Controlling Environmental Pollution

Abu Zahrim Yaser · Mohd Armi Abu Samah ·
Fazilah Ariffin · A. K. Haghi
Editors

Controlling Environmental Pollution

Practical Solutions

Springer

Editors
Abu Zahrim Yaser
Faculty of Engineering
Universiti Malaysia Sabah
Kota Kinabalu, Sabah, Malaysia

Mohd Armi Abu Samah
Kulliyyah of Science
International Islamic University Malaysia
Kuantan, Malaysia

Fazilah Ariffin
Faculty of Science and Marine Environment
Universiti Malaysia Terengganu (UMT)
Kuala Nerus, Malaysia

A. K. Haghi
Edinburgh, UK

ISBN 978-981-97-8930-6 ISBN 978-981-97-8931-3 (eBook)
https://doi.org/10.1007/978-981-97-8931-3

This Springer imprint is published by the registered company Springer Nature Singapore Pte Ltd.
The registered company address is: 152 Beach Road, #21-01/04 Gateway East, Singapore 189721, Singapore

Preface

Environmental pollution is one of the most critical issues of our time, impacting public health, ecosystems, and global economies. In Chapter "Introduction", Ariffin et al. emphasize that communities should adopt growth mindsets that enhance environmental sustainability by adapting to and profiting from severe environmental conditions. The book *Controlling Environmental Pollution—Practical Solutions* aims to give readers a comprehensive understanding of various types of pollution, particularly from the agro-based industry, medical industry, traffic, e-waste, polluted soil and water pollution. It also discusses innovative approaches to mitigating these pollutants while generating economic value. In Chapter "Environmental Pollution in Agriculture Sector; Current Trend and Future Impacts", Salikin et al. offer an in-depth examination of chemical and anthropogenic pollution in agriculture, covering air contaminants, pesticides, fertilizers, heavy metals, microplastics, antibiotics, veterinary medications, and antibiotic resistance genes (ARGs). Chapter "Environmental Pollution in Biomedical and Pharmaceutical Sectors", authored by Abd Kadir et al., addresses the various types of biomedical waste, their environmental and health impacts, relevant regulations and legislation, and the challenges of managing biomedical waste, particularly during and after the COVID-19 pandemic. In Chapter "Pollution by Microorganisms and Its Impact on the Food Safety", Ab Rashid et al. provide comprehensive insights into the diverse sources and associated risks of microbial contamination in agricultural and aquacultural products. Mohd Aminuddin et al. explore the relationship between traffic-related pollution and the number of vehicles in Chapter "Relationship Between Traffic Density and Particulate Matter Concentrations". Chapter "Indoor Air Quality Level in Mass Rapid Transit (MRT) Train in Malaysia", by Samsuddin et al., examines the correlation between CO_2 concentrations, CO levels, relative humidity, temperature, and the number of train passengers. In Chapter "Length–Weight Relationship and Condition Factor of Native Fish Population in Kenyir Lake Basin: Environmental Health Impacts and Mitigations Strategies", Abd Wahab et al. investigate the Length-Weight Relationship (LWR) and condition factor of three native species from the Kenyir Lake Basin, discussing the environmental impacts and mitigation strategies.

The world urgently needs practical solutions to its pollution challenges. In Chapter "Leveraging Environmental Epidemiology for Effective Pollution Control", Nathanael and Oginawati delve into the foundational concepts and designs of environmental epidemiology (EE) studies, discussing how to interpret these studies and the multifaceted contributions of EE. They highlight its historical significance, current applications, and future potential. Chapter "Sustainable Approaches Mitigate Environmental Pollution", by Mat Jalil, reviews various sustainable environmental protection approaches, including the use of nanotechnology, bioremediation, and green technology. Tun Ismail et al. emphasize the importance of household awareness regarding e-waste management practices in Chapter "Sustainable Electronic Waste Management Among Households in Kelantan, Malaysia", noting that understanding and knowledge of e-waste management are crucial for effective assessment. In Chapter "Qualitative Contents Analysis on Current Practices Clinical Waste Management During Pandemic in Healthcare Establishment, Kelantan", Mansor et al. employ a qualitative approach, conducting in-depth interviews with healthcare establishment (HCE) staff to explore clinical waste management (CWM) practices before and during the pandemic, and to compare how these practices have evolved over time. Chapter "Designed Landscape for Environmental Sustainability", authored by Mohd. Hussain and Abu Samah, focuses on landscape design strategies that improve land management and reduce the environmental impact of urban living. Urban and rural areas require different pollution control strategies. Urban regions, with their dense populations and limited land area, demand more compact and efficient systems. These systems are typically costly to operate and require specialized expertise, posing a significant challenge for engineers and researchers to develop high-performance systems at a reasonable cost. In contrast, rural areas, with more available land, may use less compact systems.

Precious metals and nutrients can be recovered from wastewater and solid waste, providing both environmental benefits and potential profits. Natural processes have been explored extensively for this purpose. In Chapter "Advancements in Metal Recovery from Industrial Wastes: A Comprehensive Overview", Abd Aziz et al. provide a comprehensive review of cutting-edge technologies for recovering precious metals from industrial waste streams. Mohamad and Harumain (Chapter "Harvesting Wealth: Phytomining as a Sustainable Solution for Mine Waste Management and Economic Prosperity") discuss potential solutions for managing mine waste through phytomining. Abd Aziz et al. (Chapter "Recent Advances in Processed Water Hyacinth (Eichhornia Crassipes) for the Removal of Organic Contaminants from Water") explores the complex mechanisms of adsorption using processed water hyacinth (PWH) for water remediation. Rahman (Chapter "Advancements in Constructed Wetland Technology for Enhanced Domestic Pollutants Removal: A Review") highlights the potential of constructed wetlands (CWs) as a promising technology for domestic wastewater treatment. Bukar et al. (Chapter "Effect of Different Carbon Sources on Biosurfactant Production by Bacteria Isolated from Hydrocarbon-Contaminated Soil") evaluate the impact of different carbon sources on biosurfactant production using bacteria isolated from hydrocarbon-contaminated soil. In another study, Barakwan (Chapter "Utilization of Organic Solid Waste for Hydrocarbon Contaminated Soil Remediation Through Co-composting: A Review") demonstrates that co-composting is effective in removing hydrocarbons from polluted soil. Rajin et al. (Chapter "Mesophilic Composting of Vegetable Waste Using a Passive Aerated Static Bioreactor: Simultaneous Production of Compost and Leachate") assess the performance of vegetable waste composting using passive aerated static bioreactors at both laboratory and pilot scales. Hamsun and Umor (Chapter "Utilization of Spent Mushroom Compost (SMC) from *Volvariella Volvacea* Cultivation as a Plant Substrate for *Capsicum Annuum*") explore the potential use of spent mushroom compost (SMC) from Volvariella volvacea cultivation as a plant substrate due to its high nutrient content. In Chapter "Limonite as a Feed Additive Reduces Methane Emissions During Ruminal In Vitro Fermentation", Mohd Zainudin and Maeda assess the effect of limonite and tannin on enteric methane generation during in vitro rumen fermentation.

As rural populations age and young people migrate to cities due to a lack of interest in handling unclean waste, there may be an urgent need for automation in waste management systems, even if the original systems were more affordable without it. Automation could ease the workload and attract young people to work in rural areas. In Chapter "Conclusions and Future Perspectives", Abu Samah et al. conclude that process automation using artificial intelligence (AI) will be critical in the near future. Similarly, small and medium industries require more affordable and simpler systems due to their limited financial resources.

This book is a testament to the collaborative efforts of experts from diverse fields who have contributed their knowledge and insights. We hope that the ideas and solutions presented here will inspire further innovation and action in the quest to control and mitigate environmental pollution for a more sustainable future. This book is intended to be an invaluable resource for academics, professionals, decision-makers, and students who are dedicated to finding sustainable solutions to the complex environmental challenges we face.

Kota Kinabalu, Malaysia — Abu Zahrim Yaser

Acknowledgements

The editors gratefully acknowledge the following individuals for their time and efforts in assisting the editors with the reviewing of manuscript. This book would not have been possible without the commitment of the reviewers.

- Afiza Binti Abdulah Suhaimi
- Andrea Valasek
- Anuar Mikdad Muad
- Barry N. Noller
- Fazilah Ariffin
- Husnul Azan Tajarudin
- Ibrahim A. Allamin
- Mahboob Alam
- Mariani Rajin
- Mehtab Parveen
- Mohamad Fadhli Mad Atari
- Mohammad Adam Adman
- Mohd Aizudin
- Mohd Huzairi Mohd Zainudin
- Nazrein Adrian Amaludin
- Nor Aisyah Abu Shah
- Norasiha Binti Hamid
- Nursyuhadah Othman
- Parthiban Anburajan
- Siti Balqis Zulfigar
- Siti Rohana Mohd Yatim
- Syaliza Omar
- Umar B. Ibrahim
- Zulherry Isnain
- Zulsyazwan Ahmad Khushairi

Contents

Editors and Contributors

About the Editors

Abu Zahrim Yaser is an Associate Professor in the Chemical Engineering Programme at Universiti Malaysia Sabah (UMS). Dr. Zahrim is the founder of the Kundasang Compost Living Lab and currently serves as Chairperson for the Research Cluster (Engineering and Technology). He graduated from Swansea University (UK) with a Ph.D. Since 2006, Dr. Zahrim has spearheaded multiple initiatives, and his research mostly focuses on waste processing technology. Dr. Zahrim has published 13 books, 29 book chapters, 54 journal articles, and over 50 other publications to date. With five copyrights on composting systems, his artistic creation, "Kompos" which was displayed at the Festival of Ideas (Ministry of Higher Education, Malaysia), has had over three thousand views. In addition to chairing numerous international and national seminars and conferences, he has given several keynote addresses and invited speeches on waste management and book/article publishing strategies. Dr. Zahrim has evaluated over 250 publications in his capacity as an active reviewer for multiple high-impact journals (Publon). Many communities in Sabah, Malaysia, including Nabawan and Kundasang, have embraced his inventions. Dr. Zahrim is a Professional Engineer (Ir.) registered with the Board of Engineers Malaysia and has been a visiting scientist at the University of Hull (UK). Dr. Zahrim is a recipient of the National Book Award (2023). His Scopus citations count is 1466, with an h-index of 18.

Mohd Armi Abu Samah obtained his Diploma in Forestry, B.Sc. in Environmental Science, M.Sc., and Ph.D. from Universiti Putra Malaysia (UPM), specializing in solid waste management with a focus on the composting treatment process. He previously worked as a Senior Research Officer at UPM before joining the IIUM Kuantan Campus as an Assistant Professor. Dr. Armi is actively involved in various research activities, serving as a co-researcher at UITM Shah Alam and as part of research teams at UMP, UPM, Uniza, and UMS. In addition to his academic endeavors, he writes books, magazines, proceedings, and journal papers, and serves as an invited reviewer for both international and local journals. He has published over 80 papers in various international journals, including indexed and non-indexed journals such as Scopus and ISI. Dr. Armi has received several accolades for his work, including the Malaysia Productivity Corporation Award (MPC) in the "Practitioner" category. He is also involved in knowledge transfer projects on entrepreneurship management for the community, focusing on product pitching to boost market sales and setting up composting projects for communities. Additionally, he is a member of the Malaysia Board of Technologists (MBOT). Presently, Dr. Mohd Armi Abu Samah is an Associate Professor in the Department of Chemistry, Kulliyyah of Science, IIUM Kuantan Campus.

Fazilah Ariffin is an Associate Professor at the Faculty of Science and Marine Environment, Universiti Malaysia Terengganu (UMT). She obtained her Diploma (Microbiology), B.Sc. (Hons) Microbiology, M.Sc. (Biotechnology) and Ph.D. with expertise in Environmental Microbiology. She has published more than 20 peer-reviewed technical papers (indexed by WOS and Scopus) in microbiology and one book chapter, proceedings. She has actively participated in international and local conferences. She has been granted several research grants as a Principal Investigator two from the national grants Fundamental Research Grant Scheme (FRGS) and one from the university grant, Tabung Penyelidik Muda (TPM), two industry grants from Petronas completed as a member and three grants currently ongoing as a member Talent and Publication Enhancement Research Grant Scheme, (TAPE-RG) and short-term research grant (STRG) from UniKL. She is a member of the Malaysia Board of Technologists (MBOT).

A. K. Haghi Ph.D. is a retired professor and has written, co-written, edited or co-edited more than 1000 publications, including books, book chapters, and papers in refereed journals with over 4000 citations and h-index of 33, according to Google Scholar database. He was appointed as Honorary Research Associate (HRA) of Bioprocesses and Bioproducts Group, Facultad de Ciencias Químicas, Universidad Autónoma de Coahuila, Saltillo, Coahuila, México. He has also been appointed as Honorary Research Associate (HRA) at Chemistry Centre, University of Coimbra, Portugal and held a Senior Research Fellowship (SRF) at the Tbilisi State University. He has supervised several Ph.D. and M.Sc. theses at the University of Guilan and co-supervised international doctoral projects. Professor Haghi holds a B.Sc. in urban and environmental engineering from the University of North Carolina (USA) and holds two M.Sc., one in mechanical engineering from North Carolina State University (USA) and another one in applied mechanics, acoustics, and materials from the Université de Technologie de Compiègne (France). He was awarded a Ph.D. in engineering sciences at Université de Franche-Comté (France). Professor Haghi is a frequent speaker to national and international audiences. He is the Founder and former Editor-in-Chief of the *International Journal of Chemoinformatics and Chemical Engineering* and *Polymers Research Journal*. He has acted as an editorial board member of many international journals and is a persistent reviewer of leading international journals.

Contributors

Mohd Aizudin Abd Aziz Faculty of Chemical and Process Engineering Technology, Universiti Malaysia Pahang Al-Sultan Abdullah (UMPSA), Gambang, Kuantan, Pahang, Malaysia

Norhidayah Abdull Occupational Safety, Health and Environmental Research Group, Faculty of Industrial Sciences and Technology, Universiti Malaysia Pahang Al-Sultan Abdullah, Kuantan, Pahang, Malaysia

Samsuri Abdullah Faculty of Ocean Engineering Technology and Informatics, Universiti Malaysia Terengganu, Kuala Nerus, Terengganu, Malaysia

Mohd Azmier Ahmad School of Chemical Engineering, Universiti Sains Malaysia, Nibong Tebal, Penang, Malaysia

Aminu Aliyu Department of Microbiology, Federal University Gusau, Gusau, Zamfara State, Nigeria

Binta Muhammad Aminu Department of Microbiology, Bayero University Kano, Rimin Gata, Kano State, Nigeria

Muhammad Farhan Afiq Mohd Aminuddin Faculty of Health Sciences, Centre for Environmental Health and Safety Studies, Universiti Teknologi MARA Cawangan Selangor Kampus Puncak Alam, Bandar Puncak Alam, Selangor, Malaysia;
MDC Precast Industries Sdn. Bhd, Sungai Petani, Kedah, Malaysia

Fazilah Ariffin Faculty of Science and Marine Environment, Universiti Malaysia Terengganu, Kuala Nerus, Terengganu, Malaysia

Mohd Azmir Arifin Faculty of Chemical and Process Engineering Technology, Universiti Malaysia Pahang Al-Sultan Abdullah (UMPSA), Gambang, Pahang, Malaysia

Elora Shyakilla Atinin Faculty of Engineering, Universiti Malaysia Sabah, Jalan UMS, Kota Kinabalu, Sabah, Malaysia

Rizkiy Amaliyah Barakwan Environmental Engineering Study Program, Department of Biology, Faculty of Science and Technology, Universitas Airlangga, Surabaya, Indonesia

Nur Amira Fatihah Bashari Faculty of Chemical and Process Engineering Technology, Universiti Malaysia Pahang Al-Sultan Abdullah (UMPSA), Gambang, Kuantan, Pahang, Malaysia

Suhail Bashir ENS Environmental Consultancy, Sharjah, UAE

Usman Ali Bukar Department of Microbiology, University of Maiduguri, Maiduguri, Borno State, Nigeria

Aminu Yusuf Fardami Department of Microbiology, Usmanu Danfodiyo University, Gidan Alkali, Sokoto State, Nigeria

A. K. Haghi Department of Chemistry, Institute of Molecular Sciences, University of Coimbra, Coimbra, Portugal

Muhammad Auni Hairunnaja Faculty of Chemical and Process Engineering Technology, Universiti Malaysia Pahang Al-Sultan Abdullah (UMPSA), Gambang, Kuantan, Pahang, Malaysia

Afiq Aizan Hamsun School of Biology, Faculty of Applied Science, University Technology Mara (UiTM), Kuala Pilah, Negeri Sembilan, Malaysia

Zakuan Azizi Shamsul Harumain Department of Biotechnology, Kulliyyah of Science, International Islamic University Malaysia, Pahang, Malaysia;
Research Unit in Bioinformatics and Computational Biology, Kulliyyah of Science, International Islamic University Malaysia, Pahang, Malaysia;
Environmental Biology and Biotechnology Research Unit, Kulliyyah of Science, International Islamic University Malaysia, Pahang, Malaysia

Husaini Hauwa Department of Microbiology, Sokoto State University, Dangeda, Sokoto State, Nigeria

A. Hidayah Occupational Safety and Health Research Group, Faculty of Industrial Sciences and Technology, Universiti Malaysia Pahang Al-Sultan Abdullah, Kuantan, Pahang, Malaysia

Mohd Ramzi Mohd Hussain International Islamic University Malaysia (IIUM), Kuala Lumpur, Malaysia

Khairuddin Md. Isa Faculty of Chemical Engineering Technology, Universiti Malaysia Perlis (UniMAP), Arau, Perlis, Kangar, Malaysia

HarunaYahaya Ismail Department of Microbiology, University of Maiduguri, Maiduguri, Borno State, Nigeria

Wan Nor Azilawanie Tun Ismail Faculty of Applied Social Sciences, Universiti Sultan Zainal Abidin, Gong Badak Campus, Kuala Nerus, Terengganu, Malaysia

Mohd Taufiq Mat Jalil School of Biology, Faculty of Applied Sciences, Universiti Teknologi MARA, Shah Alam, Selangor, Malaysia

Muhammad Izzat Samir Abdul Jalil School of Industrial Technology, Universiti Sains Malaysia, Gelugor, Penang, Malaysia

Nurul Huda Abd Kadir Faculty of Science and Marine Environment, Universiti Malaysia Terengganu, Kuala Nerus, Terengganu, Malaysia

Mohd Khairul Amri Kamarudin Faculty of Applied Social Sciences, Universiti Sultan Zainal Abidin, Kuala Nerus, Terengganu, Malaysia;
East Coast Environmental Research Institute (ESERI), Universiti Sultan Zainal Abidin, Kuala Nerus, Terengganu, Malaysia

Abdullahi Hassan Kawo Department of Microbiology, Bayero University Kano, Rimin Gata, Kano State, Nigeria

Toshinari Maeda Department of Biological Function and Engineering, Graduate School of Life Science and System Engineering, Kyushu Institute of Technology, Fukuoka, Japan

F. N. Mansor Occupational Safety and Health Research Group, Faculty of Industrial Sciences and Technology, Universiti Malaysia Pahang Al-Sultan Abdullah, Kuantan, Pahang, Malaysia

Mohd. Azrul Naim Mohamad Department of Biotechnology, Kulliyyah of Science, International Islamic University Malaysia, Pahang, Malaysia;
Research Unit in Bioinformatics and Computational Biology, Kulliyyah of Science, International Islamic University Malaysia, Pahang, Malaysia;
Environmental Biology and Biotechnology Research Unit, Kulliyyah of Science, International Islamic University Malaysia, Pahang, Malaysia

Rinaldy Jose Nathanael Environmental Engineering Program, Department of Biology, Faculty of Science and Technology, Airlangga University, Surabaya, Indonesia

Nur Asshifa Md Noh School of Biological Sciences, Universiti Sains Malaysia, Minden, Pulau Pinang, Malaysia

S. S. Nurud Occupational Safety and Health Research Group, Faculty of Industrial Sciences and Technology, Universiti Malaysia Pahang Al-Sultan Abdullah, Kuantan, Pahang, Malaysia

Katharina Oginawati Environmental Engineering Department, Faculty of Civil and Environmental Engineering, Bandung Institute of Technology, Bandung, Indonesia

Danar Arifka Rahman Department of Biology, Faculty of Science and Technology, Study Program of Environmental Engineering, Universitas Airlangga, Surabaya, Indonesia;
Environmental Technology and Innovation Research Group, Universitas Airlangga, Surabaya, Indonesia

Hasrul Hafizan Rahmatullah Politeknik Balik Pulau, Balik Pulau, Penang, Malaysia

Mariani Rajin Faculty of Engineering, Universiti Malaysia Sabah, Jalan UMS, Kota Kinabalu, Sabah, Malaysia

Syarifah Ab Rashid School of Biological Sciences, Universiti Sains Malaysia, Gelugor, Penang, Malaysia;
School of Biological Sciences, Universiti Sains Malaysia, Minden, Pulau Pinang, Malaysia

Nur Anniesa Farhana Mohd Roza Faculty of Science and Marine Environment, Universiti Malaysia Terengganu, Kuala Nerus, Terengganu, Malaysia

Muhamad Hafiz Saad East Coast Environmental Research Institute (ESERI), Universiti Sultan Zainal Abidin, Kuala Nerus, Terengganu, Malaysia

Sariah Saalah Faculty of Engineering, Universiti Malaysia Sabah, Jalan UMS, Kota Kinabalu, Sabah, Malaysia

Nor Hawani Salikin Renewable Biomass Transformation Cluster, School of Industrial Technology, Universiti Sains Malaysia, Gelugor, Penang, Malaysia; School of Industrial Technology, Universiti Sains Malaysia, Minden, Pulau Pinang, Malaysia

Mohd Armi Abu Samah Kuantan Campus, International Islamic University Malaysia (IIUM), Jalan Sultan Ahmad Shah, Bandar Indera Mahkota, Kuantan, Pahang, Malaysia;
Kulliyyah of Science, International Islamic University Malaysia, Kuantan, Pahang, Malaysia;
International Islamic University Malaysia (IIUM), Kuala Lumpur, Malaysia

Muhamad Armi Abu Samah International Islamic University Malaysia (IIUM), Kuantan, Pahang, Malaysia

Sufy bin Samsuddin Swift Integrated Logistics Sdn Bhd, Pelabuhan Klang, Selangor, Malaysia

Ahmad Firdaus Shafie Faculty of Health Sciences, Centre for Environmental Health and Safety Studies, Universiti Teknologi MARA Cawangan Selangor Kampus Puncak Alam, Bandar Puncak Alam, Selangor, Malaysia

Shafina Nazera Mohd Shafizan Faculty of Engineering, Universiti Malaysia Sabah, Jalan UMS, Kota Kinabalu, Sabah, Malaysia

A. A. Suhaimi Occupational Safety and Health Research Group, Faculty of Industrial Sciences and Technology, Universiti Malaysia Pahang Al-Sultan Abdullah, Kuantan, Pahang, Malaysia

Afiza Abdullah Suhaimi Occupational Safety, Health and Environmental Research Group, Faculty of Industrial Sciences and Technology, Universiti Malaysia Pahang Al-Sultan Abdullah, Kuantan, Pahang, Malaysia

Nurud Suria Suhaimi Occupational Safety, Health and Environmental Research Group, Faculty of Industrial Sciences and Technology, Universiti Malaysia Pahang Al-Sultan Abdullah, Kuantan, Pahang, Malaysia

Woei Yenn Tong Institute of Medical Science Technology, Universiti Kuala Lumpur, Kajang, Selangor, Malaysia

Mohd Ekhwan Toriman Faculty of Social Sciences and Humanities, Universiti Kebangsaan Malaysia, Bangi, Selangor, Malaysia

Noor Azrimi Umor School of Biology, Faculty of Applied Science, University Technology Mara (UiTM), Kuala Pilah, Negeri Sembilan, Malaysia

Ruqayyah Abbas Usman Department of Microbiology, Bayero University Kano, Rimin Gata, Kano State, Nigeria

Noorjima Abd Wahab Faculty of Applied Social Sciences, Universiti Sultan Zainal Abidin, Kuala Nerus, Terengganu, Malaysia

Aleena Yaaqub School of Industrial Technology, Universiti Sains Malaysia, Gelugor, Penang, Malaysia

Abu Zahrim Yaser Faculty of Engineering, Universiti Malaysia Sabah, Jalan UMS, Kota Kinabalu, Sabah, Malaysia

Siti Rohana Mohd Yatim Faculty of Health Sciences, Centre for Environmental Health and Safety Studies, Universiti Teknologi MARA Cawangan Selangor Kampus Puncak Alam, Bandar Puncak Alam, Selangor, Malaysia

Mohd Huzairi Mohd Zainudin Laboratory of Sustainable Animal Production and Biodiversity, Institute of Tropical Agriculture and Food Security, Universiti Putra Malaysia, Serdang, Selangor, Malaysia

Nurul Aili Zakaria Faculty of Applied Sciences, Universiti Teknologi MARA, Shah Alam, Selangor, Malaysia

Kamarul Zaman Zarkasi School of Biological Sciences, Universiti Sains Malaysia, Minden, Pulau Pinang, Malaysia

Introduction

Fazilah Ariffin, Mohd Armi Abu Samah, A. K. Haghi, and Abu Zahrim Yaser

Abstract Environmental pollution is an issue that exists throughout the world despite efforts to reduce it. This is due to environmental damage caused by industrial and agricultural discharges and uncontrolled waste on the ecosystem. It is important to reduce pollution in sectors such as agriculture, the pharmaceutical industry, or biomedicine through the use of environmentally friendly techniques, such as organic farming, waste recycling, or sustainable production. Bioremediation processes and environmentally friendly technologies can help prevent pollution, protect the environment, and save the environment. In this book, different efforts have been implemented in many sectors aimed at reducing pollution, because detailed environmental conservation goes hand in hand with human advancement.

1 Introduction

Environmental pollution is a growing worldwide problem that significantly affects the overall health of the environment and its people. At an alarming rate, contaminants are poisoning our air, water, and soil through industrial discharges and agricultural runoff. The deterioration of ecosystems, the decrease in biodiversity, and the rise in

F. Ariffin (✉)
Faculty of Science and Marine Environment, Universiti Malaysia Terengganu, 21030 Kuala Nerus, Terengganu, Malaysia
e-mail: fazilah@umt.edu.my

M. A. A. Samah
Kulliyyah of Science, International Islamic University Malaysia, 25200 Kuantan, Pahang, Malaysia

A. K. Haghi
Department of Chemistry, Institute of Molecular Sciences, University of Coimbra, Coimbra, Portugal

A. Z. Yaser
Faculty of Engineering, Universiti Malaysia Sabah, Jalan UMS, 88400 Kota Kinabalu, Sabah, Malaysia

A. Z. Yaser et al. (eds.), *Controlling Environmental Pollution*,
https://doi.org/10.1007/978-981-97-8931-3_1

human health issues are all clear indicators of the harmful consequences of pollution. It is critical that we look for practical solutions that may effectively reduce and regulate environmental pollution as we continue to deal with the effects of this pervasive contamination. The community should adopt growth mindsets that can enhance environmental sustainability. These mindsets not only adapt and respond to severe environmental conditions but also generate profits from them (Yaser et al. 2023; Yaser 2019).

The pharmaceutical and biomedical sectors play a crucial role in public health but also significantly affect environmental pollution. The improper disposal of biomedical waste and pharmaceutical goods can contaminate land, water, and air. Ecosystems are at risk due to the frequent discharge of dangerous chemicals and byproducts from production operations. A comprehensive approach is needed to address pollution in these areas, including creating more environmentally friendly production techniques, strict waste disposal laws, and public awareness campaigns about the correct disposal of prescription drugs. By implementing these measures, we can protect our ecosystems from future harm and reduce the environmental impact of these industries.

Additionally, agricultural activities contribute to the contamination of the environment. Although necessary for crop production, huge applications of fertilizers and pesticides can contaminate the soil and water. Agricultural pollution trends of today emphasize the necessity for environmentally friendly farming methods that minimize harm to the environment. Reducing the need for chemical inputs and minimizing pollution can be accomplished through methods like precision agriculture, organic farming by using food wastes as a beneficial biofertilizer (Razali et al. 2024), and integrated pest control. Managing agricultural waste offers a special chance to combat pollution using creative and environmentally friendly methods. One such instance is using wasted mushroom compost (SMC) from *Volvariella volvacea* cultivation as a plant substrate for *Capsicum annuum*. This method improves soil health and crop productivity in addition to recycling agricultural by-products. We may reduce agriculture's adverse impact on the environment and encourage more environmentally friendly farming methods by turning waste into a valuable resource. With this, the system of waste management will be very effective and save other sources from turning waste into wealth (Mohd Armi et al. 2011).

Therefore, the cornerstone of efficient pollution control is sustainability. Environmental pollution can be reduced through sustainable ways that use various techniques and technology that can be applied in different industries. Renewable energy sources like wind and solar power decrease greenhouse gas emissions and our dependency on fossil fuels. Green manufacturing techniques reduce waste and the industrial sector's carbon footprint. The biological method utilizing microorganisms for bioremediation presents a viable approach to pollution control. Some bacteria can break down or neutralize toxic contaminants such as pesticides, dyes, and hydrocarbons, changing them into less toxic forms (Ariffin and Anuar 2022; Vigneswari et al. 2021). Polluted water bodies, pesticide and heavy metal-filled soils, and oil spills are just a few examples of contaminated ecosystems that can be cleaned up via this natural process. Through the utilization of microbial activity, bioremediation offers a financially viable and environmentally sustainable approach to the rehabilitation of contaminated areas. This technology lowers the long-term environmental and financial costs associated with standard cleanup techniques while simultaneously restoring the health of ecosystems.

Implementing these sustainable methods might help us achieve a balance between environmental preservation and economic development. Incorporating sustainability into pollution control plans guarantees the sustainability and efficacy of our work. As inhabitants of the earth, we must maintain the sustainability of the environment as a sign of gratitude for the gift bestowed by our creator (Mohd Armi and Afiza 2024).

To sum up, combating environmental pollution necessitates a thorough and multi-dimensional strategy. "Controlling Environmental Pollution: Practical Solutions," a book, examines creative and workable approaches in various fields to reduce pollution and advance sustainability. We can make the future of our planet and its inhabitants healthier and more sustainable by looking into the causes of pollution and implementing efficient control mechanisms. Every chapter presents an analysis of particular pollution problems along with workable remedies that can be used to safeguard the environment and improve human health.

References

Ariffin F, Anuar NEMC (2022) Biodegradation of methylene blue by bacteria strains isolated from contaminated soil. Malays Appl Biol 51(3):25–35

Mohd Armi AS, Latifah AM, Wan NAS, Ahmad ZA (2011) Selection of a solid waste treatment technology using the AHP approach. Contemporary environmental quality management in Malaysia and selected countries. UPM Press, pp 308–318

Mohd Armi AS, Afiza AS (2024) Environmentally sustainable management and challenges in COVID 19 pandemic of environmental health and occupational. UPM Press, pp 167–181

Razali RM, Murshidi NZ, Ghazali MSM, Ariffin F (2024) Effect of effective microorganisms in fermentation of Rice Husk and Anchovy head using lab-scale treatment. Pertanika J Trop Agric Sci 47(1)

Vigneswari S, Sukarman MH, Ariffin F (2021) Preliminary study on thiamethoxam degrading bacteria isolated from corn plantation. Malaysian J Microbiol 17(6)

Yaser AZ (2019) Green engineering for campus sustainability. Springer, pp 1–259

Yaser AZ, Sarjadi MS, Lamaming J (2023) Cellulose: development, processing, and applications. CRC Press, pp 1–314

Environmental Pollution in Agriculture Sector; Current Trend and Future Impacts

Nor Hawani Salikin, Muhammad Izzat Samir Abdul Jalil, Aleena Yaaqub, Syarifah Ab Rashid, and Hasrul Hafizan Rahmatullah

Abstract The crucial role of the agriculture industry in feeding a growing global population cannot be overstated. However, it has been acknowledged that the agriculture sector also contributes significantly to environmental pollution. Synthetic chemicals released into the ecosystem as a result of agricultural activities pose a serious threat to the environment. This chapter comprehensively examines the various types of chemical and anthropogenic pollution associated with agriculture, including air contaminants, pesticides, fertilizers, heavy metals, microplastics, antibiotics, veterinary medications, and antibiotic resistance genes (ARGs). The adverse impacts of these pollutants on the air, water, and soil quality are discussed, focusing on contamination mechanisms throughout the entire ecosystem. The chapter also highlights the potential ecological and human health consequences, underscoring the importance of sustainable agricultural practices. Regulatory frameworks and effective mitigation approaches embraced by different stakeholders are essential to addressing these environmental challenges.

Keywords Agriculture · Environmental pollutions · Greenhouse gases (GHGs) · Agrochemical · Microplastic · Antibiotic resistance genes (ARGs)

N. H. Salikin (✉) · M. I. S. A. Jalil · A. Yaaqub
School of Industrial Technology, Universiti Sains Malaysia, 11800 Gelugor, Penang, Malaysia
e-mail: norhawani@usm.my

N. H. Salikin
Renewable Biomass Transformation Cluster, School of Industrial Technology, Universiti Sains Malaysia, 11800 Gelugor, Penang, Malaysia

S. A. Rashid
School of Biological Sciences, Universiti Sains Malaysia, 11800 Gelugor, Penang, Malaysia

H. H. Rahmatullah
Politeknik Balik Pulau, Pinang Nirai, 11000 Balik Pulau, Penang, Malaysia

A. Z. Yaser et al. (eds.), *Controlling Environmental Pollution*,
https://doi.org/10.1007/978-981-97-8931-3_2

1 Introduction

Agriculture is the term used to describe the various ways in which crop plants and domestic animals provide food and other products to the global human population. These include a wide range of practices, including cultivating soil, dairy farming, forestry, harvesting horticultural or agricultural products, managing livestock, producing honey, maple syrup or sugar, and raising or harvesting various types of seafood (Harris and Fuller 2014; Connecticut Department of Agriculture 2024). The word "agriculture" is derived from the Latin word "ager" (defined as field) and "colo" (defined as cultivate), which refers to the cultivation of land. However, the term has now expanded to cover a broad range of activities that are essential to agriculture, such as domestication, horticulture, arboriculture, cultivation, and vegeculture (Harris and Fuller 2014; Connecticut Department of Agriculture 2024). Agricultural activities also encompass the operational system, management, conservation, improvement, and maintenance of farms, and their infrastructure, tools, and equipment, including handling, planting, drying, packing, packaging, processing, freezing, grading, storing, and delivering any agricultural or horticultural commodity. Furthermore, the construction, operation, or maintenance of ditches, canals, reservoirs, or waterways used exclusively for farming purposes are also included in agricultural practices (Connecticut Department of Agriculture 2024).

Agriculture is the largest industry in the world, employing over a billion people and producing more than $1.3 trillion worth of food each year (World Wildlife 2024). Pasture and cropland cover almost half of the earth's habitable land and provide food and habitat to various species. Agriculture plays a vital role in economic growth, accounting for 4% of the global GDP and, in some of the least developed countries, more than 25% of the GDP (World Bank 2023). Since the Agricultural Revolution around 10,000–12,000 years ago, humans have utilized and managed soil and water resources, significantly impacting the development, persistence, decline, and regeneration of agriculturally supported human civilizations (Parikh and James 2012).

Sustainable agricultural operations are designed to preserve and restore critical habitats, protect watersheds, and improve soil health and water quality. Nevertheless, inappropriate agricultural practices seriously affect humans and the whole ecosystem, disturbing plant biodiversity, animal populations, soil systems, and water resources. Understanding, evaluating, and balancing the negative and positive impacts of agricultural perturbations on soil and water resources is pivotal in efforts to maintain and improve human well-being. This knowledge significantly impacts the emerging sustainability ethics and responsibility toward future human populations and ecosystems (Parikh and James 2012).

Agricultural pollution is the consequence of different inappropriate farming practices and mishandling that lead to the contamination of air, water, and soil (Du and Wang 2023). The ecological impact of these activities resonates throughout diverse ecosystems, from the use of synthetic fertilizers and pesticides to the discharge of contaminated agricultural runoff. The impacts go beyond cultivated fields, presenting risks to aquatic life, soil well-being, and overall environmental sustainability (Sharma and Singhvi 2017).

This chapter delves into the complexities of agricultural pollution by initially examining the factors that contribute to the dreadful impacts that might be endured by the communities and the environment. Understanding the origins of agricultural pollution is crucial for developing targeted mitigation strategies, as it encompasses a range of factors such as the quest for increased yields and reliance on chemical inputs. At the same time, the chapter explores how pollutants spread and the resultant effects that they may have on ecosystems, water, air, and soil quality.

2 Different Types of Agriculture Pollution

Human populations exhibited rapid expansion from low in previous centuries to high population density in recent years along with increased social complexity in just a couple of generations (Lima et al. 2024). More specifically, the human population has more than doubled between 1961 and 2016 (from 3,068,370,609 individuals in 1961 to 7,513,474,238 individuals in 2016) (Food and Agriculture Organization 2020), further validating the increasing number of individuals worldwide and food demands in parallel. These demographic shifts seem to be a notable feature of the dynamicity of *Homo sapiens* population, and they are probably facilitated by their capacity to amass cultural and technological advancements such as industrial and agricultural activities that actively alter their environments. Given the dramatic growth, demographers expect the world population to peak by the end of this century, reaching up to 10.4 billion in the mid-2080s (30% of human population increment from 2022 until the mid-2080s) (United Nations 2022). In light of the current trend, the total food demand and the global population at risk of hunger are projected to escalate by 35%–56% and change by −91% to +8%, respectively, between 2010 and 2050 (Van Dijk et al. 2021). Therefore, attempts to enhance the agriculture industries and intensive rearing of livestock are implemented vigorously around the globe to maintain sufficient food supply while enhancing food security and quality.

However, the current trend has shown a tremendous increase in forest exploitation leading to environmental pollution and debilitating threat to the globe. It was observed that from 2011 to 2015, Global Forest Change (GFC) and Tree Cover Loss (TCL) rates averaged 10.6 million hectares (Mha) per year in the tropics while the Food and Agriculture Organization (FAO)'s Forest Resources Assessment (FRA) 2020 estimated deforestation to be 10.7 Mha per year (Pendrill et al. 2022). In general, five billion hectares, or 38% of the world's land surface, are used for agricultural purposes. Of this, about one-third is farmed, and the other two-thirds are grasslands and meadows used for cattle grazing. Fruit trees, cocoa plantations, and oil palm plantations occupy around 10% of the total area of agriculture. On the other hand, an additional 21% of the land is set up for crop irrigation (Food and Agriculture Organization 2020). While the development of agriculture industries and technologies is promising, the outcome of such uplifting socioeconomic growth is associated

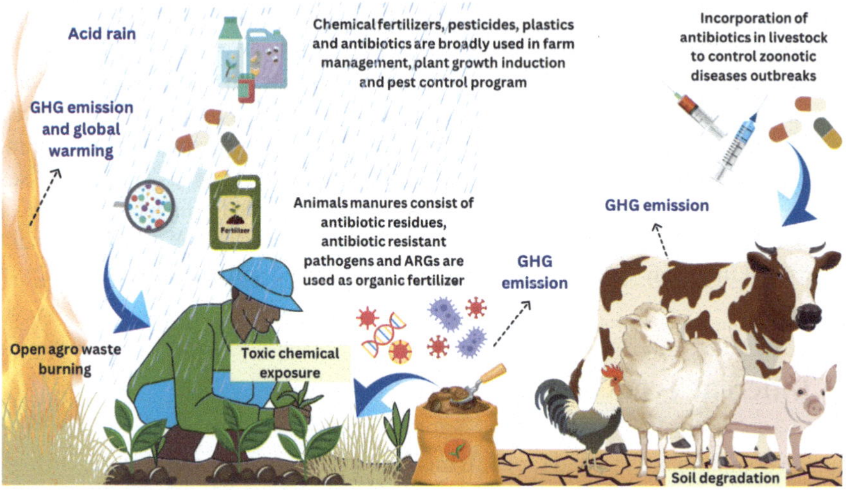

Fig. 1 Schematic diagram displaying different sources of pollutants in agriculture and the resulting impacts on humans and the environment. GHG: Greenhouse gases, ARGs: Antibiotic resistance genes

with ecological destruction and detrimental pollution to the environment, affecting the quality of air, water, soil, and biodiversity and even escalating the widespread of antibiotic resistance genes (ARGs) (Fig. 1).

2.1 Soil Degradation

Soil is the basis for human survival and development, providing humans with important ecosystem services such as food, fiber, fresh water, and biodiversity. However, approximately 33% of soils are degraded due to excessive or inappropriate human use, with agricultural systems being the main driver of soil degradation (Ma et al. 2023). Out of the 80 countries significantly impacted by land degradation worldwide, 36 are situated in Africa. In Lesotho, approximately 2% of the total land area (100 km^2) has been degraded due to overgrazing, improper farming practices, mismanagement of rangeland, and the accumulation of chemical and pesticide residues (World Health Organization 2015). Moreover, based on extensive research on soil erosion using a global dataset with over 35 million observations and a spatial regression discontinuity design, the average rate of soil erosion changes by approximately 1.4 tons per hectare per year. This reveals a significantly important impact based on the country of origin, thus requiring immediate precautionary measures to address the issue (Wuepper et al. 2020).

Soil infertility can be a result of using too many agrochemicals and burning crop residue excessively. The residue burning raises the subsoil temperatures up to nearly

33.8–42.2 °C at 10 mm depth (Gupta et al. 2004), and prolonged effects can even reach up to 15 cm of the topsoil. Frequent burning depletes soil nitrogen and carbon, kills beneficial microorganisms, and removes organic matter which ultimately causes the carbon–nitrogen equilibrium of the soil completely lost (Bhuvaneshwari et al. 2019). Consequently, soil rejuvenation processes will be hindered, and land resources will not be sustained.

Soil degradation includes the destruction of its physical, chemical, and biological quality. Degradation of soil physically leads to a decline in structural characteristics, such as pore size and connectivity. This exacerbates a soil's vulnerability to problems such as crusting, compaction, reduced water infiltration, increased surface runoff, wind and water erosion, greater soil temperature fluctuations, and a higher likelihood of desertification (Lal 2015). In addition, unfavorable changes in soil chemistry caused particularly by synthetic fertilizers and pesticides can diminish plant nutrition. This results in a decline of beneficial microbes and humus content and changing pH of the ground. Biological degradation in soil occurs when the native microbial activity is lost particularly at the top layer, resulting in a declining soil organic carbon (SOC) pool, biodiversity loss, and a surge of greenhouse gas (GHG) emissions from the soil into the atmosphere. The resulting impacts of the soil's physical, chemical, and biological degradation will eventually lead to ecological degradation when multiple ecosystem functions such as elemental cycling, water infiltration, and hydrological cycle are adversely impaired (Lal 2015; Peter Kogut 2023).

As a mitigation approach, crop rotation techniques involving the cultivation of different types of plants rather than monoculture systems have been implemented in several countries, including Romania (Stoicea et al. 2023), China (Zhao et al. 2020), and in Europe (Kollas et al. 2015). The crop rotation system is known to benefit farmers in several ways, such as by providing diversified revenue sources by encouraging the cultivation of different crops and decreasing agriculture production risk and uncertainty by minimizing the impact of highly resistant plant pests and phytopathogens infestation (Shah et al. 2021). Additionally, it promotes soil and ecological sustainability within the crop plantation through nutrient enrichment and preventing leaching into the water table (Gokhale and Sharma 2023). The implementation of crop rotation has led to an average increase of 20% in crop yields compared to continuous monoculture practices. Furthermore, legume-based rotations have resulted in a 14% higher agricultural yield when compared to rotations without legumes (Zhao et al. 2020), owing to the legume residual components that release available nitrogen for the subsequent crop and improved soil nutrients after the mineralization process facilitated by soil microorganisms (Xiao et al. 2013). However, despite those potential benefits, crop rotation practices are scarcely implemented on a global scale due to inadequate investments in machinery, infrastructure, and expertise, as well as a dearth of supporting research evidence (Shah et al. 2021). Sufficient financial resources are required for the farmers to make significant up-front investments, such as in new machinery or equipment, and impose an additional short-term cost (Shah et al. 2021). Furthermore, some farmlands are leased and strictly adhere to contracts to cultivate specific types of crops, hence limiting crop rotation practices by the farmers. In comparison to farmland owners who prioritize soil conservation and sustainable agricultural practices, short-term farmland tenancy exerts an influence on farmers, leading them to concentrate on short-term profits and unsustainable agricultural practices, i.e., excessive use of agrochemicals and biomass burning, consequently resulting in soil erosion problems (Lee 1980; Leonhardt et al. 2019). Additionally, the adoption of crop rotation practices faces obstacles due to insufficient support and coordination between public and private institutions, limited investment and concerns about production risks, and a lack of strategic studies focusing on dissemination and implementation (Shah et al. 2021). The restricted use of the crop rotation approach may eventually limit its potential benefits in the global agricultural industry, hence representing a bottleneck for sustainable agriculture system.

2.2 *Waste Generation and Air Pollution*

Approximately, 1300 million tons of agricultural waste is produced annually and the amount is possibly expanding further in a world with a growing population to feed. Despite the unprecedented surge in agricultural productivity, the amount of agricultural waste such as straw and manure increases around 7.5% annually, resulting in more than 900 million tons of biomass waste being dumped in landfills each year (Lee et al. 2022). The accumulation of moisture-rich agriculture wastes such as spoiled food waste from crops, orchards, vineyards, dairies, feedlots, farms, agricultural residues, food and meat processing, and hazardous waste in landfill leads to undesired biological degradation and in situ microbial reaction that generates greenhouse gases such as carbon dioxide, nitrous oxide, and methane (Peng et al. 2023). Methane which is a type of biogas generated under anaerobic waste fermentation by microorganisms is 25 times more potent than carbon dioxide at trapping heat in the atmosphere, thus resulting in increasing rates of global warming and unpredictable climate change.

Furthermore, agricultural waste accumulation in the dumpsites may also generate molds that can cause secondary environmental pollution and debilitating health hazards via cytotoxic fungal aerosol (Frączek et al. 2017; Lee et al. 2022). The microbial particles may be rapidly transferred to long distances depending on their respective sizes. The bigger microbe and dust aggregates (10–100 μm) descend to the ground in the vicinity of their emission sources within seconds or minutes. In contrast, smaller particles (0.5–10 μm) are typically carried farther by air since they are suspended in the atmosphere for longer periods of time (Frączek et al. 2017). As a result, the adjacent site around the landfill is commonly exposed to the airborne fungal contaminants mostly from the *Cladosporium, Penicillium, Aspergillus, Alternaria, Mucor, Fusarium, Rhizopus,* and *Trichoderma* genera (Huang et al. 2002; Abdel Hameed et al. 2015; Frączek et al. 2017) predisposing the workers and surrounding community particularly the infants, elderlies, and immunocompromised individuals to allergy or chronic fungal diseases.

Another conventional method of agricultural waste management is via thermo-chemical incineration involving the combustion of waste residues such as crop stubbles, stalks, grasses, leaves, and husks in a high-temperature and oxygen-rich environment (Lee et al. 2022). Considering the environmental issues linked to the landfill approach and the increasing rivalry for land usage, incineration has been employed as an alternative control approach of handling agricultural waste biomass. Incinerator may be a potentially powerful tool for handling agricultural wastes if combined with the right environmental treatment technology to manage incineration byproducts such as hazardous contaminants. Unfortunately, over 60% of incinerators worldwide do not currently possess these precautionary measures in place, thereby allowing hazardous wastes to leak into the ecosystems undesirably (Lee et al. 2022). In many countries such as Asia, Nepal, Taiwan, and Mexico, burning stubble is a common cost-effective and easiest practice to reduce or eradicate the abundance of combustible materials produced by agricultural activities during the end of the reaping season (Commission for Environmental Cooperation 2014). In order to prepare for the next season's sowing, farmers burn the leftover stubble on the farm, causing the emission of greenhouse gases, global warming, and deteriorated air quality (Bajracharya et al. 2021; Lee et al. 2022; Raza et al. 2022). Strong winds cause the produced smoke to broadly disseminate from the original places, hence destructively affecting other people living far from the source of pollution (Raza et al. 2022). Particulate matter (PM), acid gases, heavy metals, dioxins, furans, and greenhouse gases are among the environmental contaminants brought on by incineration. The burning biomass, e.g., woods, leaves, trees, grasses, and other agricultural wastes may result in the emission of 40% carbon dioxide (CO_2), 32% of carbon monoxide (CO), 20% of particulate matter (PM), and 50% of polycyclic aromatic hydrocarbons (PAHs) (Commission for Environmental Cooperation 2014). Based on the aerodynamic diameter and chemical constitution, PM in the air is classified as $PM_{2.5}$ and PM_{10}. $PM_{2.5}$ is small particulate matter (<2.5 μm), while PM10 is coarse particulate matter (<10 μm). Alarmingly, lightweight PM can stay floating in the air for a longer time and winds facilitate its dissemination to longer distances (Bhuvaneshwari et al. 2019). While PM such as dust, dirt, soot, and smoke can reduce visibility through smog formation, it also can readily adsorb metals, microorganisms, or other dangerous compounds like dioxins and polycyclic aromatic hydrocarbons (PAHs) and quickly being assimilated by the cells, tissues, and organs (Zhang et al. 2024). Prolonged exposure to high PM-containing environment particularly for those with vulnerabilities (infants, elderlies, and immune-deficiency individuals) may result in increasing risk of cardiovascular diseases, neurological disorders (Djebbi et al. 2021), urinary system disorders, reproductive, motor system, respiratory impairments, cancer, and even fatal (Bowe et al. 2019; Manigrasso et al. 2020).

Dioxin and dioxin-like compounds from agro waste combustion are other chlorinated organic pollutants that may potentially cause infertility, changes in thyroid levels and neurological effects in developing fetus, chloracne, diabetes, and different types of cancer. Studies have shown that dioxins represent the ligand of the aryl hydrocarbon receptor (AHR) that induce AHR-mediated biochemical and toxic responses. Furthermore, 2,3,7,8-tetrachlorodibenzo para dioxin (TCDD) is the prototypical AHR ligand and its activation may induce toxicity, immunosuppression, and tumor growth (Piwarski and Salisbury 2023).

In addition, acid gases emitted during open burning crop residues such as hydrogen chloride (HCl), hydrogen fluoride (HF), and nitrous acid (HONO) can cause allergic and inflammatory symptoms in humans while affecting the surrounding climate (Cui et al. 2021; Lee et al. 2022). The HONO photolysis resulted in the increment of hydroxyl radicals (OH) in the lower troposphere. Oxidative reactions between OH with SO_2 and NO_2 transform the radicals into sulfates and nitrates, both of which are key elements in haze formation (Cui et al. 2021). High concentrations of greenhouse gases, such as carbon dioxide (CO_2), methane (CH_4), Nitrous oxide (N_2O), and fluorinated gases in the atmosphere may result in global warming with intimidating consequences ranging from ocean acidification, desertification, biodiversity loss, and more frequent extreme events, thus threatening the well-being of humans and the globe.

2.3 *Agrochemical Pollution*

A sustainable agricultural productivity predominantly depends on continuous supply from the farm. However, insufficient nutrient supplementation remains a bottleneck for optimal plant growth and development which primarily rely on the components and concentration of mineral nutrients available in the soil (Yousaf et al. 2017). Obtaining adequate macronutrients, for example, nitrogen (N), phosphorus (P), magnesium (Mg), potassium (K), calcium (Ca), and sulfur (S) as well as micronutrients including iron (Fe), zinc (Zn), copper (Cu), boron (B), manganese (Mn), molybdenum (Mo), and chloride (Cl^-) are challenging for plants due to their relative immobility. Prolonged nutrient deficiency in crop cultivation may result in decreased agricultural productivity and quality while threatening global food security (Morgan and Connolly 2013; Kumar et al. 2021). Even worse, the damage to plants caused by competition from weeds and by other detrimental pests including viruses, bacteria, fungi, insects, and parasitic nematodes greatly impairs their yield and, in some occurrences, can totally abolish a crop (Meena et al. 2023).

Therefore, the utilization of agrochemicals, i.e., synthetic fertilizers and pesticides, is indispensable to enhance agricultural productivity while preventing crop damage due to multiple pest infestations (Penuelas et al. 2023). Modern agricultural practice primarily includes chemical fertilizers that contain three essential macro nutrients, i.e., nitrogen, phosphorus, and potassium with occasional supplementation of rock flour representing the micronutrients. Recent trend has shown that the doubled agriculture production worldwide was primarily attributed to a 6.9-fold and 3.5-fold surges in nitrogen and phosphorous fertilization, respectively (Rahman and Zhang 2018). From this rate, China has become the greatest consumer of chemical fertilizer globally, with 8.64% of the global farmland area consuming almost 28% of global chemical fertilizer (Guo et al. 2022). Moreover, to eradicate plant pests and weeds, different organic and inorganic pesticides such as dichlorodiphenyltrichloroethane (DDT), β-hexachlorocyclohexane (BHC), aldrin, dieldrin, organophosphate, endrin, chlordane, parathion, captan, 2,4-D, triazolopyrimidine, triketone, isoxazole herbicides, strobilurin, azolone fungicides, chloronicotinyl, spinosyn, and fiprole diacylhydrazine are vigorously utilized by the farmers primarily via fumigation in the farm (Sarkar et al. 2021; Tudi et al. 2021). Without the incorporation of pesticides in the cultivation process, damages due to pest invasion may result in approximately 78%, 54%, and 32% of profit losses of fruit, vegetable, and cereal production, respectively (Tudi et al. 2021). This projection further corroborates the significance of pesticide incorporation into agriculture management for sustainable food supplies with excellent quality.

2.3.1 Impacts of Pesticides Utilization in Agriculture Industries

Pesticides are generally distinguished into different categories including chemical classes, functional groups, modes of action, and toxicity. Pesticides are formulated to target specific types of plant pests, such as fungicides for fungi, insecticides for insects, and herbicides for weeds (Tudi et al. 2021). The classification of pesticides into organic and inorganic formulations is based on their chemical composition. Organic pesticides are made up of carbon as the basis of their molecular structure and may also contain oxygen, phosphorus, or sulfur (Urkude et al. 2019). Additionally, organic pesticides are classified into two distinct groups: natural organics and synthetic organics. The former, derived from natural sources such as plants, include rotenone and pyrethrum, while the latter, also known as modern pesticides, are artificially synthesized water-insoluble chemical compounds, exemplified by DDT, permethrin, malathion, and lindane (Urkude et al. 2019). In contrast, inorganic pesticides are primarily composed without carbon compounds and often contain heavy metals that are persistent and toxic to humans, such as copper sulfate, ferrous sulfate, copper, lime, and sulfur (Karuppuchamy and Venugopal 2016; Tudi et al. 2021).

Regrettably, the increased use of pesticides in fields across many nations does not accurately reflect the state of the art. Modern pesticides have been incorporated into agriculture management worldwide, yet the technology such as spraying nozzles used for their application is still inefficient, resulting in waste of pesticides and undeliberate environmental pollutions (Friedrich 1996; Azeez 2021). Furthermore, the disposal of containers with pesticide residues into ponds or open fields by uneducated and ignorant users also contributes to food poisoning, water and environmental pollutions, leading to debilitating diseases or even fatal (Obi et al. 2016).

Upon exposure to the targeted crops, pesticides can be migrated to another untargeted organism or surroundings via adsorption, leaching, volatilization, spray drift, and runoff (Robinson et al. 1999; Tudi et al. 2021). When pesticides are used, only a small amount of the applied pesticides displays a protective role to combat against plant diseases. In contrast, a large amount of pesticides are being drifted into droplets after fumigation or absorbed by the soil due to pesticides binding to the soil particles, thereby resulting in severe air and soil pollution (Xue et al. 2006). In Japan, pesticide was frequently detected in the air of residential environments and childcare facilities following its application. This observation proves that the outside application of pesticide contributes to indoor air pollution in nearby agricultural communities and residentials (Kawahara et al. 2005). Pesticides also can be easily leached in permeable soil structures, leading to water percolation that disseminates the pesticide solutes deep into the groundwater. An investigation by the United States Geological Survey (USGS) detected different types of pesticides in more than 90% of water and fish samples collected from their streams (Rose et al. 2018), validating the abundance of toxic chemicals in the environment resulted from prolonged and massive pesticides utilization in the farm.

Pesticides are also susceptible to catalytic process and degradation either by environmental microorganisms, i.e., bacteria and fungi, chemical reactions, or sunlight exposure resulting in metabolite transformation. Some of them are hazardous such as the endocrine-disrupting chemicals chlorpyrifos and its derivatives that may possibly be posing potential risks to human well-being (Yue et al. 2017). Those hazardous derivatives might possibly leach into the groundwater hence contaminating the river basins and threatening the life of aquatic inhabitants and the surrounding residentials.

2.3.2 Impact of Overfertilization in Agriculture Industries

Fertilizers are categorized into mineral fertilizers, consisting predominantly of phosphorus and potassium derived from the environment. Organic fertilizers are composed of animal excreta and decomposed plant or animal matter, while industrial fertilizers, including ammonium phosphate, urea, and ammonium nitrate are manufactured through chemical processes. Among these, industrial or chemical fertilizers are prevalent in modern agricultural practices (Sedlacek et al. 2020). The data presented in Fig. 2 indicates a 14 percent increase in the global average cropland nitrogen budget (a record of nitrogen inputs, outputs, and retention in the soil at a field scale), reaching 54 kg/ha between 2000 and 2020, despite experiencing a declining trend in the 2010s. Notably, in 2020, Asia exhibited the highest cropland nitrogen budget at 93 kg/ha, significantly surpassing the budgets of the Americas (40 kg/ha), Europe (38 kg/ha), Oceania (22 kg/ha), and Africa (12 kg/ha). Furthermore, the Asia region has reported the highest levels of synthetic fertilizer inputs, leading to the excessive accumulation of chemical components in farm soil and increased potential for agricultural pollution in the surrounding environment (Food and Agriculture Organization 2023).

While plant productivity is significantly improved via fertilization, excessive use of artificial fertilizers in the farm also contributes to dreadful consequences such as water pollution, soil nutritional imbalances, and eutrophication due to nutritional runoff (Penuelas et al. 2023). High anthropogenic nitrogen inputs in the farm soil not only lead to contaminated drinking water issues and toxic algal blooms in downstream ecosystems but also induce global climate change due to accelerated nitrous oxide emission from synthetic fertilizers that have been naturally metabolized by native microorganisms in the soil (Davidson 2009). Even for the workers, prolonged

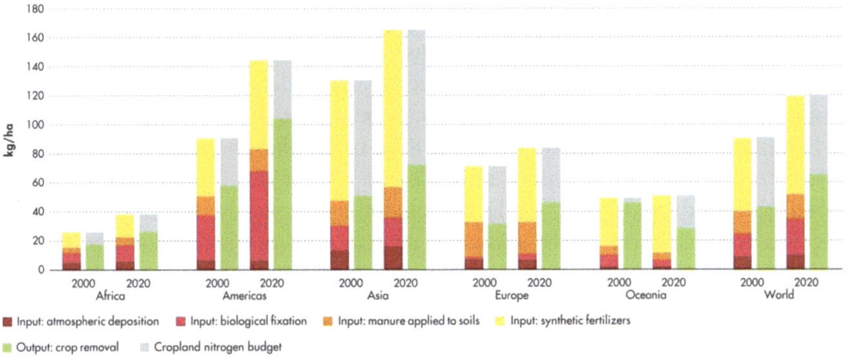

Fig. 2 The cropland nitrogen budget categorized by component and regions. *Source* FAO (2022). Adapted from Food and Agriculture Organization (2023)

occupational exposure to the dust derived from the chemical fertilizer may result in severe medical issues such as kidney and liver impairments, immune system disorders, hormonal or reproductive problems, and even cancer (Abhilash and Singh 2009; Menadi et al. 2024). The presence of xenobiotics derived from chemical fertilizers or pesticides may affect humans intracellularly, leading to alternated DNA methylation and genomic mutation (Fauteux et al. 2023).

Furthermore, heavy metals such as cadmium (Cd), lead (Pb), copper (Cu), zinc (Zn), and mercury originated from the agro waste burning, fertilizers, and pesticide fumigation also can poison agriculture products and interfere with the food supply (Mortvedt 1995; Alengebawy et al. 2021). Heavy metals can be absorbed by the plant roots through phytoextraction, phytostabilization, and rhizofiltration processes, which may cause damages to plants. Furthermore, these heavy metals can be transferred through the food chain to ultimately harm human health (Tangahu et al. 2011).

Commercial phosphate fertilizers generally contain small amounts of heavy metal contaminants which were originally minor constituents found in rock phosphate. Research has shown that rock phosphate contains the greatest level of cadmium and zinc. Chromium was the highest heavy metal in diammonium phosphate, while copper and lead were high in superphosphate samples (Carnelo et al. 1997). Cadmium (Cd) is a toxic non-essential transition metal of concern because it may affect animal and human health given its biopersistent nature (residues remain in bodies with a long half-life of about 25–30 years), causing critical and permanent damages at the cellular level (apoptosis) (Genchi et al. 2020). Exposure to cadmium may induce oxidative stress in cells due to mitochondria dysfunction, leading to irreversible damages in liver and kidney. Cadmium possesses highly toxic properties that can lead to carcinogenic effects on human health by inducing various epigenetic changes in mammalian cells (Tinkov et al. 2018). These changes can occur both in vivo (within living organisms) and in vitro (in laboratory settings), and can include alterations to DNA methylation patterns, histone modifications, and non-coding RNA expression.

Such epigenetic modifications can ultimately lead to the development of various types of cancers and other pathological conditions, such as chronic inflammation, immune dysfunction, osteoporosis, and neurodegenerative disorders such as Alzheimer's and Parkinson's diseases (Bakulski et al. 2020; Genchi et al. 2020). A report conducted by Santillán in 2023 revealed that about 80% of all phosphorus mined is utilized in fertilizer production. However, only a mere 17% of this phosphorus is used in agricultural food production, while the remaining amount is lost through various processes such as absorption by non-target crops, runoff, and soil erosion. Moreover, an alarming 46% of cadmium was found to be lost directly from farms due to soil leaching and erosion (Santillán 2023). This finding highlights the potential dangers and risks associated with cadmium exposure to neighboring communities.

2.4 Water Pollution

The 2030 Agenda for Sustainable Development acknowledges the immense importance of decent water quality via the Development Goal SDG 6. However, the problem of water pollution is revealed as a significant global issue in underdeveloped, modern, and developing countries. This issue has not only impeded economic growth but also poses a severe threat to the physical and environmental well-being of billions of people around the world (Mateo-Sagasta et al. 2017).

2.4.1 Water Pollution Due to Agricultural Activities

Worryingly, agriculture industry which accounts for 70% of water abstractions worldwide, plays a major role in water pollution. Farms' organic and inorganic contaminants such as fertilizers, pesticides, sediment, field runoff, animal feedlots, and concentrated animal feeding operations (CAFOs) can release nutrients and chemicals into water bodies, causing eutrophication, and decreased water quality, adversely affecting the aquatic ecosystem (Zahoor and Mushtaq 2023). Fertilizers and pesticides including herbicides, fungicides, and insecticides can contaminate both ground and surface water through soil runoff, leaching, and spray drift that occur during crops or commodities cultivation (Tudi et al. 2021; Zahoor and Mushtaq 2023). Heavy metals from pesticides, fertilizers, and even from animal feed additives may have serious consequences as some metals, like lead and mercury, are poisonous not only to the animal but also to consumers (Tao et al. 2020). The heavy metals can leach to the groundwater and contaminate the environment (Santillán 2023). Furthermore, in compliance of safety measurement and mishandling of agrochemical residues, i.e., disposal of pesticide containers in lakes, streams, or open fields by untrained or oblivious individuals may lead to their outflow into adjacent water sources, causing detrimental effects on aquatic inhabitants while threatening human health (Obi et al.2016).

The issue of chemical pollution caused by agricultural activities has gained global attention. The nitrogen pollution level is projected to be 150% higher by 2050 compared to 2010, of which the agricultural sector contributing to 60% of this augmentation (Martínez-Dalmau et al. 2021). According to a report by the Food and Agriculture Organization of the United Nations, nitrate from agriculture is the most widespread chemical contaminant in the world's groundwater aquifers (Mateo-Sagasta et al. 2017). Moreover, in the European Union (EU), 38% of water bodies are heavily impacted by agricultural pollution (Connor 2015). In soil, the uptake of nitrogen by plant roots occurs from a small mineral-nitrogen pool, comprised of nitrate and ammonium ions. Although applied fertilizer contributes directly to this pool, it is consistently replenished through the mineralization of organic nitrogen within the extensive soil nitrogen reservoir (Craswell 2021). A fraction of the applied fertilizer nitrogen is directly utilized by the intended crop, while another portion is lost through various mechanisms, including leaching as nitrate from the soil–plant system, and in gaseous forms such as N_2O and NH_3 (Craswell 2021). It has been estimated that, on average, only 51% of the nitrogen applied is recuperated by the above-ground components of crops, such as cereals. Furthermore, when applied at elevated rates, the average retrieval of fertilizer nitrogen by crop plants may be even lower (Dobermann and Cassman 2005). As a result, countries such as India (Sinha et al. 2022) and China (particularly the Eastern Plain, Yunnan–Guizhou Plateau, and Xinjiang) (Ji et al. 2024) have recorded increased nitrogen and phosphorous water pollution levels, hence resulting in horrendous freshwater eutrophication due to the discharge of the nutrient-overloaded aquifer to a surface body. Despite substantial financial investments by China for the restoration of lakes and rivers, the full recovery of environmental quality has not been achieved. Notably, the influx of nitrogen and phosphorus has led to significant water quality decrement in Taihu Lake (Guo 2007) and Fuxian Lake (Zhuo and Zeng 2020). Consequently, the expenses incurred for environmental remediation have exceeded more than US$30 billion for Taihu Lake (Zhang et al. 2023) and nearly US$4.21 billion for Fuxian Lake (Ji et al. 2024). Despite these efforts, the lakes' water quality remains under threat from eutrophication. The overabundance of inorganic compounds disrupts the balanced ecosystem, decreasing the biodiversity of its native inhabitants and creating favorable conditions for the uncontrolled growth of cyanobacteria, ultimately leading to the collapse of the whole habitat. Even worse, cyanobacteria produce harmful toxins such as microcystins and cyanotoxins, which are toxic to certain organisms, initiating the contamination of different levels of the food chain until it reaches human consumption (Thawabteh et al. 2023). This highlights the urgent need for sustainable agricultural practices to prevent further contamination of available water resources.

2.4.2 Water Pollution from Animal Feeding Operation

Livestock production including cattle, pigs, and chickens can be a significant source of pollutants in rivers. As the global population continues to grow and dietary habits change, the demand for animal products is also increasing. Although animal manure can serve as a natural fertilizer to promote plant growth and balance soil acidity, it's crucial to bear in mind that this substance contains more than just nitrogen and phosphate. It may also harbor dangerous pathogens and heavy metals that pose a threat to human health, leading to severe illnesses (Li et al. 2022). Studies have shown that livestock manures are commonly associated with zoonotic pathogens, for example, bacteria *Campylobacter* sp., *Escherichia coli*, *Salmonella* sp., *Providencia* sp., and *Listeria* sp. (Black et al. 2021) and parasites such as nematode *Moniezia* sp., *Trichuris* sp., *Nematodirus* sp., and *Eimeria* sp. (Davey et al. 2021) as well as pathogenic protozoa *Cryptosporidium* that can cause acute gastroenteritis and diarrhea (Li et al. 2022).

Generally, the nutrients, pathogens, and parasites present in animal manure can enter rivers via two main routes. Firstly, when manure is used to fertilize crop fields, rainwater can cause the substances to be carried along in surface runoff and eventually end up in rivers. Secondly, if manure is directly discharged into rivers, it also can introduce these substances into the water (Strokal et al. 2019). Several water pollution cases caused by animal manure have been reported within these few years. For example, the concentration of total nitrogen (TN) and total phosphate (TP) in the Huaihe River Basin, specifically in Shangqiu, Zhoukou, Heze, Zhumadian, Luohe, Jining, Xuchang, Kaifeng, Taian, and Zhengzhou, has exceeded the threshold value of 10.00 mg/L for TN and 0.08 mg/L for TP, respectively. This excess has led to critical water eutrophication (Wang et al. 2021). Moreover, an estimated 78% of coastal waters in the United States (US) are affected by eutrophication, which results in the growth of harmful algae blooms, decreased oxygen levels, and a decline in aquatic biodiversity (Malone and Newton 2020). Contamination caused by *Cryptosporidium* and bacterial pathogens were also detected in surface and groundwater for domestic water use in countries such as Haiti (Balthazard-Accou et al. 2020) and North America, causing massive water and foodborne outbreaks (Oun et al. 2014).

2.5 *Microplastic Pollution*

The International Organization for Standardization (ISO) defines microplastics as any solid plastic particles insoluble in water with dimension ranges between 1 and 1000 μm (ISO 2020). Microplastic pollution has become a global environmental concern, and its impact on aquatic ecosystems has been well-documented. However, there is growing evidence that agricultural activities also contribute significantly to microplastic pollution (Schell et al. 2020). Recent studies across 19 provinces in China and cotton fields in Shihezi, Xinjiang Uygur Autonomous Region have shown that agricultural activities are a major source of macro- and microplastic pollution, with the extensive use of plastic mulch, agrochemicals, and improper disposal of plastic waste contributing to the problem. The use of plastic mulch or irrigation pipes in agriculture, for example, has been found to degrade over time due to exposure to sunlight and mechanical stress, leading to the release of microplastics into the environment (Huang et al. 2020). Furthermore, the application of agrochemicals such as pesticides and fertilizers also plays a significant role in the release of microplastics. Plastic residues from pesticide containers and microplastic particles from fertilizers can contaminate soil and water bodies, posing a threat to terrestrial and aquatic organisms (Kumar et al. 2020). Additionally, the use of compost or biosolid in farming also can introduce plastics into the soil. Even though bigger plastic pieces are usually filtered out during the composting process, smaller fragments can still form during the milling stage and become secondary micro or nanoplastics that eventually accumulate abundantly in the soil (Kumar et al. 2020).

Understanding the long-term effects of microplastic exposure in agroecosystems is essential for evaluating its impact on agricultural sustainability. Long-term exposure to microplastics may lead to the gradual buildup of these particles in soil matrices, potentially posing ongoing challenges for agricultural productivity and environmental quality (Kumar et al. 2020; Kye et al. 2023).

Various impacts of microplastics/nanoplastics on plant growth have been identified. These include direct absorption and accumulation of these particles into plant cells via root (endocytosis) and foliar uptakes (Sun et al. 2021), as well as interference with different cellular activities in plants (Rillig et al. 2019). Microplastics can also hinder water and nutrient uptake in plants by blocking the cell wall pores, resulting in perturbation with their physiological processes (Bosker et al. 2019). This, in turn, can affect both the above- and below-ground biomass productions simultaneously, as it can hamper seed germination and root growth (Roy et al. 2023). Furthermore, the existence of microplastic can rupture the soil structure while reducing the infiltration of rainwater and irrigation water, decreasing the soil's water retention capacity, disturbing the soil microbiota while causing pH imbalance in soil and interfering with the nutrient transfer (Roy et al. 2023). Once released into the environment, microplastics come into contact with different chemicals such as organic contaminants and heavy metals, which they can transport as dangerous substances (Torres et al. 2021).

Furthermore, the potential transfer of microplastics through the food web in agroecosystems also warrants further precautionary actions. Given that microplastics are easily ingested by soil-dwelling organisms (i.e., nematodes, earthworms, isopods, collembolan, and oligochaeta) and absorbed by crop plants, there is a risk of biomagnification, where these particles may accumulate and transferred to higher trophic levels, including animals and humans (Kim and An 2019). Research has revealed that the tiny nematode *Caenorhabditis elegans* found in the soil is susceptible to the harmful effects of microplastics. When the nematodes consume these microplastics, the particles accumulate in their bodies, leading to a shorter lifespan, abnormal morphological structure, and irreversible damage to their GABAergic and cholinergic neurons which represent two critical components of their nervous system (Lei et al. 2018).

In general, there are two categories of chemicals associated with plastic that pose significant health risks to humans; bisphenols A (BPA) and phthalates. BPA is a chemical that is commonly used in the manufacturing of everyday materials and has been found to possess estrogenic activity and can disrupt the endocrine system.

Moreover, BPA can harm the human immune system by interfering with the communication of specific receptors, including the estrogen receptor, aryl hydrocarbon receptor, and peroxisome proliferator-activated nuclear receptors, while inducing abnormal immune reactions (Rogers et al. 2013; Segovia-Mendoza et al. 2020). On the other hand, phthalates, i.e., dibutyl phthalate (DBP), diethyl phthalate (DEP), and di-ethylhexyl phthalate (DEHP), are common plasticizers in the microplastic components that can leach to the environment and assimilate into the food chain. Phthalates were found affecting the zebrafish survival rate while causing morphological abnormalities and changes in genes expression (Mansuri et al. 2023). Worryingly, both BPA and phthalates can leach into the soil, hence contaminating the groundwater, drinking water, sediments, and sludge, making them ubiquitous in the environment (Zhu et al. 2022). Even worse, plants grown in BPA and phthalate-contaminated soil can absorb these chemicals through their roots. As an example, research has confirmed that phthalate esters (PAEs) are pervasive contaminants in large-scale greenhouse vegetable production in Beijing, China. The total concentrations of PAEs ranged from 0.14 to 2.13 mg/kg (mean 0.99 mg/kg) in soils and from 0.15 to 6.94 mg/kg (mean 1.49 mg/kg) in vegetables (Li et al. 2016). This potential transfer of microplastics through the agroecosystem has massive implications for food safety and human health, emphasizing the need for comprehensive risk assessments and management strategies to mitigate the long-term impacts of microplastic pollution in agricultural farms.

2.6 Antibiotic-Resistance Genes (ARGs) Contamination in Agriculture Settings

Overuse of antibiotics in agriculture has emerged as a significant concern in recent years. This practice involves the routine administration of antibiotics to livestock through feed, drinking water, or injection and the use of antibiotics as growth promoters in crops, leading to the development of antimicrobial resistance among bacteria present in agricultural farms (McEwen and Collignon 2018). Antibiotics are commonly used in livestock farming, poultry, fisheries and animal husbandry to promote animal growth and prevent infectious disease outbreaks. It has been reported that China consumes about half of the antibiotics used in animal husbandry worldwide, followed by the US, Brazil, India, and Germany (Laxminarayan et al. 2015). Without significant policy changes, the use of antimicrobials in animals for food production is expected to surge by 67% by 2030. This trend will be most pronounced in emerging economies, where demand for livestock products, particularly poultry, is growing at an accelerated pace. Such a surge poses a major global health threat by fueling the development and widespread of antimicrobial resistance, resulting in the emergence of untreatable infections (Laxminarayan et al. 2015).

Livestock feces is commonly utilized as organic soil fertilizer in farming worldwide as a substitute for chemical fertilizers. Although it contains valuable nutrients and organic-rich properties, it is frequently found to contain remnants of antibiotics, antibiotic-resistance genes (ARGs), and bacteria that are resistant to antibiotics even in manure-amended soils (Lima et al. 2020). The ARGs can cause bacteria to become resistant to antibiotics via different methods including specific mutations in genes that are linked to the antibiotic mechanisms of action (MOA) and by obtaining foreign DNA that encodes resistance determinants through conjugation, transformation or transduction; all of which are related to horizontal gene transfer (HGT) (Munita and Arias 2016). Ingestion of foods contaminated with those resistant bacteria enables its transmission, facilitating the pathogen dissemination within the community and the environment (Parkhill 2022). In the United States, it has been observed that closely related strains of *Klebsiella pneumoniae* were detected on retail meat in Flagstaff grocery stores, as well as in clinical samples from patients experiencing urinary tract infections. This indicates a possible link between the consumption of contaminated meat and the prevalence of this bacterial strain in clinical settings (Davis et al. 2015).

Furthermore, the heavy metals present in manure may exert a long-term and continuous selective pressure on ARGs via co-resistance or cross-resistance because the resistance genes for antibiotics and metals are often located together in the same plasmid or other mobile genetic element (MGE) (Yuan et al. 2020). Nowadays, ARGs are considered as emerging contaminants, considering the potential risks and their origin, mobility and elimination in the environment (Lima et al. 2020; Yuan et al. 2020). The use of antibiotics during animal growth favors the appearance of antibiotic-resistant bacteria in the gut microbiota, which consequently may be released within feces and disseminated by manure application in the farm soil (Heuer et al. 2011; Lima et al. 2020).

3 Global Agriculture Pollution Scenario

Agricultural pollution occurs not only in low-income countries but also in the urban and developing regions. These nations use different agricultural approaches, creating various environmental impacts, and embrace different regulatory responses to handle the situation.

In developed nations like the United States (US) and countries within the European Union (EU), agriculture is marked by massive utilization of chemical fertilizers, pesticides, and intensive application of antibiotics in livestock farming to enhance agricultural productivity while maintaining its excellent quality (Laxminarayan et al. 2015). In intensive-scale farming, emerging technologies such as geospatial technologies, the Internet of Things (IoT), Big Data analysis, and artificial intelligence (AI), all of which are designed for precision agriculture, could be utilized to make informed management decisions aimed at increasing crop production (Sishodia et al. 2020). In addition, sprayer-mounted drones have made pesticide application quick, efficient, and safe. Advanced sensors and high-resolution cameras on drones accurately differentiate between healthy and unhealthy plants and identify symptoms of plant diseases and weeds existence (Das et al. 2024). Despite the advancement of those sophisticated technologies, the excessive use of synthetic fertilizers and pesticides continues to be a major contributing factor to serious environmental pollution. This is because synthetic fertilizers often utilize non-renewable resources and release greenhouse gases into the atmosphere, which exacerbates climate change. It is important to consider alternative and sustainable methods of agriculture in order to mitigate the negative impact on the environment (Menegat et al. 2022). Therefore, laws and regulations are implemented as mitigation approach to control environmental pollution due to agricultural activities. This includes the EU's Council Directive 91/676/EEC by the EU countries that was introduced to address the issue of nitrates pollution from synthetic fertilizers application in agricultural practices (Food and Agriculture Organization 2022). Additionally, the European Chemicals Agency (ECHA) has enacted legislation regarding chemicals and pesticides in the EU. This legislation controls the marketing and use of specific categories of chemical products, as well as the commercialization and utilization of hazardous substances. Furthermore, protocols have been established for handling major accidents and exports of dangerous substances, indicating a high environmental safety awareness and efficient mitigation approaches against pollution (Georgios and Maria-Mirela 2023).

In some developing countries like India and China, agricultural pollution is mainly caused by conservative farming methods, poor waste management, and limited access to advanced technology (Novotny et al. 2010). Many farmers in China and India still rely on traditional irrigation methods such as flood or surface irrigation, which can lead to water waste, soil erosion, and salinization (Yang et al. 2018). Traditional farming practices frequently use excessive amounts of chemical fertilizers and pesticides, resulting in soil degradation, water pollution, and health risks. Therefore, transitioning to integrated pest management (IPM) and organic farming practices can reduce reliance on chemical inputs while improving soil health and biodiversity (Shukla et al. 2022). India's National Mission for Sustainable Agriculture (NMSA) promotes eco-friendly and sustainable agricultural practices by addressing the outdated approaches. The NMSA provides financial assistance, capacity building, and extension services to promote organic farming, soil health management, water conservation, and adoption of agroforestry (Priyadarshini and Abhilash 2020). Despite their potential, some countries still face challenges in creating and enforcing robust regulatory frameworks due to limited resources and enforcement capacity.

On the other hand, low-income regions such as those found in sub-Saharan Africa countries like Ethiopia and Kenya, rely heavily on agriculture for their livelihoods. However, they face distinct obstacles when it comes to agricultural pollution. These issues include inefficient use of resources, soil degradation, and infertility, as well as inadequate waste management. These factors contribute to pollution and possess a negative impact on human well-being. (Mateo-Sagasta et al. 2018). Regulatory initiatives in low-income countries are often in their early stages, with international aid and assistance playing a crucial role. As an example, the Comprehensive Africa Agriculture Development Program by the African Union offers assistance for the promotion of sustainable agriculture (Brüntrup 2011). In a similar vein, the Agriculture Pollution Control Project in Bangladesh, spearheaded by the World Bank, seeks to tackle the pressing issue of pollution (Faroque et al. 2011). Nevertheless, the impact of these initiatives is hampered due to a lack of resources, highlighting the importance of ongoing international cooperation to develop regulatory capabilities and promote environmentally friendly farming methods.

Despite the best efforts, finding a middle ground between agricultural productivity and environmental sustainability remains to be a persistent challenge in reality. For years, unintended consequences of modern agricultural practices have resulted in the occurrence of different types of pollution, which poses a serious threat to ecosystems, human health, and the planet's sustainability (Table 1). Several disastrous incidents related to pollution issues in agriculture industries spanning from the 1980s to 2022 are also described in Table 2. Understanding these events is crucial in highlighting the significance of sustainable agricultural practices, environmental protection, and disaster preparedness in protecting our ecosystems and the global food supply (Parr et al. 2020).

4 Conclusion and Future Perspective

Overall, the agricultural sector, while essential for sustaining human life and meeting global food demands, significantly contributes to environmental pollution through various means such as soil degradation, air, chemical, water and microplastic contamination, and antibiotic resistance gene dissemination. These pollutants pose significant threats not only to the environment but also to human health, wildlife, and ecosystem integrity. Addressing environmental pollution in agriculture requires a multifaceted approach involving policymakers, farmers, researchers, and consumers. Regulation and enforcement of environmental standards, adoption of sustainable agricultural practices, promotion of organic farming methods, and investment in innovative technologies are crucial steps toward mitigating pollution in the agricultural sector.

Furthermore, fostering awareness and education among stakeholders about the adverse effects of agricultural pollution and the importance of sustainable farming practices is paramount. Collaboration between governments, international organizations, NGOs, and private sectors is essential to implement effective policies and initiatives aimed at reducing pollution while ensuring food security and agricultural sustainability. Looking ahead, it is imperative to prioritize research and development efforts toward the development of eco-friendly farming practices, efficient waste management systems, and alternative solutions to chemical fertilizers and pesticides. Embracing emerging technologies such as precision agriculture, vertical farming, and agroecology can contribute to minimizing the environmental footprint of agriculture while enhancing productivity and resilience to climate change. Undeniably, tackling environmental pollution in the agriculture sector is a complex and pressing challenge that requires collective action and commitment from all stakeholders. By adopting sustainable practices and embracing innovation, the adverse impacts of agriculture on the environment can be mitigated paving the way for a more resilient and sustainable future for generations to come.

Table 1 Examples of global agricultural pollution disaster

Agricultural pollution disaster	Event occurrence	Common locations	References
Algal blooms and dead zones	Excessive nutrient runoff caused detrimental effects on water bodies, resulting in algal blooms that deplete oxygen levels and create dead zones where aquatic life cannot survive	Coastal areas, e.g., the Gulf of Mexico, and Chesapeake Bay	Wurtsbaugh et al. (2019)
Drinking water contamination	Groundwater contamination from agricultural runoff, which included pesticides, herbicides, and fertilizers, led to the spread of waterborne diseases, posing health risks to both humans and animals	Agricultural regions globally, e.g., Midwest USA, Punjab	Schaider et al. (2019, Syafrudin et al. (2021)
Soil erosion and desertification	Improper agricultural practices caused detrimental effects on the environment such as soil erosion, which led to the loss of fertile topsoil. This loss of topsoil was then resulted in reduced agricultural productivity and predisposed the land to become more susceptible to desertification	Global, e.g., sub-Saharan Africa, China	Becerril-Piña and Mastachi-Loza (2021)
Crop failures and food insecurity	Factors such as soil quality, contamination, and extreme weather events resulting from agricultural practices contributed to crop failures, which in turn caused food shortages and insecurity	Developing nations, e.g., parts of Africa, Southeast Asia	Onyutha (2019)
Air quality degradation	Agricultural activities, such as the burning of crop residues and the release of ammonia emissions caused a detrimental impact on air quality. This, in turn, led to respiratory problems and health issues for both humans and animals	Agricultural regions, e.g., India, Central Valley (USA)	Hill et al. (2019)
Loss of biodiversity	The decline in biodiversity caused by habitat destruction and chemical pollution from agriculture had significant consequences, disrupting ecosystems and reducing the vital services they provide	Global, e.g., Amazon Rainforest, European farmlands	Kok et al. (2018)
Resistance to pesticides and herbicides	The repeated application of pesticides and herbicides resulted in the emergence of resistant strains among pest populations, rendering these chemicals ineffective and leading to significant economic losses	Agricultural regions worldwide	Argüelles and March (2023)

Table 2 Agricultural pollution disasters tragedy (1984–2022)

Date	Disaster	Location	Description	References
1984	Bhopal Gas Tragedy	Bhopal, India	The release of methyl isocyanate gas from pesticide manufacturing plants resulted in a significant loss of life and ongoing health complications for many individuals. The catastrophe had significant impacts on the crops and soil in the neighboring regions	Palazzi et al. (2015)
2008	Kingston Fossil Plant Spill	Tennessee, USA	There was a spill of coal ash slurry that had a significant impact on water sources and farmlands. The spill resulted in the presence of heavy metals such as arsenic and mercury, causing significant and enduring harm to the environment and contamination of crops	Ramsey et al. (2019)
2010	Deepwater Horizon Oil Spill	Gulf of Mexico, USA	Although primarily an oil spill disaster, it had significant effects on fisheries and coastal agriculture. The presence of oil in the water had detrimental effects on marine life, leading to a negative impact on the seafood industry and indirectly affecting the livelihoods of farmers	(Quigg et al. 2021)
2014	Elk River Chemical Spill	West Virginia, USA	An unfortunate incident occurred when a storage tank released a chemical used in coal processing into the Elk River, resulting in the contamination of the water supply. Accessing clean water for irrigation has been a significant challenge for farmers, leading to a negative impact on crop production	Horzmann et al. (2017)
2015	Tianjin port explosion	Tianjin, China	In addition to its industrial consequences, the explosion had a detrimental effect on the surrounding agricultural regions due to the release of toxic chemicals. The contamination of soil and water had detrimental effects on the crops and livestock, greatly affecting the livelihood of the local farmers	Wang et al. (2023)

(continued)

Table 2 (continued)

Date	Disaster	Location	Description	References
2019	Amazon rainforest fires	Amazon Basin, South America	The extensive deforestation and deliberate fires have resulted in the degradation of soil, the decline of biodiversity, and the disruption of local agriculture. The fires emitted significant quantities of carbon, which had a profound effect on the Earth's climate	Costa et al. (2001)
2020	Beirut port explosion	Beirut, Lebanon	The explosion had a significant impact on urban agriculture in the surrounding area due to the release of hazardous chemicals. The contamination of soil and water presented significant challenges for local farmers, impacting the growth and health of their crops	Wang et al. (2023)
2022	Fukushima nuclear disaster	Fukushima, Japan	Radioactive contamination has a significant impact on both farmlands and fisheries, extending beyond just an agricultural disaster. The future implications for agriculture are still being revealed	Kawashima et al. (2023), Kawashima et al. (2023)

References

Abdel Hameed AA et al (2015) Airborne fungal pollution at waste application facilities. Aerobiologia 31:283–293

Abhilash PC, Singh N (2009) Pesticide use and application: an Indian scenario. J Hazard Mater 165(1–3):1–12

Alengebawy A et al (2021) Heavy metals and pesticides toxicity in agricultural soil and plants: ecological risks and human health implications. Toxics 9(3):42

Argüelles L, March H (2023) A relational approach to pesticide use: farmers, herbicides, nutsedge, and the weedy path to pesticide use reduction objectives. J Rural Stud 101:103046

Azeez OM (2021) Evaluation of pesticide application equipment efficiency for pests control and safety of the users, a review. J Exp Agric Int 43(5):54–70

Bajracharya SB, Mishra A, Maharjan A (2021) Determinants of crop residue burning practice in the Terai region of Nepal. Plos One 16(7):e0253939

Bakulski KM et al (2020) Heavy metals exposure and Alzheimer's disease and related dementias. J Alzheimer's Dis 76(4):1215–1242

Balthazard-Accou K et al (2020) Pollution of water resources and environmental impacts in urban areas of developing countries: case of the City of Les Cayes (Haiti). In: Makan A (ed) Environmental health management and prevention practice. https://doi.org/10.5772/intechopen.86951. Accessed 2 Feb 2024

Becerril-Piña R, Mastachi-Loza CA (2021) Desertification: causes and countermeasures. In: Leal Filho W, Azul A, Brandli L, Özuyar P, Wall T (eds) Life on land. Encyclopedia of the UN sustainable development goals. Springer, Cham, pp 219–231. https://doi.org/10.1007/978-3-319-71065-5_81-1. Accessed 5 Mar 2024

Bhuvaneshwari S, Hettiarachchi H, Meegoda JN (2019) Crop residue burning in India: policy challenges and potential solutions. Int J Environ Res Public Health 16(5):832

Black Z et al (2021) The fate of foodborne pathogens in manure treated soil. Front Microbiol 12:781357

Bosker T et al (2019) Microplastics accumulate on pores in seed capsule and delay germination and root growth of the terrestrial vascular plant *Lepidium sativum*. Chemosphere 226:774–781

Bowe B et al (2019) Burden of cause-specific mortality associated with PM2. 5 air pollution in the United States. JAMA Netw Open 2(11):e1915834–e1915834

Brüntrup M (2011) The comprehensive Africa Agriculture Development Programme (CAADP)—An assessment of a Pan-African attempt to revitalise agriculture. Q J Int Agric 50(1):79–106

Carnelo LGL, de Miguez SR, Marbán L (1997) Heavy metals input with phosphate fertilizers used in Argentina. Sci Total Environ 204(3):245–250

Commission for Environmental Cooperation (2014) Burning Agricultural Waste: a source of dioxins. Montreal, Canada. http://www.cec.org/files/documents/publications/11405-la-quema-de-residuos-agr-colas-es-una-fuente-de-dioxinas-en.pdf. Accessed 21 Jan 2024

Connecticut Department of Agriculture (2024) Definition of Agriculture. https://portal.ct.gov/doag/commissioner/commissioner/definition-of-agriculture. Accessed 2 Mar 2024

Connor R (2015) The United Nations world water development report 2015: water for a sustainable world. UNESCO publishing. https://unesdoc.unesco.org/ark:/48223/pf0000231823. Accessed 5 Mar 2024

Costa JBS et al (2001) Tectonics and paleogeography along the Amazon River. J S Am Earth Sci 14(4):335–347

Craswell E (2021) Fertilizers and nitrate pollution of surface and ground water: an increasingly pervasive global problem. SN Appl Sci 3(4):518

Cui L et al (2021) Nitrous acid emission from open burning of major crop residues in mainland China. Atmos Environ 244:117950

Das A et al (2024) Drone-based intelligent spraying of pesticides: current challenges and its future prospects. In: Chouhan SS, Singh UP, Jain S (eds) Applications of computer vision and drone technology in agriculture 4.0. Springer, pp 199–223. https://doi.org/10.1007/978-981-99-8684-2_12. Accessed 5 Jan 2024

Davey ML, Utaaker KS, Fossøy F (2021) Characterizing parasitic nematode faunas in faeces and soil using DNA metabarcoding. Parasit Vectors 14(1):422

Davidson EA (2009) The contribution of manure and fertilizer nitrogen to atmospheric nitrous oxide since 1860. Nat Geosci 2(9):659–662

Davis GS et al (2015) Intermingled *Klebsiella pneumoniae* populations between retail meats and human urinary tract infections. Clin Infect Dis 61(6):892–899

Djebbi E et al (2021) Effects of nickel oxide nanoparticles on survival, reproduction, and oxidative stress biomarkers in the marine calanoid copepod *Centropages ponticus* under short-term exposure. Environ Sci Pollut Res 28:21978–21990

Dobermann A, Cassman KG (2005) Cereal area and nitrogen use efficiency are drivers of future nitrogen fertilizer consumption. Sci China Ser C Life Sci 48:745–758

Du Y, Wang W (2023) The role of green financing, agriculture development, geopolitical risk, and natural resource on environmental pollution in China. Resour Policy 82:103440

Food and Agriculture Organization (2020) Sustainable Food and Agriculture. https://www.fao.org/sustainability/news/detail/en/c/1274219/. Accessed 3 Feb 2024

Food and Agriculture Organization (2022) Council Directive 91/676/EEC concerning the protection of waters against pollution caused by nitrates from agricultural sources., FAOLEX Database. https://www.fao.org/faolex/results/details/en/c/LEX-FAOC019065/. Accessed 24 Mar 2024

Food and Agriculture Organization (2023) World Food and Agriculture—Statistical Yearbook 2023. Rome, Italy. https://doi.org/10.4060/cc8166en. Accessed 2 Feb 2024

Faroque MAA, Kashem MA, Bilkis SE (2011) Sustainable agriculture: a challenge in Bangladesh. Int J Agric Res Innov Technol (IJARIT) 1:1–8

Fauteux M et al (2023) Differential effects of pesticides on dioxin receptor signaling and p53 activation. Sci Rep 13(1):21211

Frączek K et al (2017) Fungal air contamination in distinct sites within a municipal landfill area. Int J Environ Sci Technol 14:2637–2648

Friedrich T (1996) Agricultural pesticide application. FAO Agricultural Engineering Branch AGSE, FAO Rome. https://www.fao.org/fileadmin/user_upload/ags/publications/pesticide_friedrich_1.pdf. Accessed 25 Mar 2024

Genchi G et al (2020) The effects of cadmium toxicity. Int J Environ Res Public Health 17(11):3782

Georgios A, Maria-Mirela C (2023) Chemicals and pesticides, Fact Sheets on the European Union (European Parliament. https://www.europarl.europa.eu/factsheets/en/sheet/78/chemicals-and-pesticides. Accessed 24 Mar 2024

Gokhale CS, Sharma N (2023) Optimizing crop rotations via Parrondo's paradox for sustainable agriculture. Royal Soc Open Sci 10(5):221401

Guo J et al (2022) Farmland scale and chemical fertilizer use in rural China: New evidence from the perspective of nutrient elements. J Clean Prod 376:134278

Guo L (2007) Doing battle with the green monster of Taihu Lake. Science 317(5842):1166

Gupta PK et al (2004) Residue burning in rice–wheat cropping system: causes and implications. Curr Sci 1713–1717

Harris DR, Fuller DQ (2014) Agriculture: definition and overview. In: Encyclopedia of global archaeology, pp 104–113

Heuer H, Schmitt H, Smalla K (2011) Antibiotic resistance gene spread due to manure application on agricultural fields. Curr Opin Microbiol 14(3):236–243

Hill J et al (2019) Air-quality-related health damages of maize. Nat Sustain 2(5):397–403

Horzmann KA et al (2017) Comparative analytical and toxicological assessment of methylcyclohexanemethanol (MCHM) mixtures associated with the Elk River chemical spill. Chemosphere 188:599–607

Huang C-Y et al (2002) The seasonal distribution of bioaerosols in municipal landfill sites: a 3-yr study. Atmos Environ 36(27):4385–4395

Huang Y et al (2020) Agricultural plastic mulching as a source of microplastics in the terrestrial environment. Environ Pollut 260:114096

ISO U (2020) TR 21960: 2020; plastics-environmental aspects-state of knowledge and methodologies. International Organization for Standardization: Geneva, Switzerland. https://www.iso.org/standard/72300.html. Accessed 11 Nov 2023

Ji P et al (2024) Nitrogen and phosphorus trends in lake sediments of China may diverge. Nat Commun 15(1):2644

Karuppuchamy P, Venugopal S (2016) Chapter 21—Integrated pest management. In: Omkar (ed) Ecofriendly pest management for food security. Academic Press, San Diego, pp 651–684. https://doi.org/10.1016/B978-0-12-803265-7.00021-X. Accessed 5 Jan 2024

Kawahara J et al (2005) Air pollution and young children's inhalation exposure to organophosphorus pesticide in an agricultural community in Japan. Environ Int 31(8):1123–1132

Kawashima M et al (2023) Association between the deaths indirectly caused by the Fukushima Daiichi nuclear power plant accident (disaster-related deaths) and pre-disaster long-term care certificate level: a retrospective observational analysis. Int J Disaster Risk Reduct 96:103989

Kim SW, An Y-J (2019) Soil microplastics inhibit the movement of springtail species. Environ Int 126:699–706

Kok MTJ et al (2018) Pathways for agriculture and forestry to contribute to terrestrial biodiversity conservation: a global scenario-study. Biol Cons 221:137–150

Kollas C et al (2015) Crop rotation modelling—A European model intercomparison. Eur J Agron 70:98–111

Kumar M et al (2020) Microplastics as pollutants in agricultural soils. Environ Pollut 265:114980

Kumar S, Kumar S, Mohapatra T (2021) Interaction between macro-and micro-nutrients in plants. Front Plant Sci 12:665583

Kye H et al (2023) Microplastics in water systems: a review of their impacts on the environment and their potential hazards. Heliyon 9(3):e14359

Lal R (2015) Restoring soil quality to mitigate soil degradation. Sustainability 7(5):5875–5895

Laxminarayan R, Van Boeckel T, Teillant A (2015) The economic costs of withdrawing antimicrobial growth promoters from the livestock sector. OECD Food Agric Fisher Pap 78:1–42

Lee K-T et al (2022) Recover energy and materials from agricultural waste via thermalchemical conversion. One Earth 5(11):1200–1204

Lee LK (1980) The impact of landownership factors on soil conservation. Am J Agr Econ 62(5):1070–1076

Lei L et al (2018) Polystyrene (nano) microplastics cause size-dependent neurotoxicity, oxidative damage and other adverse effects in *Caenorhabditis elegans*. Environ Sci Nano 5(8):2009–2020

Leonhardt H, Penker M, Salhofer K (2019) Do farmers care about rented land? A multi-method study on land tenure and soil conservation. Land Use Policy 82:228–239

Li C et al (2016) Phthalate esters in soil, plastic film, and vegetable from greenhouse vegetable production bases in Beijing, China: concentrations, sources, and risk assessment. Sci Total Environ 568:1037–1043

Li Y et al (2022) Multi-pollutant assessment of river pollution from livestock production worldwide. Water Res 209:117906

Lima M et al (2024) Positive feedbacks in deep-time transitions of human populations. Philos Trans Royal S B 379(1893):20220256

Lima T, Domingues S, Da Silva GJ (2020) Manure as a potential hotspot for antibiotic resistance dissemination by horizontal gene transfer events. Vet Sci 7(3):110

Ma S et al (2023) Impacts of land use/land cover and soil property changes on soil erosion in the black soil region, China. J Environ Manag 328:117024

Malone TC, Newton A (2020) The globalization of cultural eutrophication in the coastal ocean: causes and consequences. Front Mar Sci 7:670

Manigrasso M et al (2020) Size resolved aerosol respiratory doses in a Mediterranean urban area: from PM10 to ultrafine particles. Environ Int 141:105714

Mansuri A et al (2023) New insight into long-term effects of phthalates microplastics in developing zebrafish: evidence from genomic alteration and organ development. Environ Toxicol Pharmacol 99:104087

Martínez-Dalmau J, Berbel J, Ordóñez-Fernández R (2021) Nitrogen fertilization. A review of the risks associated with the inefficiency of its use and policy responses. Sustainability 13(10):5625

Mateo-Sagasta J et al (2017) Water pollution from agriculture: a global review. In: Executive summary, Food and Agriculture Organization of the United Nations (FAO), Rome, Italy; International Water Management Institute (IWMI), Colombo, Sri Lanka. https://www.susana.org/en/knowledge-hub/resources-and-publications/library/details/3508. Accessed 20 Dec 2023

Mateo-Sagasta J, Zadeh SM, Turral H (2018) More people, more food, worse water?: A global review of water pollution from agriculture. Food and Agriculture Organization of the United Nations (FAO) (Italy), Water, Water, Land and Ecosystems (WLE) Program of the CGIAR, International Water Management Institute (IWMI) (Colombo). http://www.fao.org/3/CA0146EN/ca0146en.pdf. Accessed 5 Mar 2024

McEwen SA, Collignon PJ (2018) Antimicrobial resistance: a one health perspective. Microbiol Spectr 6(2):521–547

Meena SD et al (2023) Crop yield improvement with weeds, pest and disease detection. Procedia Comput Sci 218:2369–2382

Menadi S et al. (2024) Occupational exposure to the dust of chemical fertilizers (NPK 15.15. 15): effect on biochemical parameters and oxidative stress status among workers in Annaba. Comp Clin Pathol 1–8

Menegat S, Ledo A, Tirado R (2022) Greenhouse gas emissions from global production and use of nitrogen synthetic fertilisers in agriculture. Sci Rep 12(1):1–13

Morgan JB, Connolly EL (2013) Plant-soil interactions: nutrient uptake. Nat Educ Knowl 4(8):2

Mortvedt JJ (1995) Heavy metal contaminants in inorganic and organic fertilizers. Fertil Res 43:55–61
Munita JM, Arias CA (2016) Mechanisms of antibiotic resistance. In: Virulence mechanisms of bacterial pathogens, pp 481–511
Novotny V et al (2010) Comparative assessment of pollution by the use of industrial agricultural fertilizers in four rapidly developing Asian countries. Environ Dev Sustain 12:491–509
Obi FO, Ugwuishiwu BO, Nwakaire JN (2016) Agricultural waste concept, generation, utilization and management. Niger J Technol 35(4):957–964
Onyutha C (2019) African food insecurity in a changing climate: the roles of science and policy. Food Energy Secur 8(1):e00160
Oun A et al (2014) Effects of biosolids and manure application on microbial water quality in rural areas in the US. Water 6(12):3701–3723
Palazzi E, Currò F, Fabiano B (2015) A critical approach to safety equipment and emergency time evaluation based on actual information from the Bhopal gas tragedy. Process Saf Environ Prot 97:37–48
Parikh SJ, James BR (2012) Soil: the foundation of agriculture. Nat Educ Knowl 3(10):2
Parkhill J (2022) Antimicrobial resistance exchange between humans and animals: why we need to know more. Engineering (Beijing, China) 15:11
Parr JF et al (2020) Sustainable agriculture in the United States. In: Sustainable agricultural systems. CRC Press, pp 50–67
Pendrill F et al (2022) Disentangling the numbers behind agriculture-driven tropical deforestation. Science 377(6611):eabm9267
Peng X et al (2023) Recycling municipal, agricultural and industrial waste into energy, fertilizers, food and construction materials, and economic feasibility: a review. Environ Chem Lett 21(2):765–801
Penuelas J, Coello F, Sardans J (2023) A better use of fertilizers is needed for global food security and environmental sustainability. Agric Food Secur 12(1):1–9
Peter Kogut (2023) Soil degradation: harmful effects & promising solutions. EOS Data Analytics. https://eos.com/blog/soil-degradation/. Accessed 23 Mar 2024
Piwarski SA, Salisbury TB (2023) The effects of environmental aryl hydrocarbon receptor ligands on signaling and cell metabolism in cancer. Biochem Pharmacol 115771
Priyadarshini P, Abhilash PC (2020) Policy recommendations for enabling transition towards sustainable agriculture in India. Land Use Policy 96:104718
Quigg A et al (2021) Marine phytoplankton responses to oil and dispersant exposures: knowledge gained since the Deepwater Horizon oil spill. Mar Pollut Bull 164:112074
Rahman KMA, Zhang D (2018) Effects of fertilizer broadcasting on the excessive use of inorganic fertilizers and environmental sustainability. Sustainability 10(3):759
Ramsey AB, Faiia AM, Szynkiewicz A (2019) Eight years after the coal ash spill–Fate of trace metals in the contaminated river sediments near Kingston, eastern Tennessee. Appl Geochem 104:158–167
Raza MH et al (2022) Environmental and health impacts of crop residue burning: scope of sustainable crop residue management practices. Int J Environ Res Public Health 19(8):4753
Rillig MC et al (2019) Microplastic effects on plants. New Phytol 223(3):1066–1070
Robinson DE, Mansingh A, Dasgupta TP (1999) Fate and transport of ethoprophos in the Jamaican environment. Sci Total Environ 237:373–378
Rogers JA, Metz L, Yong VW (2013) Endocrine disrupting chemicals and immune responses: a focus on bisphenol-A and its potential mechanisms. Mol Immunol 53(4):421–430
Rose CE et al (2018) Holistic assessment of occurrence and fate of metolachlor within environmental compartments of agricultural watersheds. Sci Total Environ 612:708–719
Roy T, Dey TK, Jamal M (2023) Microplastic/nanoplastic toxicity in plants: an imminent concern. Environ Monit Assess 195(1):27
Santillán VS (2023) Cadmium content in fertilizers and its impacts on soil health. In: 1st meeting of the international network on soil pollution (INSOP). Food and Agriculture Organization

of the United Nations. https://www.fao.org/fileadmin/user_upload/GSP/INSOP/INSOP_2meeting_june23/Presentations/Item_2_Cd_fertilizers_Saynes.pdf. Accessed 23 Feb 2024
Sarkar S et al (2021) The use of pesticides in developing countries and their impact on health and the right to food. Policy Department, Directorate-General for External Policies, European Union, Belgium. https://www.europarl.europa.eu/cmsdata/219887/Pesticides%20health%20and%20food.pdf. Accessed 11 Jan 2024
Schaider LA et al (2019) Environmental justice and drinking water quality: are there socioeconomic disparities in nitrate levels in US drinking water? Environ Health 18:1–15
Schell T, Rico A, Vighi M (2020) Occurrence, fate and fluxes of plastics and microplastics in terrestrial and freshwater ecosystems. Rev Environ Contam Toxicol 250:1–43
Sedlacek CJ, Giguere AT, Pjevac P (2020) Is too much fertilizer a problem? Front Young Minds 8(63). https://doi.org/10.3389/frym.2020.00063. Accessed 1 June 2024
Segovia-Mendoza M et al (2020) How microplastic components influence the immune system and impact on children health: focus on cancer. Birth Defects Res 112(17):1341–1361
Shah KK et al (2021) Diversified crop rotation: an approach for sustainable agriculture production. Adv Agric 2021:1–9
Sharma N, Singhvi R (2017) Effects of chemical fertilizers and pesticides on human health and environment: a review. Int J Agric Environ Biotechnol 10(6):675–680
Shukla AK et al (2022) Fertilizer use in Indian agriculture and its impact on human health and environment. Indian J Fertil 18(3):218–237
Sinha E et al (2022) India's riverine nitrogen runoff strongly impacted by monsoon variability. Environ Sci Technol 56(16):11335–11342
Sishodia RP, Ray RL, Singh SK (2020) Applications of remote sensing in precision agriculture: a review. Remote Sens 12(19):3136
Stoicea P et al (2023) Crop rotation practiced by Romanian crop farms before the introduction of the "Environmentally Beneficial Practices Applicable to Arable Land" eco-scheme. Agronomy 13(8):2086
Strokal M et al (2019) Global multi-pollutant modelling of water quality: scientific challenges and future directions. Curr Opin Environ Sustain 36:116–125
Sun H et al (2021) Foliar uptake and leaf-to-root translocation of nanoplastics with different coating charge in maize plants. J Hazard Mater 416:125854
Syafrudin M et al (2021) Pesticides in drinking water—A review. Int J Environ Res Public Health 18(2):468
Tangahu BV et al (2011) A review on heavy metals (As, Pb, and Hg) uptake by plants through phytoremediation. Int J Chem Eng 2011:939161
Tao C et al (2020) Heavy metal content in feedstuffs and feeds in Hubei province, China. J Food Prot 83(5):762–766
Thawabteh AM et al (2023) Understanding the risks of diffusion of cyanobacteria toxins in rivers, lakes, and potable water. Toxins 15(9):582
Tinkov AA et al (2018) Cadmium and atherosclerosis: a review of toxicological mechanisms and a meta-analysis of epidemiologic studies. Environ Res 162:240–260
Torres FG et al (2021) Sorption of chemical contaminants on degradable and non-degradable microplastics: recent progress and research trends. Sci Total Environ 757:143875
Tudi M et al (2021) Agriculture development, pesticide application and its impact on the environment. Int J Environ Res Public Health 18(3):1112
United Nations (2022) Peace, dignity and equality on a healthy planet. https://www.un.org/en/global-issues/population. Accessed 22 Dec 2023
Urkude R, Dhurvey V, Kochhar S (2019) 15—Pesticide residues in beverages. In: Alexandru MG, Alina MH (eds) Quality control in the beverage industry. Academic Press, pp 529–560. https://doi.org/10.1016/B978-0-12-816681-9.00015-1. Accessed 1 Apr 2024
Van Dijk M et al (2021) A meta-analysis of projected global food demand and population at risk of hunger for the period 2010–2050. Nat Food 2(7):494–501

Wang Q et al (2023) A practical method for predicting and analyzing the consequences of ammonium nitrate explosion accidents adjacent to densely populated areas. Heliyon 9(5):e15616

Wang Y et al (2021) Effect and risk assessment of animal manure pollution on Huaihe River Basin, China. Chinese Geogr Sci 31:751–764

World Bank (2023) Agriculture and food. https://www.worldbank.org/en/topic/agriculture/overview. Accessed 25 Mar 2024

World Health Organization (2015) Section 2 Major risks resulting from human activities. https://iris.who.int/bitstream/handle/10665/177155/Synt_R_5.pdf?sequence=5&isAllowed=y. Accessed 23 Feb 2024

World Wild Life (2024) Sustainable agriculture. https://www.worldwildlife.org/industries/sustainable-agriculture. Accessed 23 Feb 2024

Wurtsbaugh WA, Paerl HW, Dodds WK (2019) Nutrients, eutrophication and harmful algal blooms along the freshwater to marine continuum. Wiley Interdiscip Rev Water 6(5):e1373

Wuepper D, Borrelli P, Finger R (2020) Countries and the global rate of soil erosion. Nat Sustain 3(1):51–55

Xiao K et al (2013) Differences in carbon and nitrogen mineralization in soils of differing initial pH induced by electrokinesis and receiving crop residue amendments. Soil Biol Biochem 67:70–84

Xue N et al. (2006) Adsorption and degradation of benfuracarb in three soils in Hunan, People's Republic of China. Bull Environ Contam Toxicol 76(4):720–727

Yang Q et al (2018) A review of soil heavy metal pollution from industrial and agricultural regions in China: pollution and risk assessment. Sci Total Environ 642:690–700

Yousaf M et al (2017) Effects of fertilization on crop production and nutrient-supplying capacity under rice-oilseed rape rotation system. Sci Rep 7(1):1270

Yuan W et al (2020) Transfer potentials of antibiotic resistance genes in *Escherichia* spp. strains from different sources. Chemosphere 246:125736

Yue L et al (2017) Adsorption–desorption behavior of atrazine on agricultural soils in China. J Environ Sci 57:180–189

Zahoor I, Mushtaq A (2023) Water pollution from agricultural activities: a critical global review. Int J Chem Biochem Sci 23(1):164–176

Zhang J et al (2023) Ancient DNA reveals potentially toxic cyanobacteria increasing with climate change. Water Res 229:119435

Zhang J et al (2024) Adverse effects of exposure to fine particles and ultrafine particles in the environment on different organs of organisms. J Environ Sci 135:449–473

Zhao J et al (2020) Does crop rotation yield more in China? A meta-analysis. Field Crops Res 245:107659

Zhu Q et al (2022) Occurrence, spatial distribution and ecological risk assessment of phthalate esters in water, soil and sediment from Yangtze River Delta, China. Sci Total Environ 806:150966

Zhuo Y, Zeng W (2020) Using stable nitrogen isotopes to reproduce the process of the impact of human activities on the lakes in the Yunnan Guizhou Plateau in the past 150–200 years. Sci Total Environ 741:140191

Environmental Pollution in Biomedical and Pharmaceutical Sectors

Nurul Huda Abd Kadir, Nur Anniesa Farhana Mohd Roza, and Suhail Bashir

Abstract COVID-19 outbreak contribute to the rising volume of biomedical waste which is harmful to both human health and the environment. There was a dramatic increase in Biomedical Waste (BMW) production in tandem with the global COVID-19 case count, which affecting global population well-being, as well as environmental health and safety. There are many alternative ways for recycling or lowering biological waste have been introduce to control BMW to environment. But yet, limited understanding on the possible mitigation worldwide to reduce burden of biological waste during this pandemic era. This article highlights the sustainability, global challenges, and solutions from the previous studies in mitigating the environmental problems from biomedical and pharmaceutical sectors. This chapter addressing the various kinds of biomedical waste, their effects on the environment and human health, regulations and legislations, and strategies, as well as challenges for managing BMW, especially during this pandemic and post-pandemic COVID-19.

Keywords Environmental pollution · Biomedical waste · Waste management and biohazard and safety

1 Introduction

Biomedical Waste (BMW) is the collective term for trash produced by companies, hospitals, and other healthcare facilities. This kind of waste is composed of a variety of potentially hazardous and infectious components. Following that, this waste is classified, sorted, and handled according to scientific procedures. Healthcare workers must inevitably possess the necessary information and have the right mindset toward BMW and its management, especially trash products containing

N. H. A. Kadir (✉) · N. A. F. M. Roza
Faculty of Science and Marine Environment, Universiti Malaysia Terengganu, 21030 Kuala Nerus, Terengganu, Malaysia
e-mail: nurulhuda@umt.edu.my

S. Bashir
ENS Environmental Consultancy, Sharjah, UAE

A. Z. Yaser et al. (eds.), *Controlling Environmental Pollution*,
https://doi.org/10.1007/978-981-97-8931-3_3

contagious or possibly infectious elements, such as research, medical, or laboratory trash, which can be produced by BMW (either solid or liquid). It is very likely that improper handling of BMW may result in infections among hospital staff, patients who are visiting the facilities, and members of the community. Additionally, BMW can be categorized as general, pathological, chemical, infectious, radioactive, sharps, biomedical, and pharmaceutical products.

BMW's management approach is based on the three Rs.: reduce, reuse, and recycle. 15% of the total quantity of BMW produced is hazardous waste, while the remaining 85% is normal (non-hazardous) waste material. The improper BMW management may result in microorganisms' infection to human body through cuts, abrasions, puncture wounds, and other openings because the product contains sharps and syringes. Additionally, there may be a possibility of ingesting or breathing airborne pathogens from BMW, which can result in illnesses such as AIDS, hepatitis A, B, and C, *Streptococcus pneumonia*, Shigella, Salmonella, *Mycobacterium tuberculosis*, COVID-19 virus, and helminthic infections.

In spite of the COVID-19 pandemic's rapid innovation for new biomedical and pharmaceutical products, long-standing BMW management issues still exist. The disproportionate number of lower-class communities and societies were affected by COVID-19 cases and there is an urgent need to address the underrepresentation of diversity in clinical research participants and practitioners to avoid unwanted pandemic scenarios. The variety of clinical, population-related, and causal intervention studies that could significantly influence the pandemic's outcomes required a broad and diverse pool of volunteers, primarily those most negatively impacted and historically underrepresented in research. Given that regulatory organizations work to guarantee the safety and effectiveness of medicines across a range of people, especially with government funding centered on community engagement in research can undoubtedly be a lever to promote diversity in study participation.

2 Commercialization and Market Demand of Biomedical and Pharmaceuticals Products

Biomedical technology industries are responsible for the development, production, marketing demands, and commercialization of biomedical and pharmaceutical products. Biomedical and pharmaceutical products are divided into five different sectors, which are diagnostic (such as lab services and test kits), hospital supplies and PPE (such as gloves and masks), medical devices (such as class II therapeutic medical technologies), therapeutics (such as drugs to treat and manage disease symptom), and vaccines (such as COVID-19 vaccines) (Mammen et al. 2022). Big data collected from biomedical research and healthcare industries as well as Artificial Intelligence (AI) encourage the utilization of biomedical and pharmaceutical products by the public, as well as the industries. European Federation of Pharmaceutical Industries and Association (EFPIA) reported that the world's pharmaceutical market worth

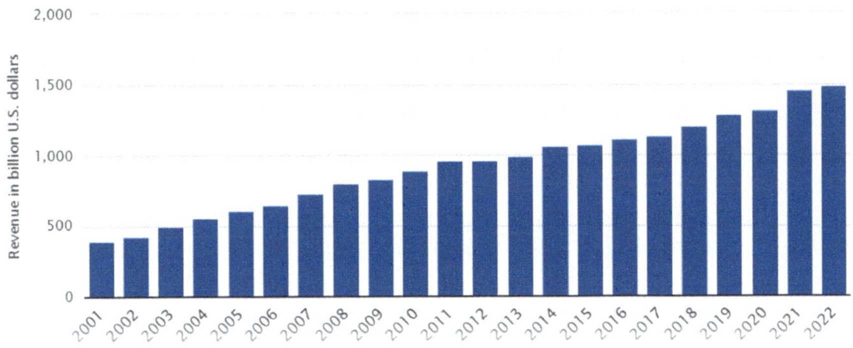

Fig. 1 Revenue of the worldwide pharmaceutical market from 2001 to 2022 (in billion U.S. dollars) (Pharmaceutical Products and Market n.d.)

approximately 1,287,736 million dollars in 2022, with North America (the USA & Canada), Europe, China, and Japan as the major contributors of the sale (52.3%, 22.4%, and 8.1%, respectively) (European Federation of Pharmaceutical Industries and Association 2023). As stated in Fig. 1, pharmaceutical industry is considered as a crisis-proof sector, as the worldwide pharmaceutical market size remains increasing for two decades (390.2 billion USD in 2001 and 1 482 billion USD market revenue in 2022) and is not affected during the economic crisis due to COVID-19 in 2019. This increasing trend of market demand for pharmaceutical products has shown that pharmaceutical industry is one of the most prominent industries which crucial in promoting societal well-being. In spite of its growing market demand, sustainability of biomedical and pharmaceutical production should be implemented to ensure the safety of the consumers and environments.

3 Sustainability of Biomedical and Pharmaceuticals Production

Sustainability holds a great significance nowadays in every industry as well as in the biomedical research and development while considering the future perspective before any development or production and is implemented by the majority of nations across the world (Rahman et al. 2020). Furthermore, the sustainability in biomedical and pharmaceutical activities demands a complete approach, where environmental stewardship goes along with societal well-being and ethical integrity, ensuring that the production remains in harmony with nature (Milanesi et al. 2020). During the recent times, the biomedical and pharmaceutical industries have found itself under increasing scrutiny and responsibilities regarding the sustainability practices in industry, holding the attention of consumers, policymakers, and other organizations alike (Chomać-Pierzecka 2023). Based on these findings, this significant focus extends across various aspects of pharmaceutical development, including the integration of sustainability principles into the formation of new delivery systems and products with lower environmental impacts, as well as efforts toward recycling of waste, reduction in water footprint, and strive for eco-friendly manufacturing methods and recyclable packaging (Haleem et al. 2023). It is evident that with the increase in the population and health issues, the biomedical and pharmaceutical industries are nowadays consuming more natural resources in the form of raw material, additives, energy, etc., intensively during the production of drugs and other healthcare products (Forezi et al. 2023). The consumption of resources for biomedical and pharmaceutical production in significantly increasing all across the world and it is significant that during the production activities sustainability approach shall be in place to use the resources efficiently and reduce the burden on the Mother Nature (Peña et al. 2021).

In addition, the significance of sustainability in the field of biomedical and pharmaceutical production extends beyond ecological considerations to cover larger societal impacts. As people around the world face with issues of health equity and access to essential medicines, sustainable production practices play a critical and significant role to guarantee that healthcare solutions are accessible and affordable for all (Alharbi 2022). By incorporating sustainability goals, pharmaceutical along with other industries can contribute to address global health disparities in a sustainable manner (Saxena et al. 2022).

The biomedical and pharmaceutical industries holding a significant contribution to pollution and resource consumption. It is important to mention that the production processes involved in manufacturing pharmaceuticals generate various forms of pollution, including air and water pollution as well as soil and groundwater pollution by generating waste in different forms (Gupta et al. 2023a). From the chemical synthesis to packaging and distribution, each stage of production results in the release of harmful substances into the environment. The waste that is generated from these industries includes solid wastes such as discarded drugs. PPE and test strips, as well as liquid wastes like contaminated chemicals, solvents, and cleaning surfactants (Sharma et al. 2023).

Meanwhile, with the upsurge in production capacity worldwide to meet the demand of growing population, there's an urgent need to prioritize the research and implementation for green initiatives and technologies within the biomedical and pharmaceutical industries (Weaver et al. 2022). This imperative arises from the necessity to reduce the waste generation and transition toward a more sustainable resource management. All these green practices and technologies, such as waste-to-resource conversion and water recycling systems will help industries to minimize their reliance on virgin natural resources and to conserve the resources for a sustainable future. It is also crucial to mention that efforts to address resource footprint and waste from biomedical, pharmaceutical production is implemented for mitigating their environmental impact (Chowdhury and Islam 2023). Implementing pollution prevention measures, such as the adoption eco-friendly and cleaner production technologies, the use of eco-friendly solvents and the waste reclamation technologies, helps to minimize the release of harmful pollutants into the environment. Meanwhile, the green chemistry principles also promote the design of pharmaceutical products and processes that minimize the use and generation of hazardous substances along with the other waste (Sapkota and Pariatamby 2023). Green chemistry practices lead to more effective drug synthesis routes, waste generation reduction, and improved safety for the environment (Asif 2021).

However, the concept of circular economy also emphasizes the sustainability approach by reusing, recycling, and repurposing of materials to minimize waste generation, reduction in ecological footprint, and increase in resource efficiency (Mahjoob et al. 2023). The sustainability discourse surrounding the biomedical and pharmaceutical sectors has also ignited interest among scholars from diverse fields such as chemistry, engineering, and environmental sciences along with business and management studies incorporating this critical topic into their studies. As the biomedical and pharmaceutical companies face a higher level of management complexity, it is imperative to balance environmental, economic, and social sustainability. By reducing waste generation, conserving resources, and implementing sustainable practices, these industries can reduce their environmental footprint while concurrently fostering innovation, economic growth, and resilience in an increasingly resource-constrained world (Ang et al. 2021).

It is not a moral imperative to implement sustainability approach in biomedical and pharmaceutical industries along with other sectors but it is now a strategic and necessary step to safeguard collective future and promote sustenance (Mishra et al. 2021). During these times, environmental challenges assumed unprecedented proportions with resource scarcity that threatens global stability (Sharma and Tiley 2021). However, integrating sustainability into industry practices is no longer a choice but the need of the hour.

4 Global Challenges of Biomedical and Pharmaceutical Industries

4.1 COVID-19 Pandemic

Coronavirus disease, COVID-19 is a transmittable disease caused by Severe Acute Respiratory Syndrome Coronavirus 2 (SARS-CoV-2) virus. The first case was reported in Wuhan, China, in December 2019, where pneumonia-like disease cases were reported between the adults involved in a commercial and trading activity of livestock and poultries such as snakes and bats at the Huanan seafood marketplace (Siddiqui et al. 2022). Examination of the respiratory samples of infected individuals leads to a discovery of a novel beta coronavirus that possesses a nucleotide homology to the SARS-CoV (79% similarity) and existing bat-borne SARS-like coronaviruses isolated from Chinese horseshoe bats, *Rhinolophus sinicus,* and Intermediate horseshoe bat, *Rhinolophus affinis* (89% and 96% similarity, respectively) (Mackenzie and Smith 2020; Roberts et al. 2021). As of December 2023, a cumulative total of over 773 million COVID-19 cases was reported worldwide, with a significant amount of almost 7 million of reported deaths (World Health Organization 2023). COVID-19 disease can be transmitted by close contact interactions, where the infectious liquid particle spreads to healthy people via activities like coughing, sneezing, laughing, talking, or sharing foods. Symptoms of SARS-CoV-2 infections include fever, dry cough, headache, weakness, and diarrhea. Study revealed that elderly patients (>65 years old) are at a higher risk of fatality, and individuals with underlying diseases (lung, hypertension, and cardiovascular) possess severe COVID-19 outcomes (Yuki et al. 2020).

4.2 *Biomedical and Pharmaceutical Product Disposal*

Equipment and devices such as Personal Protective Equipment (PPE), test kits, and sanitizers used in medical and pharmaceutical activity are called biomedical and pharmaceutical products. Biomedical Waste (BMW) is defined as the remaining items (wastes) disposed of during research activities, and/or medical practices such as; diagnosis and disease immunization, as well as medicinal and pharmaceutical residues (Datta et al. 2018; Ilyas et al. 2020). According to the World Health Organization (WHO), BMW falls into eight classifications: infectious waste (such as autopsies of infected patients and discarded samples), pathological waste (such as discarded body parts of infected patients), sharp waste (such as needles and blades), chemical waste (such as expired chemicals and disinfectants), cytotoxic waste (such as cytotoxic drugs), radioactive waste (such as radiotherapeutic materials), pharmaceutical waste (such as expired medicines), and non-hazardous/general waste (such as household wastes) (World Health Organization 2018). Biomedical and pharmaceutical product disposal requires a safe and responsible management of waste generated from healthcare facilities, research institutions, and pharmaceutical manufacturing.

Proper biomedical and pharmaceutical product disposal methods need to be done in order to comply with the Sustainable Development Goals (SDGs) introduced by the United Nations, particularly Goal 3 (ensuring the well-being of all ages) and Goal 6 (sustainable management of water and sanitation). Globally, methods in disposal of biomedical and pharmaceutical products are created by complying the guidelines by WHO and The Basel, Rotterdam and Stockholm conventions. There are multiple biomedical and pharmaceutical product disposal methods such as incineration, autoclaving, landfill disposal, chemical, thermal, ion radiation, deep burial, and microwave (Hegde et al. 2007). Incineration and deep burial are the most familiar methods in disposing the BMWs. During the incineration process, solid wastes are combusted into by-products such as ashes, gases, heats, steams, and ashes. In addition, deep burial method is a method where BMWs are buried deeply, (approximately 2 m) at a specified, allocated burial site (Jaseem et al. 2017). In the autoclaving method, BMWs are broken down into smaller sizes, before being immersed in an optimum temperature and period to kill pathogens, meanwhile, in microwaving methods apply the idea of utilizing an electromagnetic field, triggering the water molecules presented in BMWs to heat up, causing the pathogen in the BMWs to be destroyed (Ghasemi and Yusuff 2016). Land disposal method is a method where BMWs are dumped into pits that are made in low-leveled areas that are usually far away from flooded and landslide areas.

During the COVID-19 outbreak, the total of BMWs was reported to be approximately 16 649.48 tons/day, globally (Ngoc et al. 2022). The United Nations Environment Programme (UNEP) reported that, in developing countries, about 2.5 kg/bed of medical wastes (MWs) were produced daily (United Nations Environment Programme 2020a). During the initial outbreak of COVID-19 in Wuhan (China), the total of BMW increased exponentially from 50 to 247 tons daily, meanwhile, in the United States, an estimated increased amount of 2.5 million tons of BMW were yielded per month, compared to 5 million tons per year (Ilyas et al. 2020). This is mainly due to the increased number of PPEs used by not only health professionals but also public such as gloves, head covers, masks, and gowns that are used as protective gears to minimize the risk of exposure to the SARS-CoV-2 virus. These BMWs might be infectious as the SARS-CoV-2 virus can last for 3 h on tissues, 2 days on cloth fabrics, and 4 days on glass.

Considering the hazard of COVID-19 disease, WHO suggested compliance to the Water, Sanitation, Hygiene, and Waste Management (WASH) guidelines on handling general waste of COVID-19 BMWs. Specified, marked bins or boxes need to be prepared for disposal of COVID-19 BMWs, and these need to be treated on-site before being disposed of. In addition, the infected BMWs should be stored away from scavengers, and the trash bags need to be sealed before it is 70% full in order to avoid direct contact with BMWs (United Nations Environment Programme 2020b). Transports that mobilized the BMWs must be labeled as infectious and biohazard, and documentations such as weight, waste class, and number of volumes need to be prepared (Capoor and Parida 2021). The existing methods in handling BMWs before COVID-19 are effective in killing the SARS-CoV-2 virus thus, these methods still can be used post-COVID-19.

4.3 *Biohazard and Safety*

The abundant increasing amounts of BMWs after COVID-19 has resulted in more problems in BMWs management, as few countries are already struggling in BMWs management even before COVID-19 hits. WHO reported that only 58% of the sampled facilities of 24 countries have adequate systems in order to properly manage BMWs in 2015 (World Health Organization 2018). Disposal of biomedical and pharmaceutical products may cause concerns, as improper BMWs management can lead to the release of toxic pollutants to the environment.

One of the prominent safety issues caused by improper handling of BMWs is the emission of harmful pollutants as a by-product of the incineration process. Examples of the pollutants emitted due to BMWs incineration are particulate matter (PM), metals, and greenhouse gases (GHG) such as nitrous oxide and carbon monoxide

(Belkhir and Elmeligi 2018). PM released in the environment can cause air pollution, and is a lung carcinogen as reported by studies (Mudu et al. 2020; Pehnec et al. 2020; Valavanidis et al. 2013). GHGs are the main source of global warming as these gases trap the infrared radiation of the sun in the atmosphere, creating the greenhouse effect (Darkwah et al. 2018). Climate change, sea level rise, life-threatening natural disaster are the examples of events that will arise if global warming is not seriously taken into consideration. Disposal of BMWs by landfilling method may cause biohazard as the pathogenic BMWs can spread parasites to plants that later damaging the ecosystem. Under-constructed landfill may cause a landfill liner system that can cause release of harmful leachate that contaminates groundwater (Parida et al. 2022). In addition, a study has reported that landfill areas are potent with microbial dust and pathogenic microbes that can cause chronic pulmonary disease. Improper sharp waste may cause injury and may cause infection. For example, needle injury can cause infection of blood-carrying pathogens like human immunodeficiency virus (HIV) and hepatitis C virus (HCV) (Gupta et al. 2023b). Heavy metals and radioactive materials of BMWs can cause soil and water contamination. Heavy metal accumulation is undesirable because they can affect the ecosystem safety by disrupting organisms' food chain (Li et al. 2019).

4.4 Regulation and Legislation

Legislation is a description of a legal obligation and its associated penalty and punishment, which may include a fine, an arrest, or rehabilitation. Meanwhile, regulation is the administrative framework that the government has put in place to enforce the legislation. Regulation and legislation of BMWs may vary from country to country, based on the condition of the country itself. Regulation and legislation regarding BMWs should be practical, applicable, and comprehensive in order to ensure a proper handling procedure to be done (Al-Shahi Salman et al. 2014),

United Nations Environment Programme (UNEP) plays a pivotal role in setting international standards for biomedical waste management. In 2020, a comprehensive report entitled "*Waste Management during the COVID-19 Pandemic: from response to recovery*" was released in order to provide guidelines on practices that need to be taken to handle BMWs properly (United Nation Environment Programme 2020). This report provides strategies for efficient management of BMWs including proper segregation, storage, collection, and treatment of waste, along with the development of innovative waste disinfection technologies.

In addition, U.S. Environmental Protection Agency has released a report to assist the existing Resource Conservation and Recovery Act (RCRA) formed by the United States of America authority in 2015 (United State Environment Protection Agency 2021a, b). This report highlights guidelines in executing regulatory oversight of premises, ensuring the existing requirement and enforcement of RCRA to be in line with the changes due to COVID-19. Changes such as reduction of on-site inspections, virtual audit implementation, and regulatory flexibility were provided to ensure compliance while addressing health and safety concerns. RCRA 2015 law addresses the management and disposal of hazardous waste, including certain types of biomedical waste in the United States (United States Environmental Protection Agency 2015). Even though BMWs are not specifically mentioned and defined in the RCRA section of the United States 40 Code of Federal Regulations (CFR), the BMWs meet the criteria of the hazardous waste mentioned in part 261–part 265 of CFR (Office of Federal Register 2016). Parts 261–265 of Title 40 CFR as well as regulatory framework of RCRA includes identification of infectious and toxic biochemical waste as hazardous waste, as well as standards application toward BMWs' generating facilities to comply with the waste identification, labeling, storage, and inventory of RCRA. Transportation of hazardous waste needs to comply with the RCRA guidelines on proper packaging, labeling, and transportation procedure. RCRA requires all facilities that handle biohazard to have a permit issued by the Environmental Protection Agency (EPA) or any of its authorized state agencies. This permit includes guidelines on the facility design, waste operational plans, and treatment method.

"Guidelines for handling, treatment and disposal waste generated during treatment, diagnostics and quarantine of COVID-19 patients" was issued by the Ministry of Environment, Forest and Climate Change of India in March 2020 (the Ministry of Environment Forest and Climate Change of India 2020). These guidelines are applicable to every individual in India who is involved in the biomedical and pharmaceutical disposal institutions (such as hospitals, laboratories, and blood banks). This rule highlights the duty of occupants and operators to apply all the precaution and preventive steps in order to avoid adverse effects toward citizens and the environment. The duty includes wearing PPE during COVID-19 BMWs management, as well as having recorded data of the quantity of generated, transported, treated, and disposed of COVID-19 wastes.

COVID-19 BMWs must be segregated into a labeled and dedicated container prior to being stored, transported, and disposed. The bags must be labeled as biohazard or bio-waste, and infectious BMWs must be sterilized by microwave or autoclave method before the final disposal. In BMWs transportation, the vehicle must comply with the Motor Vehicles Act, 1988 (59 of 1988) requirements. Grant authorization of the occupier and operator must be applied to the State Pollution Control Board and Pollution Control Committee.

In Malaysia, The COVID-19 Management Guidelines were introduced by the Ministry of Health (MOH). In these guidelines, COVID-19 BMWs segregated must comply with the existing healthcare facility policies and local regulations (Ministry of Health Malaysia 2020). The Environmental Quality (Scheduled Wastes) Regulations 2005 introduced on 15 August 2005 were one of the regulations applied in handling COVID-19 BMWs. BMWs segregated must comply with the color coding standard classification. Blue colored plastic bags are used to indicate BMWs that need to be autoclaved, yellow bags are for BMWs to be incinerated, and black plastic bags are for general waste (Ambali et al. 2013). Waste generators must submit an inventory of generation, treatment, and disposal to the Department of Environment (DOE), as well as filling out the "e-Consignment Note," which serves as a monitoring tool for any transaction between waste generator and waste receiver (Department of Environment

2009). For transportation of BMWs, dedicated vehicles and internal routes must be used, and the transport must be sensitized, cleaned, and disposed of as biohazard chemicals.

5 Solution and Challenges

It is essential to prioritize both solution and preventive measures for a sustainable future in every sector including the biomedical and pharmaceutical industries (Ignacz et al. 2022), where the main concern raises from these industries including the production of waste that raises a significant concern due to their persistent and significant threat to human and environmental health along with sustainability (Ray et al. 2022). There are many solutions available that can be adopted to achieve sustainability among these sectors including awareness, reduction in ecological and carbon footprint (Daughton and Ruhoy 2010), integrated waste reduction management practices, recycling, and research program to explore more options toward sustainability (Sapkota and Pariatamby 2023). However, during the recent times, the advanced and emerging technologies in the field of biomedical and pharmaceutical industries are among the primary solutions to achieve sustainability by precise production as well as reduction in waste generation along with other benefits like to reduce the significant impact on environment, human health and also to maintain the efficacy along with economic viability. (Weaver et al. 2022).

6 Conclusion

In conclusion, the result of waste production and resource reliance from the biomedical and pharmaceutical industries have substantial environmental and safety issues. In these industries, sustainability entails minimizing resource use, cutting waste production, and implementing environmentally friendly practices, especially during the pandemic. By using green chemistry practices like solvent replacement and catalysis, pharmaceutical manufacture can drastically lessen its environmental impact which helps to cater the waste problems. In addition, recycling and reusing waste goods potentially decrease environmental damage, and this may help the industry's dependency on limited resources by putting circular economy ideas into practice. The biomedical and pharmaceutical industries are vital in tackling global issues like pollution, resource depletion, and climate change. These industries may support a healthier society by putting sustainability first and taking preventative action based on the guidelines and regulations provided by authorities.

References

Al-Shahi Salman R, Beller E, Kagan J, Hemminki E, Phillips RS, Savulescu J, Macleod M, Wisely J, Chalmers I (2014) Increasing value and reducing waste in biomedical research regulation and management. Lancet 383:176–185. https://doi.org/10.1016/S0140-6736(13)62297-7

Alharbi MF (2022) Impact of green supply chain management practices on sustainability of healthcare organizations: mediating role of environmental responsibility. Gomal Univ J Res 38(2):145–159. https://doi.org/10.51380/gujr-38-02-02

Ambali AR, Bakar AN, Merican FM (2013) Environmental policy in Malaysia: biomedical waste, strategies and issues (1)

Ang KL, Saw ET, He W, Dong X, Ramakrishna S (2021) Sustainability framework for pharmaceutical manufacturing (PM): a review of research landscape and implementation barriers for circular economy transition. J Clean Prod 280:124264. https://doi.org/10.1016/j.jclepro.2020.124264

Asif M (2021) Green synthesis, green chemistry, and environmental sustainability: an overview on recent and future perspectives of green chemistry in pharmaceuticals. Green Chem Technol Lett 7:18–27. https://doi.org/10.18510/Gctl.2021.713

Belkhir L, Elmeligi A (2018) Carbon footprint of the global pharmaceutical industry and relative impact of its major players. J Clean Prod 214. https://doi.org/10.1016/j.jclepro.2018.11.204

Capoor MR, Parida A (2021) Biomedical waste and solid waste management in the time of COVID-19: a comprehensive review of the national and international scenario and guidelines. J Lab Phys 13(2):175–182. https://doi.org/10.1055/s-0041-1729132

Chomać-Pierzecka E (2023) Pharmaceutical companies in the light of the idea of sustainable development—An analysis of selected aspects of sustainable management. Sustainability 15(11):8889. https://doi.org/10.3390/SU15118889

Chowdhury MA, Islam MZ (2023) Eco-friendly Bio-medical waste management technology practice in Bangladesh for sustainable development. Malaysian J Sustain Environ 10:1–30. https://doi.org/10.24191/myse.v10i1.21247

Darkwah WK, Odum B, Addae M, Koomson D, Kwakye Danso B, Oti-Mensah E, Asenso T, Buanya B (2018) Greenhouse effect: greenhouse gases and their impact on global warming. J Sci Res Rep 17:1–9. https://doi.org/10.9734/JSRR/2017/39630

Datta P, Mohi GK, Chander J (2018) Biomedical waste management in India: critical appraisal. J Lab Phys 10(1):6–14. https://doi.org/10.4103/JLP.JLP_89_17

Daughton CG, Ruhoy IS (2010) Reducing the ecological footprint of pharmaceutical usage: linkages between healthcare practices and the environment. https://doi.org/10.1007/978-3-642-05199-9_6

Department of Environment (DOE) (2009) Guideline on the handling and management of clinical wastes in Malaysia

European Federation of Pharmaceutical Industries and Association (2023) The pharmaceutical industry in figures. https://www.efpia.eu/media/rm4kzdlx/the-pharmaceutical-industry-in-figures-2023.pdf

Forezi L, da SM, Ferreira PG, Carvalho AS, de Silva F, de C, da Ferreira VF (2023) Medicinal chemistry for sustainable development. Curr Top Med Chem 23(11):957–969. https://doi.org/10.2174/1568026623666230517114621

Ghasemi MK, Yusuff RM (2016) Advantages and disadvantages of healthcare waste treatment and disposal alternatives: Malaysian scenario. Polish J Environ Stud 25(1):17–25. https://doi.org/10.15244/pjoes/59322

González Peña OI, López Zavala MÁ, Cabral Ruelas H (2021) Pharmaceuticals market, consumption trends and disease incidence are not driving the pharmaceutical research on water and wastewater. Int J Environ Res Publ Health 18:2532. https://doi.org/10.3390/ijerph18052532

Gupta PP, Bankar NJ, Mishra VH, Sanghavi S, Badge AK, Gupta PP, Bankar NJ, Mishra VH, Sanghavi S, Badge A (2023a) The efficient disposal of biomedical waste is critical to Public

Health: insights from the Central Pollution Control Board Guidelines in India. Cureus 15(10). https://doi.org/10.7759/cureus.47303

Gupta N, Kirar J, Deswal Y, Deswal L, Chandra K (2023b) Pollution by pharma industries: an overview

Haleem A, Javaid M, Singh R, Suman R, Qadri M (2023) A pervasive study on Green Manufacturing towards attaining sustainability. Green Technol Sustain 1:100018. https://doi.org/10.1016/j.grets.2023.100018

Hegde V, Kulkarni RD, Ajantha GS (2007) Biomedical waste management. J Oral Maxillofac Pathol 11(1):5. https://doi.org/10.4103/0973-029X.33955

Ignacz G, Orkenyi R, Konczol A, Szekely G (2022) Sustainable separations in pharmaceutical manufacturing. In: Contemporary chemical approaches for green and sustainable drugs, pp 155–200. https://doi.org/10.1016/B978-0-12-822248-5.00008-5

Ilyas S, Srivastava RR, Kim H (2020) Disinfection technology and strategies for COVID-19 hospital and bio-medical waste management. Sci Total Environ 749:141652. https://doi.org/10.1016/j.scitotenv.2020.141652

Jaseem M, Kumar P, John RM (2017) An overview of waste management in pharmaceutical industry. Pharma Innov 6(3, Part C):158

Kleber J, Cohen B (2020) Reducing waste and increasing sustainability in health care settings. Am J Nurs 120:45–48. https://doi.org/10.1097/01.NAJ.0000660032.02514.ec

Lalor F, Fitzpatrick J, Sage C, Byrne E (2019) Sustainability in the biopharmaceutical industry: seeking a holistic perspective. Biotechnol Adv 37. https://doi.org/10.1016/j.biotechadv.2019.03.015

Li C, Zhou K, Qin W, Tian C, Qi M, Yan X, Han W (2019) A review on heavy metals contamination in soil: effects, sources, and remediation techniques. Soil Sediment Contam: Int J 28:1–15. https://doi.org/10.1080/15320383.2019.1592108

Mahjoob A, Alfadhli Y, Omachonu V (2023) Healthcare waste and sustainability: implications for a circular economy. Sustainability 15:7788. https://doi.org/10.3390/su15107788

Mammen M, Narasimhan V, Kuntz R, Lewis-Hall F, Poul M, Schechter A (2022) Health product manufacturers and innovators COVID-19 impact assessment: lessons learned and compelling needs. NAM Perspectives. https://doi.org/10.31478/202201b

Mackenzie JS, Smith DW (2020) COVID-19: a novel zoonotic disease caused by a coronavirus from China: what we know and what we don't. Microbiology Australia, MA20013. https://doi.org/10.1071/MA20013

Milanesi M, Runfola A, Guercini S (2020) Pharmaceutical industry riding the wave of sustainability: review and opportunities for future research. J Clean Prod 261:121204. https://doi.org/10.1016/j.jclepro.2020.121204

Ministry of Environment Forest and Climate Change India (2020) Guidelines for handling, treatment and disposal of waste generated during treatment/diagnosis/quarantine of COVID-19 patients central pollution control board

Ministry of Health Malaysia (2020) Guidance on the use of masks with regards to Covid-19 pandemic. https://covid-19.moh.gov.my/garis-panduan/garis-panduan-kkm/Annex_8a_RECOMENDATIONS_ON_FACE_MASK.pdf

Mishra M, Sharma M, Dubey R, Ranjan V, Pandey J (2021) Green synthesis interventions of pharmaceutical industries for sustainable development. Curr Res Green Sustain Chem 4:100174. https://doi.org/10.1016/j.crgsc.2021.100174

Moors E, Cohen A, Schellekens H (2014) Towards a sustainable system of drug development. Drug Discov Today 19. https://doi.org/10.1016/j.drudis.2014.03.004

Mudu P, Pérez Velasco R, Zastenskaya I, Jarosinska D (2020) The importance and challenge of carcinogenic air pollutants for health risk and impact assessment. Euro J Public Health 30(Supplement_5):ckaa165.841. https://doi.org/10.1093/eurpub/ckaa165.841

Ngoc S-MV, Nguyen M-A, Nguyen T-L, Thi HV, Dao TL, Bui TMP, Hoang VT, Chu D-T (2022) COVID-19 and environmental health: a systematic analysis for the global burden of biomedical waste by this epidemic. Case Stud Chem Environ Eng 6:100245. https://doi.org/10.1016/j.cscee.2022.100245

Office of Federal Register (2016) Code of Federal Regulations. Title 40: protection of environment. https://www.ecfr.gov/current/title-40

Parida VK, Sikarwar D, Majumder A, Gupta AK (2022) An assessment of hospital wastewater and biomedical waste generation, existing legislations, risk assessment, treatment processes, and scenario during COVID-19. J Environ Manag 308:114609. https://doi.org/10.1016/j.jenvman.2022.114609

Pehnec G, Jakovljević I, Godec R, Sever Štrukil Z, Žero S, Huremović J, Džepina K (2020) Carcinogenic organic content of particulate matter at urban locations with different pollution sources. Sci Total Environ 734:139414. https://doi.org/10.1016/j.scitotenv.2020.139414

Pharmaceutical Products & Market (n.d.) Statista—The statistics portal. https://www.statista.com/markets/412/topic/456/pharmaceutical-products-market/#overview. Accessed 11 Jan 2024

Rahman MM, Ghoshal U, Ragunath K, Jenkins G, Rahman M, Edwards C, Hasan M, Taylor-Robinson S (2020) Biomedical research in developing countries: opportunities, methods, and challenges. Indian J Gastroenterol 39. https://doi.org/10.1007/s12664-020-01056-5

Ray M, Njoku A, Nimo-Sefah L (2022) Comparing environmental policies to reduce pharmaceutical pollution and address disparities. Int J Environ Res Public Health 19:8292. https://doi.org/10.3390/ijerph19148292

Roberts DL, Rossman JS, Jarić I (2021) Dating first cases of COVID-19. Plos Pathog 17(6):e1009620. https://doi.org/10.1371/journal.ppat.1009620

Sapkota B, Pariatamby A (2023) Pharmaceutical waste management system—Are the current techniques sustainable, eco-friendly and circular? A review. Waste Manag (New York, N.Y.) 168:83–97. https://doi.org/10.1016/j.wasman.2023.05.052

Saxena K, Balani S, Srivastava P (2022) The role of pharmaceutical industry in building resilient health system. Front Public Health 10:964899. https://doi.org/10.3389/FPUBH.2022.964899/BIBTEX

Sharma S, Tiley A (2021) Environmental sustainability in the pharmaceutical industry. https://doi.org/10.4324/9781003190516-24

Sharma N, Kaur B, Singh CP, Shivani S, Thakur K (2023) Modern structured waste disposal techniques in pharma industries. AIP Conf Proc 2558(1). https://doi.org/10.1063/5.0120817/2873998

Siddiqui S, Alhamdi HWS, Alghamdi HA (2022) Recent Chronology of COVID-19 pandemic. Front Public Health 10. https://www.frontiersin.org/articles/10.3389/fpubh.2022.778037

Sreekanth K, Gupta NV, Raghunandan HV, Kashyap U (2014) A review on managing of pharmaceutical waste in industry. Int J PharmTech Res 6:899–907

United State Environment Protection Agency (2021) Authorized state hazardous waste program inspections and operations were impacted during coronavirus pandemic. https://www.epaoig.gov/reports/evaluation/authorized-state-hazardous-waste-program-inspections-and-operations-were

United State Environment Protection Agency (2021) Authorized state hazardous waste program inspections and operations were impacted during coronavirus pandemic. https://www.epaoig.gov/sites/default/files/documents/2021-11/_epaoig_20211201-22-e-0009.pdf

United Nation Environment Programme (2020) Waste management during the COVID-19 pandemic from response to recovery. www.un.org/Depts/Cartographic/english/htmain.htm

United Nations Environment Programme (2020a) Waste management during the COVID-19 pandemic from response to recovery. https://wedocs.unep.org/bitstream/handle/20.500.11822/33416/WMC-19.pdf

United Nations Environment Programme (2020b) COVID-19 Waste management Factsheets 7: household medical waste management strategies. UNEP—UN Environment Programme. https://wedocs.unep.org/bitstream/handle/20.500.11822/32775/FS7.pdf?sequence=1&isAllowed=y

United States Environmental Protection Agency (2015) History of the Resource Conservation and Recovery Act [Other Policies and Guidance]. https://www.epa.gov/rcra/history-resource-conservation-and-recovery-act

Valavanidis A, Vlachogianni T, Fiotakis K, Loridas S (2013) Pulmonary oxidative stress, inflammation and cancer: respirable particulate matter, fibrous dusts and ozone as major causes of lung carcinogenesis through reactive oxygen species mechanisms. Int J Environ Res Public Health 10(9):Article 9. https://doi.org/10.3390/ijerph10093886

Weaver E, O'Hagan C, Lamprou D (2022) The sustainability of emerging technologies for use in pharmaceutical manufacturing. Expert Opin Drug Deliv. https://doi.org/10.1080/17425247.2022.2093857

World Health Organisation (2018) Health-care waste. https://www.who.int/news-room/fact-sheets/detail/health-care-waste

World Health Organization (2023) WHO COVID-19 dashboard. https://data.who.int/dashboards/covid19/cases

Yuki K, Fujiogi M, Koutsogiannaki S (2020) COVID-19 pathophysiology: a review. Clin Immunol 215:108427. https://doi.org/10.1016/j.clim.2020.108427

Pollution by Microorganisms and Its Impact on the Food Safety

Syarifah Ab Rashid, Nur Asshifa Md Noh, Kamarul Zaman Zarkasi, Nor Hawani Salikin, Mohd Taufiq Mat Jalil, Nurul Aili Zakaria, and Woei Yenn Tong

Abstract Nutrition is vital for sustaining life, making food safety a fundamental human right. Microbial pollution poses a significant threat to food safety as it can result in a broad spectrum of foodborne diseases. Consumers worldwide are at risk of consuming fresh produce and animal products due to food spoilage microbes, leading to reported cases of foodborne illnesses. Apart from that, food spoilage microbes can also reduce the shelf life of the products which contributes to significant economic losses for the food production industry. The early stages of the food chain, involving the cultivation or harvesting of fresh produce and animal products, are recognized as sources of contamination, particularly in agricultural and aquacultural products. A multitude of global organizations and governmental entities, including the FDA, FAO, USDA, and WHO, are dedicated to upholding food safety and security through regulatory guidelines and the advancement of cutting-edge technologies. Innovative techniques such as microbe-free contact surfaces, hurdle technology, and new packaging materials have been identified as potential solutions to enhance food safety and preservation. These methods ensure both the safety and quality of food from production to consumer delivery. This review provides comprehensive insights into the diverse sources and associated risks of microbial contamination in agricultural and aquacultural products. It also emphasizes the pivotal role of international bodies in addressing food safety issues and highlights the application of both conventional and modern techniques in securing the food supply.

S. A. Rashid (✉) · N. A. M. Noh · K. Z. Zarkasi
School of Biological Sciences, Universiti Sains Malaysia, 11800 Minden, Pulau Pinang, Malaysia
e-mail: syarifahabrashid@usm.my

N. H. Salikin
School of Industrial Technology, Universiti Sains Malaysia, 11800 Minden, Pulau Pinang, Malaysia

M. T. M. Jalil · N. A. Zakaria
Faculty of Applied Sciences, Universiti Teknologi MARA, 40450 Shah Alam, Selangor, Malaysia

W. Y. Tong
Institute of Medical Science Technology, Universiti Kuala Lumpur, 1-1, Jalan TKS 1, Taman Kajang Sentral, 43000 Kajang, Selangor, Malaysia

A. Z. Yaser et al. (eds.), *Controlling Environmental Pollution*,
https://doi.org/10.1007/978-981-97-8931-3_4

Keywords Foodborne pathogen(s) · Food safety · Microbial pollution · Agriculture · Aquaculture

1 Introduction

Microorganisms are essential for the functioning of various ecosystems including nutrient cycling, decomposition, and fermentation. Nevertheless, certain microorganisms have the potential to inflict harm upon humans, animals, and plants, thereby engendering diseases and presenting notable threats to public health and safety. A critical domain where microbial contamination can precipitate severe consequences is in safeguarding food safety. The presence of pathogenic microorganisms in food can instigate foodborne diseases, impacting millions of individuals globally and incurring considerable economic burdens. Foodborne diseases are illnesses caused by consuming contaminated food or water containing harmful pathogens in sufficient quantities. These diseases manifest through various gastrointestinal symptoms such as nausea, vomiting, diarrhea, abdominal pain, or fever. In severe cases, complications such as meningitis, sepsis, abortions, Reiter syndrome, Guillain–Barré syndrome, or death may occur (Soto Varela et al. 2016; Finger et al. 2019). Most of these cases are caused by diarrhea. According to the World Health Organization (WHO 2023), the consumption of contaminated food most commonly results in diarrhea, affecting approximately 550 million individuals and causing 230,000 deaths worldwide annually. Highly vulnerable individuals encompass the elderly, infants, children, and those with compromised immune systems due to weakened immunity.

Persistent exposure to foodborne diseases is attributed to several factors, including substandard hygiene practices, insufficient sanitation, and limited access to secure and uncontaminated food supplies. The prevalence of foodborne diseases is particularly pronounced in Africa and South-East Asia, where individuals in developing countries confront a heightened risk of exposure. This concern is accentuated by documented annual deaths, totalling 1.8 million, resulting from diarrheal diseases. Alarmingly, more than 80% of these fatalities affect children under the age of five, emphasizing the urgent need for comprehensive interventions in these regions (Hodge et al. 2016; Bakir et al. 2017). The comprehensive interventions include vaccination and supplementation, education, breastfeeding program, etc. Non-typhoidal *Salmonella* sp. ranks second in causing illnesses, but first in causing hospitalizations and deaths. In addition to *Salmonella* sp., the Centers for Disease Control and Prevention (CDC) have identified *Toxoplasma gondii*, *Listeria monocytogenes*, *Norovirus* sp., and *Campylobacter* sp. as the top five germs responsible for deaths from foodborne illnesses (CDC 2023a). These bacteria are associated with hard-to-treat diseases and often exhibit resistance to multiple available antibiotics. Pyrimethamine (PYR) and sulfadiazine (SDZ) represent the predominant medications utilized in treating toxoplasmosis caused by *T. gondii*; however, several cases documented treatment failures, indicating the presence of drug-resistant strains (Montazeri et al. 2018). As for *L. monocytogenes*, it has been reported to resist oxacillin, kanamycin, levofloxacin, and teicoplanin, amoxicillin/clavulanic acid, rifampicin, and ciprofloxacin (Şanlibaba et al. 2018). Campylobacter demonstrates a notable propensity for genetic mutation when subjected to fluoroquinolone treatment, and acquiring resistance does not diminish its capacity to thrive and replicate (Shen et al. 2018). Noroviruses, as with all viral infections, persist without susceptibility to antibiotic therapy (Nordgren et al. 2016). Hence, staying hydrated during illness is crucial, as vomiting and diarrhea can result in dehydration.

The excessive and improper use of drugs in human medicine has been linked to the development and propagation of antibiotic-resistant bacteria, making it challenging to treat foodborne diseases in humans. For instance, *Salmonella* sp., particularly those causing diarrhoeal and invasive disease, has demonstrated resistance to quinolones, fluoroquinolones, sulfonamides, trimethoprim, β-lactams, and chloramphenicol (Wang et al. 2022). Invasive disease, especially caused by non-typhoidal *Salmonella*, poses a severe and life-threatening risk, impacting the bloodstream, meninges, and other typically sterile areas of the body (Marchello et al. 2021). The sources of infections that result in cases of disease are not fully understood, with two primary theories: transmission from a zoonotic reservoir, as seen in other non-typhoidal salmonellosis, or person-to-person transmission (Marchello et al. 2022). Due to this resistance profile and threat, *Salmonella* sp. has been categorized under "Priority 2: High" class by WHO which requires urgent antibiotic alternative (WHO 2017).

Foodborne microbial hazards are not limited to bacteria, as they encompass a diverse range of pathogens, including fungi, viruses, parasites, and worms. These pathogens could potentially contaminate both animal and non-animal products at multiple stages of the food production chain, viz., (1) primary production, which takes place in farms where plants are grown or animals are raised for food; (2) harvest and slaughter of food produce and animals; (3) transportation; (4) food processing; (5) storage; (6) distribution; and (7) preparation and serving, including outside or inside the home. Among these, the most common source of environmental pollution with microorganisms occurs in the primary production of the pre-harvest ones (Fig. 1). Their transmission routes or agents are through avenues beyond food consumption. These include vectors (such as insects or rodents), fomites (inanimate objects that can carry pathogens), humans, animals, waterborne, and airborne routes (Hoffman and Scallan 2017; Stein and Chirilã 2017). Thus, it is imperative to highlight the necessity of implementing effective strategies and mechanisms to safeguard agricultural and aquacultural practices, ensuring they remain free from the risks associated with microbial contamination and foodborne illnesses.

Fig. 1 Microorganisms are predominantly produced during the pre-harvest stage, especially within farm environments. This occurs through the presence of microorganisms in soil, water sources, or on the surfaces of plants and animals

2 Microbial Pollution in Agriculture and Aquaculture Products

2.1 Animals and Their Products

Poultry meat consumption has undergone a substantial rise, nearly doubling from 67 million tons in the year 2000 to 131 million tons in the year 2020, surpassing all other meat sectors and emerging as the largest meat sector (OECD 2021). The figures indirectly orchestrate a risk to public health if the meats become contaminated. The microbiological contamination of poultry meat is influenced by the conditions under which the animals are raised, slaughtered, and processed at slaughterhouses (Motarjemi and Leliveld 2014). Bacteria are found on surfaces, in the air as aerosols, and in liquids within these facilities. As a result, carcasses and cuts obtained after the slaughter of animals can be contaminated by the microbiota from both the animals and the slaughterhouse environment (Rouger et al. 2017). The process of poultry slaughtering encompasses various stages, including killing, scalding, defeathering, eviscerating, washing, and chilling (Marmion et al. 2021). After killing stage, poultry carcasses undergo scalding to ease feather removal via hot water dipping at 50–60 °C in a continuous-flow water tank. Throughout the scalding process, dust and fecal matter from the skin, feathers, as well as intestinal tract are liberated into the scalding water and unintentionally spread intestinal contents to subsequent carcasses' skin (Cox and Paciv 2010; Behravesh et al. 2012; Rouger et al. 2017).

In the subsequent step of defeathering, bacteria that thrive in moderate temperatures (mesophiles) may spread between the carcasses or from the equipment used for defeathering, potentially leading to an increase in their numbers on the carcasses (Ahmed et al. 2013). According to research by Russel (2012) and Seliwiorstow et al. (2015), defeathering has been identified as a significant contributor to the contamination of carcasses with pathogens such as *Staphylococcus aureus*, *Salmonella* sp., *Campylobacter* sp., and *Escherichia coli*. A report by Perez-Arnedo et al. (2020) revealed that *Salmonella* sp. is actively present after defeathering, evisceration, and washing processes, with the highest prevalence (40%) occurring after the defeathering process. The presence of the bacteria is equally distributed at 36% after evisceration and after washing processes. Similarly, *L. monocytogenes* was observed at 36% after washing and chilling processes in poultry slaughtering. A lower prevalence (12%) was detected after evisceration, while no detection was reported for this bacterium after defeathering. Another actively contaminating bacteria in poultry is *Campylobacter* sp., predominantly present at each stage of the slaughtering process. The highest prevalence was recorded after evisceration and after chilling, both at 96%. This was followed by after defeathering step (92%) and after washing step (88%). *Salmonella* sp. and *Campylobacter* sp. are recognized for their ability to endure on chicken skin. These bacteria can trap in the deeper layers, crevices, or feather follicles.

In addition, foodborne bacteria are capable of producing biofilms. The biofilms are not only responsible for food spoilage and outbreaks but also for damage to slaughtering processing apparatus. These substances can develop on surfaces, whether living (biotic) or non-living (abiotic), facilitated by the secretion of extracellular polymeric substances (EPS). In general, biofilms comprise proteins, polysaccharides, nucleic acids, and lipids (Ashrafudoulla et al. 2020). Initially, the formation of biofilm can result in fouling—a buildup of microorganisms, organic matter, and various materials on surfaces. This accumulation can lead to blockages, corrosion, and equipment damage across multiple sectors, including healthcare, water distribution systems, food processing and packaging, industrial manufacturing, marine industries, and sanitation (Shineh et al. 2023).

Pseudomonas biofilms pose a serious risk of contaminating subsequent batches of food passing over the biofilm region or through aerosols originating from the contaminated surface. According to Zarei et al. (2023), surfaces of equipment within poultry slaughterhouses act as significant reservoirs for *Pseudomonas* strains, leading to the spoilage of poultry meat during subsequent refrigerated storage. This bacterial genus is recognized for its remarkable diversity, encompassing the highest number of species. A total of 14 *Pseudomonas* sp. isolates were obtained from chilling tanks, 5 isolates from the surfaces of hanging shackles, 18 isolates from viscera sorting tables, and 21 isolates from packing tables. Null *Pseudomonas* sp. was found on the rubber fingers of the plucking machine. These bacteria can form dual-species biofilms with *Salmonella enterica* serovar Enteritidis. However, this ability depends on whether *S. enterica* serovar Enteritidis forms the biofilm concurrently with the *Pseudomonas* strains or if the *Pseudomonas* strains have already established a biofilm, as illustrated in Fig. 2.

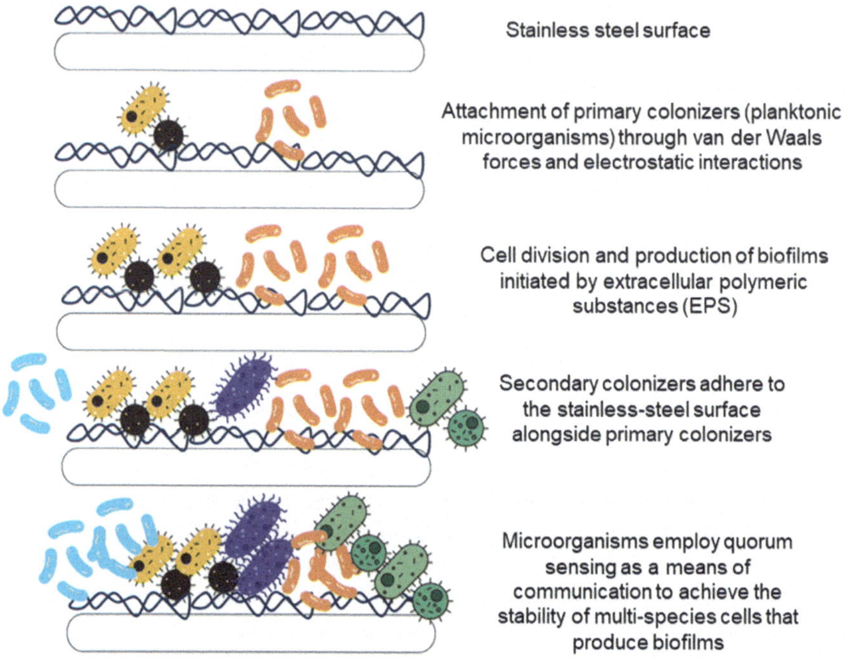

Fig. 2 Biofilm formation on stainless steel surfaces by dual or multi-species microorganisms

Multi-species biofilms tend to demonstrate greater stability in comparison to biofilms composed of a single species. Enterobacteriaceae was identified on camel, cattle, and sheep meats, exhibiting a total viable count ranging from 4.91 to 5.37 log10 CFU/g. This data was gathered from both slaughterhouses and butcher shops in Saudi Arabia. Among 120 meat samples, *E. coli* was prevalent in the majority, accounting for 70% (Al-Asmari et al. 2023). On the other hand, Bakhtiary et al. (2016) found that the bacterial diversity in environmental samples from the sheep slaughter line was higher than that observed in cattle. The variation in contamination rates can be attributed to heightened transmission from animals to the environment and the production line, potentially stemming from the manual slaughtering of sheep. Conversely, in the cattle slaughter line, all processes were conducted on a production line featuring vertical rail dressing and automatic hide pullers. The hygienic conditions during bleeding in the cattle slaughter line were superior, as animals were hoisted by one leg, and bleeding continued until the blood flow was minimal. Overall, the use of automatic hide removal contributed to a reduction in carcass contamination, as it involved less handling and fewer knife applications. This information explains the detection of 25 Enterobacteriaceae strains from 15 genera in sheep meat, representing the highest microbial prevalence compared to cattle and camel meats, which exhibited only 14 different genera (Al-Asmari et al. 2023).

Salmonella species, another important pathogen, can contaminate both the shells and the interior of eggs. It can potentially contaminate eggshells through various pathways, including the deposition by birds during the egg-laying process and contact with bird droppings after laying. Significantly, handling eggs from retail establishments is generally considered a minor risk factor for illness, as these eggs undergo thorough washing procedures before reaching the stores. However, the transmission of *Salmonella* sp. extends beyond the external surface of eggs, as the bacterium can infiltrate the interior during the egg's developmental stage within the chicken before the formation of the protective shell (CDC 2023b). Other sources of *Salmonella* sp. contamination encompass the nest, hens, feet, fecal matter, wash water, handling materials, hands of the person, and soiled containers. Fecal matter, particularly, serves as a significant source of *S. enterica* serovar Enteritidis and Enterococcus. Additional primary contaminants of eggs involve Gram-positive bacteria of the genera *Arthrobacter*, *Bacillus*, *Micrococcus*, and *Staphylococcus*, as well as Gram-negative bacteria of the genera *Acetobacter*, *Aeromonas*, *Alcaligenes*, *Enterobacter*, *Escherichia*, *Flavobacterium*, *Proteus*, *Pseudomonas*, *Serratia*, and coliforms (Erkmen and Bozuglu 2016). A recent study by Abdelhamid and Yousef (2023) found that *S. enterica* serovar Enteritidis found in eggs and poultry share a close genetic similarity. In contrast, these strains exhibit notable genomic differences when compared to those associated with beef. This distinct genetic correlation highlights the significance of the vertical transmission route of *S. enterica* serovar Enteritidis, particularly from breeding chickens to hatching eggs, as highlighted by previous research by Liu et al. (2022). Additionally, bacteria can undergo the vertical transmission route, also referred to as trans-ovarian contamination, during the egg formation process within either the ovary or the oviduct (Gast et al. 2019).

Effective egg storage preserves their quality; however, both physical and chemical transformations occur during the deterioration process. Physically, the egg white becomes waterier due to a reduction in viscosity, and water migrates from the egg white to the yolk, causing thinning. Evaporation of water occurs through the shell, and the escape of carbon dioxide leads to an elevation in the pH value. Consequently, protein breakdown commences, accompanied by various other alterations (Obi and Igbokwe 2007; Siripongvutikorn et al. 2024). Fungal growth on the eggshell occurs under conditions of relatively high humidity (<80%). Once fungi colonize the surface of the shell, hyphae can rapidly penetrate the pores. Increased condensation on the eggshell elevates the risk of fungal growth (Erkmen and Bozuglu 2016). According to Rajmani et al. (2011), at least six fungal genera, viz., *Aspergillus* sp., *Penicillium* sp., *Fusarium* sp., *Mucor* sp., *Rhizopus* sp., and *Alternaria* sp. were isolated from eggshell. The presence of *Aspergillus* sp. in egg samples from poultry farms indicates potential contamination, either in the poultry feeds, the raw materials used in feed production, or due to a general lack of hygiene on these farms. This is a significant health concern as the fungus can produce aflatoxins, known to be carcinogenic, teratogenic, and mutagenic in both humans and animals (Sarma et al. 2017).

2.2 Fruits, Vegetables, and Cereals

Fruits and vegetables are commonly consumed in a ready-to-use or ready-to-eat state and seldom undergo any heat processing before consumption. The initial stage of fresh produce contamination involves the attachment of pathogens to phyllosphere or rhizosphere tissue. This process involves cellular mobility facilitated by structures such as curli, fimbriae, and flagella, ultimately culminating in the formation of biofilms (Ölmez 2016). *Salmonella* sp. and *E. coli* are two prevalent microorganisms linked to fresh produce. In general, pathogenic bacteria linked to both fruits and vegetables can rapidly multiply when deposited onto injured portions of the produce. This phenomenon was demonstrated when *E. coli* O157:H7 exhibited an 11-fold increase in population during a 4-h incubation period after being introduced to cut lettuce stems (Brandl 2008). Subsequently, a comprehensive transcriptomic analysis of the entire genome unveiled an upregulation of genes linked to *E. coli* O157:H7 attachment, virulence, DNA repair, resistance to oxidative stress, and resistance to antimicrobials when the bacterium was exposed to the exudates of injured plant tissue (Kyle et al. 2010). Notably, *E. coli* O157:H7 demonstrated its capacity to endure low temperatures on food over a specific duration. This was substantiated by its persistence in fresh-cut mango and papaya stored at −20 °C for 180 days (Strawn and Danyluk 2010). Furthermore, at 5 °C, *E. coli* O157:H7 exhibited survival in fresh-cut pineapple and melon for 15 days (Abadias et al. 2012). Intriguingly, in "Rocha" pears, this bacterium thrived, reaching a population exceeding 8 log CFU/g when exposed to temperatures of 12 and 20 °C (Graça et al. 2017). Conversely, in peaches, under conditions of 20 and 25 °C, *E. coli* O157:H7 cell counts were recorded at 2 log CFU/g (Alegre et al. 2010). The majority of *E. coli* strains are not harmful. However, some types, viz., Shiga toxin-producing *E. coli* (STEC), can lead to serious foodborne illness. When aggressive strains like *E. coli* O157:H7, the main serotype within the STEC group, are present in fruits, they can significantly raise the risk of foodborne illness outbreaks due to their ability to survive and spread in various environments, their genetic flexibility, and their preference for certain types of fruit (WHO 2018).

From 2006 to 2023, the CDC documented approximately 70 foodborne salmonellosis outbreaks in humans. These include infections associated with the consumption of fresh and fresh-cut fruits in the United States (CDC 2024a). According to He et al. (2021), tomatoes, cantaloupes melon, and cucumbers were implicated in roughly 80% of salmonellosis cases related to the consumption of fruits. RNA sequencing depicted both *Salmonella enterica* serotypes Newport and Typhimurium exhibited resilient survival on diverse fresh-cut fruit items under refrigeration storage conditions, regardless of the initial inoculation levels. After a 7-day storage period at 4 °C, *S. enterica* populations were detected in fresh-cut tomato and cantaloupe melon, registering an approximate 0.2 log reduction. In contrast, a growth in population was observed in fresh-cut cucumber and apple, showing a 0.5 log increase. Furthermore, the analysis indicated that *S. enterica* orchestrates its survival on fresh-cut fruits by regulating genes associated with specific carbon utilization and metabolic pathways.

Notably, the bacterium up-regulated Pag genes, including PagC, PagD, and PagN, serving as virulence factors for survival on both cantaloupe melon and tomato. This finding suggests a link between events of food contamination and the potential for human infection. *Salmonella* serotype Poona is also under concern, contaminating cantaloupe melon. Washing cantaloupe at ambient temperature proves ineffective in eliminating microbial contaminants. Cells of *Salmonella* Poona can attach and form biofilms inside the netting of the rind, rendering them inaccessible to the washing solution. Storing untreated, inoculated cantaloupes at room temperature for 24–72 h post-inoculation resulted in a notable rise in populations of *Salmonella* Poona and *E. coli* compared to storage at 4 °C (Annous et al. 2004).

Listeria monocytogenes is extensively found in plant vegetation, including raw vegetables. The presence of this bacterium on plant materials is likely a result of contamination from feces of livestock, wildlife, humans, and surface waters (Gartley et al. 2022). It was reported that outbreaks of *Listeria* associated with fresh produce occur less frequently compared to outbreaks caused by *Salmonella* sp. and *E. coli* (Ölmez 2016). Similar with *S. enterica* serovar Enteritidis and *E. coli* O157:H7, *L. monocytogenes* emerged as a significant microorganism posing a threat to the safety of fruits and vegetables. Concurrently, there was a reported outbreak associated with *L. monocytogenes*-contaminated stone fruits such as peaches, plums, and nectarines. *L. monocytogenes* exhibited sustained viability throughout the entire storage period, persisting until day 26, with no significant deviation in levels compared to those observed on day 19–21. This observation suggests the robust resilience of this bacterium under the specified storage conditions, where psychrotolerant pathogenic bacteria like *L. monocytogenes* thrive at or below 7 °C (Greig et al. 2005). Notably, the types of stone fruit utilized in the study did not exert a significant influence on the survival of *L. monocytogenes*, indicating a consistent bacterial response across different stone fruit varieties (De Jesus et al. 2020). Demonstratively, cases of *L. monocytogenes* influencing public health via contamination of different fruits and vegetables were evident. For instance, it played a role in a 2010 listeriosis outbreak associated with chopped celery in Texas, leading to the deaths of 10 individuals (Gaul et al. 2013). Similarly, in 2011, 30 people contracted listeriosis from melons contaminated with *Listeria* in Colorado (CDC 2011). Additionally, a 2014 listeria outbreak in California was linked to caramel apple contamination (CDC 2015). To that, CDC (2024b) estimated 1,600 people experience listeriosis annually, leading to around 260 fatalities. This consistent recurrence of incidents related to *Listeria* highlights the ongoing challenge of preventing its contamination across various fresh fruits, vegetables, and associated produce. This challenge is emphasized not only by the stone fruit-related outbreak mentioned earlier but also by incidents involving chopped celery, melons, and caramel apples.

Cereal grains hold a pivotal position in global nutrition, playing a significant role in meeting the caloric requirements of the worldwide population. However, the presence of diverse pathogenic microbes poses a potential threat, leading to food spoilage, mycotoxin production, or both. In the field, cereals are frequently subjected to contamination by fungi such as *Fusarium* sp., *Alternaria* sp., *Cladosporium* sp., and *Helmintosporium* sp. Conversely, during storage, there is a higher incidence of contaminants such as *Penicillium* sp., *Aspergillus* sp., *Eurotium* sp., and *Rhizopus* sp. (Minutillo et al. 2022). *Bacillus cereus* is commonly associated with intricate food items that incorporate rice, such as rice-based products, as well as starchy foods like noodles, which have been reported to be linked to *B. cereus* intoxication (Wang et al. 2023). Cooked rice, which typically has a pH near 7 and comprises 79% carbohydrates, 7% protein, 2% fat, along with essential vitamins and minerals, serves as an optimal growth medium for *B. cereus*. This is attributed to the specific moment in the cooking process when the substrate's humidity reaches water activity values conducive to the microorganism's proliferation (Rodrigo et al. 2021). *B. cereus* spores, along with active or vegetative cells found in the intestines of insects, may contaminate crops near the soil, such as rice, and subsequently be transferred to humans (Jensen et al. 2003; Rodrigo et al. 2021). Despite cooking rice at temperatures above 55 °C being effective in killing *B. cereus* cells, spores can endure the cooking process. Later, under favorable growth conditions, these spores can germinate and develop into bacteria (Albaridi 2022).

2.3 *Aquaculture*

Aquatic food items rank as some of the most nourishing and precious foods globally. They boast a significant protein content, supplying over 20% of the average per capita animal protein intake for approximately 3.3 billion people. In 2017, fish alone accounted for about 17% of the total animal protein and 7% of overall protein consumption worldwide (FAO 2020). The spoilage dynamics of fish are influenced by various factors. Firstly, the type and size of the fish play a crucial role, with flat fish, high in lipids and trimethylamine oxide, being more susceptible compared to cylindrical, fatty fish, which spoil at a slower rate. Secondly, whether the fish is native or cultivated, along with the capture conditions, contributes to the spoilage process. Fish subjected to exhaustion through shaking, oxygen deprivation, and improper handling, as well as those with a full digestive tract, tend to spoil more rapidly. Thirdly, the type and microbiological load are significant determinants, with a higher load accelerating the deterioration process. Lastly, storage temperature plays a crucial role, with a higher temperature leading to a shorter shelf life (Naef et al. 2023). Moreover, research indicates that microbial metabolism of carbon and nitrogen sources plays a significant role in contributing to the spoilage of aquaculture products during storage (Zhuang et al. 2021).

In healthy fish, all the complex metabolic processes are in balance, and the flesh remains free of contaminants. However, after capture, the fish will eventually die and undergo irreversible changes that lead to spoilage (Adeyeye 2016). Fish possesses an almost neutral pH, high water activity, and serves as a rich source of proteins, lipids, vitamins, and minerals (Ali et al. 2022). However, these characteristics render fish highly susceptible to contamination and spoilage, primarily attributed to autolysis, lipid oxidation, and microbiological activity (Mei et al. 2019). Spoilage in fresh and processed fish as well as fishery products comes from yeast, fungi, and bacteria. Fish harvested for human consumption undergo processes like drying, salting, brining, and smoking to enhance their quality. While bacteria struggle to grow in dried and salted fish due to very low water activity, the issue of fungal growth arises when fish lose 15% of their water content, leading to spoilage. Thus, the Codex Standard sets a maximum allowable water activity of 0.75 a_w (containing only 10% moisture or less) for fish products. This standard is in place to control the growth of bacteria pathogens and spoilage fungi, as these microorganisms require a certain level of water to thrive (Fasuan et al. 2022).

Based on a study by Deng et al. (2021), a total of 12 genera was isolated from dried fish, with *Fusarium* sp., *Penicillium* sp., and *Aspergillus* sp. as predominant fungal isolates. Among these, the contamination caused by *Aspergillus flavus* was notably higher, ranging from 1.10×10^3 to 2.40×10^4 CFU/g. Other fungal genera identified included *Neoscytalidium* sp., *Cutaneotrichosporon* sp., *Trichoderma* sp., *Naganishia* sp., *Kodamaea* sp., *Phialemoniopsis* sp., *Nigrospora* sp., *Ceriporia* sp., and *Phellinus* sp. It has been reported that the primary spoilage agent for dried fish produced in temperate regions was *Wallemia sebi.* This fungus was also found to contaminate salt-based fish along with *Sporendonema* sp. and *Oospora* sp. (Galaviz-Silva et al. 2008; Gram 2009). In another study by Dutta et al. (2018) on smoked sardines (*Sardinella aurita*), *Aspergillus fumigatus* and *A. flavus* were identified, both of which have mycotoxin potential. Fungal growth not only jeopardizes the quality of fishery products but can also lead to the production of mycotoxins, posing health risks. This was substantiated by the presence of aflatoxin B1, T-2 toxin, ochratoxin A, and deoxynivalenol alongside dried fish (Deng et al. 2021). Most of these toxins exhibit carcinogenic properties, particularly affecting the liver and kidneys. Additionally, they are implicated in disrupting intestinal barrier function, modulating the immune system, and impairing fetal development, collectively influencing human health (Awuchi et al. 2022).

Pseudomonas sp. is recognized as a specific spoilage organism in fish stored aerobically at both low and ambient temperatures (Naef et al. 2023). Bacteria from the Pseudomonadaceae family are implicated in the formation and accumulation of biogenic amines, such as histamine, putrescine, cadaverine, tyramine, agmatine, and tryptamine. Researchers have emphasized that histamine, when present at elevated concentrations, poses a substantial risk for poisoning, particularly considering its greater abundance in fish products (Visciano et al. 2014; Ruiz-Capillas and Herrero 2019). Histamine synthesis within fish tissue occurs through the enzymatic decarboxylation of free histidine, a process facilitated by exogenous decarboxylases released by microorganisms, including *Pseudomonas* sp. The challenge is exacerbated by the fact that the levels of histamine cannot be reduced through cooking, canning, or freezing, as this compound exhibits heat stability (Visciano et al. 2014). Furthermore, the occurrence of *Pseudomonas* in protein-rich foods has been observed to stimulate the proliferation of *L. monocytogenes*, attributed to its proteolytic activity and the release of readily metabolizable peptides in contaminated products. A study by Maggio et al. (2021) demonstrated that *L. monocytogenes* exhibited significantly higher attachment numbers on surfaces with preexisting *Pseudomonas* biofilms compared to surfaces without *Pseudomonas* presence.

Shrimps and prawns constituted the primary categories of crustaceans, contributing to approximately 17% of the overall value of globally traded aquaculture products (FAO 2019). Crustaceans are prone to spoilage during storage. In a study conducted by Broekaert et al. (2013), both *Pseudoalteromonas* sp. and *Psychrobacter* sp., isolated from brown shrimp, have been identified as potential contributors to spoilage due to their capacity to break down lipids and degrade amino acids and proteins. Significantly, *Pseudoalteromonas* sp. exhibited a greater responsibility for generating off-odors in comparison to *Psychrobacter* sp. Additionally, Qian et al. (2018) detected different bacteria, including *Shewanella putrefaciens*, *Aeromonas hydrophila*, and *Aeromonas sobria*, which were identified as potential spoilage agents in Pacific white shrimp. The coexistence of *S. putrefaciens* and *A. sobria* was found to elevate the levels of total volatile basic nitrogen (TVB-N) and putrescine in shrimp juice. These two substances are known contributors to shrimp spoilage. In another study, fungi were identified in samples of dried salted shrimp, with approximately 1.0×10^2 and 1.02×10^6 CFU/g of fungi. The predominant microorganisms isolated were *Aspergillus* sp., *Penicillium* sp., and *Cladosporium* sp. (Freitas et al. 2021). The majority of fungi identified in dried salted shrimp samples may pose a potential human health risk. Moreover, the consequences of consuming contaminated food on the gastrointestinal tract remain under-explored.

3 Control, Prevention, and Management Options for Spoilage Microorganisms

3.1 Key Players in Food Security

As the global population continues to grow, the expanding scale and industrialization of agriculture and aquaculture to meet the escalating demand for food present both opportunities and challenges in ensuring food safety. Numerous agencies and organizations are actively engaged in the regulation, mitigation, and supervision of microbial agents contributing to food spoilage within the realm of food security. In many countries, the main goal of having a Food and Drug Administration (FDA) or a similar agency is to ensure compliance with food safety laws, aiming to protect public health and safety. This involves informing citizens about the nutritional content of important food products, enforcing existing laws in the food industry to ensure the availability of safe products, and monitoring the food supply chain to eliminate potential toxic contaminants (Fung et al. 2018).

The Food and Agriculture Organization (FAO), a specialized agency of the United Nations, spearheads global endeavors to eradicate hunger. Collaborating closely with the World Health Organization (WHO), the World Organisation for Animal Health (OIE), and the United Nations Environment Programme (UNEP), FAO ensures food safety across the entire food supply chain. Additionally, FAO evaluates the effectiveness of national food control systems and assists in establishing the necessary infrastructure to address food safety hazards and swiftly respond to food-related emergencies. In the FAO Strategic Framework for 2022–2031, four key areas have been identified under Programme Priority Areas (PPAs) to enhance the food safety system. These include (i) strengthening intergovernmental and multi-stakeholder engagement for coordinated food safety governance at all levels, (ii) providing robust scientific advice and evidence to inform food safety decision-making, (iii) enhancing and continually improving national food control systems, and (iv) encouraging collaboration between public and private stakeholders to ensure effective food safety management and controls across agri-food systems (FAO 2023).

The USDA is tasked with safeguarding consumers from foodborne illnesses and promoting food safety. They offer guidance and materials on food spoilage, such as the FoodKeeper app, which assists in extending the freshness and quality of food products by detailing proper storage techniques. This app is a convenient tool for individuals to ensure the safety of their food by offering helpful guidance on how to store various foods and drinks effectively to enhance their freshness and reduce food wastage. Through educating users on proper food storage techniques, the app enables consumers to choose methods that prolong the shelf life of their items and maintain their freshness for a longer period compared to improper storage practices (USDA 2024).

The World Health Organization (WHO) takes the lead in addressing global health issues, including ensuring food safety. They advocate for safe food practices through preventive measures and awareness campaigns, monitor the worldwide impact of foodborne and zoonotic diseases, and revise the WHO Global Strategy for Food Safety to assist Member States in enhancing their domestic food safety mechanisms. The WHO collaborates closely with the Food and Agriculture Organization (FAO), the World Organization for Animal Health (OIE), the UN Environment Programme (UNEP), and other international organizations to guarantee food safety across the entire food chain, from production to consumption (WHO 2022).

3.2 Innovative Techniques and Technologies for Enhancing Food Safety and Preservation

Conventional and modern techniques, including innovations such as microbe-free contact surfaces, hurdle technology, and novel packaging materials with antimicrobial properties, have become indispensable in ensuring food safety and security. These technologies not only extend shelf life but also preserve nutritional integrity and enhance overall product safety.

3.2.1 Microbe-Free Contact Surfaces

Responding to effective preservation, consistent cleaning and disinfection of equipment surfaces are paramount. Surfaces that come into contact with food, including packaging materials, cutting devices, and conveyor belts, pose a persistent risk of contaminating food products during transportation, processing, and storage (Hertwig 2017). Microbe-free contact surfaces are necessary to manage microbial presence and prevent cross contamination. The use of chemicals in the food industry for disinfection purposes is not new but remains effective. These chemicals can be categorized into oxidative and non-oxidizing disinfectants. Oxidative disinfectants, such as chlorine compounds and peracetic acid, are commonly employed. On the other hand, representative non-oxidizing disinfectants include Quaternary Ammonium Compounds (QACs or Quats) (Mendoza et al. 2022). Chlorine compounds are commonly used in the food industry for disinfecting various agricultural products such as animal products, fruits, and vegetables. During the production process of animal products, disinfectants such as low levels of chlorine and peracetic acid (also known as peroxyacetic acid) are utilized to cleanse both the interior and exterior of the poultry washer. These disinfectants play a crucial role during pre-chilling, chilling, and post-chill treatments (Wideman et al. 2016). Their application at concentrations ranging from 20 to 50 ppm aims to minimize contamination from blood, tissue, or fecal matter. Similar chemicals are introduced into cold water, particularly at the final stage of poultry processing, to sanitize poultry products effectively (Keener et al. 2004; Wang et al. 2020). Their primary function lies in the rapid cooling of poultry and inhibition of bacterial growth. Notably, they also serve as agents for carcass sanitation (Wang et al. 2020).

There are concerns regarding the use of chlorine compounds in food because their interaction with organic components can potentially form carcinogenic compounds. As a result, there are restrictions on their application. In the European Union (EU), the use of disinfectants such as chlorine or peracetic acid on poultry carcasses is not prohibited. Conversely, in the United States, the United States Department of Agriculture (USDA) Food Safety and Inspection Service (FSIS) allows the addition of chlorine to processing waters at levels up to 50 ppm, calculated as free available chlorine, for carcass wash applications, and chiller make-up water. FSIS mandates that chlorinated water with a minimum of 20 ppm available chlorine must be applied to all carcass surfaces when inner surfaces have been re-processed (due to carcass contamination), besides trimming alone (USDA Food Safety and Inspection Service 2017). Japan adopts a similar approach, utilizing sodium hypochlorite at concentrations ranging from 35 to 200 ppm as total chlorine concentration to disinfect poultry carcasses. Despite this measure, numerous cases of cross contamination persist, particularly during the chilling process at slaughterhouses. Consequently, bacteria are detected in processed poultry meats, even at retail markets (Sasaki et al. 2013).

3.2.2 Hurdle Technology

Recent advancements in food preservation highlight hurdle technology's role in combatting microorganisms by creating a sterile environment, thwarting their growth or survival. Emerging technologies like high-pressure process (HPP), pulsed electric field (PEF), ultrasound, and irradiation have been innovated to uphold the nutritional and sensory attributes of foods. These advancements simultaneously cater to consumer preferences for natural preservation methods.

HPP stands out as a prominent hurdle technology, employing elevated pressure to deactivate microorganisms while disrupting non-covalent bonds within enzymes, ribosomes, and cellular membranes, thus achieving microbial inactivation (Putnik et al. 2020). HPP is also known as high-hydrostatic pressure processing (HHP) or ultra-high-pressure processing (UHP). It typically exposes liquid and solid foods to pressures ranging from about 400 to 600 MPa at refrigeration or mild process temperatures (<45 °C). This method upholds food safety while preserving the freshness and quality of food products, which minimally affects taste, texture, appearance, or nutritional value (Muntean et al. 2016). It has the capacity to eliminate pathogens such as *S. enteritidis, E. coli, L. monocytogenes*, and *Vibrio* sp. (Silva and Evelyn 2023), along with spoilage microorganisms like lactic acid bacteria and coliforms (Balamurugan et al. 2018), thereby meeting the standards set by food authorities. According to Wu et al. (2022), the adoption of the HPP hurdles strategy demonstrates efficacy in mitigating microbial proliferation and enhancing the longevity of packaged food items within the cold chain. This methodological approach presents synergistic advantages when integrated with various hurdles, including temperature modulation, pH optimization, antimicrobial interventions, and modified atmosphere packaging. By leveraging the HPP-based hurdle strategy, it is possible to lessen the reliance on additional hurdles, decrease the need for additives, and uphold the sensory attributes and nutritional quality of the food.

Pulsed electric fields are frequently utilized in the processing of fluid foods since the electrical current moves more efficiently within the liquid. With electric field intensities typically falling between 20 and 80 kV/cm, this method efficiently destroys microorganism cell membranes in less than one second (Bhavya and Umesh Hebbar 2017). In terms of mechanism of action, there are a few different theories available but the most recent one relates to the free radical theory. According to this theory, PEF treatment leads to the formation of free radicals and hydrogen peroxide. These reactive species penetrate the cell membranes, initiating an oxidation process that affects crucial cell components such as lipids and proteins. Furthermore, enzymes within the cell are adversely affected, ultimately resulting in cell death and a reduction in microbial populations within the treated sample. The effectiveness of PEF treatment hinges greatly on factors such as the characteristics of the cell's outer membrane, its orientation to the electric field, as well as parameters including cell dimensions and behavior in response to the electric field (Buckow et al. 2013). Moreover, Ghoshal (2023) later confirmed that the development of an electrical field and electro-mechanical rigidity, which transmit to the formation of pores in the membrane, is the primary mechanism behind microbial inactivation. For instance, *E. coli, Salmonella sp., L. monocytogenes*, mold spores (Li et al. 2021), *Saccharomyces cerevisiae* (Emanuel et al. 2020), and *Bacillus subtilis* (Siemer et al. 2014) represent some of the microorganisms effectively inactivated by this method.

Ultrasonication is a versatile process that operates across three distinct frequency ranges: low frequency (10–100 kHz), medium frequency (100–1 MHz), and high frequency (1–10 MHz). The specific frequency used can be adjusted based on the intended purpose of the treatment. Ultrasonication is widely utilized for a variety of applications, including extraction, emulsification, filtration, encapsulation, and curing of meat (Jadhav and Choudhary 2024). However, it is important to note that medium frequency sonication has been reported to generate undesirable chemical radicals, which can lead to unwanted changes in food products. As a result, medium frequency ultrasonication is less common in the food industry (Ranjha et al. 2021). Ultrasonication can be classified into direct or indirect sonication methods. In direct sonication, a sonication horn is submerged directly into the liquid food, whereas in indirect sonication, the food is contained within a vessel, which is then immersed in an ultrasonic bath. The cavitation effect generated during the ultrasound process can be harnessed for various purposes, including intensified extraction of bioactive compounds, deactivation of enzymes and microbes, enhanced heat and mass transfer during processing, intensified synthesis of lipids, alteration of rheological properties, separation of compounds, and formation of stable emulsions (Chemat et al. 2011; Jadhav and Choudhary 2024). Multiple recent studies have affirmed the efficacy of this approach against *E. coli* O157:H7 contamination on fresh produce (Ha and Kang 2015; Lin et al. 2019; Bai et al. 2023).

Irradiation does not render foods radioactive, alter their nutritional quality, or significantly affect their taste, texture, or appearance. The changes induced by irradiation are typically so subtle that distinguishing irradiated food from non-irradiated counterparts is difficult. It delivers effective results on agricultural and aquacultural products (FDA 2024). The primary forms of irradiation utilized in food industries include electromagnetic gamma rays, derived from radioactive isotopes of cobalt (^{60}Co) or cesium (^{137}Cs), and corpuscular beta rays, consisting of electrically generated electrons within a cathodic tube. These electrons are then accelerated by an electron accelerator to produce a high-speed stream directed at foods (Bouzarjomehri et al. 2020; FDA 2024). As per Lung et al. (2015) and Federighi (2019), the energy of beta rays should not surpass 10 million electron-volts (MeV) and under 5 MeV for gamma rays. These radiations can induce both direct and indirect damage to microbial DNA. Direct damage occurs when the rays disrupt the bonding between base pairs in DNA or RNA, resulting in cell reproductive death, which is observable. Indirect damage to DNA involves the generation of free radicals and reactive oxygen species through the breakage of water molecules. *Listeria* sp., *Salmonella* sp., *Clostridium* sp., *Cronobacter* sp., and *S. aureus* from agricultural and aquacultural products are the common targets of irradiation (Munir and Federighi 2020).

3.2.3 Novel Packaging Materials

Employing microbicidal or microbiostatic approaches on products and packaging is crucial, as effective packaging not only physically shields food but also ensures its safety and freshness throughout transportation and storage. Prolonging the shelf life of food necessitates measures to prevent the ingress of moisture and gases, including oxygen and carbon dioxide, mitigate exposure to light, and impede microbial decay (Ahankari et al. 2021). Polyethylene (PE), polyethylene terephthalate (PET), polystyrene (PS), polyvinyl chloride (PVC), polypropylene (PP), polyvinyl alcohol (PVA), ethylene vinyl alcohol (EVOH), polytetrafluoroethylene (PTFE), and poly (methyl methacrylate) (PMMA) are among conventionally employed polymers in food packaging applications. These polymers serve as barriers against chemical, biological, and physical threats, providing a preservation system for the food, particularly safeguarding the original aroma, flavor, and nutrients (Vallejos et al. 2022).

Interest in integrating nanotechnology into polymeric packaging is growing. Utilizing nanotechnology-based food packaging for natural products entails emphasizing its multitude of advantages compared to conventional packaging. These include heightened preservation, enhanced safety, increased sustainability, efficient delivery of bioactive compounds, and innovative properties. In contrast, conventional packaging methods may lack preservation, pose environmental concerns, lack advanced features, and raise safety issues, thereby making nanotechnology-based solutions a more appealing choice for packaging natural products. Ab Rashid et al. (2022) found that a film made of naringenin-polyvinyl alcohol (PVA) nanoparticles reduced bacterial load in beef by 100% and showed bactericidal effects against various foodborne pathogens. In another study, the same researchers discovered that thymol-loaded PVA nanoparticles inhibited MRSA, *Y. enterocolitica*, and *S. typhi* on blueberries, achieving a 100% reduction in *Y. enterocolitica.* Additionally, hyperbranched polyamide-amine (HPAMAM) combined with silver nanoparticles effectively reduced silver leakage and curtailed the growth of *E. coli* and *S. aureus* when used as packaging for cherry tomatoes (Gu et al. 2019). Furthermore, a study by Youssef et al. (2014) showcased that the incorporation of chitosan, nanosilver, and nanogold demonstrated a comprehensive antimicrobial spectrum, effectively inhibiting both Gram-positive and Gram-negative bacteria, as well as fungi and yeasts. The microorganisms targeted encompassed *S. aureus*, *P. aeruginosa*, *A. niger*, and *C. albicans*, respectively. Yu et al. (2019) fabricated cellulose nanofibrils (CNFs) and silver nanoparticles (AgNPs) composite via a reduction method using sodium borohydride. The utilization of this composite in food packaging demonstrated a more effective inhibitory effect on *E. coli* O157:H7 and *L. monocytogenes*, without any discernible adverse impact on the viability of colon cells at concentrations up to 1000 μg/mL. These findings suggest the significant potential of utilizing metal sources in food packaging. However, their toxicity, particularly concerning leaching issues, necessitates thorough evaluation before widespread implementation in industrial applications. Thus, researchers are recognizing considerably safer alternatives by combining edible natural products with less toxic or non-toxic polymers. Lotfinia et al. (2023) successfully developed antibacterial food packaging that can inhibit *S. aureus* and *E. coli* by applying honey and red cabbage anthocyanin with sodium alginate. The susceptibility of both bacteria was represented by a diameter zone of more than 17 mm. Similarly, Md Abu et al. (2020) identified honey as an antibiotic against common foodborne pathogens, including *B. subtilis, S. aureus*, and *P. mirabilis*. In their experiment, nanocellulose derived from medical-grade cotton wool was used as a platform for the controlled release of honey loaded in polymeric polyvinylpyrrolidone. All of the mentioned reports point out the potential of nanotechnology in food packaging.

Nevertheless, its implementation is not without challenges. These hurdles encompass the need for cost-effective methods in nanomaterial characterization for food, toxicity assessment, scalability solutions, and stability assurance in field conditions. The perception and acceptance of nanotechnology in food packaging by consumers are critical, necessitating effective communication about the benefits and safety of nanomaterials to establish consumer trust. Since nanotechnology is a novel approach, meeting regulatory requirements is another challenge. This requires ongoing research on existing regulations and the development of new ones. Once marketed, it is essential to provide customers with clear and standardized labeling for products. This information not only offers details about the product but also contributes to transparency in the market.

Modified atmosphere packaging (MAP) is a method utilized for prolonging the shelf life of food by modifying the air surrounding the product. This involves carefully adjusting the balance of atmospheric gases, including oxygen (O_2), carbon dioxide (CO_2), and nitrogen (N_2), to inhibit oxidation, respiration, and slow down microbial growth, thereby preserving the quality of the food for an extended period. This method is appropriate for various food items across typical food categories, such as sausages, various meats, vegetables, fruits, dairy products, bread, fish, seafood, and other similar products (Czerwiński et al. 2021). Spoilage in meat products like sausages and raw meat primarily results from microbial growth, high moisture, and nutrient content, beginning shortly after butchering. MAP is an effective method for extending the shelf life of various meat products, with CO_2 playing a critical role as a protective gas. To maintain the appealing red color of red meat, MAP with high O_2 levels prevents oxidation. Additionally, CO_2 within modified atmosphere packaging (MAP) serves to inhibit bacterial growth, thereby extending shelf life. However, excessive levels of CO_2 during MAP may induce a sour taste in meat. Notably, the N_2 application can also result in a more pronounced sour taste (Hansen et al. 2023). Furthermore, elevated oxygen concentrations within MAP have been associated with lipid oxidation and subsequent rancidity in meat products (Domínguez et al. 2019).

Aquaculture products are highly prone to spoilage due to their rapid decline in quality post-harvest, primarily due to their neutral pH level and the presence of enzymes that facilitate microbial growth, resulting in unpleasant taste and odor. Moreover, fish rich in fatty acids are susceptible to rapid rancidity, highlighting the need for stringent cooling measures at temperatures of 0–4 °C to extend their shelf life (Suárez-Medina et al. 2024). Additionally, it is important to note that MAP gases containing CO_2 levels exceeding 20% effectively inhibit the growth of bacteria.

Specifically, when oxygen is absent and CO_2 is present, there is a bacteriostatic effect on aerobic microflora, preventing the proliferation of Gram-negative bacteria and bacteria producing hydrogen sulfide (Speranza 2013; Czerwiński et al. 2021). It is common practice to utilize a gas mixture with a 50% CO_2 level to preserve freshness (De la Caba et al. 2019), although higher concentrations may lead to undesirable sour tastes and dryness. Occasionally, combining elevated CO_2 levels with O_2 can maintain color while inhibiting specific bacteria. N_2 is frequently employed as an inert gas to prevent CO_2 absorption and packaging collapse (Czerwiński et al. 2021). In MAP for agricultural products like fruits and vegetables, the suggested oxygen levels for optimal safety and quality range from 1 to 5%. However, in practice, the oxygen concentration in MAP production often drops below 1% (Singh 2010). For example, the ideal CO_2 concentration of 5% and O_2 concentration of 5% effectively postpone the ripening of tomatoes (Fagundes et al. 2015).

4 Conclusions

In conclusion, microorganism pollution poses a significant threat to food safety. Vigilant attention and comprehensive mitigation strategies are crucial, given the contamination risks in production, especially in agriculture and aquaculture, and the potential health risks for consumers. Implementing stringent measures, such as employing microbe-free contact surfaces, hurdle technology, and novel packaging materials, can provide a robust solution for food safety. However, researchers play a crucial role in developing effective monitoring and safety protocols. Achieving this requires a collaborative effort among consumers, researchers, regulatory bodies, and industries to ensure the integrity of our food supply.

References

Ab Rashid S, Rosli NSM, Hway TS, Yenn TW, Ring LC, Md Yusof FA, Wen-Nee T (2022) Naringenin-grafted polyvinyl alcohol (Na/PVA) nanoparticles: synthesis, characterisation and in vitro evaluation of its antimicrobial efficiency on fresh beef. Trop Life Sci Res 33(1):143–161. https://doi.org/10.21315/tlsr2022.33.1.9

Abadias M, Alegre I, Oliveira M, Altisent R, Viñas I (2012) Growth potential of *Escherichia coli* on fresh-cut fruits (melon and pineapple) and vegetables (carrot and escarole) stored under different conditions. Food Control 27:37–44. https://doi.org/10.1016/j.foodcont.2012.02.032

Abdelhamid AG, Yousef AE (2023) Egg-associated *Salmonella enterica* serovar Enteritidis: comparative genomics unveils phylogenetic links, virulence potential, and antimicrobial resistance traits. Front Microbiol 14. https://doi.org/10.3389/fmicb.2023.1278821

Adeyeye SAO (2016) Traditional fish processing in Nigeria: a critical review. Nutr Food Sci 46(3):321–335. https://doi.org/10.1108/NFS-11-2015-0148

Ahankari SS, Subhedar AR, Bhadauria SS, Dufresne A (2021) Nanocellulose in food packaging: a review. Carbohydr Polym 255:117479. https://doi.org/10.1016/j.carbpol.2020.117479

Ahmed TA, Saeed SAM, Hussien HA (2013) Evaluation of poultry meat safety based on ISO 22000 as food safety management system. Pak J Nutr 12:121–129

Al-Asmari F, Hamad SH, Al Hashedi SA (2023) Prevalence of Enterobacteriaceae in camel, cattle, and sheep carcasses at slaughterhouses and butcher shops. Appl Sci 13(20):11495. https://doi.org/10.3390/app132011495

Albaridi N (2022) Risk of *Bacillus cereus* contamination in cooked rice. Food Sci Technol 42. https://doi.org/10.1590/fst.108221

Alegre I, Abadias M, Anguera M, Usall J, Viñas I (2010) Fate of *Escherichia coli* O157:H7, *Salmonella* and *Listeria innocua* on minimally-processed peaches under different storage conditions. Food Microbiol 27(7):862–868. https://doi.org/10.1016/j.fm.2010.05.008

Ali A, Wei S, Ali A, Khan I, Sun Q, Xia Q, Wang Z, Han Z, Liu Y, Liu S (2022) Research progress on nutritional value, preservation and processing of fish—A review. Foods 11(22):3669. https://doi.org/10.3390/foods11223669

Annous BA, Burke A, Sites JE (2004) Surface pasteurization of whole fresh cantaloupes inoculated with *Salmonella poona* or *Escherichia coli*. J Food Prot 67(9):1876–1885. https://doi.org/10.4315/0362-028x-67.9.1876

Ashrafudoulla M, Mizan MFR, Ha AJ, Park SH, Ha SD (2020) Antibacterial and antibiofilm mechanism of eugenol against antibiotic resistance Vibrio parahaemolyticus. Food Microbiol 91:103500. https://doi.org/10.1016/j.fm.2020.103500

Awuchi CG, Ondari EN, Nwozo S, Odongo GA, Eseoghene IJ, Twinomuhwezi H, Ogbonna CU, Upadhyay AK, Adeleye AO, Okpala COR (2022) Mycotoxins' toxicological mechanisms involving humans, livestock and their associated health concerns: a review. Toxins 14(3):167. https://doi.org/10.3390/toxins14030167

Bai M, Dai J, Li C, Cui H, Lin L (2023) Antibacterial and antibiofilm performance of low-frequency ultrasound against *Escherichia coli* O157:H7 and its application in fresh produce. Int J Food Microbiol 400:110266. https://doi.org/10.1016/j.ijfoodmicro.2023.110266

Bakhtiary F, Sayevand HR, Remely M, Hippe B, Hosseini H, Haslberger AG (2016) Evaluation of bacterial contamination sources in meat production line. J Food Qual 39(6):750–756. https://doi.org/10.1111/jfq.1224

Bakir H, Hadi M, Jurdi M (2017) Towards a renewed public health regulatory and surveillance role in water, sanitation and hygiene. Eastern Mediterr Health J 23(8):525–526. https://doi.org/10.26719/2017.23.8.525

Balamurugan S, Inmanee P, Souza JD, Strange P, Pirak T, Barbut S (2018) Effects of high pressure processing and hot water pasteurization of cooked sausages on inactivation of inoculated *Listeria monocytogenes*, natural populations of lactic acid bacteria, *Pseudomonas* spp., and coliforms and their recovery during storage at 4 and 10 °C. J Food Protect 81(8):1245–1251

Behravesh CB, Williams IT, Tauxe RV (2012) Emerging foodborne pathogens and problems: expanding prevention efforts before slaughter or harvest. Improving food safety through a one health approach. National Academics Press, Washington, USA, pp 307–331

Bhavya ML, Umesh Hebbar H (2017) Pulsed light processing of foods for microbial safety. Food Qual Saf 1:187–201

Bouzarjomehri F, Dad V, Hajimohammadi B, Shirmardi SP, Yousefi-Ghaleh SA (2020) The effect of electron-beam irradiation on microbiological properties and sensory characteristics of sausages. Radiat Phys Chem 168:108524

Brandl MT (2008) Plant lesions promote the rapid multiplication of Escherichia coli O157:H7 on postharvest lettuce. Appl Environ Microbiol 74(17):5285–5289. https://doi.org/10.1128/AEM.01073-08

Broekaert K, Noseda B, Heyndrickx M, Vlaemynck G, Devlieghere F (2013) Volatile compounds associated with *Psychrobacter* spp. and *Pseudoalteromonas* spp., the dominant microbiota of brown shrimp (*Crangon crangon*) during aerobic storage. Int J Food Microbiol 166(3):487–493. https://doi.org/10.1016/j.ijfoodmicro.2013.08.013

Buckow R, Ng S, Toepfl S (2013) Pulsed electric field processing of orange juice: a review on microbial, enzymatic, nutritional, and sensory quality and stability. Compr Rev Food Sci Food Saf 12:455–467

CDC (2023a) Foodborne germs and illnesses. https://www.cdc.gov/foodsafety/foodborne-germs.html

CDC (2011) Multistate outbreak of listeriosis linked to whole cantaloupes from Jensen farms, Colorado (FINAL UPDATE). https://www.cdc.gov/listeria/outbreaks/cantaloupes-jensen-farms/index.html

CDC (2015) Multistate outbreak of listeriosis linked to commercially produced, prepackaged caramel apples. http://www.cdc.gov/listeria/outbreaks/caramel-apples-12-14/index.html

CDC (2023b). Salmonella and eggs. https://www.cdc.gov/foodsafety/communication/salmonella-and-eggs.html

CDC (2024a) List of multistate foodborne outbreak notices. https://www.cdc.gov/foodsafety/outbreaks/lists/outbreaks-list.html

CDC (2024b) Listeria (Listeriosis). https://www.cdc.gov/listeria/index.html

Chemat F, Zill-E-Huma, Khan MK (2011) Applications of ultrasound in food technology: processing, preservation and extraction. Ultrason Sonochem 18:813–35

Cox JM, Pavic A (2010) Advances in enteropathogen control in poultry production. J Appl Microbiol 108(3):745–755

Czerwiński K, Rydzkowski T, Wróblewska-Krepsztul J, Thakur VK (2021) Towards impact of modified atmosphere packaging (MAP) on shelf-life of polymer-film-packed food products: challenges and sustainable developments. Coatings 11(12):1504. https://doi.org/10.3390/coatings11121504

De Jesus AJ, Sheth I, Kwon HJ, Gao Z, Palmer J, Hur M, Hammack TS, Macarisin D, Chen Y (2020) Survival of a serotype 4b strain and a serotype 1/2a strain of *Listeria monocytogenes*, isolated from a stone fruit outbreak investigation, on whole stone fruit at 4 °C. Int J Food Microbiol 334:108801. https://doi.org/10.1016/j.ijfoodmicro.2020.108801

De la Caba K, Guerrero P, Trung TS, Cruz-Romero M, Kerry J-P, Fluhr J, Maurer M, Kruijssen F, Albalat A, Bunting S, Burt S, Little D, Newton R (2019) From seafood waste to active seafood packaging: an emerging opportunity of the circular economy. J Clean Prod 208:86–98

Deng Y, Wang Y, Deng Q, Sun L, Wang R, Ye L, Tao S, Liao J, Gooneratne R (2021) Fungal diversity and mycotoxin contamination in dried fish products in Zhanjiang market, China. Food Control 121:107614. https://doi.org/10.1016/j.foodcont.2020.107614

Domínguez R, Pateiro M, Gagaoua M, Barba FJ, Zhang W, Lorenzo JM (2019) A comprehensive review on lipid oxidation in meat and meat products. Antioxidants (Basel, Switzerland) 8(10):429

Dutta M, Majumdar PR, Rakeb-Ul-Islam MD, Saha D (2018) Bacterial and fungal population assessment in smoked fish during storage period. J Food: Microbiol Saf Hyg 3(1):1–7. https://doi.org/10.4172/2476-2059.1000127

Emanuel E, Dubrovin I, Hanya E, Pinhasi GA, Pogreb R, Cahan R (2020) Eradication of *Saccharomyces cerevisiae* by pulsed electric field treatments. Microorganisms 8(11):1684

Erkmen O, Bozoglu TF (2016) Chapter 17. Spoilage of eggs and egg products. In: Food microbiology: principles into practice, pp 296–300. https://doi.org/10.1002/9781119237860.ch17

Fagundes C, Moraes K, Pérez-Gago MB, Palou L, Maraschin M, Monteiro AR (2015) Effect of active modified atmosphere and cold storage on the postharvest quality of cherry tomatoes. Postharvest Biol Technol 109:73–81

FAO (2023) FAO Strategic Priorities for food safety within the FAO strategic framework 2022–2031. https://doi.org/10.4060/cc4040en

FAO (2019) FAO yearbook: Fishery and aquaculture statistics 2017. Food and Agriculture Organization of the United Nations, Rome, pp 50–52

FAO (2020) The state of world fisheries and aquaculture 2020. https://doi.org/10.4060/ca9229en

Fasuan AA, Akin-Obasola B, Abiodun BO (2022) Water activity relations of spoilage fungi associated with smoke-dried catfish (*Clarias gariepinus*) sold in some open markets in Nigeria. J Food Sci Technol 59(6):2168–2176. https://doi.org/10.1007/s13197-021-05229-8

FDA (2024) Food irradiation: what you need to know. https://www.fda.gov/food/buy-store-serve-safe-food/food-irradiation-what-you-need-know

Federighi M (2019) Les rayonnements ionisants comme méthode de préservation des aliments: État de l'art. Revue de Medecine Veterinaire 4–6:58–72

Finger JA, Baroni WS, Maffei DF, Bastos DH, Pinto UM (2019) Overview of foodborne disease outbreaks in Brazil from 2000 to 2018. Foods 8(10):434. https://doi.org/10.3390/foods8100434

de Freitas J, Pereira Neto LM, da Silva TIB, de Oliveira TFL, da Rocha JHL, Souza MD, de Marchi PGF, de Araújo ÁV (2021) Counting and identification of molds and yeasts in dry salted shrimp commercialized in Rio Branco, Acre, Brazil. Food Sci Technol 41(suppl 1):284–289. https://doi.org/10.1590/fst.16720

Fung F, Wang HS, Menon S (2018) Food safety in the 21st century. Biomed J 41(2):88–95. https://doi.org/10.1016/j.bj.2018.03.003

Galaviz-Silva L, Goméz-Anduro G, Molina-Garza ZJ, Ascencio-Valle F (2008) Food safety issues and the microbiology of fish and shellfish. In: Heredia NL, Wesley IV, Garcia JS (eds) Microbiologically safe foods. Wiley, Hoboken, NJ, USA, pp 227–273

Gartley S, Anderson-Coughlin B, Sharma M, Kniel KE (2022) *Listeria monocytogenes* in irrigation water: an assessment of outbreaks, sources, prevalence, and persistence. Microorganisms 10(7):1319. https://doi.org/10.3390/microorganisms10071319

Gast RK, Regmi P, Guraya R, Jones DR, Anderson KE, Karcher DM (2019) Contamination of eggs by *Salmonella Enteritidis* in experimentally infected laying hens of four commercial genetic lines in conventional cages and enriched colony housing. Poult Sci 98:5023–5027

Gaul LK, Farag NH, Shim T, Kingsley MA, Silk BJ, Hyytia-Trees E (2013) Hospital-acquired listeriosis outbreak caused by contaminated diced celery—Texas, 2010. Clin Infect Dis 56:20–26. https://doi.org/10.1093/cid/cis817

Ghoshal G (2023) Comprehensive review on pulsed electric field in food preservation: gaps in current studies for potential future research. Heliyon 9(6):e17532. https://doi.org/10.1016/j.heliyon.2023.e17532

Graça A, Santo D, Quintas C, Nunes C (2017) Growth of *Escherichia coli*, *Salmonella enterica* and *Listeria* spp., and their inactivation using ultraviolet energy and electrolyzed water, on 'Rocha' fresh-cut pears. Food Control 77:41–49. https://doi.org/10.1016/j.foodcont.2017.01.017

Gram L (2009) Microbiological spoilage of fish and seafood products. In: Sperber WH, Doyle MP (eds) Compendium of the microbiological spoilage of foods and beverages. Springer, New York, USA, pp 87–119. https://doi.org/10.1007/978-1-4419-0826-1_4

Greig J, Nesbakken T, Stephan R (2005) Meat and meat products. In: Roberts TA, Pitt JI (eds) Microorganisms in foods, 2nd edn. Kluwer Academic/Plenum Publishers, New York, pp 1–88

Gu R, Yun H, Chen L, Wang Q, Huang X (2019) Regenerated cellulose films with amino-terminated hyperbranched polyamic anchored nanosilver for active food packaging. ACS Appl Bio Mater 3(1):602–610. https://doi.org/10.1021/acsabm.9b00992

Ha J, Kang D (2015) Enhanced inactivation of food-borne pathogens in ready-to-eat sliced ham by near-infrared heating combined with UV-C irradiation and mechanism of the synergistic bactericidal action. Appl Environ Microbiol 81. https://doi.org/10.1128/AEM.01862-14

Hansen AÅ, Langsrud S, Carlehög M, Haugen J-E, Moen B (2023) CO2 packaging increases shelf life through reduction of off-odor production by CO2 tolerant bacteria in addition to growth inhibition of the spoilage bacteriota. Food Control 144:109390. https://doi.org/10.1016/j.foodcont.2022.109390

He Y, Chen R, Qi Y, Salazar JK, Zhang S, Tortorello ML, Deng X, Zhang W (2021) Survival and transcriptomic response of *Salmonella enterica* on fresh-cut fruits. Int J Food Microbiol 348:109201. https://doi.org/10.1016/j.ijfoodmicro.2021.109201

Hertwig C (2017) Cold atmospheric pressure plasma processing for decontamination of solid food and food contact surfaces. In: Innovative food processing technologies, pp 656–661

Hodge J, Chang HH, Boisson S, Collin SM, Peletz R, Clasen T (2016) Assessing the association between thermotolerant coliforms in drinking water and diarrhea: an analysis of individual-level data from multiple studies. Environ Health Perspect 124(10):1560–1567. https://doi.org/10.1289/EHP156

Hoffmann S, Scallan E (2017) Epidemiology, cost, and risk analysis of foodborne disease. Foodborne Dis 2017:31–63. https://doi.org/10.1016/b978-0-12-385007-2.00002-4

Jadhav HB, Choudhary P (2024) Emerging techniques for the processing of food to ensure higher food safety with enhanced food quality: a review. Discov Food 4:20. https://doi.org/10.1007/s44187-024-00089-5

Jensen GB, Hansen BM, Eilenberg J, Mahillon J (2003) The hidden lifestyles of *Bacillus cereus* and relatives. Environ Microbiol 5(8):631–640. https://doi.org/10.1046/j.14622920.2003.00461.x. PMid:12871230

Keener KM, Bashor MP, Curtis PA, Sheldon BW, Kathariou S (2004) Comprehensive review of *Campylobacter* and poultry processing. Compr Rev Food Sci Food Saf 3:105–116. https://doi.org/10.1111/j.1541-4337.2004.tb00060.x

Kyle JL, Parker CT, Goudeau D, Brandl MT (2010) Transcriptome analysis of *Escherichia coli* O157:H7 exposed to lysates of lettuce leaves. Appl Environ Microbiol 76:1375–1387. https://doi.org/10.1128/AEM.02461-09

Li Z, Yang Q, Du H, Wu W (2021) Advances of pulsed electric field for foodborne pathogen sterilization. Food Rev Intl 39(7):3603–3619. https://doi.org/10.1080/87559129.2021.2012798

Lin L, Wang X, Li C, Cui H (2019) Inactivation mechanism of *E. coli* O157:H7 under ultrasonic sterilization. Ultrason Sonochem 59:104751. https://doi.org/10.1016/j.ultsonch.2019.104751

Liu B, Zhang X, Ding X, Bin P, Zhu G (2022) The vertical transmission of *Salmonella* Enteritidis in a one-health context. One Health (Amsterdam, Netherlands) 16:100469

Lotfinia F, Norouzi MR, Ghasemi-Mobarakeh L, Naeimirad M (2023) Anthocyanin/honey-incorporated alginate hydrogel as a bio-based pH-responsive/antibacterial/antioxidant wound dressing. J Funct Biomater 14(2):72. https://doi.org/10.3390/jfb14020072

Lung H-M, Cheng Y-C, Chang Y-H, Huang H-W, Yang BB, Wang C-Y (2015) Microbial decontamination of food by electron beam irradiation. Trends Food Sci Technol 44:66–78

Maggio F, Rossi C, Chaves-López C, Serio A, Valbonetti L, Pomilio F, Chiavaroli AP, Paparella A (2021) Interactions between *L. monocytogenes* and *P. fluorescens* in dual-species biofilms under simulated dairy processing conditions. Foods (Basel, Switzerland) 10(1):176

Marchello CS, Birkhold M, Crump JA, Vacc-iNTS Consortium Collaborators (2022) Complications and mortality of non-typhoidal salmonella invasive disease: a global systematic review and meta-analysis. Lancet Infect Dis 22(5):692–705. https://doi.org/10.1016/s1473-3099(21)00615-0

Marchello CS, Fiorino F, Pettini E, Crump JA, Vacc-iNTS Consortium Collaborators (2021) Incidence of non-typhoidal Salmonella invasive disease: a systematic review and meta-analysis. J Infect 83(5):523–532. https://doi.org/10.1016/j.jinf.2021.06.029

Marmion M, Ferone MT, Whyte P, Scannell AGM (2021) The changing microbiome of poultry meat; from farm to fridge. Food Microbiol 99:103823

Md Abu T, Zahan KA, Rajaie MA, Leong CR, Ab Rashid S, Mohd Nor Hamin NS, Tan WN, Tong WY (2020) Nanocellulose as drug delivery system for honey as antimicrobial wound dressing. Mater Today: Proc 31:14–17. https://doi.org/10.1016/j.matpr.2020.01.076

Mei J, Ma X, Xie J (2019) Review on natural preservatives for extending fish shelf life. Foods 8(10):490. https://doi.org/10.3390/foods8100490

Mendoza IC, Luna EO, Pozo MD, Vásquez MV, Montoya DC, Moran GC, Romero LG, Yépez X, Salazar R, Romero-Peña M, León JC (2022) Conventional and non-conventional disinfection methods to prevent microbial contamination in minimally processed fruits and vegetables. LWT 165:113714. https://doi.org/10.1016/j.lwt.2022.113714

Minutillo SA, Ruano-Rosa D, Abdelfattah A, Schena L, Malacrinò A (2022) The fungal microbiome of wheat flour includes potential mycotoxin producers. Foods 11:676. https://doi.org/10.3390/foods11050676

Montazeri M, Mehrzadi S, Sharif M, Sarvi S, Tanzifi A, Aghayan SA, Daryani A (2018) Drug resistance in *Toxoplasma gondii*. Front Microbiol 9:2587

Motarjemi Y, Lelieveld H (2014) Chapter 1—Fundamentals in management of food safety in the industrial setting: challenges and outlook of the 21st century. In: Lelieveld YM (ed) Food safety management. Academic Press, San Diego, USA. pp 1–20

Munir MT, Federighi M (2020) Control of foodborne biological hazards by ionizing radiations. Foods (Basel, Switzerland) 9(7):878

Muntean M-V, Marian O, Barbieru V, Cătunescu GM, Ranta O, Drocas I, Terhes S (2016) High pressure processing in food industry—Characteristics and applications. Agric Agric Sci Procedia 10:377–383. https://doi.org/10.1016/j.aaspro.2016.09.077

Naef E, Piaggio M, Glodowsky A, Sacchi C, Broggi E, Barreto L, Lound L (2023) Aplicación de la impregnación a vacío en la biopreservación de carne de pescado de río. Ciencia, Docencia y Tecnología Suplemento 13(14). https://pcient.uner.edu.ar/index.php/Scdyt/article/view/1564

Nordgren J, Sharma S, Kambhampati A, Lopman B, Svensson L (2016) Innate resistance and susceptibility to *Norovirus* Infection. Plos Pathog 12(4):e1005385. https://doi.org/10.1371/journal.ppat.1005385

Obi CN, Igbokwe AJ (2007) Microbiological analysis of freshly laid and stored domestic poultry egg in selected poultry farms in Umuahia, Abia State, Nigeria. Res J Biol Sci 4(12):1297–1303. https://doi.org/10.3923/rjbsci.2009.1297.1303

OECD (2021) Meat consumption. https://data.oecd.org/agroutput/meat-consumption.htm

Ölmez H (2016) Foodborne pathogenic bacteria in fresh-cut vegetables and fruits. Food Hygiene Toxicol Ready-to-Eat Foods 151–166. https://doi.org/10.1016/b978-0-12-801916-0.00009-1

Perez-Arnedo I, Cantalejo MJ, Martínez-Laorden A, Gonzalez-Fandos E (2020) Effect of processing on the microbiological quality and safety of chicken carcasses at slaughterhouse. Int J Food Sci Technol 56(4):1855–1864. https://doi.org/10.1111/ijfs.14815

Putnik P, Pavlić B, Šojić B, Zavadlav S, Žuntar I, Kao L, Kitonić D, Kovačević DB (2020) Innovative hurdle technologies for the preservation of functional fruit juices. Foods (Basel, Switzerland) 9(6):699

Qian Y, Ye J, Yang S, Lin Z, Cao W, Xie J (2018) Evaluation of the spoilage potential of *Shewanella putrefaciens*, *Aeromonas hydrophila*, and *Aeromonas sobria* isolated from spoiled Pacific white shrimp (*Litopenaeus vannamei*) during cold storage. J Food Saf 38(6). https://doi.org/10.1111/jfs.12550

Rajmani RS, Singh AP, Singh PK, Doley J, Verma SP (2011) Fungal contamination in eggs. J Vet Public Health 9(1):59–61

Ranjha MMAN, Irfan S, Lorenzo JM, Shafique B, Kanwal R, Pateiro M, Arshad RN, Wang L, Nayik GA, Roobab U, Aadil RM (2021) Sonication, a potential technique for extraction of phytoconstituents: a systematic review. Processes 9:1–21

Rodrigo D, Rosell CM, Martinez A (2021) Risk of *Bacillus cereus* in relation to rice and derivatives. Foods 10(2):302. https://doi.org/10.3390/foods10020302

Rouger A, Tresse O, Zagorec M (2017) Bacterial contaminants of poultry meat: sources, species, and dynamics. Microorganisms 5(3):50. https://doi.org/10.3390/microorganisms5030050

Ruiz-Capillas C, Herrero AM (2019) Impact of biogenic amines on food quality and safety. Foods 8(2):62. https://doi.org/10.3390/foods8020062

Russel SM (2012) Controlling *Salmonella* in poultry production and processing. CRC Press, Florida, USA, pp 1–277

Şanlıbaba P, Tezel BU, Çakmak GA (2018) Prevalence and antibiotic resistance of *Listeria monocytogenes* isolated from ready-to-eat foods in Turkey. J Food Qual 2018:1–9

Sarma UP, Bhetaria PJ, Devi P, Varma A (2017) Aflatoxins: implications on health. Indian J Clin Biochem 32(2):124–133. https://doi.org/10.1007/s12291-017-0649-2

Sasaki Y, Maruyama N, Zou B, Haruna M, Kusukawa M, Murakami M, Asai T, Tsujiyama Y, Yamada Y (2013) *Campylobacter* cross-contamination of chicken products at an abattoir. Zoonoses Public Health 60:134–140

Seliwiorstow T, Baré J, Van Damme I, Uyttendaele M, De Zutter L (2015) *Campylobacter* carcass contamination throughout the slaughter process of *Campylobacter*-positive broiler batches. Int J Food Microbiol 194:25–31. https://doi.org/10.1016/j.ijfoodmicro.2014.11.004
Shen Z, Wang Y, Zhang Q, Shen J (2018) Antimicrobial resistance in *Campylobacter* spp. Microbiol Spectr 6. https://doi.org/10.1128/microbiolspec.arba-0013-2017
Shineh G, Mobaraki M, Perves Bappy MJ, Mills DK (2023) Biofilm formation, and related impacts on healthcare, food processing and packaging, industrial manufacturing, marine industries, and sanitation—A review. Appl Microbiol 3(3):629–665. https://doi.org/10.3390/applmicrobiol3030044
Siemer C, Toepfl S, Heinz V (2014) Inactivation of *Bacillus subtilis* spores by pulsed electric fields (PEF) in combination with thermal energy—I. Influence of process- and product parameters. Food Control 39:163–171
Silva FVM, Evelyn (2023) Pasteurization of food and beverages by high pressure processing (HPP) at room temperature: inactivation of *Staphylococcus aureus, Escherichia coli, Listeria monocytogenes, Salmonella,* and other microbial pathogens. Appl Sci 13(2):1193. https://doi.org/10.3390/app13021193
Singh S (2010) Modified atmosphere packaging of fresh produce: current status and future needs. LWT-Food Sci Technol 3:381–392
Siripongvutikorn S, Usawakesmanee W, Sud-on P (2024) Effect of egg type and storage conditions on the quality of crispy Thai pancake (Thong Phap). Italian J Food Sci 36(1):68–79. https://doi.org/10.15586/ijfs.v36i1.2433
Soto Varela Z, Pérez Lavalle L, Estrada Alvarado D (2016) Bacteria causing of foodborne diseases: an overview at Colombia. Salud Uninorte 32(1):105–122. https://doi.org/10.14482/sun.32.1.8598
Speranza B (2013) Use of desirability approach to predict the inhibition of *Pseudomonas fluorescens, Shewanella putrefaciens* and *Photobacterium phosphoreum* in fish fillets through natural antimicrobials and modified atmosphere packaging. Food Bioprocess Technol 6:2319–2330
Stein RA, Chirilã M (2017) Routes of transmission in the food chain. Foodborne Dis 2017:65–103. https://doi.org/10.1016/B978-0-12-385007-2.00003-6
Strawn LK, Danyluk MD (2010) Fate of *Escherichia coli* O157:H7 and *Salmonella* spp. on fresh and frozen cut mangoes and papayas. Int J Food Microbiol 138:78–84. https://doi.org/10.1016/j.ijfoodmicro.2009.12.002
Suárez-Medina MD, Sáez-Casado MI, Martínez-Moya T, Rincón-Cervera MÁ (2024) The effect of low temperature storage on the lipid quality of fish, either alone or combined with alternative preservation technologies. Foods 13(7):1097
USDA Food Safety and Inspection Service (2017) Safe and suitable ingredients used in the production of meat, poultry, and egg products. FSIS Directive 7120.1, Rev 39
USDA (2024) USDA's FoodKeeper app uses open data to keep consumers safe and food fresh. https://www.usda.gov/media/blog/2016/11/29/usdas-foodkeeper-app-uses-open-data-keep-consumers-safe-and-food-fresh
Vallejos S, Trigo-López M, Arnaiz A, Miguel Á, Muñoz A, Mendía A, García JM (2022) From classical to advanced use of polymers in food and beverage applications. Polymers (Basel) 14(22):4954. https://doi.org/10.3390/polym14224954
Visciano P, Schirone M, Tofalo R, Suzzi G (2014) Histamine poisoning and control measures in fish and fishery products. Front Microbiol 5:500. https://doi.org/10.3389/fmicb.2014.00500
Wang H, Qin X, Li X, Wang X, Lei Y, Zhang C (2020) Effect of chilling methods on the surface color and water retention of yellow-feathered chickens. Poult Sci 99(4):2246–2255
Wang Y, Ge H, Wei X, Zhao X (2022) Research progress on antibiotic resistance of *Salmonella.* Food Quality Saf 6. https://doi.org/10.1093/fqsafe/fyac035
Wang X, Huang Q, Huang R (2023) Quantitative risk assessment of *Bacillus cereus* in wet rice noodles from raw material to marketing phase. Heliyon 9(3):e14354. https://doi.org/10.1016/j.heliyon.2023.e14354

WHO (2017) WHO publishes list of bacteria for which new antibiotics are urgently needed. https://www.who.int/news/item/27-02-2017-who-publishes-list-of-bacteria-for-which-new-antibiotics-are-urgently-needed

WHO (2018) *E. coli*. https://www.who.int/news-room/fact-sheets/detail/e-coli

WHO (2022) Food safety. https://www.who.int/news-room/fact-sheets/detail/food-safety

WHO (2023) Assessing microbiological risks in food. https://www.who.int/activities/assessing-microbiological-risks-in-food

Wideman N, Bailey M, Bilgili S, Thippareddi H, Wang L, Bratcher C, Sanchez-Plata M, Singh M (2016) Evaluating best practices for *Campylobacter* and *Salmonella* reduction in poultry processing plants. Poult Sci 95(2):306–315

Wu C-P, Wu S-M, Lin Y-H, Wu Y-H, Huang B-C, Huang H-W, Wang C-Y (2022) High pressure processing-based hurdle strategy for microbial shelf life of packed food in the Cold Chain. Food Packag Shelf Life 34:100983

Youssef AM, Abdel-Aziz MS, El-Sayed SM (2014) Chitosan nanocomposite films based on Ag–NP and Au–NP biosynthesis by *Bacillus subtilis* as packaging materials. Int J Biol Macromol 69:185–191

Yu Z, Wang W, Kong F, Lin M, Mustapha A (2019) Cellulose nanofibril/silver nanoparticle composite as an active food packaging system and its toxicity to human colon cells. Int J Biol Macromol 129:887–894. https://doi.org/10.1016/j.ijbiomac.2019.02.084

Zarei M, Rahimi S, Fazlara A, Anvari SE (2023) High biofilm-forming *Pseudomonas* strains isolated from poultry slaughterhouse surfaces: Their importance in the persistence of *Salmonella enteritidis* in slaughterhouses. Int J Food Microbiol 390:110126. https://doi.org/10.1016/j.ijfoodmicro.2023.110126

Zhuang S, Liu X, Li Y, Zhang L, Hong H, Liu J, Luo Y (2021) Biochemical changes and amino acid deamination & decarboxylation activities of spoilage microbiota in chill-stored grass carp (*Ctenopharyngodon idella*) fillets. Food Chem 336:127683. https://doi.org/10.16/j.foodchem.2020.127683

Relationship Between Traffic Density and Particulate Matter Concentrations

Muhammad Farhan Afiq Mohd Aminuddin, Siti Rohana Mohd Yatim, Ahmad Firdaus Shafie, Samsuri Abdullah, and Mohd Armi Abu Samah

Abstract This study examines particulate matter (PM) trends and their relationship with motor vehicle numbers in a selected area. Air sampling using Dust Trax TM DRX Aerosol Monitor 8534 was conducted from 8 a.m. to 12 p.m. PM concentrations (PM_1, $PM_{2.5}$, PM_{10}) showed similar distribution patterns, with levels starting low at 8 a.m., peaking at 10 a.m., and declining by 11 a.m. The average concentrations over 4 h were PM_1: 0.054 mgm^{-3}, $PM_{2.5}$: 0.055 mg m^{-3}, and PM_{10}: 0.064 mg m^{-3}. $PM_{2.5}$ levels exceeded daily National Ambient Air Quality Standards, while PM_{10} levels were below. Weak correlations were found between PM concentrations and motor vehicle numbers ($R = 0.154$ for PM_{10}, $R = 0.157$ for $PM_{2.5}$, $R = 0.152$ for PM_1). Overall, PM distribution trended similarly across all sizes, peaking at 10 a.m., and showing a positive but weak relationship with motor vehicle numbers. Future studies in other Malaysian regions could enhance understanding of traffic-related pollution.

Keywords Air pollution · Particulate matter · Traffic density

M. F. A. M. Aminuddin · S. R. M. Yatim (✉) · A. F. Shafie
Faculty of Health Sciences, Centre for Environmental Health and Safety Studies, Universiti Teknologi MARA Cawangan Selangor Kampus Puncak Alam, 42300 Bandar Puncak Alam, Selangor, Malaysia
e-mail: sitirohana@uitm.edu.my

A. F. Shafie
e-mail: firdausshafie@uitm.edu.my

M. F. A. M. Aminuddin
MDC Precast Industries Sdn. Bhd, PT 48877, Jalan, Persiaran Amanjaya 5, Bandar Amanjaya, 08000 Sungai Petani, Kedah, Malaysia

S. Abdullah
Faculty of Ocean Engineering Technology and Informatics, Universiti Malaysia Terengganu, Public Health, 21030 Kuala Nerus, Terengganu, Malaysia
e-mail: samsuri@umt.edu.my

M. A. A. Samah
Kuliyyah of Science, International Islamic University Malaysia, Kuantan, Malaysia
e-mail: marmi@iium.edu.my

A. Z. Yaser et al. (eds.), *Controlling Environmental Pollution*,
https://doi.org/10.1007/978-981-97-8931-3_5

1 Introduction

Air pollution is defined as the presence of one or more pollutants in the atmosphere at concentration and duration that exceed natural degrees. Ozone (O_3), airborne lead (Pb), carbon monoxide (CO), sulfur oxides (SO_x), and nitrogen oxides (NO_x) are examples of chemicals that are considered pollutants in the air (Saxena and Sonwani 2019). Due to these pollutants' capability to permeate lung tissue and trigger local and systemic consequences, particulate matter (PM) that is less than 2.5 μm has recently been the focus for outdoor air pollution investigations (Sweileh et al. 2018). The city's urbanization and the expanding development in the surrounding areas have substantially enhanced the people's quality of life. This, however, has caused a rise in pollutant-emitting sources such as traffic vehicles and industrial activities (Lee et al. 2020).

Air pollution has been shown in many studies to cause countless health adverse effects. Short-term exposures to $PM_{2.5}$, NO_2, and O_3 were shown to be strongly related to asthma-related mortality (Liu et al. 2019). Upon exposures with lower concentrations of $PM_{2.5}$, NO_2, and O_3, cases of airway inflammation have been recorded, which is regarded to be a significant indication of asthma. It has been proposed that NO_2 and O_3 exposure might also increase airway hyperresponsiveness, which is another indication of asthma (Peden 2001). Furthermore, $PM_{2.5}$, NO_2, and O_3 exposures have been connected to oxidative stress, which in turn has been linked to severe asthma (Liu et al. 2019).

Even on days with mild air pollution, the health of vulnerable and sensitive persons might be impacted. Short-term air pollution exposure is linked to COPD (chronic obstructive pulmonary disease), cough, shortness of breath, wheezing, asthma, respiratory disease, and high hospitalization rates, which is known as one of the measurements of morbidity (Manisalidis et al. 2020). Long-term impacts of air pollution are chronic asthma, pulmonary insufficiency, cardiovascular illnesses, and cardiovascular mortality (Manisalidis et al. 2020).

On-street vehicles such as cars, lorries, and buses contribute significantly to serious air pollution emissions. The use of fossil fuels such as gasoline and diesel has raised several adverse effects. Certain $PM_{2.5}$ pollutants are produced directly during combustion from vehicle exhausts, and there are also certain $PM_{2.5}$ pollutants that are formed indirectly due to the interactions of polluting gases in the atmosphere and combined with sunlight to form even more dangerous and complex pollutants, i.e., ammonium, nitrogen oxides, sulfur oxides, and volatile organic compounds. Although most of these pollutants are generated by vehicle exhaust, some volatile organic compounds are also emitted through gasoline evaporation during refilling and leaks in cars' fuel tanks and lines (Environmental Research Kang et al. 2022). Additional $PM_{2.5}$ particles are produced from the process of braking and wearing out of tires (Matthaios et al. 2022). This study will investigate the relationship between traffic density and particulate matter concentration. Furthermore, this study will also be able to assist the authority in taking the necessary actions to reduce the exposure of particulate matter, especially among pedestrians, hawkers, and sensitive groups.

2 Materials and Methods

This is a cross-sectional study and generally consists of five steps. The first step is to set up instruments. The researcher obtained the equipment needed for this study and inspected the instruments for any damage that may affect the results. Secondly, the pre-sampling process which happened a day before the actual data collection where the researcher stood by the roadside and observed the traffic conditions to determine traffic density and peak hours. Next, the actual sampling where the data on the concentration of PM_{10}, $PM_{2.5}$, and PM_1 along with the number of motor vehicles are recorded. After the data were obtained, data analysis was done to achieve the desired objective which is to determine the trend of particulate matters and to establish the relationship between concentrations of particulate matters and number of motor vehicles. Lastly, all the results derived from this study are written in this report.

Ambient air sampling is conducted using Dust Trax TM DRX Aerosol Monitor 8534 as shown in Fig. 1. This instrument is used to measure the concentration of PM_{10}, $PM_{2.5}$, and PM_1. It is placed on a tripod 1.5 m above the ground, 3.5 m away from the roadside. Furthermore, it should be placed 20 m or twice the height of the tree above the inlet. Moreover, the distance from any air flow obstacles such as buildings should be two times the height of obstacles above the inlet. Airflow restriction should be unrestricted in at least three of four wind quadrants. Local source interferences were considered when choosing the site for ambient air sampling. The data interval is stored in detailed 5 min intervals in the sampler's microprocessor with the flow rate 16.67 L min^{-1} or 1 m^3 h^{-1}. Data of concentration of PM_{10}, $PM_{2.5}$, and PM_1 was collected from 8.00 a.m. to 12.00 p.m. since this is the peak hours for the selected road as determined from pre-sampling procedures.

The descriptive statistics and the trend of PM_{10}, $PM_{2.5}$, and PM_1 concentration at study areas plots and the number of vehicles is conducted by Microsoft Excel Spreadsheet 2020. The descriptive statistics are summarized in box plot and table showing the temporal variation of PM_{10}, $PM_{2.5}$, and PM_1 concentrations throughout the study period to see the trend of the concentration of particulate matter. Meanwhile, the number of vehicles and concentration of PM_{10}, $PM_{2.5}$, and PM_1 is summarized using correlation in linear graph.

Regression was used in this study as a statistical method, and to investigate the relationship between dependent and independent variables. This analysis is used to determine the relationship of traffic density on PM_{10}, $PM_{2.5}$, and PM_1 using Microsoft Excel 2020. The dependent variable is concentration of PM_{10}, $PM_{2.5}$, and PM_1 and the independent variable is traffic density. This method is one of the most used methods in environmental particulate exposure evaluation (Latif et al. 2021). The mathematical model of simple linear regression can be described in Equation (1).

$$Y = \beta 0 + \beta 1x + \varepsilon \quad (1)$$

The intercept and slope are the unknown regression coefficients, while the random error, ε, has a mean of zero and an unknown variance of σ^2 where the intercept and the slope are, which are unknown regression coefficients, while the random error has a mean of zero and an unknown variance of σ2.

3 Results and Discussion

The daily and mean mass concentrations of the particulate pollutants (PM_{10}, $PM_{2.5}$, and PM_1) were calculated hourly as shown in Figs. 2, 3, and 4 in terms of temporal pattern. During the study period, the overall average (mean of 4 h data) concentrations of PM_1 (ultra-fine), $PM_{2.5}$ (fine), and PM_{10} (coarse) were 0.054 mg m^{-3}, 0.055 mg m^{-3}, and 0.064 mg m^{-3}, respectively (Table 1). The mean of $PM_{2.5}$ concentrations (55 μg m^{-3}) was higher than the daily National Ambient Air Quality Standards while the PM_{10} concentrations (64 μg m^{-3}) are lower than the NAAQS (NAAQS: $PM_{2.5} = 35$ μg m^{-3} and $PM_{10} = 150$ μg m^{-3}).

One previous study shows that a tropical environment with high temperatures and natural emissions such as soil dust and other natural sources affect the concentration of PM_{10} in ambient air (Halim et al. 2020). Meteorological conditions, such as wind speed and direction, temperature, humidity, and atmospheric pressure, can all affect the dispersion of particulate matter in the atmosphere (Li et al. 2022). Particulate matter can come from both natural and human-made sources. Natural sources of particulate matter include wind-blown dust and sand, which can be generated by natural processes such as erosion and weathering (e.g., winds, storms). Another example of a natural source is pollen, which is released into the air by plants as part of their reproductive process. Meanwhile, human-made particulate matter can be derived from emissions from vehicles such as cars, trucks, and buses. These emissions include particulates such as soot and other particles. The burning of biomass (e.g., wood, animal dung) for cooking or heating, which can release particulate matter into

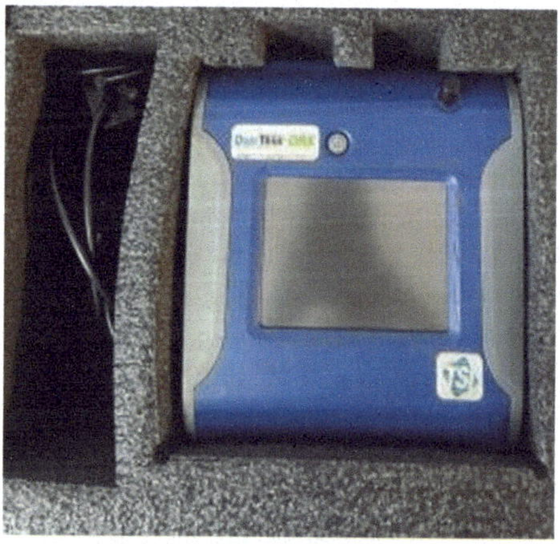

Fig. 1 Dust Trax TM DRX Aerosol Monitor 8534

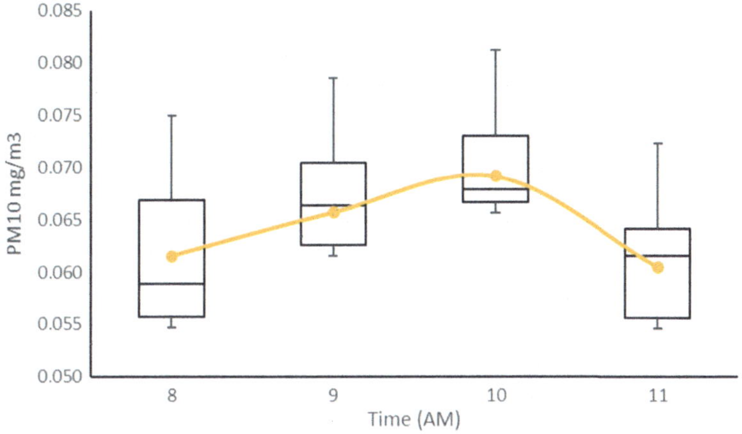

Fig. 2 Trend of PM_{10}

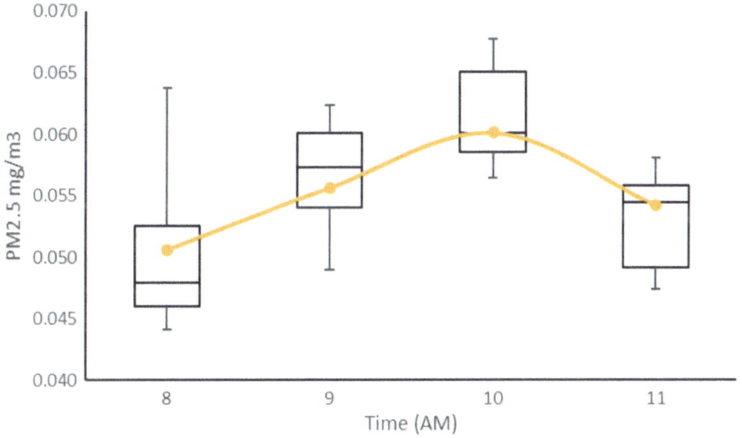

Fig. 3 Trend of $PM_{2.5}$

the air as smoke (Kirešová and Guzan 2022; *Particulate Matter (PM) Basics | US EPA* 2022).

3.1 Trend of Coarse Particulate Matter (PM_{10})

Figure 2 shows the box and whisker plots of the temporal distributions of PM_{10}, from 8 a.m. to 12 p.m. During 8:00 to 9:00 a.m., the distribution is positively skewed, since the whisker and half-box are longer on the upper side of the median than on the

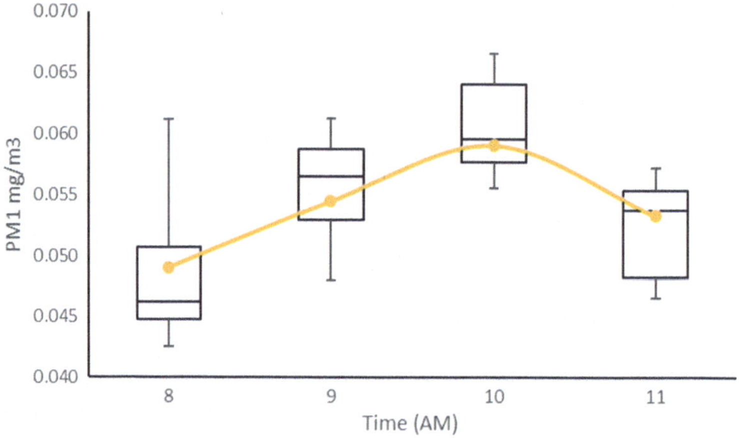

Fig. 4 Trend of PM_1

Table 1 Summary of descriptive statistics for the particulate matter measured at Jalan Tengku Ampuan Bariah

Descriptive statistical indices	PM_{10} (mg/m^3)	$PM_{2.5}$ (mg^3)	$PM_{1.0}$ (mg/m^3)	Moto vehicle (units)
Mean	0.064	0.055	0.054	17
Median	0.054	0.048	0.046	17
Standard deviation	0.022	0.020	0.020	5
Kurtosis	0.042	0.278	0.140	2.659
Skewness	1.072	1.108	1.068	0.750
Minimum	0.039	0.031	0.030	5
Maximum	0.127	0.116	0.111	43

bottom. Distribution during 9:00 a.m. to 10:00 a.m. is symmetrical, since both half-boxes are the same length (0.004 on both sides). The center of distribution during 10:00 to 11:00 a.m. is the highest with median 0.067, due to the impact of vehicular emissions during morning rush hours (Latif et al. 2021). PM_{10} concentrations were the lowest during (08:00–09:00 a.m.) possibly due to people having not started their daily activity and the limited number of vehicles around as shown in Fig. 7. This shows that the concentration of particulate matter is directly proportional to the number of vehicles. Furthermore, a decrease in concentration can be seen from 11:00 a.m. to 12:00 p.m. The major source of the $PM_{2.5}$ and PM_{10} all over the city is due to the exponential growth of the number of vehicles followed by construction activities, domestic pollution, paved and unpaved road dust, increased use of diesel generator sets, and open trash burning (Devaraj et al. 2019).

3.2 Trend of Fine Particulate Matter ($PM_{2.5}$)

Figure 3 shows box and whisker plot with a mean line of the temporal distributions of $PM_{2.5}$, from 8 a.m. to 12 p.m. During 8:00 to 9:00 a.m. the distribution is positively skewed, since the whisker and half-box are longer on the upper side of the median than on the bottom. It also shows the lowest distribution with 0.048 median. Distribution during 9:00 a.m. to 10:00a.m. is symmetrical, since both half-boxes are the same length (0.003 on both sides). The center of distribution during 10:00 to 11:00 a.m. is the highest with a median 0.059 and mean 0.06. Furthermore, a decreasing concentration trend can be seen from 11:00 a.m. to 12:00 p.m. with mean 0.054. According to a study published in the journal of Environmental Research in 2020 by Manisalidis et al., it was found that short-term exposure to $PM_{2.5}$ was associated with an increased risk of acute nasopharyngitis in a sample of children in Athens, Greece.

3.3 Trend of Ultra-Fine Particulate Matter (PM_1)

Figure 4 shows box and whisker plot with a mean line of the temporal distributions of PM_1, from 8 a.m. to 12 p.m. During 8:00 to 9:00 a.m. the distribution is positively skewed, since the whisker and half-box are longer on the upper side of the median than on the bottom. It also shows the lowest distribution with a mean of 0.049. Distribution from 9:00 a.m. to 10:00 a.m. increases slightly to a mean 0.055. The center of distribution from 10:00 to 11:00 a.m. is the highest with mean 0.059 and median 0.056. Furthermore, a decreased concentration trend can be seen from 11:00 a.m. to 12:00 p.m. with a mean 0.054.

Generally, the distribution trends across all PM_{10}, $PM_{2.5}$, and PM_1 are almost the same, which is low at 8 a.m. and slowly increasing and peaked at 10 a.m. and start to decrease at 11 a.m. This may be due to the traffic pattern of the studied area. PM_1 concentrations can be influenced by several factors, including weather patterns, land use, and the presence of other pollutants in the air. For example, PM_1 concentrations may be higher in areas with high levels of traffic or industrial activity, or during times of low wind speeds or high humidity (Kirešová and Guzan 2022).

3.4 Relationship Between Number of Motor Vehicles and Concentration of Particulate Matter

Figures 5, 6 and 7 show correlation coefficients for PM_{10}, $PM_{2.5}$, and PM_1 with the relation of motor vehicles, respectively. Results show that the concentration of particulate matter has a positive relation with the number of motor vehicles. This is in line with a study which shows an increase in air pollution emissions with the increase of revolution per minutes (RPM) of the engine. These findings support the hypothesis that most pollutants are derived from the fuel combustion happening in the engine (Macedo et al. 2022). Non-exhaust emissions have become a bigger part of the total emissions and now outweigh exhaust emissions in many countries (Harrison et al. 2021). Particle emissions that don't come from the engine are caused by wear and tear of the brakes, tires, and road surface, as well as the re-suspension of road dust. Organization for Economic Cooperation and Development (OECD) says that the amount of non-exhaust particulate matter ($PM_{2.5}$) released by passenger vehicles around the world is likely to rise by 53% by 2030 compared to a baseline year of 2017. This is mostly because people are driving their cars more (OECD 2020).

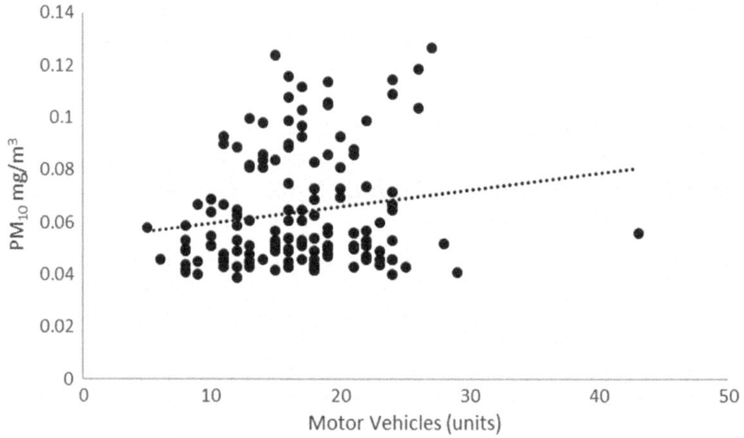

Fig. 5 Relationship between number of motor vehicles and concentration of PM_{10}

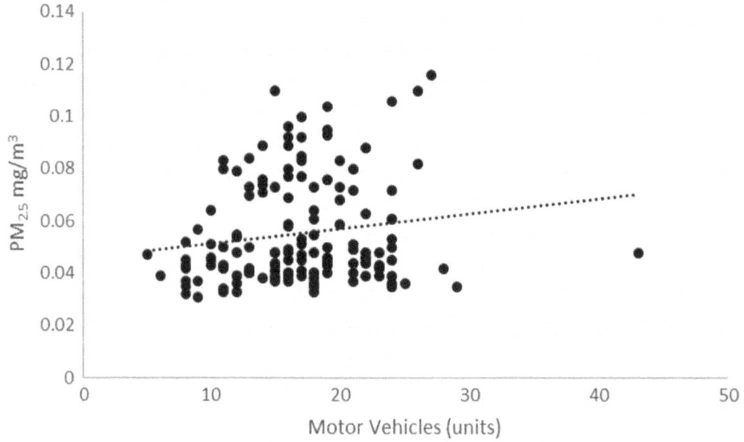

Fig. 6 Relationship between number of motor vehicles and concentration of $PM_{2.5}$

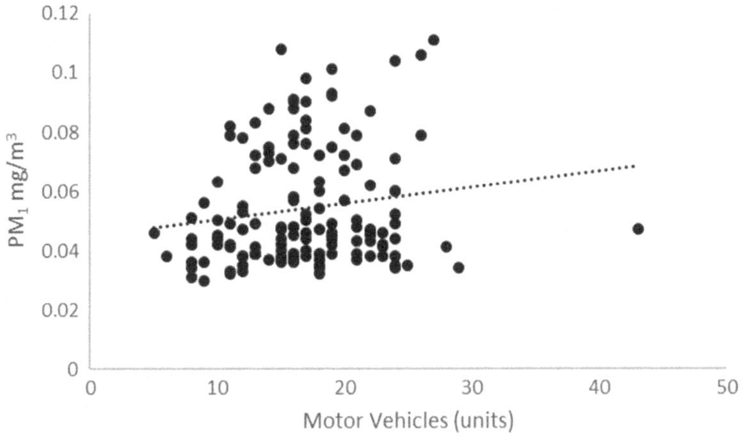

Fig. 7 Relationship between number of motor vehicles and concentration of PM_{10}

However, results show a weak correlation with R values 0.154, 0.157, and 0.152 for PM_{10}, $PM_{2.5}$, and PM_1, respectively. This weak correlation may be due to other factors that also can affect the concentration of particulate matter like meteorological factors. Previous studies have shown a strong correlation between concentration of particulate matters with meteorological factors such as air pressure, temperature, and precipitation (Li et al. 2015). The study observed higher concentrations of PM_{10}, $PM_{2.5}$, and $PM_{1.0}$ in winter compared to summer in Chengdu from 2009 to 2011. Peak concentrations occurred in January, while minimum values were recorded in September. This indicates an improvement in atmospheric PM pollution with annual decreases in PM_{10}, $PM_{2.5}$, and $PM_{1.0}$ concentrations. Moreover, wind environment factors such as wind speed and turbulence have a direct impact on air pollution, whereas changes in wind speed and turbulence in the urban wind environment are influenced by urban spatial form. As a result, air pollution in cities can be attributed to the effects of wind environment and urban spatial form (Yang et al. 2020).

Table 2 shows the equation retrieved from regression analysis. The model can be used to predict the future concentration of particulate matter in the study area. Model 1 is used to predict concentration of PM_{10}, C_{10} by inserting number of motor vehicles, n. Meanwhile, model 2 is used to predict concentration of $PM_{2.5}$, $C_{2.5}$ and Model 3 is used to predict concentration of PM_1.

Table 2 Equation retrieved from regression analysis

Model	Equation	R
1	$C10 = 0.0006\,(n) + 0.0536$	0.154
2	$C2.5 = 0.0006\,(n) + 0.0454$	0.157
3	$C1 = 0.0006\,(n) + 0.0448$	0.152

4 Conclusion

It can be concluded that the general distribution trend across all PM_{10}, $PM_{2.5}$, and PM_1 in the study areas is almost the same: low at 8 a.m., slowly increasing, peaking at 10 a.m., then starting to decrease at 11 a.m. Meanwhile, the relationship between the number of motor vehicles and particulate matter concentrations indicates that the number of motor vehicles manipulated the PMs. When the number of motor vehicles increases, the concentrations of particulate matter also increase. Based on these findings, it can be concluded that traffic density plays a significant role in determining the concentration of particulate matter in the study areas. The study suggests that efforts to reduce traffic density, such as promoting alternative forms of transportation, may be effective in reducing particulate matter concentrations in the area. Additionally, this study's findings also highlight the importance of monitoring and controlling vehicular emissions in order to mitigate the negative impacts of traffic on air quality.

References

Devaraj S, Tiwari S, Ramaraju HK, Dumka UC, Sateesh M, Parmita P, Shivashankara GP (2019) Spatial and temporal variation of atmospheric particulate matter in Bangalore: a technology-intensive Region in India. Arch Environ Contam Toxicol 77(2):214–222. https://doi.org/10.1007/S00244-019-00643-8

Kang G, Cho K, Shin J, Lee S, Lee S-B, Woo S (2022) Real-time detection of vehicle- originated condensable particulate matter through thermodenuder integrated aerosol measurement method at tailpipes. Environ Res.https://doi.org/10.1016/j.envres.2022.113487

Halim NDA, Latif MT, Mohamed AF, Maulud KNA, Idrus S, Azhari A, Othman M, Sofwan NM (2020) Spatial assessment of land use impact on air quality in mega urban regions, Malaysia. Sustain Cities Soc 63(January):102436. https://doi.org/10.1016/j.scs.2020.102436

Harrison RM, Allan J, Carruthers D, Heal MR, Lewis AC, Marner B, Murrells T, Williams A (2021) Non-exhaust vehicle emissions of particulate matter and VOC from road traffic: a review. Atmos Environ 262:118592. https://doi.org/10.1016/J.ATMOSENV.2021.118592

Kirešová S, Guzan M (2022) Determining the correlation between particulate matter PM10 and meteorological factors. Eng 3(3):343–363. https://doi.org/10.3390/eng3030025

Latif MT, Dominick D, Hawari NSSL, Mohtar AAA, Othman M (2021) The concentration of major air pollutants during the movement control order due to the COVID-19 pandemic in the Klang Valley, Malaysia. Sustain Cities Soc 66:102660. https://doi.org/10.1016/J.SCS.2020.102660

Li Y, Chen Q, Zhao H, Wang L, Tao R (2015) Variations in PM10, PM2.5 and PM1.0 in an urban area of the Sichuan basin and their relation to meteorological factors. Atmosphere 6(1):150–163. https://doi.org/10.3390/ATMOS6010150

Lee CC, Tran MV, Choo CW, Tan CP, Chiew YS (2020) Evaluation of air quality in Sunway City, Selangor, Malaysia from a mobile monitoring campaign using air pollution micro-sensors. Environ Pollut 265. https://doi.org/10.1016/J.ENVPOL.2020.115058

Li Q, Li X, Li H (2022) Factors influencing PM2.5 concentrations in the Beijing–Tianjin–Hebei urban agglomeration using a geographical and temporal weighted regression model. Atmosphere 13(3). https://doi.org/10.3390/atmos13030407

Liu Y, Pan J, Zhang H, Shi C, Li G, Peng Z, Ma J, Zhou Y, Zhang L (2019) Short-term exposure to ambient air pollution and asthma mortality. Am J Respir Crit Care Med 200(1):24–32. https://doi.org/10.1164/RCCM.201810-1823OC/SUPPL_FILE/DISCLOSURES.PDF

Macedo E, Pascale A, Ferreira E, Guarnaccia C, Coelho MC (2022) Experimental evaluation of gear-shift and internal-combustion engine variables on fuel consumption, noise and pollutant emissions. Transport Res Procedia 62:703–710. https://doi.org/10.1016/J.TRPRO.2022.02.087

Manisalidis I, Stavropoulou E, Stavropoulos A, Bezirtzoglou E (2020) Environmental and health impacts of air pollution: a review. Front Public Health 8:14. https://doi.org/10.3389/FPUBH.2020.00014

Matthaios VN, Lawrence J, Martins MAG, Ferguson ST, Wolfson JM, Harrison RM, Koutrakis P, Chan HTH (2022) Quantifying factors affecting contributions of roadway exhaust and non-exhaust emissions to ambient $PM_{10-2.5}$ and $PM_{2.5-0.2}$ particles. Sci Total Environ 835:155368. https://doi.org/10.1016/J.SCITOTENV.2022.155368

OECD (2020) Non-exhaust particulate emissions from road transport. In: Non-exhaust particulate emissions from road transport. https://doi.org/10.1787/4A4DC6CA-EN

Particulate Matter (PM) Basics | US EPA (2022) United States Environmental Protection Agency. https://www.epa.gov/pm-pollution/particulate-matter-pm-basics

Peden DB (2001) Air pollution in asthma: effect of pollutants on airway inflammation. Annals Allergy Asthma Immunol 87(6):12–17. https://doi.org/10.1016/S1081-1206(10)62334-4

Saxena P, Sonwani S (2019) Criteria air pollutants and their impact on environmental health, vol 1, no 1. Springer, Singapore

Sweileh WM, Al-Jabi SW, Zyoud SH, Sawalha AF (2018) Outdoor air pollution and respiratory health: a bibliometric analysis of publications in peer-reviewed journals (1900–2017). Multidiscip Respir Med 13(1):1–12. https://doi.org/10.1186/S40248-018-0128-5/TABLES/5

Yang J, Shi B, Shi Y, Marvin S, Zheng Y, Xia G (2020) Air pollution dispersal in high density urban areas: research on the triadic relation of wind, air pollution, and urban form. Sustain Cities Soc 54:101941. https://doi.org/10.1016/J.SCS.2019.101941

Indoor Air Quality Level in Mass Rapid Transit (MRT) Train in Malaysia

Sufy bin Samsuddin, Norhidayah Abdull, Afiza Abdullah Suhaimi, and Nurud Suria Suhaimi

Abstract This study aims to determine the correlation between CO_2 concentrations, CO concentration, relative humidity, and temperature to the number of train passengers. Even though the number of passengers is growing, some reports assert that the quality of the indoor air of Malaysian passenger trains is controversial, particularly during peak hours. The survey included 28 major stations and was carried out over 2 h on weekdays at peak and non-peak hour travel times. The direct-reading instrumentations were in the middle of the train compartment and used for temperature, humidity, air movement, carbon monoxide (CO), and carbon dioxide (CO_2). The number of passengers in and out of the train was also counted manually. The finding indicates that the concentration of CO_2 (up to 1380 ppm) during peak hours exceeded the standard limits (1000 ppm) set by the American Society of Heating, Refrigerating and Air-Conditioning Engineers (ASHRAE), and Department of Occupational Safety and Health (DOSH) Malaysia. A positive correlation exists between the number of passengers and CO_2 concentration. However, the other parameters such as relative humidity, temperature, air movement, and CO concentration have weak correlations to the number of passengers. This study can provide initial data to the train management team on the requirement to improve indoor air quality by applying suitable control measures such as a proper ventilation system.

Keywords Indoor air quality · Peak hours · Non-peak hours · Number of passengers

---

S. Samsuddin
Swift Integrated Logistics Sdn Bhd, Lot 23, Lebuh Sultan Mohamed 1, Kawasan Perusahaan Fasa 2, Bandar Sultan Suleiman, Port Klang, 42000 Pelabuhan Klang, Selangor, Malaysia

N. Abdull (✉) · A. A. Suhaimi · N. S. Suhaimi
Occupational Safety, Health and Environmental Research Group, Faculty of Industrial Sciences and Technology, Universiti Malaysia Pahang Al-Sultan Abdullah, 26300, Kuantan, Pahang, Malaysia
e-mail: hidayahabdull@umpsa.edu.my

A. Z. Yaser et al. (eds.), *Controlling Environmental Pollution*,
https://doi.org/10.1007/978-981-97-8931-3_6

1 Introduction

The urban transportation system is developing, resulting in increased traffic congestion. As a result of this issue, people rely more on public transportation to avoid traffic congestion while traveling to work (Chiu 2014). Additionally, these individuals spent approximately 90 minutes commuting to work, which compelled them to spend time indoors (Li 2006). While interest in indoor air quality (IAQ) investigations has increased, relatively few studies have focused on the IAQ of subway and railway car cabins (Hidayatullah 2015). It was hypothesized that extended exposure during commuting could have a detrimental influence on the occupants' health and well-being. Several poor train air quality symptoms include headaches, sleepiness, and eye and nose irritation (Maryam et al. 2018; Zuurbier 2010). Regardless of the stated unfavorable health effects of continuous exposure, another significant worry of public transportation Mass Rapid Transit (MRT) is comfort, including congestion and inconvenience with current units and locomotives (Syahirah 2014). Therefore, indoor air quality is a matter that needs attention in public transportation. Apart from breakdowns and technical difficulties, a prominent concern that can lead to poor indoor air quality in MRT is air conditioning. It is a significant factor in determining most of the occupant's exposure to air pollution. Indoor air quality can hurt the health and comfort of MRT users in Malaysia (Batutay 2020). Most exposed would be transport workers such as traffic enforcers, drivers, and daily MRT users. Previous research has established that air pollutants generated by various train cabin and ventilation system components constitute essential sources of dangerous air contaminants that may cause health and safety threats to passengers (Penney 2014).

On the stated issue, this study aimed to assess the level of indoor air quality (CO_2 concentrations, CO concentration, relative humidity, and temperature) and its correlation to the number of train passengers. Besides, the independent variable is the change in the number of passengers on MRT Kajang Lines. The study was conducted twice during peak and off-peak hours to determine the correlation between CO_2 concentrations, CO concentrations, relative humidity, and temperature to the number of train passengers along the Kajang Lines from Kajang Station to Kwasa Sentral Station.

2 Methodology

A cross-sectional study is a quantitative approach to determining indoor air quality in MRT Kajang Lines between non-peak and peak hours. This study was conducted with the measurement of CO_2, CO, temperature, and relative humidity data in MRT. This study also aims to determine the relationship between the number of passengers and CO_2, CO, temperature, and relative humidity. Furthermore, the data was collected using a simple random sampling method, and the data was analyzed by using correlation and descriptive statistics.

3 Sampling Location

This was statistical dichotomy research of level indoor air quality along the MRT Kajang Line from Sungai Buloh to Kajang. The study area consisted of MRT cabins and users on a selected cabin in the line that begins from Sungai Buloh, located to the north-west of Kuala Lumpur, and runs through the city center of Malaysia's capital before terminating in Kajang, a rapidly rising town to the city's southeast. This area has enormous urban development and a high population density. The Klang Valley MRT Project implemented is the 51 km Kajang Line.

4 Data Collection

The data was collected using monitoring instruments to find indoor air quality exposure in MRT. IAQ status in the MRT was collected using instrument monitors such as an Air Quality Meter (AIRFLOW Instrument Model 9565-P-NB) for CO_2 concentrations, CO, relative humidity, air movement, and temperature. The checklist was done before conducting the study to determine the current situation in the MRT environment. The things on the checklist are human exposure and comfort level, potential contaminant sources, ventilation, and air conditioning. The checklist was taken from the Industrial Code of Practice (ICOP) on Indoor Air Quality (IAQ), Department of Occupational Safety and Health, Malaysia (Industrial Code of Practise on Indoor Air Quality, Department of Occupational Safety and Health 2010).

5 Instrumentation

An Air Quality Meter (AIRFLOW Instrument Model 9565-P-NB) was used to monitor the CO_2 concentrations, CO, relative humidity, air movement, and temperature in a selected coach during an 84-minute journey on the MRT Kajang Line between Sungai Buloh and Kajang, which is 51 km. According to (Cheng 2016), The Air Quality Meter was positioned 1.0–1.5 meters above the train car's floor. For the duration of the sample period, the Air Quality Meter reading was recorded upon train departure from each station. On weekdays, the air quality was tested inside the train cabin at each station throughout the non-peak and peak hours. The non-peak hour starts from 2:00 to 4:00 p.m. while the peak hour starts from 5:00 pm to 7:00 pm. According to Loh (2021), the evening peak hour in MRT is between 5.00 pm and 8.00 pm, while the non-peak hour is between 12 pm and 4 pm. All equipment was calibrated before use, ensuring no obstructions.

A method of observation was used to count the number of MRT passengers. The passengers inside the train cabin were counted and IAQ data was acquired using a single direct-reading sample approach while the door was closed, and the train was moving toward the next station. Only 58 passengers could fit in a single train cabin while everyone else had to stand to get to their destination.

6 Data Analysis

(a) Descriptive statistics: Using descriptive statistics, the result was shown by a graph of air quality.
(b) Correlation analysis: The Pearson correlation was used to study the relationship between CO_2 concentrations, CO concentrations, and relative humidity with the number of passengers during peak and non-peak hours. While the Spearman correlation was used to study the relationship between relative humidity and temperature with the number of passengers during peak and non-peak hours.

7 Results and Discussion

Number of MRT Users, Concentration of CO_2 and CO, Relative Humidity, Temperature, and Air Velocity During Peak and Non-peak Hours along Kajang Lines During Weekdays.

Figure 1 and Table 1 show an overview of passenger traffic at various train stations (from Kajang to Kwasa Damansara Stations) during peak and non-peak hours within the Klang Valley, Malaysia. Overall, a notable trend observed across all stations is the significant increase in passenger numbers during peak hours compared to non-peak hours, indicating the rush hour phenomenon common in urban transit systems. Stations like Bukit Bintang, Pasar Seni, and Tun Razak Exchange stand out with high passenger volumes during peak hours, suggesting their pivotal roles as major transit hubs or strategic locations within commercial districts. Conversely, stations like Kwasa Sentral and Kota Damansara experience a considerable decrease in passenger numbers during peak hours, serving specific demographic segments. Stations such as Taman Mutiara, Taman Midah, and Taman Pertama consistently demonstrate high passenger traffic throughout peak and non-peak hours, highlighting their significance as key transit points or serving densely populated residential areas.

Starting from Taman Pertama, the graph was significantly inclined to Semantan Station during peak hours while the graph for the non-peak hours increased but not as fast as peak hours. The rate of increase in the number of passengers during peak hours was higher than during non-peak hours. The maximum value for the non-peak hour is 63 people at Bandar Utama Station while for peak hour is 178 at Bukit Bintang Station. The minimum value of passengers was from the first station (Kajang Station) which is 6 and 18 for non-peak and peak hours, respectively. The average value for non-peak hours was 35 passengers while peak hours were 92 passengers, and it was 2.63 times greater than non-peak hours.

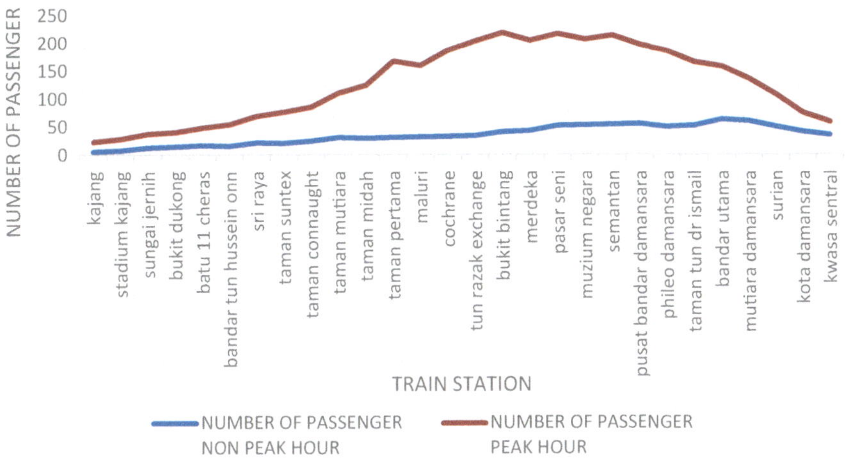

Fig. 1 Number of passengers for peak and non-peak hours

Table 1 Average, minimum, and maximum number of passengers

Parameter	Peak hour (passenger)	Non-peak hour (passenger)
Average	92	35
Minimum	18	6
Maximum	178	63

The non-peak hour starts from 2:00 to 4:00 p.m., departing from Kajang Station to Kwasa Sentral Station. Only a few passengers could not sit, and the others were sitting because the population of passengers was high, and it was during working hours during the time from Kajang to Kwasa Sentral. During this non-peak hour journey, there are no issues with crowded people in one cabin. The peak hour started from 5:00 pm to 7:00 pm during travel also from Kajang to Kwasa Sentral. Many passengers use the train as a transportation to go home to avoid traffic on the road at peak hours. The population density of the area and workplace in the city where the station is located is a factor that may have an impact on the number of passengers. As previously mentioned, there were only 58 seats available for passengers in one cabin on an MRT coach. In this situation, passengers may begin standing up at Taman Connaught Station and continue until they reach their destination.

During peak hours, Bukit Bintang Station experiences a surge in passenger numbers, attributed to its strategic location within Kuala Lumpur's bustling commercial and entertainment district. Serving as a major transportation hub, it connects various modes of transport and attracts a diverse crowd of commuters, tourists, and shoppers. The area's renowned shopping centers, hotels, and vibrant nightlife contribute to the high foot traffic during peak hours, along with the presence of corporate offices and business centers. Bukit Bintang also hosts events and festivals, further boosting passenger numbers. With a large variety of malls, eateries, bars, hotels, and hostels dotting every street, it's one of KL's top commercial and leisure districts (Cheng 2021). Conversely, during non-peak hours, Bandar Utama Station sees a peak in passenger traffic due to its primarily residential nature. Residents commute for daily errands, school, and recreational activities, reflecting the suburban lifestyle of the area. With its proximity to educational institutions and local amenities, Bandar Utama is a focal point for community activities during off-peak times.

Table 2 presents a summary of indoor air quality (IAQ) parameters within MRT cabins during peak and non-peak hours. Additionally, Fig. 2 illustrates how CO_2 concentration levels increase as passenger numbers rise on MRT trains traveling from Kajang to Kwasa Sentral. The red line represents the safety limits outlined by ASHRAE standard 62-2001 (1000 ppm) and the ICOP on IAQ 2010 (Malaysia). Notably, only during peak hours did CO_2 concentrations surpass the safety threshold. During peak hours, CO_2 concentration readings exceeded the limit starting from Taman Midah Station (1028 ppm) with 95 passengers on board, continuing to Phileo Damansara Station (1043 ppm) with 178 passengers. CO_2 levels peaked at Bukit Bintang Station (1380 ppm) with 178 passengers. Conversely, non-peak hour sessions remained within safe CO_2 concentration limits, with the highest recorded level at Surian Station (912 ppm) with 50 passengers.

Figures 3 and 4 depict the temperature and relative humidity data for both peak and non-peak hours. The plots reveal fluctuating trends in these parameters, unaffected by variations in passenger numbers. Despite an increasing passenger count, there is no discernible impact on temperature and relative humidity levels. At the initial station, the maximum temperatures recorded were 26.8 °C and 24.8 °C for non-peak and peak hours, respectively. This disparity can be attributed to the delayed spread of the ventilation system's temperature throughout the train cabin. Referring to Table 2, during non-peak hours, the average temperature was 21.5 °C with an average relative humidity of 69.1%, while during peak hours, the average temperature was 20.6 °C with an average relative humidity of 69.4%. Despite fluctuations in passenger numbers and crowded conditions within the train cabin, the MRT maintained controlled and cold temperatures ranging between 23.0 °C and 26.0 °C.

Figure 5 reveals varying levels of air movement at different stations. Stations such as Taman Midah and Maluri demonstrate relatively high air movement values, reaching 0.35 m/s and 0.31 m/s, respectively. This suggests adequate ventilation and air circulation within the cabins at these locations. Conversely, stations like Bandar Tun Hussein Onn and Taman Mutiara exhibit lower air movement values, around 0.19 m/s and 0.26 m/s, respectively, indicating potentially less efficient ventilation. Despite passenger numbers and fluctuations in environmental conditions, the overall trend shows relatively consistent air movement values across most stations. Ensuring adequate air movement is crucial for maintaining indoor air quality and passenger comfort during travel. Stations with higher air movement values likely offer better ventilation, contributing to a more pleasant commuting experience for passengers.

Table 2 Summary of IAQ parameters in MRT train cabin from Kajang to Kwasa Sentral for peak and non-peak hours

Train station	Number of passenger		Temperature (°C)		Relative humidity (%rh)		CO_2 (ppm)		Air movement (m/s)		CO (ppm)	
	1	2	1	2	1	2	1	2	1	2	1	2
Kajang	6	18	26.8	24.8	66.7	58.3	900	746	0.22	0.33	0.6	0.2
Stadium Kajang	8	21	24	22.8	63.1	66.3	939	814	0.14	0.19	0	0
Sungai Jernih	13	25	23.2	21.2	67.9	69.5	742	757	0.15	0.21	0.1	0
Bukit Dukong	15	26	22.4	20.8	72.9	65.9	696	612	0.14	0.22	0	0
Batu 11 Cheras	17	32	22	20.4	66.4	71	643	629	0.18	0.17	0	0
Bandar Tun Hussein Onn	16	39	21.8	20.9	65.7	81.8	516	595	0.19	0.22	0	0
Sri Raya	22	48	21	20.1	65.6	69.3	612	561	0.19	0.16	0	0
Taman Suntex	21	56	20.8	19.8	64.3	67.9	794	590	0.21	0.15	0	0
Taman Connaught	25	61	20.3	19.7	69	71.7	858	545	0.15	0.22	0	0
Taman Mutiara	31	80	20.9	20.8	74.9	71.9	985	519	0.26	0.21	0	0
Taman Midah	30	95	20.3	20.3	68.1	62	1028	564	0.19	0.35	0	0
Taman Pertama	31	137	20.9	20.4	76.5	69.8	1195	510	0.17	0.2	0	0
Maluri	32	128	20.6	20.2	67.2	68.8	1225	593	0.14	0.31	0	0
Cochrane	33	154	20.3	21	63.3	73.7	1250	643	0.19	0.24	0	0
Tun Razak Exchange	34	170	19.9	20.9	65.4	67.6	1304	691	0.14	0.3	0	0
Bukit Bintang	41	178	20.8	20.3	75.2	61.1	1380	654	0.15	0.24	0	0
Merdeka	43	162	21.3	19.7	67.6	61.3	1235	701	0.15	0.21	0	0
Pasar Seni	52	165	20.7	19.7	64.9	70.9	1132	760	0.19	0.18	0	0
Muzium Negara	53	154	20.7	19.7	66.7	68.9	1096	764	0.31	0.17	0	0
Semantan	54	160	21.8	19.9	79.6	78.3	1038	638	0.17	0.17	0	0
Pusat Bandar Damansara	55	142	21.3	19.9	67.7	67.8	1044	757	0.2	0.18	0	0

(continued)

Table 2 (continued)

Train station	Number of passenger		Temperature (°C)		Relative humidity (%rh)		CO_2 (ppm)		Air movement (m/s)		CO (ppm)	
	1	2	1	2	1	2	1	2	1	2	1	2
Phileo Damansara	50	136	20.1	20.1	65.7	65.6	1043	796	0.19	0.18	0	0
Taman Tun Dr Ismail	52	114	21.2	20.6	80.7	67.5	965	721	0.18	0.15	0	0
Bandar Utama	63	95	22.4	21	71.9	75	745	702	0.2	0.17	0	0
Mutiara Damansara	60	76	22.5	20.6	69	67.9	793	865	0.24	0.27	0	0
Surian	50	58	21.6	20.6	65.7	76.7	688	912	0.23	0.24	0	0
Kota Damansara	41	34	21.2	20.3	69.1	74.1	592	865	0.15	0.24	0	0
Kwasa Sentral	35	24	21.8	20.1	74.8	72.4	595	739	0.19	0.28	0	0

1 = peak hour, 2 = non-peak hour

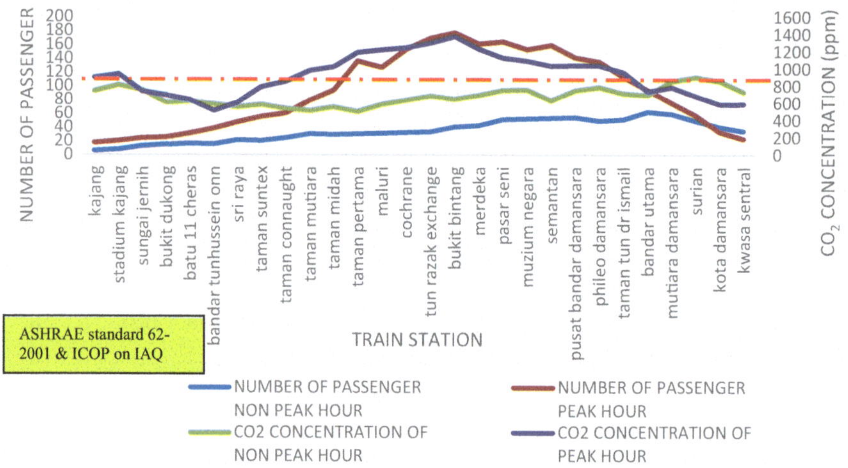

Fig. 2 CO_2 concentration (ppm) and MRT station

Figure 6 shows the graph of carbon monoxide along the train station from Kajang to Kwasa Sentral. There is a value of carbon monoxide only at the beginning of the train journey which is at the Kajang Station for both peak and non-peak hours and there are also readings at Sungai Jernih Station on the non-peak hour. The readings are 0.6 ppm and 0.2 ppm for non-peak and peak hours, respectively. When the train arrives at Sungai Jernih during the non-peak hour, the reading of CO is 0.1 ppm which

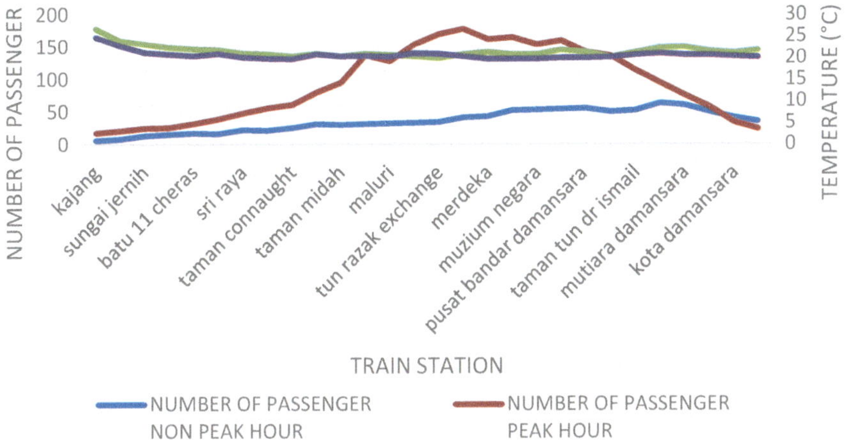

Fig. 3 Temperature (°C) and MRT station

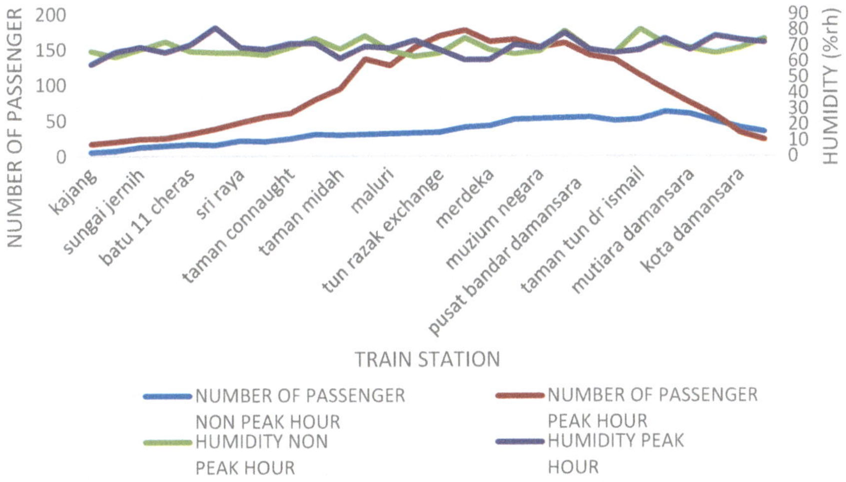

Fig. 4 Relative humidity (%RH) and MRT station

is very small in value. Then, the data had no reading until the end of the train journey, which was 0 ppm, and it is good for ventilation and safety for the passengers in the cabin. This is because carbon monoxide was from the outside train at the beginning of the journey, but the ventilation system in the train helps to improve the air quality in a space by removing other pollutants and contaminants from the air, which can help to reduce the overall level of CO in the air by removing sources of the gas. In addition, air conditioners can help to circulate and freshen the air, which can help to reduce the concentration of CO and other gases.

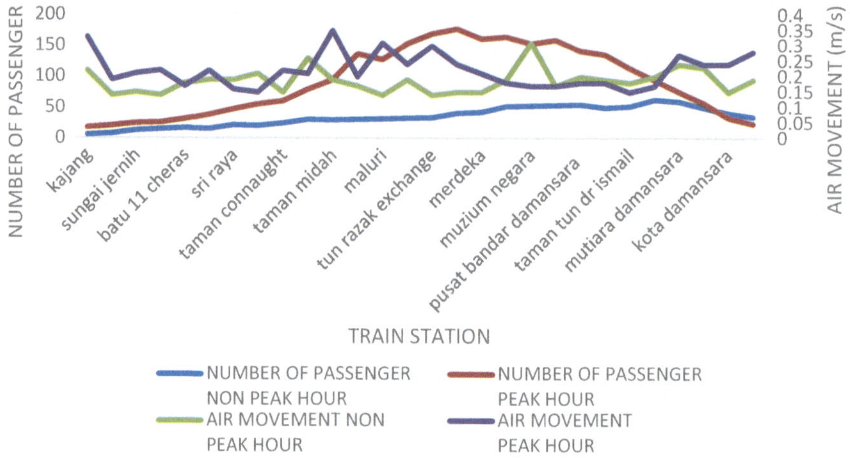

Fig. 5 Air movement (m/s) and MRT station

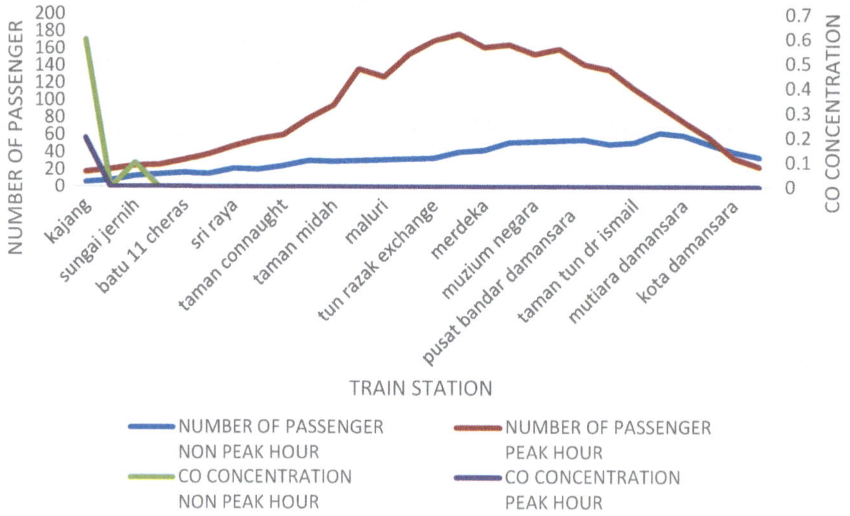

Fig. 6 CO concentration (ppm) and MRT station

8 Correlation Between CO_2 Concentrations to the Number of Passengers

The indoor air quality (IAQ) within MRT cabins significantly influenced passenger numbers during peak hours, when trains were densely populated, resulting in CO_2 concentrations surpassing the ASHRAE standard 62-2001 threshold of 1000 ppm. A parametric Pearson analysis assessed the correlation between CO_2 concentrations and passenger numbers during peak and non-peak hours. The correlation coefficient during peak hours was 0.860, indicating a strong correlation between CO_2 concentrations and passenger numbers, as depicted in Table 3 and Fig. 7. Conversely, the correlation coefficient for non-peak hours was 0.389, representing a moderate correlation. The p-values for CO_2 concentrations during peak and non-peak hours were 0.00 and 0.041, respectively, signifying significant differences in the correlation between CO_2 concentrations and passenger numbers. Fluctuations in passenger numbers, attributed to inconsistent data distribution and the changing air circulation resulting from the opening and closing of train doors at each station, further compounded these effects.

During peak hours when MRT cabins were crowded, the indoor air quality (IAQ) significantly impacted passenger volumes, leading to carbon dioxide (CO_2) levels exceeding the ASHRAE standard 62-2001 threshold of 1000 ppm. The Pearson analysis revealed a strong correlation coefficient of 0.860 between CO_2 levels and passenger volumes during peak hours, as shown in Table 3 and Fig. 7. Conversely, the correlation coefficient was 0.389 during non-peak hours, indicating a moderate relationship. The p-values for CO_2 levels during peak and non-peak hours were 0.00

Table 3 Pearson correlation between the number of passengers and CO_2 concentration

		CO_2 concentration of non-peak hour	CO_2 concentration of peak hour
Number of passengers non-peak hour	Pearson correlation	0.389*	0.270
	Sig. (2-tailed)	0.041	0.164
	N	28	
Number of passengers peak hour	Pearson correlation	−0.084	0.860**
	Sig. (2-tailed)	0.671	0.000
	N	28	

**Correlation is significant at the 0.01 level (2-tailed)
*Correlation is significant at the 0.05 level (2-tailed)

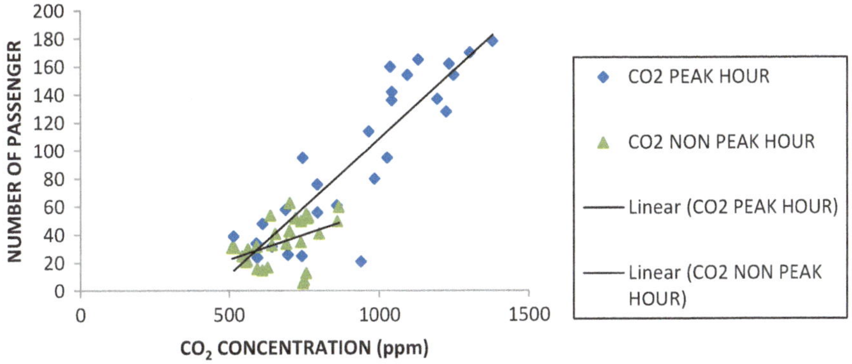

Fig. 7 Correlation line value for number of passengers and CO_2 concentration

and 0.041, respectively, indicating significant disparities in the association between CO_2 levels and passenger volumes. Variations in passenger volumes, influenced by irregular data distribution and the fluctuating air circulation resulting from the opening and closing of train doors at each station, further exacerbated these effects.

This study utilized the Spearman rho test to evaluate the correlation regarding temperature, as the data for both peak and non-peak hours did not exhibit normal distribution, thus falling into the non-parametric category. The non-normality of the data could be attributed to fluctuations in external environmental factors, such as varying weather conditions and passenger activity, which could influence cabin temperature irregularly throughout the day. The correlation analysis revealed a weak correlation with coefficient of −0.419 between temperature and passenger numbers during peak hours, as depicted in Table 4 and Fig. 8. Conversely, the correlation coefficient was −0.187 during non-peak hours, indicating a negligible correlation. Significant differences were observed between temperature and passenger numbers during peak hours ($P = 0.027 < 0.05$), while no such significance was found during non-peak hours ($P = 0.341 > 0.05$), as shown in Table 4. Despite the air circulation facilitated by air conditioning within the cabin, the temperature gradually decreased with the increasing number of passengers.

Passenger-induced temperature rise is often observed in enclosed spaces, such as vehicles, where body heat accumulation can elevate the internal temperature. However, the impact of this effect varies based on factors such as vehicle size, insulation, and external ambient temperature (Horak 2017). Larger vehicle spaces may mitigate temperature rises despite a high passenger density. Adequate air conditioning systems effectively manage this heat influx by maintaining the thermostat's setpoint temperature, thereby counteracting the passenger-induced temperature rise (Zhu 2020). However, the internal temperature may increase if the air conditioning system is insufficient. Thus, a vehicle's number of passengers can influence its internal temperature. Therefore, the MRT has done its best where the temperature in the train remains cool despite the increase in passengers for comfort and health. Thus, MRT needs to maintain special interest and effort in air quality maintenance

Table 4 Spearman correlation between the number of passengers and temperature during peak and non-peak hour

		Temperature during peak hour	Temperature during non-peak hour
Number of passengers peak hour	Spearman correlation	−0.419*	
	Sig. (2-tailed)	0.027	
	N	28	
Number of passengers non-peak hour	Spearman correlation		−0.187
	Sig. (2-tailed)		0.341
	N	28	

**Correlation is significant at the 0.01 level (2-tailed)
*Correlation is significant at the 0.05 level (2-tailed)

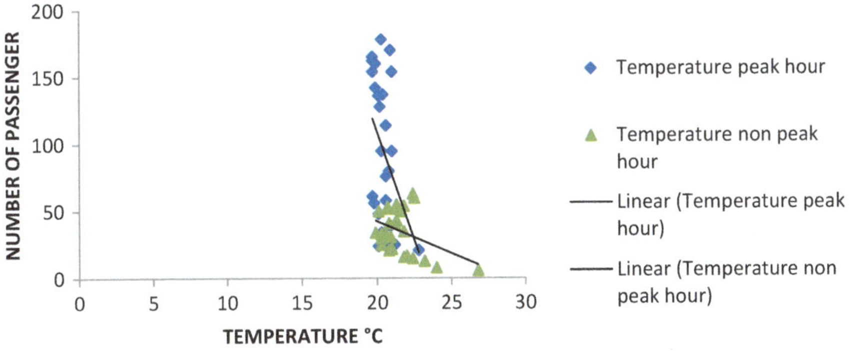

Fig. 8 Correlation line for number of passengers and CO_2 concentration

during peak hours with a lot of passengers inside one cabin. MRT states they have control setting electronic thermostats for desired temperature settings from 16 °C to 28 °C or other temperature ranges as per the manufacturer's standard to maintain the temperature in the MRT (MRT Corp Guideline 2021).

A Pearson correlation, a parametric test, was utilized to investigate the correlation between relative humidity and passenger numbers during both peak and non-peak hours. The correlation coefficient for peak hours was found to be −0.126, indicating a negligible correlation between humidity and passenger numbers, as presented in Table 5 and Fig. 9. Similarly, during non-peak hours, the correlation coefficient was 0.296, also suggesting a negligible correlation. The p-values for humidity during peak and non-peak hours were 0.522 and 0.126, respectively.

Figure 9 demonstrates the fluctuating nature of relative humidity findings, consistent with recent studies highlighted by Bong (2013). These fluctuations indicate the influence of various factors on the results, making it challenging to discern the relationship between passenger numbers and humidity. The presence of occupants in a space can impact humidity levels through increased heat and moisture production.

Table 5 Pearson correlation between the number of passengers and humidity

		Relative humidity non-peak hour (%RH)	Relative humidity peak hour (%RH)
Number of passengers non-peak hour	Pearson correlation	0.296	0.186
	Sig. (2-tailed)	0.126	0.343
	N	28	
Number of passengers peak hour	Pearson correlation	0.152	−0.126
	Sig. (2-tailed)	0.440	0.522
	N	28	

**Correlation is significant at the 0.01 level (2-tailed)

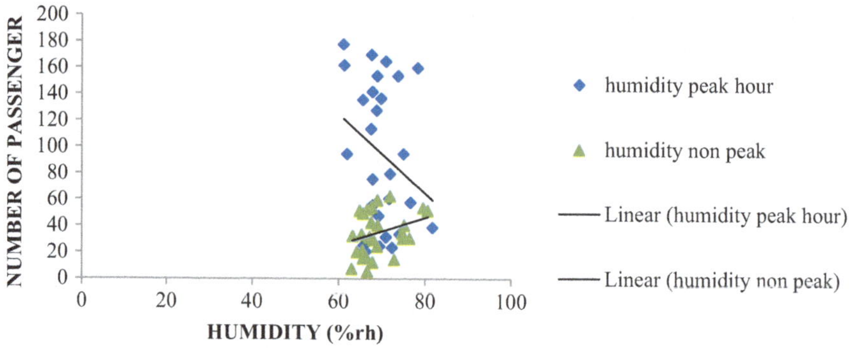

Fig. 9 Correlation line for number of passengers and relative humidity

Conversely, air conditioning systems can reduce humidity levels by extracting moisture from the air. Additionally, factors such as space size and ventilation systems can affect humidity accumulation. The heating, ventilation, and air conditioning (HVAC) system in buildings or vehicles is designed to regulate temperature and indoor air quality. Ventilation systems utilize fans to circulate air, promoting even distribution of heat or cool air and enhancing occupant comfort. Moreover, ventilation systems facilitate the exchange of indoor and outdoor air, improving air quality by removing stale air and introducing fresh air. However, in this study, the opening and closing of train doors at each station introduce varying external conditions, making it challenging to ascertain the true relative humidity within the train. Furthermore, external weather conditions, such as rain during non-peak hours and hot weather during peak hours, further influence humidity measurements.

A parametric test was used to study the relationship between air movement to the number of passengers for peak hours which is the Pearson correlation. The correlation coefficient for peak hour is −0.090 which was the negligible relationship between humidity and the number of passengers as shown in Table 6 and Fig. 9. P-value (significant 2 tails) air movement for peak hour is ($P = 0.649$). So, there is no significant difference between CO_2 concentrations and the number of passengers for peak hour.

Table 6 Pearson correlation between the number of passengers and air movement peak hour

		Air movement peak hour	Air movement during non-peak hour
Number of passengers peak hour	Pearson correlation	−0.090	
	Sig. (2-tailed)	0.649	
	N	28	
Number of passengers non-peak hour	Pearson correlation		0.319
	Sig. (2-tailed)		0.099
	N	28	

For non-peak hours, the Spearman's rho test was used to assess the relationship, as the data for non-peak hours did not follow a normal distribution and was therefore categorized as non-parametric, as shown in Table 8. The relationship coefficient for non-peak hours was 0.319 which was a low positive relationship between air movement and the number of passengers as shown in Table 6. P-value (significant 2 tails) air movement for non-peak hours is ($P = 0.099 =< 0.05$). So, there is a significant difference between air movement and the number of passengers for non-peak hours.

There are several ways in which air movement is managed in trains. Firstly, some trains have mechanical ventilation systems that use fans to circulate air. Air movement can also happen when having air conditioning systems that help to cool the air inside the car and improve the comfort of the passengers. Next, the motion of the train can create natural convection currents, causing air to flow through the car. This can help to circulate fresh air and improve the air quality. Then, the open and closing of doors can also create air movement in a train car, as air is drawn in or expelled through the doorway. Overall, the air movement in a train is designed to help regulate the temperature and air quality inside the car and improve the comfort of the passengers.

The study found there was a weak correlation between air movement and the number of passengers because most air movement in the train was created from the ventilation system. By providing negative or positive pressures in the system, the fan in ventilation systems overcomes all component resistances (ductwork and fittings, dampers, air filters, and heating or cooling coils) to the air movement. The modern air conditioning system in the train cabin removes the stale air from within and replaces it six to eight times an hour with fresh air from outside. Only 20% of the air would be fresh after this operation, leaving up to 80% of the old air within (Schrecengost 2022). Air temperature and velocity in train cabins that use the ventilation mode with air intake from the floor and exhaust from the roof were discovered to be higher but more comfortable for people (Konstantinov 2015). When the airflow organization of the two ventilation systems is compared, it is discovered that the ventilation system with supply air diffusers located at the roof of the train cabin and the door being used as an exhaust air outlet has a more uniform temperature distribution and a higher ventilation efficiency (Wang 2014).

The correlation coefficient for peak hours was −0.262, indicating a negligible relationship between CO concentrations and passenger numbers, as presented in Table 7 and Fig. 10. Similarly, during non-peak hours, the correlation coefficient was −0.387, suggesting a negligible relationship. The p-values for CO concentrations during peak and non-peak hours were 0.178 and 0.042, respectively. It is believed that the CO in train does not contribute by the human respiration, but due to external factors, particularly emissions from traffic. As trains often traverse urban or industrial areas with heavy vehicular traffic, they can be exposed to CO emissions from surrounding vehicles. In urban settings, automobiles, trucks, and buses are common sources of CO, with internal combustion engines producing CO as a byproduct of incomplete fuel combustion. Traffic congestion, stop-and-go driving patterns, and inefficient engine performance can exacerbate vehicle CO emissions. As trains travel alongside roadways or pass through urban centers, they may draw in air contaminated with CO from nearby traffic emissions, leading to CO within the train cabin.

The result of the IAQ measurement of peak and non-peak hours was compared with the ICOP on IAQ 2010 and ASHRAE standard 62-2001 to ensure passengers are safe and the MRT train follows the regulation as stated in Table 7. The value of CO_2 concentration during peak hour does not meet the standards of ASHRAE standard 62-2001 and ICOP on IAQ 2010 which exceeds 1000 ppm when the train starts crowded with the passenger. The train station that does not comply with the regulation is at

Table 7 Pearson correlation between the number passengers and CO concentration

		CO non-peak hour (ppm)	CO peak hour (ppm)
Number of passengers peak hour	Pearson correlation	−0.299	−0.262
	Sig. (2-tailed)	0.122	0.178
	N	28	
Number of passengers non-peak hour	Pearson correlation	−0.387*	−0.346
	Sig. (2-tailed)	0.042	0.071
	N	28	

**Correlation is significant at the 0.01 level (2-tailed)
*Correlation is significant at the 0.05 level (2-tailed)

Table 8 Acceptable range for IAQ parameters according to ICOP on IAQ 2010 (DOSH, Malaysia)

Parameter	Acceptable range
Air temperature	23–26
Relative humidity	40–70%
Air movement	0.15–0.50 m/s
Carbon monoxide	10 ppm
Carbon dioxide	1000 ppm

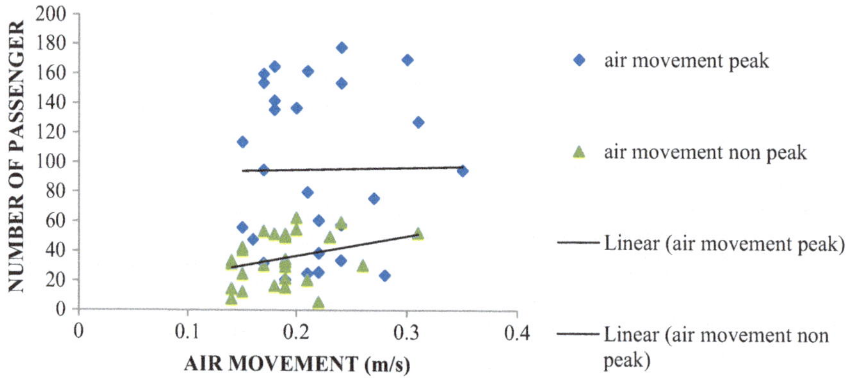

Fig. 10 Correlation line for number of passengers and air movement

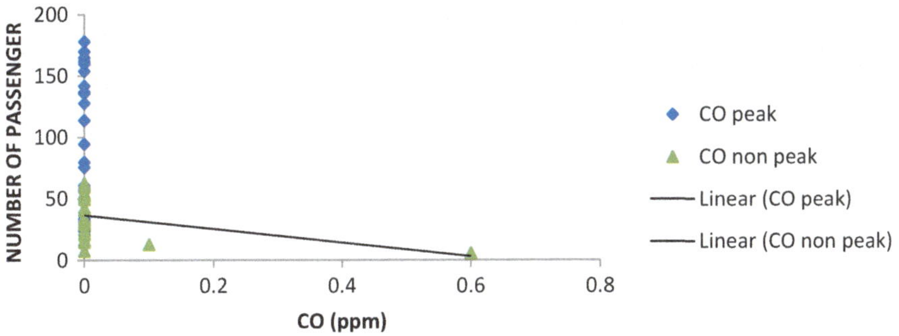

Fig. 11 Correlation line for number of passengers and CO

Taman Midah until Phileo Damansara because of the pack of passengers in one cabin. The next physical contaminant is carbon monoxide, and the reading in the MRT cabin follows the ICOP on IAQ 2010 regulation which is 10 ppm because the maximum value is only 0.2 ppm and 0.6 ppm for peak and non-peak, respectively. The mean for CO concentration during peak hour was 0.007 while the non-peak hour was 0.025 which was very low.

The acceptable range for air temperature in ICOP on IAQ 2010 is between 23 °C and 26 °C. The average temperature for peak hour is 20.6 °C while the non-peak hour is 21.5 °C. So, it complied with the ICOP on IAQ 2010 and is good for MRT passenger health and comfort. Extreme heat is a serious health risk. Besides, high temperatures can cause insufficient humidity. There is some reason why the temperature reading was lower despite of increased number of passengers. The air conditioning system may be overly powerful for the number of people in the cabin because of its high cooling capacity. This indicates that the body heat of the passengers is not reducing the cooling impact of the air conditioning, resulting in a lower temperature reading. Then, the insulation inside the cabin can be overly effective, which would keep an excessive amount of cold air inside. Despite an increase in passengers, this can cause the temperature to read lower. The average relative humidity is 69.4% and 69.1% for peak and non-peak, respectively. Based on the ICOP on IAQ 2010 (Table 8), the average humidity reading value in the MRT train has passed the acceptable limit which is between 40 and 70%. Nevertheless, the maximum humidity value at peak hours is 81.8% and 80.7 on the non-peak hour is exceeding the safe IAQ limit for ICOP on IAQ 2010. Passengers who spend too much time in a high humidity may become ill, particularly if they have respiratory illnesses. In air with a relative humidity of 60% or more, the bacteria and viruses that cause disease thrive and multiply.

According to the Industry Code of Practice on Indoor Air Quality (ICOP IAQ) Malaysia 2010, optimal air movement is defined within the range of 0.15 to 0.50 meters per second (m/s). The collected data indicates that the average air movement was 0.22 m/s during peak periods and 0.19 m/s during non-peak periods. These measurements fall within the acceptable range, suggesting adequate ventilation and air circulation in the monitored environment. Controlled airflow comes from the

MRT ventilation system for ventilation in the train. Reducing the risk of air leaks and ensuring that cooling equipment is properly placed and well maintained can help improve the ventilation system's energy efficiency. Because particles can more easily gather to deadly concentrations indoors than outdoors, where wind swiftly disperses the aerosols, the virus spreads more quickly. Building occupants are more prone to contract the virus through inhalation when the virus is present at higher quantities. Therefore, one of the best strategies to lower the danger of spreading airborne viruses is to properly ventilate.

References

Batutay TJG, H S (2020) Quantitative and qualitative analysis of indoor air quality inside the MRT 3 train cabins. IEOM Society International

Bong C, K Y (2013) Study on air quality characteristics of subway stations using sensor module. Int J Environ Sci Dev

Cheng HD (2016) Characteristics of PM2.5, CO_2 and particle-number concentration in mass transit railway carriages in Hong Kong. Environ Geochem Health 1–12

Cheng TL (2021) Better times for malls only in 2023. Free Malaysia Today I FMT

Chiu CO, K M (2014) Mode choice between private and public transport in. Sci World J 1–14

Hidayatullah (2015) The analysis of indoor air quality inside the Kelana Jaya light rail transit train in Kuala Lumpur. Master thesis, Johor, Universiti Tun Hussein Onn Malaysia

Horak J (2017) Cabin air temperature of parked vehicles in summer conditions: life-threatening environment for children and pets calculated by a dynamic model. Springer

Industrial Code of Practise on Indoor Air Quality, Department of Occupational Safety and Health, Malaysia (2010) Department of Occupational Safety and Health (DOSH)

Konstantinov M (2015) Numerical simulation of the thermal comfort in train cabin. Int J Railway Technol 4(3):69–88. https://doi.org/10.4203/ijrt.4.3.3

Li TT, B Y (2006) Air quality in passenger cars of the ground railway transit system in Beijing, China. Sci Total Environ 89–95

Loh (2021, Nov 4) What time is the peak hours in Singapore? Mrtsgt. https://blog.sgtrains.com/2021/11/morning-train-fares-50-cents-discount/

Maryam Z, Hazrin AH, Hizrri A, Norhidayah AH, Samsuddin NS, Shukri M (2018) Association of particulate matter (PM) with respiratory symptoms among children in selected primary schools in Pahang. J Clean WAS (JCleanWAS) 2:11–15. Zibeline International Publishing. https://doi.org/10.26480/jcleanwas.01.2018.11.15

Penney D, B V (2014) MRTC. "Maintenance". Metro Rail Transit Corporation. https://mrt3.com/index.php/maintenance.html

Schrecengost R (2022) Air movement and air handling systems engineer

Syahirah B, O. M. (2014) An assessment of commuters' perceptions of safety and comfort levels of 'Women-Only Coach': the case study of KTM Komuter Malaysia. Procedia Environ Sci 20:197–205

Wang H (2014) Performance evaluation of air distribution systems in three different China railway high-speed train cabins using numerical simulation. Cabins. Building Simulation

Zhu H-C (2020) Fast prediction for multi-parameters (concentration, temperature and humidity) of indoor environment towards the online control of HVAC system. Building Simulation

Zuurbier M, H. G. (2010) Commuters' exposure to particulate matter air pollution is affected by mode of transport, fuel type, and route. Environ Health Perspect 118(6):783–789.

Length–Weight Relationship and Condition Factor of Native Fish Population in Kenyir Lake Basin: Environmental Health Impacts and Mitigations Strategies

Noorjima Abd Wahab, Mohd Khairul Amri Kamarudin, Mohd Ekhwan Toriman, Muhamad Hafiz Saad, and Muhamad Armi Abu Samah

Abstract Introduction: The vegetation and wildlife that live around water bodies are included in freshwater ecosystems. Each of them is part of a larger biological and hydrological landscape that also includes upland terrestrial ecosystems, riparian zones, and subsurface groundwater aquifers. Objective: The aim of this study is to identify the Length–Weight Relationship (LWR) and condition factor of three native species from the Kenyir Lake Basin, as well as the environmental impacts and mitigation strategies. Methodology: LWR and condition factor of *Channa micropeltes* (Toman), *Hampala macrolepidata* (Sebarau) and *Chitala lopis* (Belida) were conducted at two main rivers (Terengganu River and Petuang River) during wet season, normal season and dry season. Result and Discussion: Among these three native species generally showed the equilibrium allometric growth recorded values

N. A. Wahab · M. K. A. Kamarudin (✉)
Faculty of Applied Social Sciences, Universiti Sultan Zainal Abidin, Gong Badak Campus, 21300 Kuala Nerus, Terengganu, Malaysia
e-mail: mkhairulamri@unisza.edu.my

N. A. Wahab
e-mail: noorjimaabdwahab@unisza.edu.my

M. K. A. Kamarudin · M. H. Saad
East Coast Environmental Research Institute (ESERI), Universiti Sultan Zainal Abidin, Gong Badak Campus, 21300 Kuala Nerus, Terengganu, Malaysia
e-mail: hafiz@koperasisiswazah.com

M. E. Toriman
Faculty of Social Sciences and Humanities, Universiti Kebangsaan Malaysia, 43600 Bangi, Selangor, Malaysia
e-mail: ikhwan@ukm.edu.my

M. A. A. Samah
International Islamic University Malaysia (IIUM), Kuantan Campus, Jalan Sultan Ahmad Shah, Bandar Indera Mahkota, 25200 Kuantan, Pahang, Malaysia
e-mail: marmi@iium.edu.my

A. Z. Yaser et al. (eds.), *Controlling Environmental Pollution*,
https://doi.org/10.1007/978-981-97-8931-3_7

of 3.87 ± 2.201 and condition factor (K) as 6.25 ± 0.97 during wet, normal, and dry seasons at Terengganu River and Petuang River. Overall, the growth level of these three native species showed the ideal growth at the Terengganu River compared to the Petuang River. Conclusion: Management and monitoring programs will need to be initiated, or modifications to an existing program will need to be made so that the ecological effects of implementing the flow recommendations can be evaluated adequately.

Keywords Length–Weight Relationship (LWR) · *Channa micropeltes* (Toman) · *Hampala macrolepidata* (Sebarau) · *Chitala lopis* (Belida) · Terengganu river · Petuang river

1 Introduction

The presence of biodiversity is a sign of healthy ecosystems. The loss of biodiversity has been linked to high rates of species extinction. The effects of invasive non-native species should not be disregarded, despite the fact that additional threats to ecosystem health have been discovered, including habitat loss, overexploitation, pollution, endemism, and anthropogenic causes. According to Reid et al. (2019), freshwater ecosystems have been significantly impacted by anthropogenic change due to damming, ecosystem modification for navigating, the agricultural sector, hydroelectricity, water extraction, acidification, eutrophication, pollution, and the release of non-native animals. Freshwater fishes are among the most endangered vertebrate species worldwide, which warrants particular management actions and restoration measures to stop the catastrophic decrease. The body of knowledge on inland waterbodies' mitigation, rehabilitation, and restoration is extensive and varied. Even while most ecological management initiatives will also impact fish populations, they are often driven by a larger, legally sanctioned environmental framework (Keat-Chuan et al. 2017; Birk et al. 2020; Khan et al. 2022).

Stressors frequently interact with one another as well as with ecosystem processes. Hydrology and hydrodynamics, or how tides and freshwater flow interact to define the temporal and geographical variation of the estuary's physical environment, impact things like nutrient enrichment, harmful compounds, and temperature. This makes evaluating and assessing whether a given stressor will have positive, negative, or neutral overall consequences for the ecosystem and its properties more difficult. Additionally, animals have different personal responses to the majority of stresses. This confluence of factors leads to a complex biological, geographical, and chronological mosaic of impacts (Taniwaki et al. 2017; Palmer & Ruhi 2019; Comte et al. 2021).

The assessment of the impacts is somewhat influenced by the ecological services and demands that are of significance or concern, such as the availability of safe and useable water, commercial and recreational fisheries, habitat quality, or public access to the delta. The process of ranking stresses in order to prioritize society's investments in their improvements is therefore far ,more complicated than it might initially appear, despite the fact that doing so is politically acceptable (Scholte et al. 2015; Ehrlich et al. 2017; Bratman et al. 2019). The environmental services and demands that are important or of concern, such as the accessibility of safe and usable water, corporate and recreational aquaculture, the condition of habitat, or public access to the delta, have some influence on the assessment of the consequences. Therefore, although doing so is acceptable in society, evaluating stresses in order to prioritize society's efforts in their improvements is significantly more complex than it might initially appear (Muhamad Khair et al. 2018; Nitanan et al. 2020). Figure 1 shows the conceptual diagram of interactive stressors affecting the San Francisco Bay-Delta based on water quality, habitat condition, and overall ecosystem structure and functioning. This research concept focused on the key fish species, such as *salmonids*, which are closely related to other biotic elements of the ecosystem, such as primary producers that are planktonic and benthic, grazers, larvae, and juvenile and mature invertebrate and fish species (National Academies of Sciences, Engineering, and Medicine 2012).

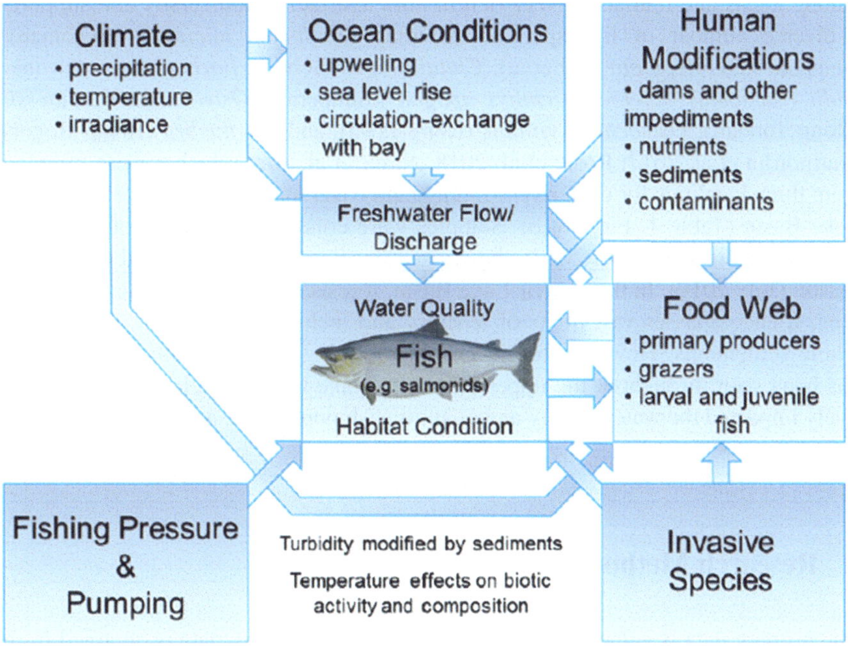

Fig. 1 The conceptual diagram showing the interactive stressors affecting the water quality, habitat condition, and overall ecosystem structure and functioning. *Source* National Academies of Sciences, Engineering, and Medicine (2012)

Table 1 Research location in Terengganu river and Petuang river

Sampling station	River	Longitude	Latitude
Station 1	Terengganu River I	102°55′37.6″T	5° 01′57.3″U
Station 2	Terengganu River II	102°56′12.051″T	5° 03′04.559″U
Station 3	Terengganu River III	102°56′28.354″T	5° 04′00.363″U
Station 4	Petuang River I	102°39′36.0″T	5° 15′21.1″U
Station 5	Petuang River II	102°39′43.8″T	5° 15′11.7″U
Station 6	Petuang River III	102°39′37.4″T	5° 14′14.9″U

2 Study Area and Research Methodology

An important part of Malaysia's economic development is the development of man-made lakes such as Kenyir Lake, which are used to store water, supply water for agriculture, provide areas for cage aquaculture, regulate flooding, and produce hydropower. Kenyir Lake Basin, which is a sustainable ecotourism hotspot among locals and tourists, is very rich in flora and fauna biodiversity and supports a diverse number of fish species, including the *hanna micropeltes* (Toman), *Hampala macrolepidata* (Sebarau), *Chitala lopis* (Belida), *Barbonymus schwanefeldii* (Lampam), *Cyclocheilichthys apogon* (Temperas), *Osteochilus waandersii* (Rong Toboali), *Osteochilus vittatus* (Rong Jawa), and *Tor tambra* (Kelah Bogor) (Farinordin et al. 2017; Rosle et al. 2018; Naser et al. 2021). Fishes were sampled from three localities for each river (Terengganu River and Petuang River) in Kenyir Lake Basin (Table 1, Fig. 2a, b). Samples were collected at each location during the wet season (December 2017), the dry season (September 2018), and the normal season (July 2019). In the Kenyir Lake Basin, forests serve as the primary land use type. It encompasses vast tracts of territory and includes agricultural, tourist, and whole settlements. However, recent considerable logging for plantation activities has been seen throughout the upper part of the lake basin, which may have negatively impacted the water quality and quantity, fish population, and natural beauty in the area.

3 Research Methodology

The primary data is gained during fieldwork while secondary data is obtained from existing resources. Fishes were sampled for each sampling station using the size 2 cm × 2 cm fish net. The fish net was used as it is the best option to catch fish in deeper pools. The fish caught were randomly sampled. The specimens were preserved in 10% formalin and brought back to the laboratory. The fish caught were identified in the laboratory for identification. The fish are then measured in total length and weighted. The specimens were brought out of the deep freezer and allowed to thaw,

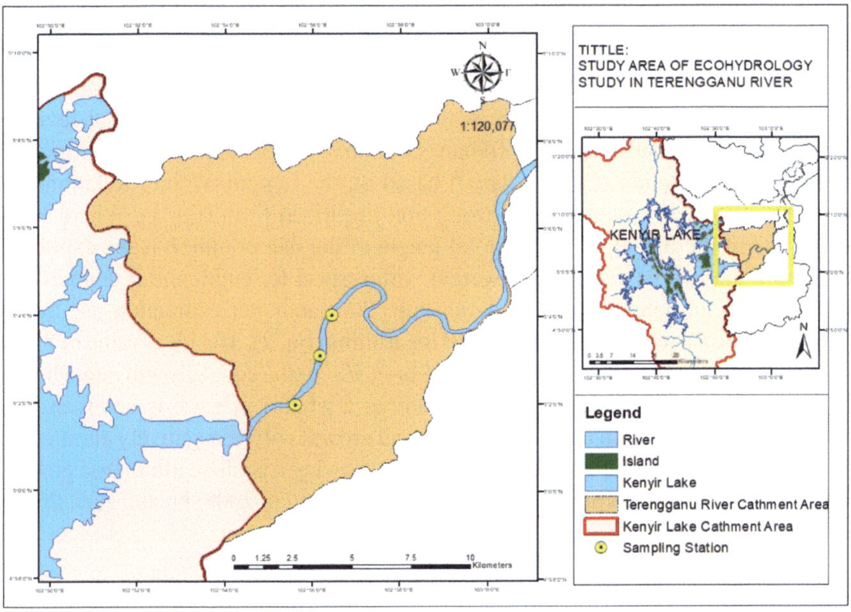

(a) The research location map in Terengganu River, Kenyir Lake

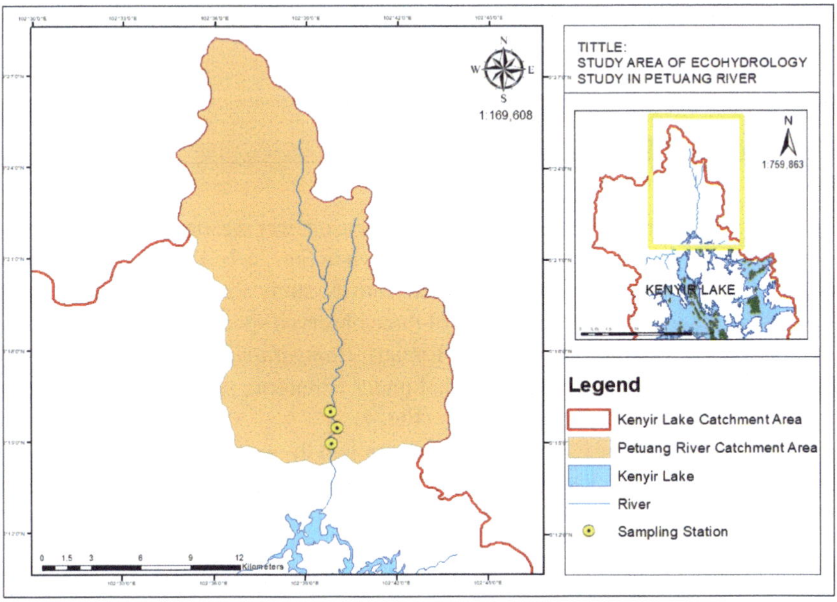

(b) The research location map in Petuang River, Kenyir Lake

Fig. 2 **a** The research location map in Terengganu River, Kenyir Lake, **b** The research location map in Petuang River, Kenyir Lake

and their body length and weight were measured. The total length was measured using a one-meter measuring board graduated in mm. The fish was wiped with a dry napkin before weighing and body weight was measured using a weighing balance. In this study, in the determination of the length–weight relationship and relative condition factor of native species (Toman (*Channa macropeltes*), Sebarau (*Hampala macrolepidata*), Belida (*Chitala lopis*)) based on the Length–Weight Relationship (LWR) method was calculated following non-linearized form (Eq. 1). Where W is the weight of the fish in g, L is the total length of the fish in mm, b is the exponent describing the rate of variation in weight with respect to length and a denotes the coefficient of the length–weight relationship. For parameter estimation use linear regression of data transformed to base 10 logarithms (Eq. 2). The ideal value of b is 3 that represents an isometric growth (b $\geq$ 2.5 or $\approx$ 3/> 3) (the equal growth rates shown by different body parts with that of the body as a whole), when b is less than 2.5, fish become slimmer with increasing length and growth will be negatively allometric. When b is greater than 2.5, fish become heavier showing a positive allometric growth and reflecting optimum conditions for growth (Samat et al. 2008; Nyanti et al. 2018; Awang et al. 2022).

$$W = aL^b \tag{1}$$

$$\text{Log W} = \text{log a} + \text{b} * \text{log L} \tag{2}$$

4 Result and Discussion

The list of fish species collected during the dry season, wet season and normal season at Terengganu River and Petuang River are shown in Fig. 3a, b, Tables 2, 3, 4, 5, 6 and 7. 21 species (seven families) were recorded from the study area during sampling time (along three seasons respectively). Out of three observed species from 21 selected as native species as *Channa micropeltes* (Toman), *Hampala macrolepidata* (Sebarau), and *Chitala lopis* (Belida) were recorded under isometric, positive allometric, and negative allometric growth, respectively (Fig. 4).

This study sampled 21 fish species, but the length–weight relationship was not calculated for three species. The age sizes of fish covered in this study varied from young to adult stages with different growth rates. This proved there is high variability due to the collection of fish species based on several mesh sizes (Connor et al. 2017; Farinordin et al. 2017; Wilkinson et al. 2018). All the age sizes were used as the determination tools to identify the length–weight relationship for each species. Linear regressions were significant for all species during three seasons respectively with R^2 values > 0.5. Figures 5a–d, 6a–e and 7a–e indicates the results from analysis on the key species *Channa micropeltes* (Toman), *Hampala macrolepidata* (Sebarau) and *Chitala lopis* (Belida) from the Terengganu River and Petuang River during dry

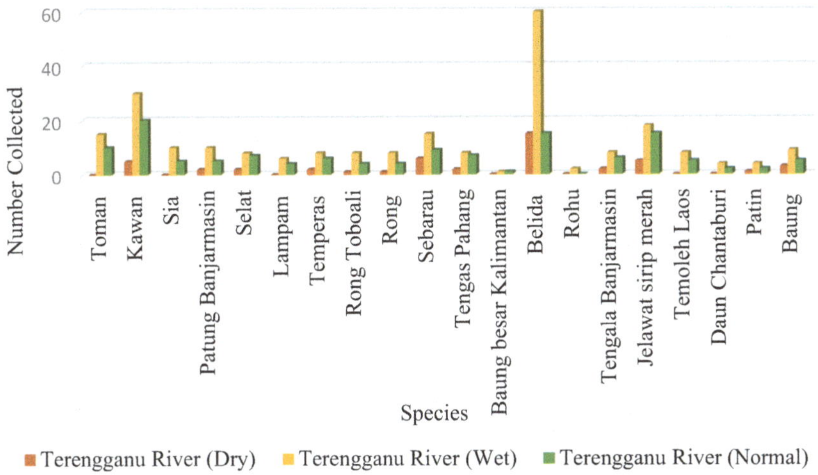

(a) The number collected of fishes's species in the Terengganu River

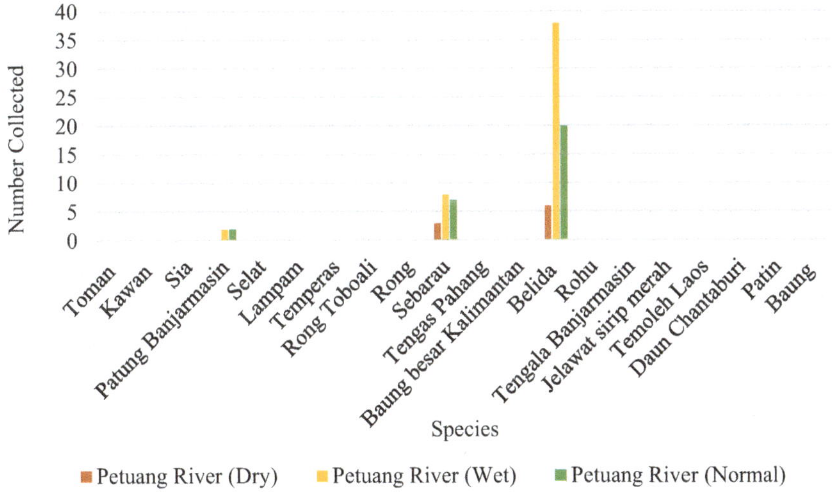

(b) The number collected of fishes's species in Petuang River

Fig. 3 **a** The number collected of fish species in the Terengganu River, **b** The number collected of fish species in the Petuang River

season, normal season and wet season. All three native species from the Terengganu River area show the fish grew (the b value was between 2.7877 and 3.87) and the Petuang River area shows the fish grew (the b value was between 2.2005 and 3.2067) during three seasons respectively. The b values of LWR were close to 3.0, hence indicating isometric growth and representing the optimum shape of fish species. Sebarau and Belida recorded negative allometric growth at Petuang River during the

Table 2 The distribution of fish collected during dry season in the Terengganu River

Bil	Species name (Origin name)	Species name (Scientific name)	The number collected	Average of width (cm)	Average of length (cm)	Average of weight (g)
1	Toman	*Channa micropeltes*	0	0.00	0.00	0.00
2	Kawan	*Labiobarbus leptocheilus*	5	4.74	13.76	69.8
3	Sia	*Mystacoleucus obtusirostris*	0	0.00	0.00	0.00
4	Patung Banjarmasin	*Pristolepis fasciata*	2	4.15	7.25	26.80
5	Selat	*Notopterus notopterus*	2	4.95	14.8	84.00
6	Lampam	*Barbonymus schwanefeldii*	0	0.00	0.00	0.00
7	Temperas	*Cyclocheilichthys apogon*	2	6.4	11.6	55.85
8	Rong Toboali	*Osteochillus waandersii*	1	4.5	15.3	27.8
9	Rong	*Osteochilus vittatus*	1	5.3	12.5	66.5
10	Sebarau	*Hampala macrolepidata*	6	4.90	14.02	56.67
11	Tengas Pahang	*Neolissochilus soroides*	2	4.80	14.50	46.00
12	Baung besar Kalimantan	*Hemibagrus capitulum*	0	0.00	0.00	0.00
13	Belida	*Chitala lopis*	15	4.04	15.46	40.48
14	Rohu	*Labeo rohita*	0	0.00	0.00	0.00
15	Tengala Banjarmasin	*Puntioplites bulu*	2	4.25	12.70	62.70
16	Jelawat sirip merah	*Leptobarbus rubripinna*	5	4.90	14.70	73.26
17	Temoleh Laos	*Probarbus julleni*	0	0.00	0.00	0.00
18	Daun Chantaburi	*Poropuntius normani*	0	0.00	0.00	0.00
19	Patin	*Pangasius sp.*	1	5.8	15.1	90.2
20	Baung	*Hemibagrus sp.*	3	4.17	16.03	92.50
			Maximum	6.4	16.03	92.5
			Depth (m)	2.5 ± 1.1		
			Velocity (m/s)	0.045 ± 0.010		
			Discharge, Q (m^3/s)	8.5202 ± 4.8939		

Table 3 The distribution of fish collected during normal season in the Terengganu River

Bil	Species name (Origin name)	Species name (Scientific name)	The number collected	Average of width (cm)	Average of length (cm)	Average of weight (g)
1	Toman	*Channa micropeltes*	10	4.36	27.12	280.97
2	Kawan	*Labiobarbus leptocheilus*	20	4.33	15.04	68.11
3	Sia	*Mystacoleucus obtusirostris*	5	3.32	15.22	44.20
4	Patung Banjarmasin	*Pristolepis fasciata*	5	3.20	33.60	21.50
5	Selat	*Notopterus notopterus*	7	4.87	31.10	59.00
6	Lampam	*Barbonymus schwanefeldii*	4	5.23	11.43	51.75
7	Temperas	*Cyclocheilichthys apogon*	6	6.72	11.80	51.55
8	Rong Toboali	*Osteochillus waandersii*	4	3.88	16.93	26.68
9	Rong	*Osteochilus vittatus*	4	5.18	11.60	54.08
10	Sebarau	*Hampala macrolepidata*	9	4.43	13.63	63.52
11	Tengas Pahang	*Neolissochilus soroides*	7	5.01	16.24	46.06
12	Baung besar Kalimantan	*Hemibagrus capitulum*	1	5.1	20.4	142.65
13	Belida	*Chitala lopis*	15	4.63	15.35	40.44
14	Rohu	*Labeo rohita*	0	0.00	0.00	0.00
15	Tengala Banjarmasin	*Puntioplites bulu*	6	4.85	14.38	68.47
16	Jelawat sirip merah	*Leptobarbus rubripinna*	15	5.66	15.84	92.87
17	Temoleh Laos	*Probarbus julleni*	5	3.94	17.54	77.40
18	Daun Chantaburi	*Poropuntius normani*	2	3.65	9.25	18.8
19	Patin	*Pangasius sp.*	2	5.8	18.5	138.8
20	Baung	*Hemibagrus sp.*	5	3.26	18.40	124.14
			Maximum	6.72	33.6	280.97
			Depth (m)		2.8 ± 1.1	
			Velocity (m/s)		0.060 ± 0.011	
			Discharge, Q (m^3s)		18.7415 ± 12.5884	

Table 4 The distribution of fish collected during the wet season in the Terengganu River

Bil	Species name (Origin name)	Species name (Scientific name)	The number collected	Average of width (cm)	Average of length (cm)	Average of weight (g)
1	Toman	*Channa micropeltes*	15	4.91	29.55	315.73
2	Kawan	*Labiobarbus leptocheilus*	30	4.97	15.03	70.37
3	Sia	*Mystacoleucus obtusirostris*	10	4.11	15.84	47.79
4	Patung Banjarmasin	*Pristolepis fasciata*	10	3.49	8.57	33.79
5	Selat	*Notopterus notopterus*	8	5.44	15.59	73.10
6	Lampam	*Barbonymus schwanefeldii*	6	5.50	11.08	51.13
7	Temperas	*Cyclocheilichthys apogon*	8	6.85	11.54	49.60
8	Rong Toboali	*Osteochillus waandersii*	8	4.54	16.31	33.60
9	Rong	*Osteochilus vittatus*	8	5.53	11.19	43.36
10	Sebarau	*Hampala macrolepidata*	15	4.94	15.21	87.63
11	Tengas Pahang	*Neolissochilus soroides*	8	5.44	16.15	49.45
12	Baung besar Kalimantan	*Hemibagrus capitulum*	1	5.00	22.15	173
13	Belida	*Chitala lopis*	60	4.82	15.88	41.05
14	Rohu	*Labeo rohita*	2	6.75	16.55	52.15
15	Tengala Banjarmasin	*Puntioplites bulu*	8	4.91	14.10	75.09
16	Jelawat sirip merah	*Leptobarbus rubripinna*	18	5.81	16.23	90.86
17	Temoleh Laos	*Probarbus julleni*	8	4.74	17.09	85.21
18	Daun Chantaburi	*Poropuntius normani*	4	4.63	9.63	35.25
19	Patin	*Pangasius sp.*	4	6.73	19.63	147.58
20	Baung	*Hemibagrus sp.*	9	4.28	19.39	125.30
			Maximum	6.85	26.12	268.99
			Depth (m)		6.1 ± 0.7	
			Velocity (m/s)		0.11 ± 0.010	
			Discharge, Q (m^3s)		9.89 ± 0.41	

Table 5 The distribution of fish collected during the dry season in the Petuang River

Bil	Species name (Origin name)	Species name (Scientific name)	The number collected	Average of width (cm)	Average of length (cm)	Average of weight (g)
1	Toman	*Channa micropeltes*	0	0.00	0.00	0.00
2	Kawan J	*Labiobarbus leptocheilus*	0	0.00	0.00	0.00
3	Sia	*Mystacoleucus obtusirostris*	0	0.00	0.00	0.00
4	Patung Banjarmasin	*Pristolepis fasciata*	0	0.00	0.00	0.00
5	Selat	*Notopterus notopterus*	0	0.00	0.00	0.00
6	Lampam	*Barbonymus schwanefeldii*	0	0.00	0.00	0.00
7	Temperas	*Cyclocheilichthys apogon*	0	0.00	0.00	0.00
8	Rong Toboali	*Osteochillus waandersii*	0	0.00	0.00	0.00
9	Rong	*Osteochilus vittatus*	0	0.00	0.00	0.00
10	Sebarau	*Hampala macrolepidata*	3	4.83	13.40	55.37
11	Tengas Pahang	*Neolissochilus soroides*	0	0.00	0.00	0.00
12	Baung besar Kalimantan	*Hemibagrus capitulum*	0	0.00	0.00	0.00
13	Belida	*Chitala lopis*	6	3.52	14.83	31.48
14	Rohu	*Labeo rohita*	0	0.00	0.00	0.00
15	Tengala Banjarmasin	*Puntioplites bulu*	0	0.00	0.00	0.00
16	Jelawat sirip merah	*Leptobarbus rubripinna*	0	0.00	0.00	0.00
17	Temoleh Laos	*Probarbus julleni*	0	0.00	0.00	0.00
18	Daun Chantaburi	*Poropuntius normani*	0	0.00	0.00	0.00
19	Patin	*Pangasius sp.*	0	0.00	0.00	0.00
20	Baung	*Hemibagrus sp.*	0	0.00	0.00	0.00
			Maximum	4.83	14.83	55.37
			Depth (m)		2.3 ± 0.9	
			Velocity (m/s)		0.200 ± 0.010	
			Discharge, Q (m^3s)		6.25 ± 4.9214	

Table 6 The distribution of fish collected during normal season in the Petuang River

Bil	Species name (Origin name)	Species name (Scientific name)	The number collected	Average of width (cm)	Average of length (cm)	Average of weight (g)
1	Toman	*Channa micropeltes*	0	0.00	0.00	0.00
2	Kawan	*Labiobarbus leptocheilus*	0	0.00	0.00	0.00
3	Sia	*Mystacoleucus obtusirostris*	0	0.00	0.00	0.00
4	Patung Banjarmasin	*Pristolepis fasciata*	2	4.05	6	26.45
5	Selat	*Notopterus notopterus*	0	0.00	0.00	0.00
6	Lampam	*Barbonymus schwanefeldii*	0	0.00	0.00	0.00
7	Temperas	*Cyclocheilichthys apogon*	0	0.00	0.00	0.00
8	Rong Toboali	*Osteochillus waandersii*	0	0.00	0.00	0.00
9	Rong	*Osteochilus vittatus*	0	0.00	0.00	0.00
10	Sebarau	*Hampala macrolepidata*	7	4.64	11.87	38.90
11	Tengas Pahang	*Neolissochilus soroides*	0	0.00	0.00	0.00
12	Baung besar Kalimantan	*Hemibagrus capitulum*	0	0.00	0.00	0.00
13	Belida	*Chitala lopis*	20	4.48	13.54	155.23
14	Rohu	*Labeo rohita*	0	0.00	0.00	0.00
15	Tengala Banjarmasin	*Puntioplites bulu*	0	0.00	0.00	0.00
16	Jelawat sirip merah	*Leptobarbus rubripinna*	0	0.00	0.00	0.00
17	Temoleh Laos	*Probarbus julleni*	0	0.00	0.00	0.00
18	Daun Chantaburi	*Poropuntius normani*	0	0.00	0.00	0.00
19	Patin	*Pangasius sp.*	0	0.00	0.00	0.00
20	Baung	*Hemibagrus sp.*	0	0.00	0.00	0.00
			Maximum	4.64	13.54	155.23
			Depth (m)		2.1 ± 0.1	
			Velocity (m/s)		0.48 ± 0.081	
			Discharge, Q (m^3s)		16.4242 ± 10.4007	

Table 7 The distribution of fish collected during the wet season in Petuang River

Bil	Species name (Origin name)	Species name (Scientific name)	The number collected	Average of width (cm)	Average of length (cm)	Average of weight (g)
1	Toman	*Channa micropeltes*	0	0.00	0.00	0.00
2	Kawan	*Labiobarbus leptocheilus*	0	0.00	0.00	0.00
3	Sia	*Mystacoleucus obtusirostris*	0	0.00	0.00	0.00
4	Patung Banjarmasin	*Pristolepis fasciata*	2	4.15	6.5	26.45
5	Selat	*Notopterus notopterus*	0	0.00	0.00	0.00
6	Lampam Palembang	*Barbonymus schwanefeldii*	0	0.00	0.00	0.00
7	Temperas Indonesia	*Cyclocheilichthys apogon*	0	0.00	0.00	0.00
8	Rong Toboali	*Osteochillus waandersii*	0	0.00	0.00	0.00
9	Rong	*Osteochilus vittatus*	0	0.00	0.00	0.00
10	Sebarau	*Hampala macrolepidata*	8	4.41	14.15	85.35
11	Tengas Pahang	*Neolissochilus soroides*	0	0.00	0.00	0.00
12	Baung besar Kalimantan	*Hemibagrus capitulum*	0	0.00	0.00	0.00
13	Belida	*Chitala lopis*	38	4.21	15.3	42.65
14	Rohu	*Labeo rohita*	0	0.00	0.00	0.00
15	Tengala Banjarmasin	*Puntioplites bulu*	0	0.00	0.00	0.00
16	Jelawat sirip merah	*Leptobarbus rubripinna*	0	0.00	0.00	0.00
17	Temoleh Laos	*Probarbus julleni*	0	0.00	0.00	0.00
18	Daun Chantaburi	*Poropuntius normani*	0	0.00	0.00	0.00
19	Patin	*Pangasius sp.*	0	0.00	0.00	0.00
20	Baung	*Hemibagrus sp.*	0	0.00	0.00	0.00
			Maximum	4.26	11.97	51.48
			Depth (m)		2.4 ± 0.8	
			Velocity (m/s)		0.337 ± 0.081	
			Discharge, Q (m^3s)		9.89 ± 0.408	

Fig. 4 Native species such as *Channa micropeltes* (Toman), *Hampala macrolepidata* (Sebarau), and *Chitala lopis* (Belida)

dry season and Belida Jakarta recorded a b value of 2.3713 at Terengganu River during the normal season. This is likely due to the existence of a slight pressure from environmental factors such as physical factors, and predation is the reason for the decrease in the equilibrium of these two parameters. All three native species from the Terengganu River area during the wet season show the fish grew well (the b value was between 2.7877 and 3.87), thus suggesting that the existing aquatic environmental condition was good. The size (length and weight) of the fish caught varies from one species to the other and also between the areas. From this study, it was found that *Channa micropeltes* are among the larger native species (Average length: 27.16 cm and average weight: 284.513 g during wet season; Average length: 27.12 cm and average weight: 280.97 g during normal season). According to Panchan (2010) and Tongnunui et al. (2016), *Channa micropeltes* species can grow to as large as 2 m in total length and 20 kg in body weight. In principle, if *Channa micropeltes* species could do better, it is believed that other species that share other habitats could do the same. For most local species, spawning may have occurred more than once a year. If any changes occurred to water level, or flow discharge, it may interrupt their spawning stages or ruin their spawning process. It is also worth noticing that fish recorded in the study area are all native to the region. Thus, they may be using their habitat for one or more spawning, rearing, feeding, and local migration. Stream flow plays a central role in stream ecology. It controls several important stream characteristics such as depth, width, current velocity, and substrate composition. Flow variability has also been found to have profound effects on aquatic fauna, such as in fish abundance and diversity.

The condition factor (K) is a quantitative parameter to determine the best condition of the fish and show the level of balance of food resources. This factor varies according to the influence of physiological factors, fluctuating according to different stages of development. A good fish growth rate is shown in the increase in length and weight, especially for population analysis at a particular time (Hamid et al. 2015). The relationship between length and weight shows the condition factor (condition factor, c.f.) of the fish species applied to see the actual growth rate of the species in a river basin or lake. Furthermore, the condition factor is measured by the fish species' good condition based on the fish sample's weight measurement. Based on the sampling results recorded in this study for the three seasons, the selection of fish species as biological indicators is based on the strength of the highest length and weight relationship, which is the best condition of fish species that can inhabit the habitats of the two rivers studied. The average value of K for the dry season in both rivers of the study recorded a slightly lower value compared to the normal season and the wet season, which shows the spawning and recovery conditions for the fish. The calculation of condition factors in this study describes the health condition of the fish population in general in a certain habitat. The results of these findings can be used in estimating the status of fishery resources by fish biologists and fisheries resource managers in Malaysia (Hashim et al. 2017; Chai et al. 2021).

5 Environmental Health Impacts and Mitigation Strategies

An estimated 4600 billion cubic meters of water are utilized annually worldwide, with a 1% annual growth rate. Agriculture, energy generation, industry utilization, and human beings all anticipate needing a lot of water. Globally, more than 3856 km^3 of freshwater were removed for domestic, commercial, and agricultural uses (FAO 2016; Flörke et al. 2013). Agriculture (70%) and industry (19%) were the two sectors that used most of the water that was extracted, while home usage made up 11% of total use (Doungmanee 2016). Asia uses an estimated 72% more water for agriculture than any other continent (FAO 2021). Due to the strong demand for livestock products, freshwater consumption in agriculture is already high and is predicted to rise in the upcoming years (Udimal et al. 2017). Figure 8 shows the percentage and level of fresh water usage in Malaysia based on sectors based on the data provided by AQUASTAT from 2010 until 2018 (FAO 2021). Only approximately 61.6% of the wastewater produced by the agriculture sector, which includes aquaculture, was effectively treated. Due to the ecology of fish farming, aquaculture, a subsector of the fisheries industry, is one of those sectors that uses a lot of water. With the rising pace of the global population, which is anticipated to reach 8.3–10.9 billion people

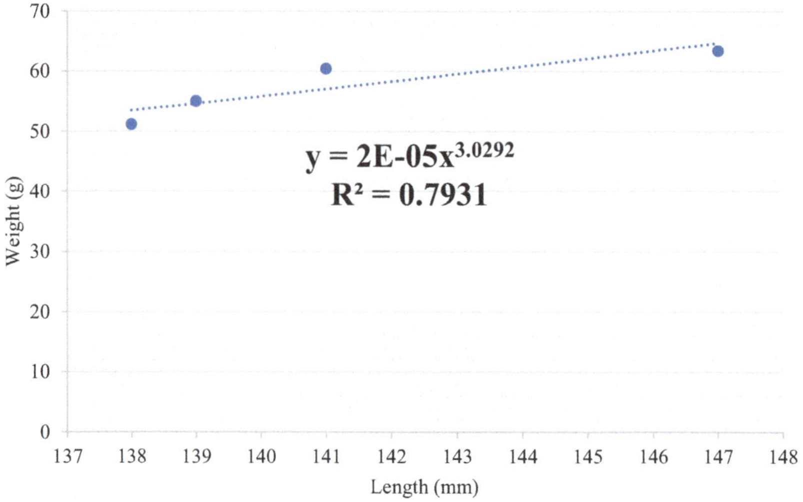

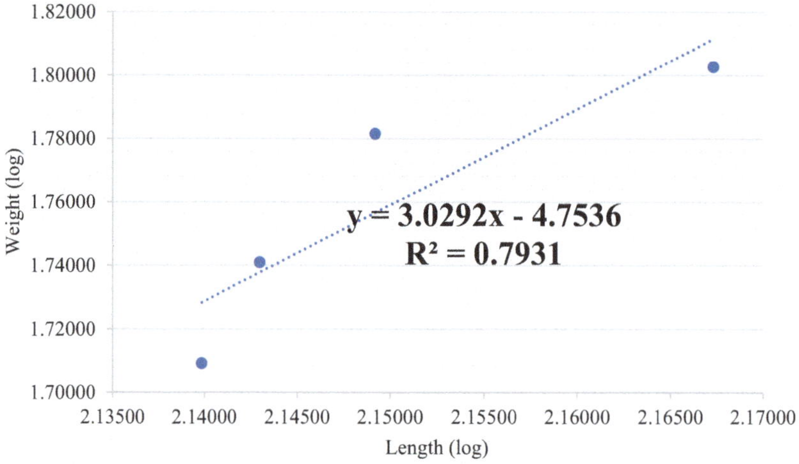

Model	Length-Weight Relationship of Sebarau (*Hampala macrolepidata*)
R^2	0.7931
Coefficient a	1.7636×10^{-5}
Coefficient b	3.0292 (> 3)
Condition Factor	2.056

(a)

Fig. 5 **a** Length–weight relationship of Sebarau (*Hampala macrolepidata*) (*analysis of well-being*) in Terengganu River during the dry season, **b** Length–weight relationship of Belida (*Chitala lopis*) (*analysis of well-being*) in Terengganu River during the dry season, **c** Length–weight relationship of Sebarau (*Hampala macrolepidata*) (*analysis of well-being*), in Petuang River during the dry season, **d** Length–weight relationship of Belida (*Chitala lopis*) in Petuang River during the dry season

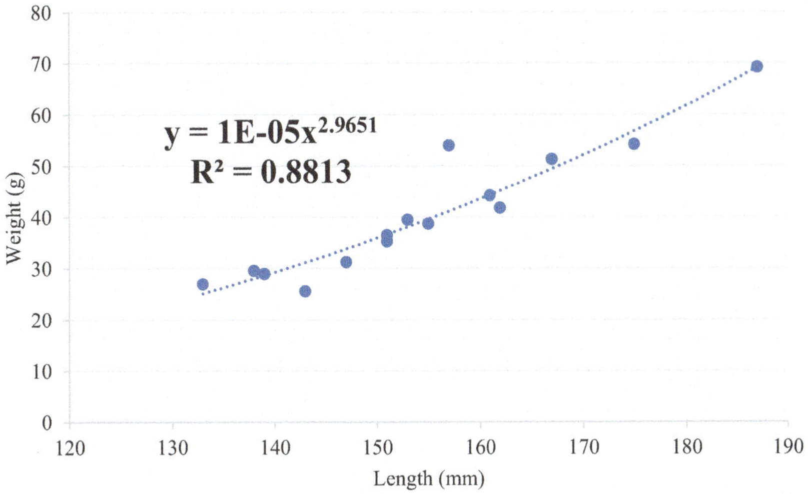

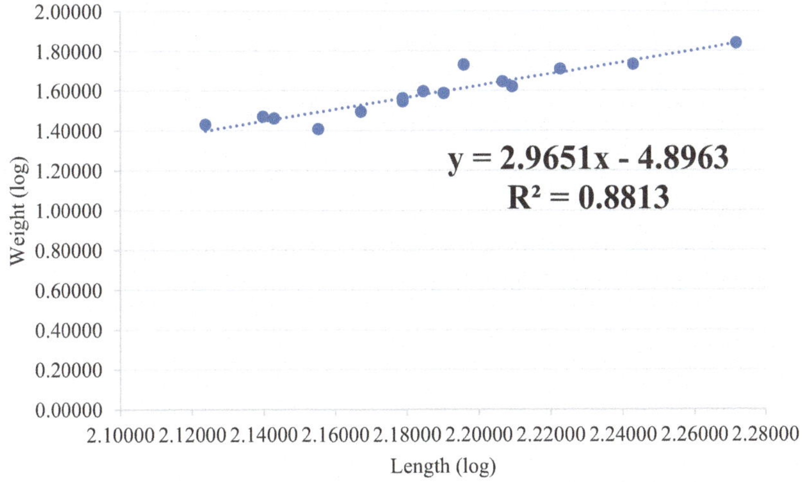

Model	Length-Weight Relationship of Belida (*Chitala lopis*)
R^2	0.8813
Coefficient a	1.2697 x 10^{-5}
Coefficient b	2.965 (≈ 3)
Condition Factor	1.095

(b)

Fig. 5 (continued)

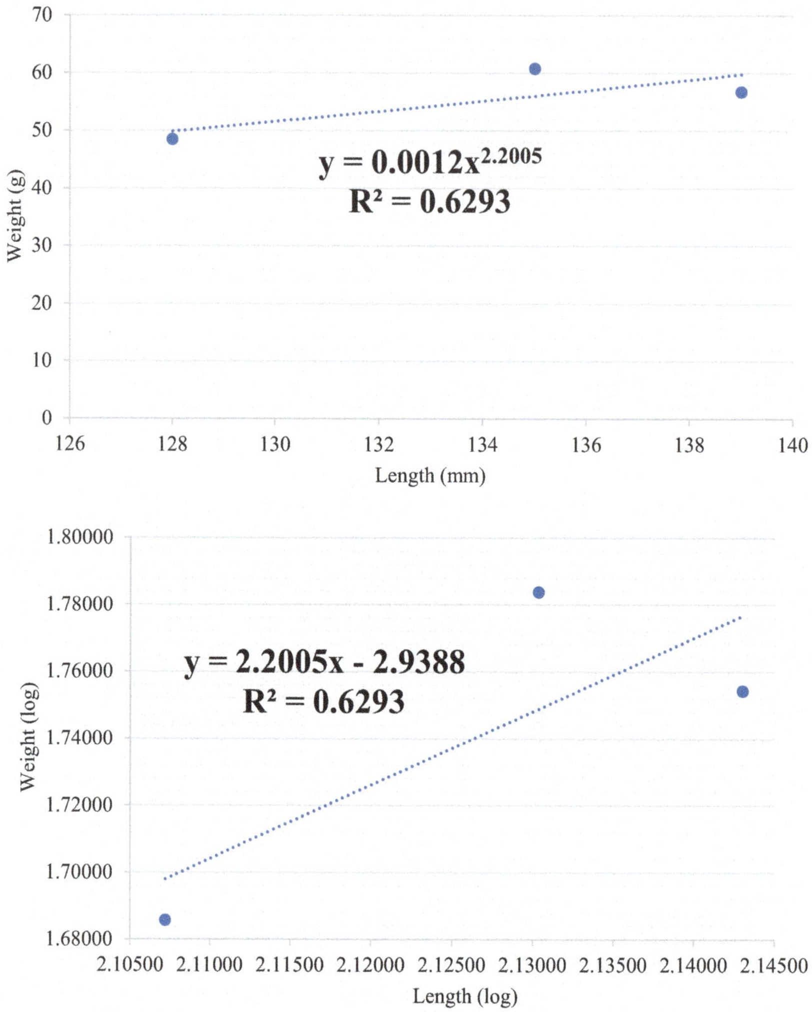

Model	Length-Weight Relationship of Sebarau (*Hampala macrolepidata*)
R^2	0.6293
Coefficient a	1.1513 x 10^{-3}
Coefficient b	2.2005 (< 3)
Condition Factor	2.301

(c)

Fig. 5 (continued)

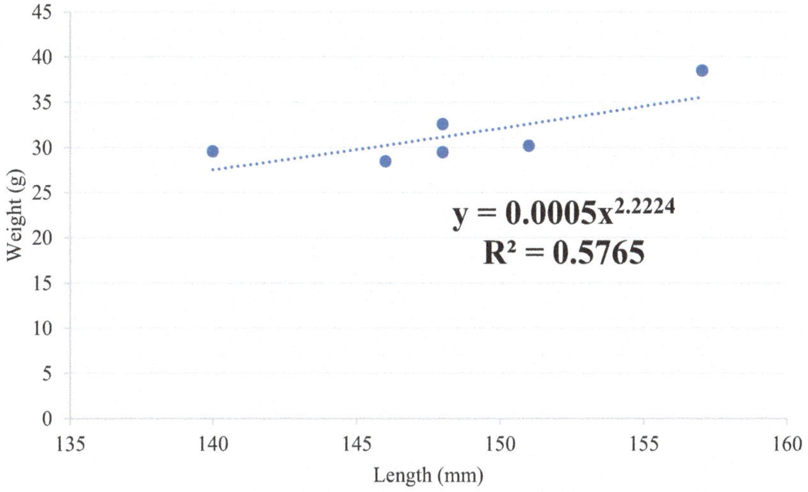

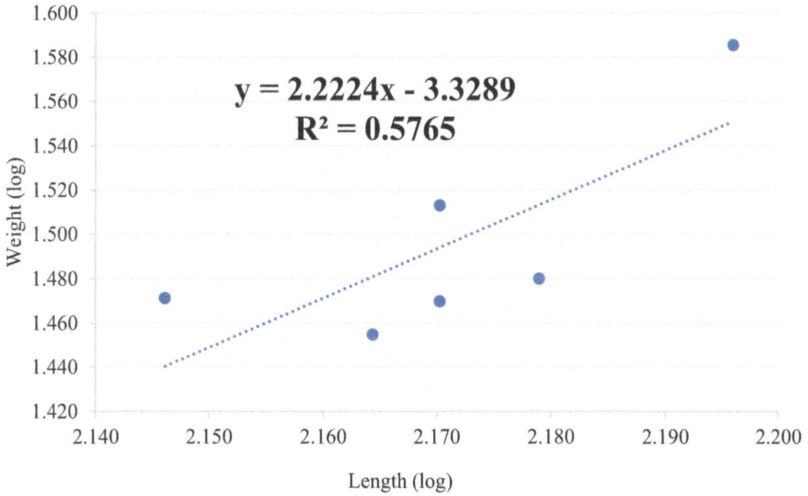

Model	Length-Weight Relationship of Belida (*Chitala lopis*)
R^2	0.5765
Coefficient a	4.6892×10^{-4}
Coefficient b	2.2224 (< 3)
Condition Factor	0.965

(d)

Fig. 5 (continued)

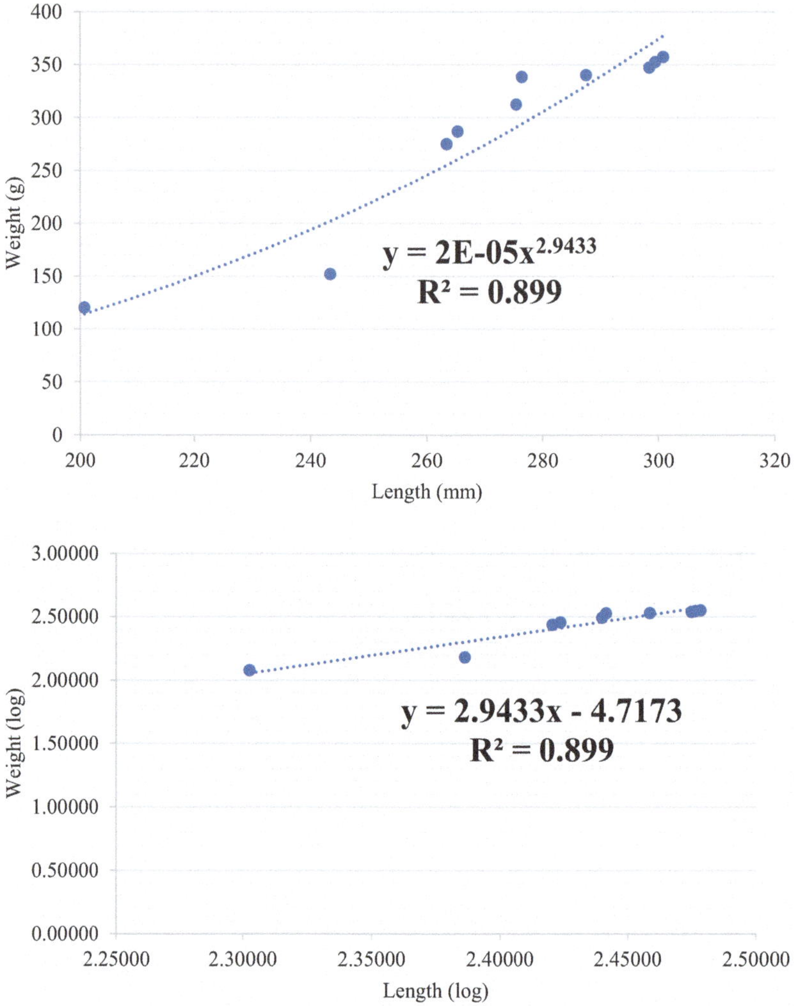

Model	Length-Weight Relationship of Toman (*Channa macropeltes*)
R^2	0.899
Coefficient a	1.9173×10^{-5}
Coefficient b	2.9433 (≈ 3)
Condition Factor	1.409

(a)

Fig. 6 **a** Length–weight relationship of Toman (*Channa macropeltes*) in Terengganu River during normal season, **b** Length–weight relationship of Sebarau (*Hampala macrolepidata*) in Terengganu River during normal season, **c** Length–weight relationship of Belida (*Chitala lopis*) in Terengganu River during normal season, **d** Length–weight relationship of Sebarau (*Hampala macrolepidata*) in Petuang River during normal season, **e** Length–weight relationship of Belida (*Chitala lopis*) in Petuang River during normal season

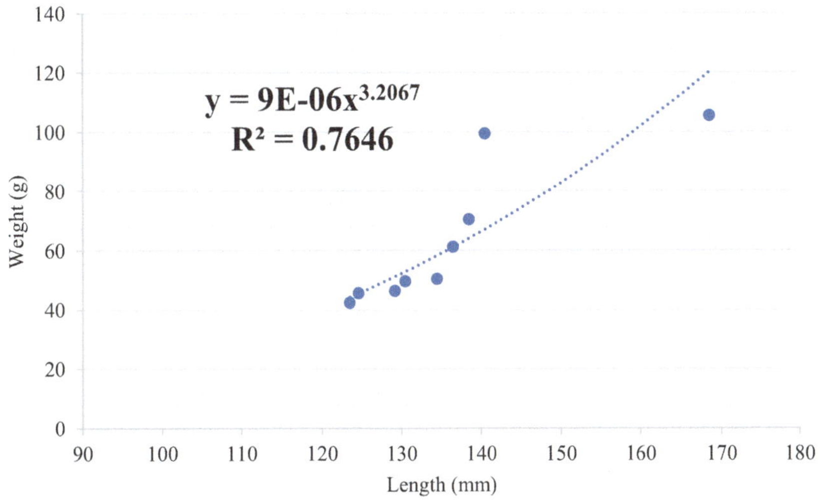

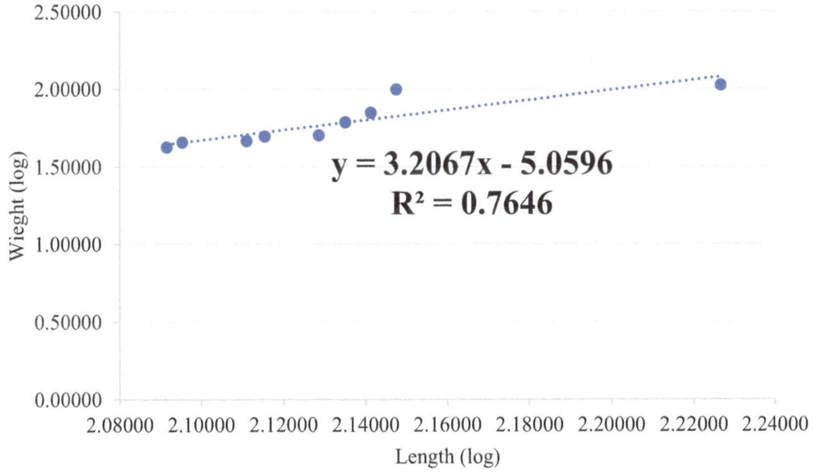

Model	**Length-Weight Relationship of Sebarau (*Hampala macrolepidata*)**
R^2	0.7646
Coefficient a	8.7177×10^{-6}
Coefficient b	3.2067 (> 3)
Condition Factor	2.509

(b)

Fig. 6 (continued)

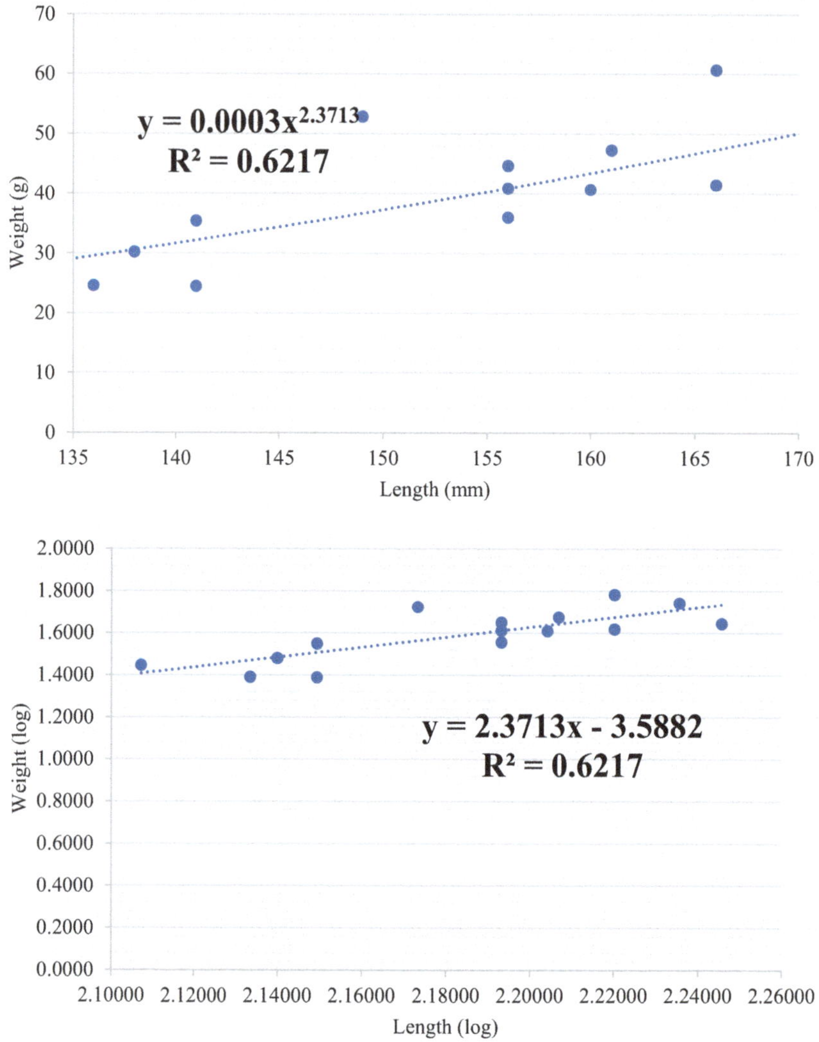

Model	Length-Weight Relationship of Belida (*Chitala lopis*)
R^2	0.6217
Coefficient a	2.5811×10^{-54}
Coefficient b	2.3713 (<3)
Condition Factor	1.118

(c)

Fig. 6 (continued)

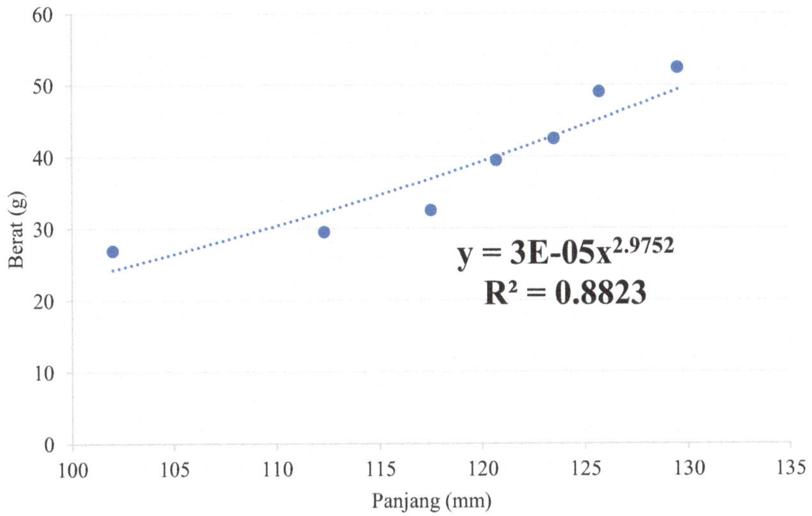

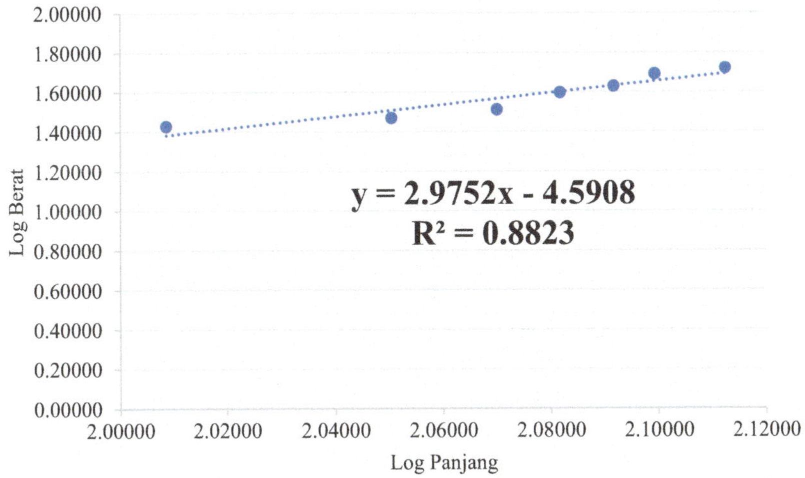

Model	Length-Weight Relationship of Sebarau (*Hampala macrolepidata*)
R^2	0.8823
Coefficient a	2.5657×10^{-5}
Coefficient b	2.9752 (≈ 3)
Condition Factor	2.326

(d)

Fig. 6 (continued)

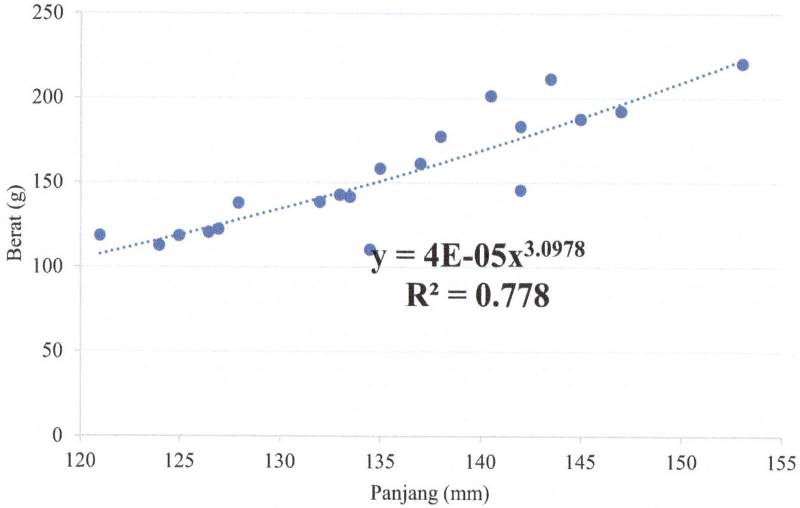

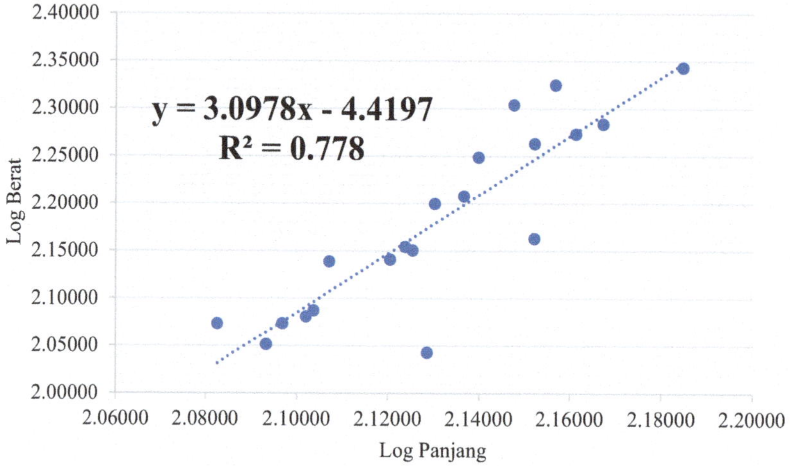

Model	Length-Weight Relationship of Belida (*Chitala lopis*)
R^2	0.778
Coefficient a	3.8203×10^{-5}
Coefficient b	3.0978 (> 3)
Condition Factor	6.253

(e)

Fig. 6 (continued)

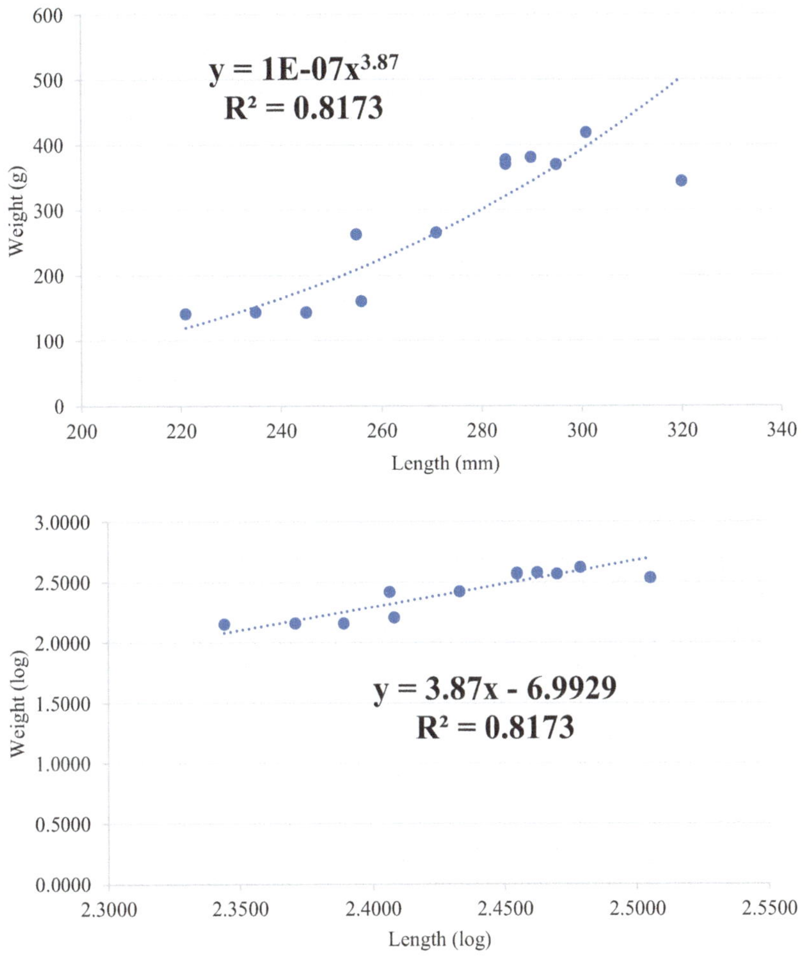

Model	**Length-Weight Relationship of Toman (*Channa macropeltes*)**
R^2	0.8173
Coefficient a	1.0165×10^{-7}
Coefficient b	3.87 (> 3)
Condition Factor	1.22

(a)

Fig. 7 **a** Length–weight relationship of Toman (*Channa macropeltes*) in Terengganu River during the wet season, **b** Length–weight relationship of Sebarau (*Hampala macrolepidata*) in Terengganu River during the wet season, **c** Length–weight relationship of Belida (*Chitala lopis*) in Terengganu River during the wet season, **d** Length–weight relationship of Sebarau (*Hampala macrolepidata*) in Petuang River during the wet season, **e** Length–weight relationship of Belida (*Chitala lopis*) in Petuang River during the wet season

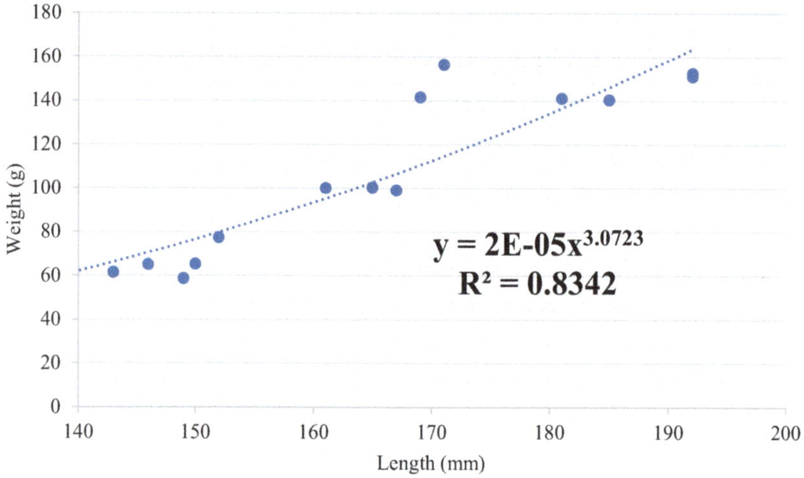

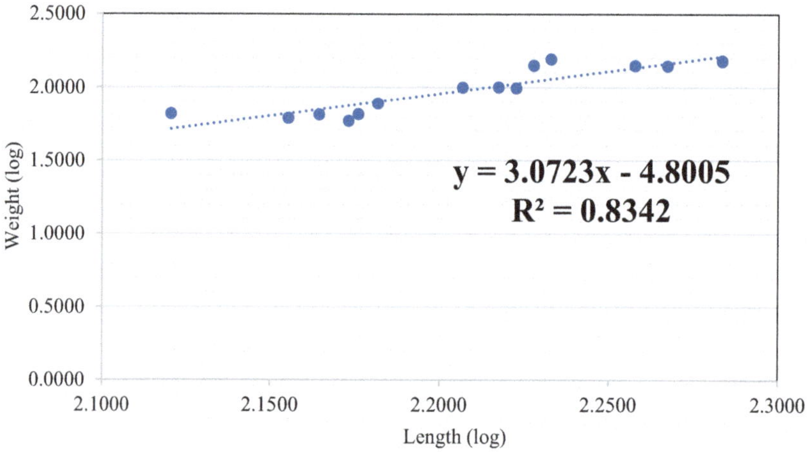

Model	**Length-Weight Relationship of Sebarau (*Hampala macrolepidata*)**
R^2	0.8342
Coefficient a	1.5831×10^{-5}
Coefficient b	3.0723 (> 3)
Condition Factor	2.49

(b)

Fig. 7 (continued)

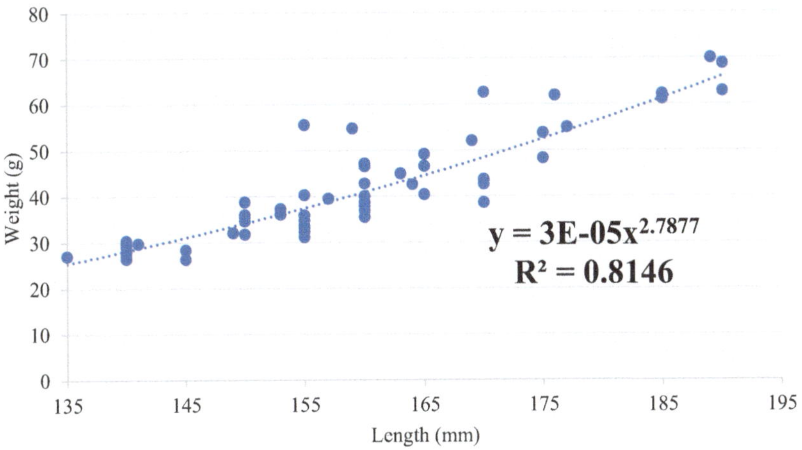

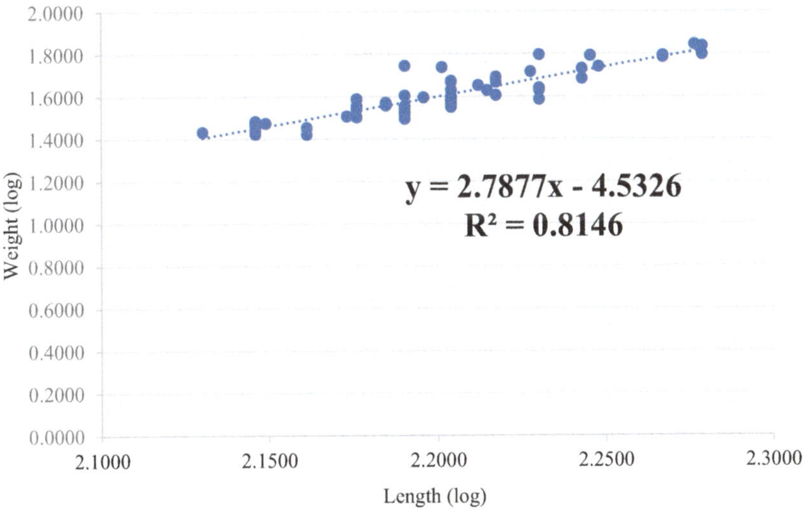

Model	Length-Weight Relationship of Belida (*Chitala lopis*)
R^2	0.8146
Coefficient a	2.934×10^{-5}
Coefficient b	2.7877 (≈3)
Condition Factor	1.03

(c)

Fig. 7 (continued)

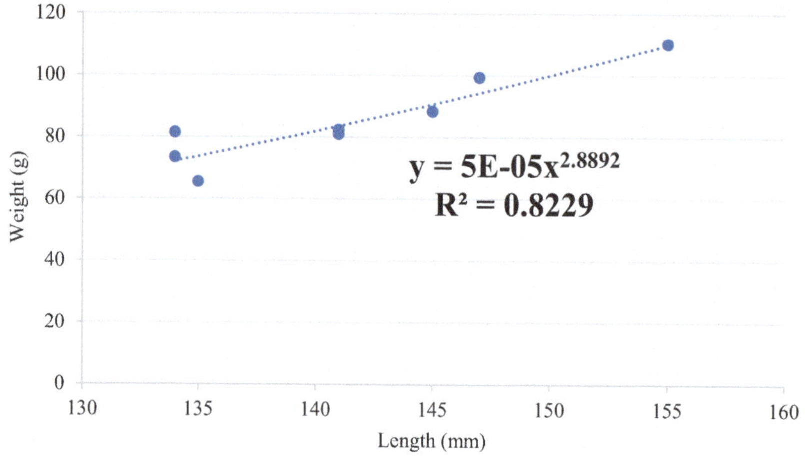

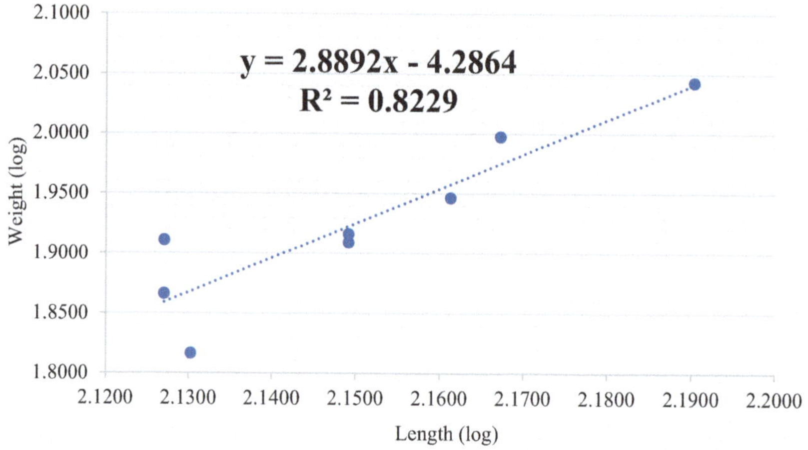

Model	**Length-Weight Relationship of (*Hampala macrolepidata*)**
R^2	0.8229
Coefficient a	5.1713×10^{-5}
Coefficient b	2.8892 (≈ 3)
Condition Factor	3.01

(d)

Fig. 7 (continued)

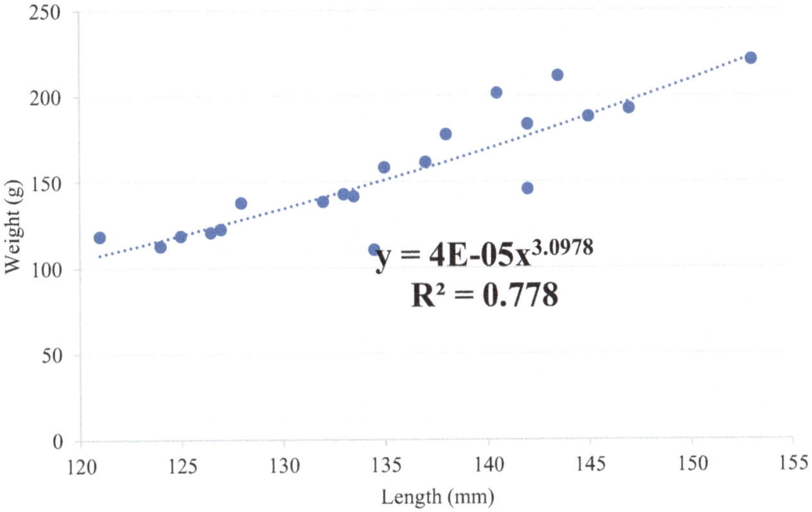

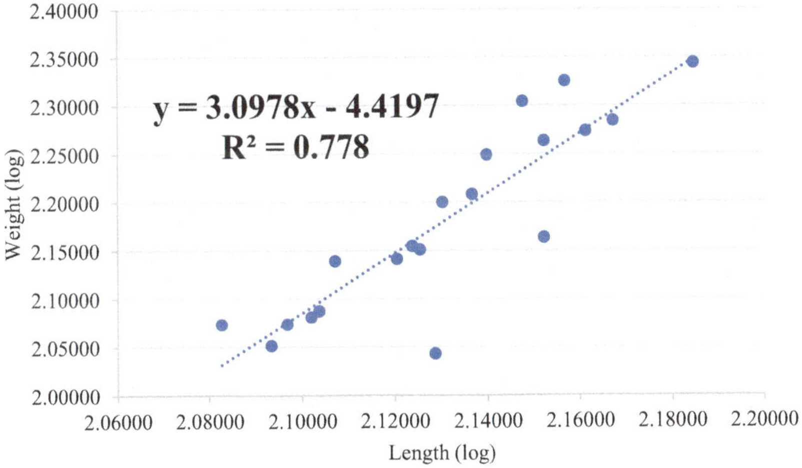

Model	Length-Weight Relationship of Belida (*Chitala lopis*)
R^2	0.778
Coefficient a	3.8203×10^{-5}
Coefficient b	3.0978 (> 3)
Condition Factor	1.19

(e)

Fig. 7 (continued)

by the year 2050 (Washington 2015), aquaculture has been expanding quickly. The demand for protein consumption will rise along with advancements in the population forecast. According to estimates, human fish consumption increased on average by 3.1% yearly between 1961 and 2017 (FAO 2020), which is more than twice as fast as the annual global population growth rate of 1.6% during the same period. To meet the subsequently rising demand caused by these factors, the aquaculture industry will be industrialized. This industry uses a lot of water and is similar to conventional agriculture in that it does so. According to Dauda et al. (2019); Goddard and Delghandi (2020), common ponds, flow-through raceways, and recirculation are among the fish production systems used. A 46.0% increase from the overall production of aquaculture in 2000 and 51.3 million tons (62.5%) from in-land aquaculture contributed to the 82.1 million tons of aquaculture production in 2018. Therefore, in-land freshwater aquacultures, which use about 16,316.3 m^3 of water per kilogram of production (Hussan et al. 2019), are more highly demanded water-intensive efforts.

Figure 9 illustrates that, up to three decades after aquaculture activities started, marine capture fish dominated the entire production. Beginning in the 2000s, terrestrial cultivation expanded quickly, surpassing marine catch fish production in 2010. While the national output eventually became independent of the captured sea fish aquaculture, in-land systems for aquaculture are still expanding quickly at the moment. According to estimates, the majority of aquaculture production (45.9% and 46.9% of the overall output, respectively) was based on the in-land cultivation of freshwater and brackish water commodities. Based on the growth of the in-land culture system in 2010, the produced wastewater also showed a considerable increase. With the expansion of the in-land culture system, it is predicted that the amount of

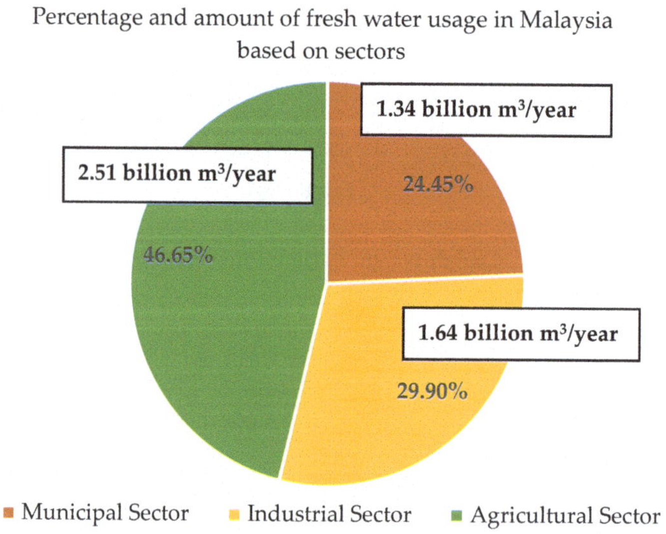

Fig. 8 The percentage and amount level of fresh water usage in Malaysia based on sectors. *Source* FAO (2013)

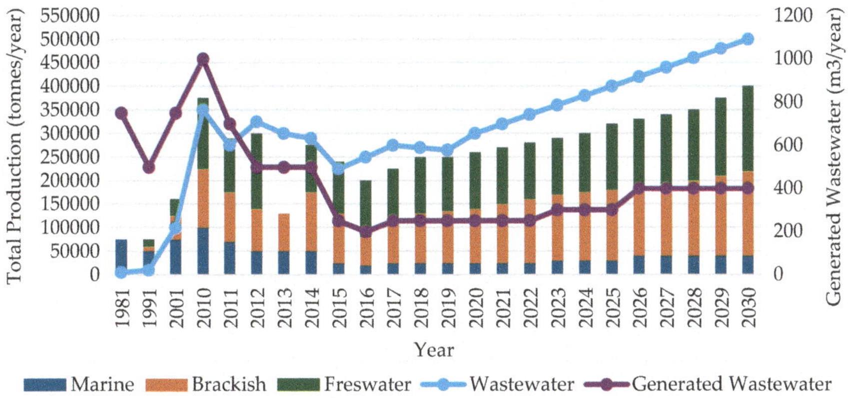

Fig. 9 The distribution of total fish production in Malaysia and estimated wastewater generated from 1981 until 2030. *Source* Department of Fisheries Malaysia (2020) with the latest available data for 2019, and prediction until 2030 and adapted from Kurniawan et al. (2021)

wastewater produced by the aquaculture industry will rise. The aquaculture industry will produce wastewater in the amount of 7.519 km^3 by 2030.

Table 8 summarizes the characteristics of aquaculture wastewater from several countries worldwide for comparison with various fish cultures. According to the criteria of chemical oxygen demand (COD), nitrogen, and phosphorus contents, Table 9 shows that aquaculture wastewater was frequently reported to have a neutral pH level and had large amounts of organic matter. Wastewater effluent pollution is a global problem, but tropical and subtropical regions, such as Malaysia, require additional attention since they contain habitats for delicate coral reefs and communities that are already at risk from ocean acidification and global warming (Hozumi et al. 2018). The features of wastewater effluent vary based on the species that are grown and the methods that are used, but the majority of these traits require further attention due to their organic and nutritional contents (Huang et al. 2018; John et al. 2020; Zhang et al. 2014). The concentration of organic compounds in the ecosystem could be considerably impacted by these dissolved organic materials, which are mostly composed of polysaccharides, lipids, nutrition, proteins, and humic substances (Huang et al. 2018). Antibiotics, which are regarded as Contaminants of Emerging Concern (CEC), are also present in aquaculture effluent, as evidenced by signs of their use in fish tanks to treat or prevent infections and foster the growth of cultured species (Huang et al. 2019). There is currently little analysis of CECs in aquaculture in Malaysia. Despite its tiny percentage, the survival of CEC in relation to bacterial development and the expansion of the antibiotic resistance gene needs additional consideration (Huang et al. 2019). In order to effectively produce fish based on sexual dimorphism, hormones, primarily estrogens, and androgens, are also used. This adds yet another chemical to the list of CECs in aquaculture effluents. Hormones can influence endocrine systems in addition to contaminating the environment (Hoga et al. 2018) (Table 5).

Table 8 Summarizes the characteristics of aquaculture wastewater from several countries in the world for comparison with various fish cultures

pH	Salinity	COD (mg/L)	Total phosphorus (mg/L)	Total nitrogen (mg/L)	Ammoniacal nitrogen (mg/L)	Nitrite nitrogen (mg/L)	Nitrate nitrogen (mg/L)	Dissolved oxygen (mg/L)	Culture	Country	References
7.78	2.80%	–	0.42	6.81	4.24	0.13	2.00	–	*Penaeus vannamei Boone* fish	China	Gao et al. (2016)
–	–	66.6	0.23	3.60	2.35	0.13	0.51	2.45	Yellow catfish	China	Huang et al. (2019)
7.8	–	–	–	–	1.07	0.28	0.40	6.7	Mixed-culture fish and prawn	India	John et al. (2020)
7.4	29.9 PSU	<3	<0.1	1	0.8	–	–	–	Olive flounder fish	Korea	Choi et al. (2020)
6.72	–	205	32.5	–	9.45	25	30.17	3.83	Mixed-culture fish	Malaysia	Akinbile and Yusoff (2012)
7.9		155	1.4	8.5	1.8	0.2	0.18	3	*Penaeus monodon* shrimp	Malaysia	Nyanti et al. (2010)
7.78	–	125	2.21	–	24.75	<0.01	<0.3	–	Catfish	Malaysia	Ng and Chan (2021)
7.9	–	758	0.91	0.83	–	–	0.83	9.6	–	Nigeria	Igwegbe and Onukwuli (2019)
6.68	–	108.70	2.50	–	0.08	–	8.50	4.17	–	Nigeria	Omotade et al. (2019)
6.97	–	253	~2.5	18.5	–	–	18.1	–	Tilapia fish	Norway	Halfhide et al. (2014)
7.67	44%	–	0.07	1.09	0.63	0.17	0.38	–	Mixed-culture fish	Spain	Villar-Navarro et al. (2019)

Table 9 Contaminants of Emerging Concerns (CECs) compounds existing in aquaculture effluents as a reference to further study the existence of CECs in Malaysian aquaculture wastewater

Type of aquaculture wastewater	Species	Summary	References
Fish culture	*Paralichthys olivaceus* (Olive flounder)	i. Out of 177 wastewater samples from 16 facilities, the percentage of different antibiotics found ii. 52.5% contains oxytetracycline (OTC) iii. 20.3% contains oxalinic acid (OXO) iv. 19.2% contains florfenicol (FLO) v. 12.4% contains amoxicillin (AMX)	Choi et al. (2020)
Fish culture	Yellow catfish	i. Concentration of two common antibiotics in aquaculture: Enrofloxacin (ENR) 25.7–67.4 ng/L and sulfamethoxazole (SMZ)—64.0–211.2 ng/L ii. A large range of antibiotic concentrations related to the temporary variation of antibiotic dosage iii. ENR is a fluoroquinolone, which has a low biodegradation rate and high photolysis iv. SMZ is a sulfonamide, which has low sorption to solid substrate and high mobility v. Respective effluent also contains different types of antibiotic resistance genes (RG): Quinolone RG: qepA, qnrA, qnrB, and qnrD; Sulfonamide RG: sul1 and sul; Tetracycline RG: tet(34), tetB, tetL, tetM, and tetV, where quinolone RG presence is the largest percentage	Huang et al. (2019)

(continued)

Table 9 (continued)

Type of aquaculture wastewater	Species	Summary	References
Fish culture	*Paralichthys olivaceus* (Olive flounder)	i. A total of 22 antibiotic RGs found in aquaculture effluents (sampled from 12 sites) in South Jeolla province and Jeju Island, South Korea ii. tetracycline resistance: *tetA, tetB*, tetD*, tetE, tetG, tetH, tetM, tetQ, tetX, tetZ, tetBP* iii. sulfonamide resistance: *sul1, sul2* iv. quinolone resistance: *qnrD, qnrS, aac(6′)-Ib-cr* v. *β*-lactams resistance: bla_{TEM}, bla_{CTX}, bla_{SHV} vi. macrolide resistance: *ermC* vii. florfenicol resistance: *floR* viii. multidrug resistance: *oqxA* ix. class 1 integrons–integrase gene: *intI1* ranged from 4.24×10^3 to 1.46×10^2 copies/16S rRNA gene with two major communities found out of all antibiotic RG (74.8%–98.0%)	Jang et al. (2018)
Not specified	–	A total of eight residues of antibiotics are found in aquaculture wastewater: amoxicillin (AMOX), ampicillin (AMPI), doxycycline (DXC), oxytetracycline (OTC), lincomycin (LCM), vancomycin (VCM), sulfamethazine (SMT), and sulfamethoxazole (SMZ)	Do et al. (2019)
Not specified	–	i. Only 20% to 30% of tetracycline antibiotics are used by the fishes, while approximately 75% is released to the environment ii. Leads to selective pressure toward bacteria and alters microbial community, eutrophication, and health hazards toward humans and animals	

The best way to describe environmental ecological evaluation and management is as an interconnected cycle of different feedbacks, particularly with regard to adaptive management or an ecosystem-based framework. Designing, implementing, and evaluating biological assessments, creating models (such as habitats and environmental model management), and assessing numerous environmental ecology conservation and management methods all depend on spatial ecology. Information is arranged according to a topical framework corresponding to the assessment and management cycle. Figure 10 illustrates a conceptual diagram of the fisheries' mitigation strategies cycle based on aspects of spatial ecology. Collectively, biological evaluation, when supplemented by an understanding of spatial ecology, can give managers the ability to comprehend how and when fish and their habitats may be vulnerable to various hazards. Naturally, having this knowledge makes it easier to assess or create measures to safeguard the long-term viability of fishing production (Koushlesh et al. 2018). Failure to comprehend the spatial ecology of fish and to include spatiotemporal data can bias abundance assessments and projections and perhaps result in inefficient or counterproductive management efforts. Integrated Water Resources Management (IWRM) is a process that promotes the coordinated development and management of water, land, and related resources, in order to maximize the economic and social welfare in an equitable manner without compromising the sustainability of vital ecosystems. IRBM is an integral part of IWRM, where river basins constitute the management unit. Each river basin or sub-basin is unique and may have a different focus, with IWRM providing the generic and common policy. IWRM involves equity, ecological integrity or environmental sustainability, and economic efficiency. In order to implement IWRM, it will involve three elements that need to move concurrently. The three elements are enabling environment through policies and legislation, institutional framework with emphasis on functions of the various administrative levels and stakeholders, and management instruments including operational instruments for effective regulation, monitoring, and enforcement. Therefore, IWRM should address and accommodate the use of the river in a sustainable manner by reducing conflict and competing effects. The plan should aim to enhance river conditions by employing the best available techniques while maximizing environmental control strategies or best management practices for river basin management (Elfithri et al. 2012; Leong and Lai 2017).

6 Conclusion

This study provides the first basic information for three major native species collected from Kenyir Lake Basin (inlet river basin: Petuang River and outlet river basin: Terengganu River Basin). Mostly three species namely *Channa micropeltes* (Toman), *Hampala macrolepidata* (Sebarau), and *Chitala lopis* (Belida) exhibited a trend of positive growth at Terengganu River Basin compared to the Petuang River Basin during three seasons, respectively. These growth patterns denote that this lake basin could provide a favorable environment and suitable habitat for the growth and population dynamics along the lake basin. This study also has contributed some information and knowledge on sustainable ecological resources in Kenyir Lake Basin.

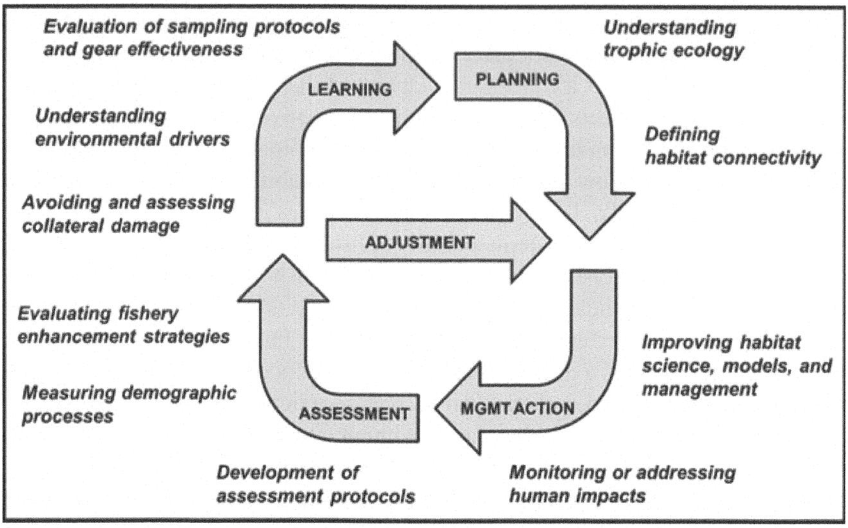

Fig. 10 The illustration of a conceptual diagram of the fisheries mitigation strategies cycle based on aspects of spatial ecology. *Source* Cooke et al. (2016)

References

Akinbile CO, Yusoff MS (2012) Assessing water hyacinth (Eichhornia crassopes) and lettuce (Pistia stratiotes) effectiveness in aquaculture wastewater treatment. Int J Phytoremediat 14:201–211

Awang MM, Alfitri A, Ahmad AR, Careemdeen JD, Ahmad J (2022) Socio-ecological support and physical facilities satisfaction: how they link to social participation and well-being among urban residents in Malaysia. Sustainability 14(3):1–18

Birk S, Chapman D, Carvalho L, Spears BM, Andersen HE, Argillier C, Hering D (2020) Impacts of multiple stressors on freshwater biota across spatial scales and ecosystems. Nat Ecol Evol 4(8):1060–1068

Bratman GN, Anderson CB, Berman MG, Cochran B, De Vries S, Flanders J, ... Daily GC (2019) Nature and mental health: an ecosystem service perspective. Sci Adv 5(7):eaax0903

Chai XL, Hashim R, Rahmat A, Nur M, Kamaludin MH, Habullah NA, Kamaruddin IS (2021) Length-weight relationship and relative condition factor of two dominant species (*Cyclocheilithys apogon* and *Notopterus notopterus*) at Subang Lake, Selangor, Malaysia. Pertanika J Sci Technol 29(3):1557–1566

Choi S, Sim W, Jang D, Yoon Y, Ryu J, Oh J, Woo JS, Kim YM, Lee Y (2020) Antibiotics in coastal aquaculture waters: occurrence and elimination efficiency in oxidative water treatment processes. J Hazard Mater 396

Comte L, Olden JD, Tedesco PA, Ruhi A, Giam X (2021) Climate and land-use changes interact to drive long-term reorganization of riverine fish communities globally. Proc Natl Acad Sci 118(27):e2011639118

Connor L, Matson R, Kelly FL (2017, January) Length-weight relationships for common freshwater fish species in Irish lakes and rivers. In: Biology and environment: proceedings of the Royal Irish Academy, vol 117, no 2. Royal Irish Academy, pp. 65–75

Cooke SJ, Martins EG, Struthers DP, Gutowsky LF, Power M, Doka SE, ... Krueger CC (2016) A moving target—incorporating knowledge of the spatial ecology of fish into the assessment and management of freshwater fish populations. Environ Monit Assess 188:1–18

Dauda AB, Ajadi A, Tola-Fabunmi AS, Akinwole AO (2019) Waste production in aquaculture: sources, components and managements in different culture systems. Aquac Fish 4:81–88

Department of Fisheries Malaysia (2020). Annual Report 2019

Do TCMV, Nguyen DQ, Nguyen KT, Le PH (2019) Tio2 and Au-TiO2 nanomaterials for rapid photocatalytic degradation of antibiotic residues in aquaculture wastewater. Materials (Basel) 12

Doungmanee P (2016) The nexus of agricultural water use and economic development level. Kasetsart J Soc Sci 37:38–45

Ehrlich O, Bi X, Borisova T, Larkin S (2017) A latent class analysis of public attitudes toward water resources with implications for recreational demand. Ecosyst Serv 28:124–132

Elfithri R, Beghum RA, Mokhtar M (2012) Instruments for integrated water resource management in Malaysia. J Appl Sci Res 8(12):5599–5607

FAO (2013) National aquaculture sector overview: Malaysia. FAO Ctry. Notes

FAO (2016) The state of world fisheries and aquaculture 2016. Contributing to Food Security and Nutrition for All. Food and Agriculture Organization of the United Nations, Rome

FAO (2020) The state of world fisheries and aquaculture 2020, The state of world fisheries and aquaculture 2020. https://doi.org/10.4060/ca9229en

FAO (2021). Aquastat [WWW Document]. Aquastat. http://www.fao.org/nr/water/aquastat/data/query/index.html?lang=en. Accessed 09 Apr 20

Farinordin FA, Nilam WSW, Husin SM, Samat A, Nor SM (2017) Scale morphologies of freshwater fishes at Tembat forest reserve, Terengganu, Malaysia. Sains Malaysiana 46(9):1429–1439

Flörke M, Kynast E, Bärlund I, Eisner S, Wimmer F, Alcamo J (2013) Domestic and industrial water uses of the past 60 years as a mirror of socio-economic development: a global simulation study. Glob Environ Chang 23:144–156

Gao F, Li C, Yang Z-HH, Zeng G-MM, Feng L-JJ, Liu JZ, Liu M, Cai HW (2016) Continuous microalgae cultivation in aquaculture wastewater by a membrane photobioreactor for biomass production and nutrients removal. Ecol Eng 92:55–61

Goddard S, Delghandi M (2020) Importance of the conservation and management of freshwater to aquaculture. In: Encyclopedia of the World's Biomes, pp 35–44

Halfhide T, Åkerstrøm A, Lekang OI, Gislerød HR, Ergas SJ (2014) Production of algal biomass, chlorophyll, starch and lipids using aquaculture wastewater under axenic and non-axenic conditions. Algal Res 6:152–159

Hamid M, Mansor MASHHOR, Nor SAM (2015) Length-weight relationship and condition factor of fish populations in Temengor Reservoir: indication of environmental health. Sains Malaysiana 44(1):61–66

Hashim M, Abidin DAZ, Das SK, Mazlan AG (2017) Length-weight relationship, condition factor and TROPH of Scatophagus argus in Malaysian coastal waters. Aquac Aquar Conserv Legis 10(2):297–307

Hoga CA, Almeida FL, Reyes FGR (2018) A review on the use of hormones in fish farming: analytical methods to determine their residues. CYTA—J Food 16:679–691

Hozumi A, Hong PY, Kaartvedt S, Røstad A, Jones BH (2018) Water quality, seasonality, and trajectory of an aquaculture-wastewater plume in the Red Sea. Aquac Environ Interact 10:61–77

Huang L, Li M, Ngo HH, Guo W, Xu W, Du B, Wei Q, Wei D (2018) Spectroscopic characteristics of dissolved organic matter from aquaculture wastewater and its interaction mechanism to chlorinated phenol compound. J Mol Liq 263:422–427

Huang XF, Ye GY, Yi, NK, Lu LJ, Zhang L, Yang LY, Xiao L, Liu J (2019) Effect of plant physiological characteristics on the removal of conventional and emerging pollutants from aquaculture wastewater by constructed wetlands. Ecol Eng 135:45–53

Hussan A, Adhikari S, Adhikari S, Das A, Hoque F, Pillai BR (2019) Fish culture without water!!! Does aquaculture contribute to water scarcity? Is aquaculture without water possible? Int J Fish Aquat Stud 7:39–48

Igwegbe CA, Onukwuli OD (2019) Removal of total dissolved solids (TDS) from aquaculture wastewater by coagulation-flocculation process using Sesamum indicum extract: effect of operating parameters and coagulation-flocculation kinetics. Pharmaceut Chem J 6(4):32–45

Jang HM, Kim YB, Choi S, Lee Y, Shin SG, Unno T, Kim YM (2018) Prevalence of antibiotic resistance genes from effluent of coastal aquaculture, South Korea. Environ Pollut 233:1049–1057

John EM, Krishnapriya K, Sankar TV (2020) Treatment of ammonia and nitrite in aquaculture wastewater by an assembled bacterial consortium. Aquaculture 526

Keat-Chuan NC, Aun-Chuan OP, Wong WL, Khoo G (2017) An overview of the status, trends and challenges of freshwater fish research and conservation in Malaysia. J Survey Fish Sci 7–21

Khan S, Naushad M, Govarthanan M, Iqbal J, Alfadul SM (2022) Emerging contaminants of high concern for the environment: current trends and future research. Environ Res 207:112609

Koushlesh SK, Sinha A, Kumari K, Borah S, Chanu TN, Baitha R, ... Das BK (2018) Length-weight relationship and relative condition factor of five indigenous fish species from Torsa River, West Bengal, India. J Appl Ichthyol 34(1):169–171

Kurniawan SB, Ahmad A, Rahim NFM, Said NSM, Alnawajha MM, Imron MF, ... Hasan HA (2021) Aquaculture in Malaysia: water-related environmental challenges and opportunities for cleaner production. Environ Technol Innov 24:101913

Leong WK, Lai SH (2017, June) Application of water evaluation and planning model for integrated water resources management: case study of Langat River Basin, Malaysia. IOP Conf Series: Mater Sci Eng 210(1):012024). IOP Publishing

Muhamad Khair NK, Lee KE, Mokhtar M, Goh CT (2018) Integrating responsible care into quality, environmental, health and safety management system: a strategy for Malaysian chemical industries. J Chem Health Saf 25(5):10–18

Naser MA., Fahmi-Ahmad M, Ahmad A (2021) Length-weight relationship and condition factor of river sprats, Clupeichthys sp. From Tasik Kenyir, Terengganu, Malaysia. Malays Appl Biol 50(3):69–75

National Academies of Sciences, Engineering, and Medicine (2012) Sustainable water and environmental management in the California Bay-Delta. The National Academies Press, Washington, DC. https://doi.org/10.17226/13394.

Ng YS, Chan DJC (2021) The enhancement of treatment capacity and the performance of phytoremediation system by fed batch and periodic harvesting. RSC Adv 11:6049–6059

Nitanan KM, Shuib A, Sridar R, Kunjuraman V, Zaiton S, Herman MA (2020) The total economic value of forest ecosystem services in the tropical forests of Malaysia. Int Rev 22(4):485–503

Nyanti L, Noor-Azhar NI, Soo CL, Ling TY, Sim SF, Grinang J, Lee KSP (2018) Physicochemical parameters and fish assemblages in the downstream river of a tropical hydroelectric dam subjected to diurnal changes in flow. Int J Ecol 1:1–10

Nyanti L, Nyanti L, Berundang G, Ling TY (2010) Short term treatment of shrimp aquaculture wastewater using water hyacinth (Eichhornia crassipes). World Appl Sci J 8:1150–1156

Omotade IF, Alatise MO, Olawale O (2019) Growth and yield performance of hot pepper using aquaculture wastewater. Agricultural Engineering International: CIGR J 21(2):18–25

Palmer M, Ruhi A (2019) Linkages between flow regime, biota, and ecosystem processes: implications for river restoration. Science 365(6459):eaaw2087

Panchan R (2010) Fish assemblage patterns in littoral zone of Nam Oun reservoir, Sakon Nakhon province, Thailand. J Fish Environ 34(2):27–39

Reid AJ, Carlson AK, Creed IF, Eliason EJ, Gell PA, Johnson PT, ... Cooke SJ (2019) Emerging threats and persistent conservation challenges for freshwater biodiversity. Biol Rev 94(3):849–873

Rosle S, Ibrahim S, Yusof MN (2018) Relationship of water quality parameters and depth with fish density in Kenyir Lake, Malaysia. Int J Adv Sci Res 22(November):22–25

Samat A, Shukor MN, Mazlan AG, Arshad A, Fatimah MY (2008) Length-weight relationship and condition factor of Pterygoplichthys pardalis (Pisces: Loricariidae) in Malaysia Peninsula. Res J Fish Hydrobiol 3(2):48–53

Scholte SS, Van Teeffelen AJ, Verburg PH (2015) Integrating socio-cultural perspectives into ecosystem service valuation: a review of concepts and methods. Ecol Econ 114:67–78

Taniwaki RH, Piggott JJ, Ferraz SF, Matthaei CD (2017) Climate change and multiple stressors in small tropical streams. Hydrobiologia 793:41–53

Tongnunui S, Beamish FW, Kongchaiya C (2016) Fish species, relative abundances and environmental associations in small rivers of the Mae Klong River basin in Thailand. Agric Nat Resour 50(5):408–415

Udimal TB, Jincai Z, Ayamba EC, Mensah Owusu S (2017) China's water situation; the supply of water and the pattern of its usage. Int J Sustain Built Environ 6:491–500

Villar-Navarro E, Levchuk I, Rueda-Márquez JJ, Manzano M (2019) Combination of solar disinfection (SODIS) with H2O2 for enhanced disinfection of marine aquaculture effluents. Sol Energy 177:144–154

Washington H (2015) Overshoot. Ref. Modul. Earth Syst. Environ. Sci. https://doi.org/10.1016/b978-0-12-409548-9.09517-8.

Wilkinson CL, Yeo DC, Tan HH, Fikri AH, Ewers RM (2018) Land-use change is associated with a significant loss of freshwater fish species and functional richness in Sabah, Malaysia. Biol Conserv 222:164–171

Zhang Q, Achal V, Xu Y, Xiang WN (2014) Aquaculture wastewater quality improvement by water spinach (Ipomoea aquatica Forsskal) floating bed and ecological benefit assessment in ecological agriculture district. Aquac Eng 60:48–55

Leveraging Environmental Epidemiology for Effective Pollution Control

Rinaldy Jose Nathanael and Katharina Oginawati

Abstract Environmental epidemiology (EE) plays a crucial role in understanding the health impacts of pollutants and guiding effective pollution control strategies. This chapter explores the basic concept, basic designs of EE studies, how to interpret them, and the multifaceted contributions of EE, highlighting its historical significance, current applications, and future potential. The chapter is structured to provide environmental scientists, engineers, and interdisciplinary practitioners with a foundational understanding of EE. Key epidemiological measures such as incidence, prevalence, risk ratio (RR), and odds ratio (OR) are explained to equip readers with the tools to interpret EE study results. Furthermore, we discuss the essential role of EE in supporting pollution prevention and remediation efforts, using historical study and case like the Harvard Six Cities study and Love Canal tragedy. Key challenges and opportunities for EE are highlighted. The emergence of new contaminants and complex chemical mixtures requires innovative methods to assess their combined health effects. Technological advances in big data analytics, biomonitoring, and remote sensing enhance the precision and scope of EE studies. We also stress the importance of interdisciplinary collaborations, especially with engineers, to develop targeted technologies for pollution treatment and management. This chapter aims to initiate and spark collaborations by sharing the fundamentals of EE with scientists and practitioners from diverse disciplines, fostering a comprehensive and effective approach to pollution control.

R. J. Nathanael (✉)
Environmental Engineering Program, Department of Biology, Faculty of Science and Technology, Airlangga University, Surabaya, Indonesia
e-mail: jose.nathanael@fst.unair.ac.id

K. Oginawati
Environmental Engineering Department, Faculty of Civil and Environmental Engineering, Bandung Institute of Technology, Bandung, Indonesia

A. Z. Yaser et al. (eds.), *Controlling Environmental Pollution*,
https://doi.org/10.1007/978-981-97-8931-3_8

1 Introduction

It is not uncommon to start the discussion on environmental pollution with human activities. The increasing activities and varieties of industry, agriculture, and recreation have increased environmental pollution across the globe and in multiple environmental compartments (Edo et al. 2024). However, the challenge of ever-increasing human activities in the twenty-first century pertains not only to the increased pollutant concentrations across environmental compartments but also to their increasing varieties and effects on the environment. Emerging contaminants and their wide range of health effects, for example, have taken center stage in discussions about the advancement of pollution control and environmental protection (Yadav et al. 2021).

In controlling environmental pollution, there are several basic concepts underlying the discussions and practice, from the most preferred to the least: (1) pollution prevention, (2) pollution reduction, (3) waste treatment and management, and (4) remediation. Pollution prevention seeks to eliminate pollutant generation at the source through regulations and cleaner production processes. Pollution reduction minimizes the amount and toxicity of pollutants produced by improving resource use efficiency and implementing regulatory measures. When prevention and reduction are insufficient, waste treatment and management involve treating pollutants to neutralize harmful substances before environmental discharge, using physical, chemical, and biological treatments. Remediation addresses existing contamination, using methods like soil excavation and groundwater treatment to clean up and restore polluted sites.

In controlling environmental pollution using these concepts, the actions often start with awareness of the specific pollutant's effects on the environment and human health. Historical examples highlight how understanding the impacts of certain pollutants has driven regulatory action and pollution control measures. For instance, the devastating effects of methylmercury were brought to global attention through the epidemic neurological disease incident in Minamata, Japan. Discovered in the 1950s, this disease caused severe neurological damage and death among residents due to industrial wastewater discharge into Minamata Bay, leading to the bioaccumulation of methylmercury in fish and shellfish. This tragic event spurred stricter regulations on mercury discharge and heightened awareness of industrial pollution's potential impacts on human health and the environment (Semionov 2018). Another pivotal moment in environmental pollution control was the publication of Rachel Carson's Silent Spring in 1962. This groundbreaking book highlighted the adverse effects of pesticides, particularly DDT, on wildlife and ecosystems. Carson's work raised public awareness and prompted policy changes, eventually leading to the regulation and banning of several harmful pesticides (Epstein 2014).

It is interesting to look back at those cases and think: What did the government and scientists in Minamata think and do when they were faced with the devastating case that killed and disabled many people? Also, why and how did Carson link the unusual silence of the spring that year with the excessive use of pesticides? How exactly did they finally conclude that methylmercury was the main cause of the poisoning in Minamata and that DDT was the main cause of the death of many birds? What if they had never identified and confirmed the exact pollutants causing those devastating effects? The control and solution to the pollution might have come later, and the damage could have been worse than what actually happened.

We are then convinced that historically, even until now, awareness of the adverse effects of pollutants on the environment and human health, whether locally or globally, has initiated or at least boosted pollution control measures. In this age, the harmful effects of many environmental pollutants from wastewater, gas emissions, or toxic wastes have been well documented and rapidly researched. However, determining and prioritizing the control of the most critical pollutants requires more than knowledge of their individual harmful effects or safe limits. In pollution control measures, we are faced with various kinds of pollutants across environmental compartments and limited resources to control everything at the same time. As environmental engineers, scientists, or policymakers, we often have to prioritize and focus on specific scopes or cases one by one. Spending resources and time on pollutants, sites, or compartments that are relatively not critical is deemed ineffective. Using environmental health methods to guide the determination of "hot spots," the critical pollutants, and environmental compartments that need immediate attention and action is one way to effectively control environmental pollution.

This chapter is designed to summarize an environmental health research method that has historically been, and still is, widely utilized to support effective pollution control: environmental epidemiology. This chapter will explore the brief history, concept overview, and basic designs of this method, as well as examples of how findings from environmental epidemiology studies can help achieve effective pollution control. It is primarily aimed at environmental scientists, engineers, and policymakers who are not closely related to or actively engaged in environmental health methods, to help them understand and interpret research that utilizes these methods. By providing this foundational knowledge, we aim to bridge the gap between environmental health research and practical applications in pollution control, fostering interdisciplinary collaboration and informed decision-making.

2 Environmental Epidemiology

2.1 History and Overview

Human experimentation is considered the most reliable form of scientific evidence, but it is often impractical and unethical for studying potentially harmful substances. As a result, researchers use alternative methods such as animal and in vitro studies to evaluate environmental health risks. These methods, however, have limitations due to biological differences between species, which affect how toxins are processed and their impacts (Voisin et al. 1990). Furthermore, using genetically uniform test populations does not reflect the genetic diversity found in humans, and laboratory exposure scenarios often differ from real-world conditions. Therefore, to study the effects of environmental exposure on human health, a "natural experimentation" or observation method is commonly used.

Epidemiology, the study of disease distribution and determinants in populations, has evolved significantly over centuries and is a crucial branch of medicine and public health. Its roots trace back to ancient times with Hippocrates, who proposed that environmental factors and lifestyle influenced disease occurrence. During the Renaissance, Girolamo Fracastoro introduced the concept of contagion, and John Graunt's seventeenth-century analysis of London mortality data marked the beginning of systematic statistical analysis in public health. The eighteenth century saw James Lind's controlled trial on scurvy and Edward Jenner's development of the smallpox vaccine, demonstrating the power of empirical evidence and preventive measures (Tulchinsky and Varavikova 2014a). Environmental epidemiology, as a distinct discipline, emerged in the mid-twentieth century, though its roots can be traced back to the nineteenth century with notable public health investigations. One of the earliest and most famous examples is John Snow's study during the 1854 cholera outbreak in London. Snow used epidemiological methods to identify a contaminated water pump as the source of the outbreak, laying the groundwork for modern environmental epidemiology (Vineis 2018). In the early twentieth century, the field began to take shape as industrialization and urbanization led to increased awareness of environmental hazards. The Great Smog of London in 1952, which caused thousands of deaths due to severe air pollution, highlighted the need to study the health impacts of environmental factors. This event, among others, spurred the development of air pollution epidemiology and regulatory measures.

Environmental epidemiology studies the relationship between environmental exposures and human health. It focuses on identifying and quantifying the health effects of environmental hazards, such as pollutants in air, water, soil, and food. By documenting actual conditions, including diverse exposures and their effects, these observational studies provide a more accurate picture of how environmental factors influence human health. This approach maintains ethical standards and takes advantage of the complexity and diversity of human environments. Environmental epidemiology uses these natural experiments to offer a more practical and thorough way to study the health effects of environmental exposures.

Epidemiology studies can be broadly categorized into two main types: descriptive and inferential. Descriptive epidemiology is primarily used to describe the distribution of diseases and health outcomes by examining patterns based on person, place, and time. It focuses on identifying the "what," "who," "when," and "where" of health-related events and is fundamental in generating initial hypotheses about potential causes of health issues (Tulchinsky and Varavikova 2014b). This approach is particularly useful in initial investigations of disease outbreaks or unusual health events, providing essential data that can inform further, more detailed studies. These studies are foundational as they provide essential information about the occurrence and patterns of health events, which can then generate hypotheses about possible environmental causes. Analytical (inferential) epidemiology, on the other hand, delves deeper into the "why" and "how" by testing these hypotheses through observational or experimental studies. This method aims to identify and quantify the relationship between exposures (such as pollutants) and health outcomes, often using comparison groups to establish associations or causality. Due to its capacity to provide more detailed insights into the determinants of health and disease, inferential studies are more commonly employed in environmental epidemiology, especially when precise risk assessments are needed to inform public health interventions and policy decisions.

Given the complexities and specific requirements of environmental epidemiology, inferential studies are typically more prevalent than descriptive ones. They offer detailed analyses that are crucial for understanding the impacts of environmental exposures and for developing effective pollution control strategies. The next subchapters will overview the basic study designs of inferential environmental epidemiology commonly conducted to guide effective pollution control.

2.2 *Cohort Design*

A cohort study is an observational study that follows a group of individuals (the "cohort") who share a common characteristic or experience within a defined period (Bloom 2019). These studies aim to investigate the incidence and natural history of a disease by identifying and following a group of people who do not have the disease at the start of the study. Cohort studies can be prospective (following participants forward in time) or retrospective (using historical data to follow participants from the past to the present).

The primary difference between prospective and retrospective cohort designs is the timing of the exposure relative to the start of the study. In a prospective cohort study, the exposure begins at the present time (the start of the study), and we follow the cohort into the future to observe outcomes. In a retrospective cohort study, however, the exposure began in the past, and we use historical data to observe outcomes up to the present time (the start of the study). See Fig. 1 for illustration of cohort design.

For example, a prospective cohort study design might be used to investigate the impact of air pollution on respiratory health. We enroll 5,000 participants from various cities, all of whom are initially free of chronic respiratory diseases. Over the next 10 years, we regularly collect data on their exposure to air pollutants such as particulate matter (PM2.5) and nitrogen dioxide (NO_2), along with their respiratory health status. Throughout the study, we find that higher long-term exposure to these pollutants is associated with a significant increase in the incidence of asthma and chronic bronchitis. This evidence helps policymakers implement stricter air quality standards to protect public health.

Now, suppose we conduct a retrospective cohort study to examine the health effects of a chemical spill that occurred a decade ago. We identify individuals who lived near the spill site and use historical health records to assess their exposure to the leaked chemicals. We then compare the health outcomes of these individuals to a similar population that was not exposed to the spill. Our analysis reveals a higher incidence of liver and kidney diseases among those exposed to the chemicals. This retrospective study provides crucial evidence that leads to stricter regulations and call to immediate actions, such as site remediation.

From these two examples, we can get an impression of how cohort environmental epidemiology studies may help make pollution control measures more effective. Cohort design is primarily used when we are interested in evaluating the health impact of specific pollutants. Prospective cohort studies are particularly useful for evaluating the long-term effects of regulations or policies, such as air quality or water quality standards, as they allow for the continuous collection of exposure and health data over time. Retrospective cohort studies, on the other hand, are beneficial for detecting the health impacts of past exposures, such as industrial accidents, chemical spills, occupational exposure, or historical environmental contamination. This design helps uncover long-term health effects that may not have been immediately apparent, providing evidence to support the need for remediation efforts and stricter regulations to prevent future occurrences. Together, these study designs offer robust methods for linking environmental exposures to health outcomes, guiding effective pollution control strategies based on empirical evidence.

2.3 Case–Control Design

Case–control studies are a fundamental research design in environmental epidemiology, particularly useful for investigating the causes of rare diseases or outcomes. These studies are retrospective, meaning they look back in time to examine the relationship between exposure and outcome. Unlike cohort studies, which follow participants over time, case–control studies begin with individuals who already have the outcome of interest (cases) and compare them to individuals without the outcome (controls) (Bloom 2019). The primary difference between cohort design and case–control design is thus the grouping of individuals. The cohort design groups individuals based on their exposure status and compares the outcomes, while the case–control design groups individuals based on their disease status (case or control) and compares their past exposure to any environmental factors in question (Fig. 2).

Many people often confuse case–control studies with retrospective cohort design. To better understand the distinction, let's consider this example. Suppose we conduct a case–control study to investigate the link between multiple environmental factors (such as drinking water source and quality, occupational chemical exposure, and pesticide exposure) and a rare type of cancer, such as non-Hodgkin lymphoma. We select 200 individuals diagnosed with non-Hodgkin lymphoma as cases and 200 healthy individuals without the disease as controls. Using detailed questionnaires, we gather data on their historical exposure to various factors, their duration and frequency of exposure, and other potential risk factors. After analyzing the data, we find that cases are significantly more likely to have had high levels of exposure to a specific type of pesticide compared to controls. The analysis then shows a strong association between pesticide exposure and the development of non-Hodgkin lymphoma. These findings can then inform policymakers to prioritizing and considering stricter regulations on pesticide use and encourage further research into safer alternatives.

Case–control studies, with their ability to uncover associations in a time-efficient and cost-effective manner, play a crucial role in identifying potential environmental risk factors to a certain outcome. This information can be instrumental in guiding public health interventions and shaping environmental policies aimed at reducing exposure to harmful substances. Case–control design is also cost-effective and time-efficient because they use existing data and are retrospective, avoiding the need for prolonged follow-up periods and large sample sizes required in cohort studies. This efficiency allows for quicker, more resource-efficient studies when initial exploratory research or preliminary evidence is needed.

However, case–control studies face significant limitations compared to cohort design. They are susceptible to recall bias, as participants' memories of past exposures can be unreliable, potentially skewing the results. Selection bias is another concern, as improper matching of cases and controls can lead to inaccurate associations. Moreover, case–control studies cannot establish causality due to their retrospective nature; they can only identify associations between exposures and outcomes. Cohort studies, particularly prospective ones, mitigate these issues by collecting exposure data in real time, reducing recall bias, and providing a clearer temporal relationship between exposure and outcome, thereby offering a stronger basis for inferring causality.

Nonetheless, both cohort and case–control designs are crucial in effective pollution control. The exposure-focused perspective of cohort studies and the outcome-driven approach of case–control studies are essential for collecting, synthesizing, and communicating evidence about environmental pollutants that need attention due to their detrimental effects on human health.

2.4 Cross-Sectional Design

Cross-sectional studies are a type of observational study that analyzes data from a population, or a representative subset, at a specific point in time. Unlike cohort and case–control studies, which follow participants over a period to examine the progression or history of exposure and outcome, cross-sectional studies provide a snapshot of the population's health status and environmental exposures at a single moment (Bloom 2019). These studies are particularly useful for assessing the prevalence of health outcomes or risk factors in a given population. By simultaneously measuring exposures and outcomes, cross-sectional studies can identify associations between environmental factors and health conditions, although they cannot establish causality due to the simultaneous collection of data on both variables (Fig. 3).

Cross-sectional studies offer several advantages, especially when compared to cohort and case–control designs. One major strength is their efficiency and cost-effectiveness; they do not require long-term follow-up, making them quicker and less expensive to conduct. Cross-sectional studies provide valuable prevalence data on health outcomes and exposures, essential for public health planning and resource allocation, similar to case–control studies, which also use existing data for analysis. Additionally, cross-sectional studies can assess multiple outcomes and exposures simultaneously, providing a comprehensive snapshot of the population's health and environmental status, a feature that cohort studies also share but on a longitudinal basis.

A notable drawback of cross-sectional studies is their inability to determine causality; since exposure and outcome data are collected simultaneously, it's impossible to establish whether the exposure preceded the outcome. This limitation is shared with case–control studies, which also struggle with temporal ambiguity, though cohort studies generally avoid this issue by following subjects over time. Furthermore, cross-sectional studies do not capture changes over time, hindering the assessment of exposure and disease progression dynamics, unlike cohort studies, which are designed to observe these changes. Cross-sectional studies are also prone to biases, such as selection bias and recall bias, particularly if participants are asked to remember past exposures, a challenge that is also common in case–control studies. These limitations highlight the importance of carefully choosing the study design that best fits the research question and the resources available.

Cross-sectional design shares several strengths and limitations with case–control design when compared with cohort design, but they serve their own distinct purpose. Cross-sectional studies are generally quicker and less resource-intensive than case–control studies, as it involves a single point of data collection. They also allow for the examination of multiple outcomes and exposures simultaneously within a population, making them distinct from case–control studies, which are more focused on identifying and comparing risk factors associated with a specific outcome.

Cross-sectional studies are also particularly useful as preliminary studies to explore potential, less-known links between pollutants and health outcomes. Because these studies require less time to complete than longitudinal designs, they can provide rapid insights that guide future research directions and inform immediate pollution control measures and public health responses. Suppose we conduct a cross-sectional study to investigate the potential association between exposure to microplastics in drinking water and gastrointestinal issues among residents of a small town. We randomly select 1000 residents and collect data on their gastrointestinal health, such as the frequency of symptoms like stomach pain, bloating, and indigestion, as well as their current exposure to microplastics through their drinking water. By analyzing these data, we might find that residents with higher levels of microplastics in their drinking water report more gastrointestinal issues. While this finding would indicate an association, it would not confirm that microplastics cause these issues, as the study design does not account for temporal relationships or potential confounding factors. However, the finding could prompt more detailed cohort studies or experimental research to investigate causality and support the development of policies and measures to limit pollutants in the environment.

2.5 Interpreting Environmental Epidemiology Studies

After learning the basic designs of environmental epidemiology studies—cohort, case–control, and cross-sectional—it is crucial to understand how to interpret the results derived from these studies. Interpreting the results is key to understanding the significance and implications of the findings. This subchapter will cover important, foundational terms and measures commonly used in reporting epidemiology study results.

1. Incidence (I) and Prevalence (P)

In health communication, effectively conveying the magnitude and impact of diseases within a population is crucial not only for informed decision-making and public awareness. Therefore, measures that quantify the number of cases relative to the population size are essential tools for this purpose. These metrics provide a clear picture of how widespread a disease is, helping health professionals, policymakers, and the public understand the severity and reach of health issues.

Incidence refers to the rate at which new cases of a particular disease or condition occur in a specified population during a defined time period. It is a measure of the probability or risk of individuals developing the disease over that period. To calculate incidence, epidemiologists divide the number of new cases by the number of individuals at risk in the population, often expressed per 1,000 or 100,000 people. This measure is crucial for identifying trends and patterns in disease occurrence, evaluating the effectiveness of public health interventions, and allocating resources for prevention and treatment efforts. Incidence is particularly useful for studying acute conditions or outbreaks where the goal is to understand the onset and spread of new cases.

Prevalence, on the other hand, indicates the proportion of individuals in a population who have a specific disease or condition at a particular point in time or over a specified period. It encompasses all cases, both new and pre-existing, within the population. Prevalence is calculated by dividing the total number of cases (both new and existing) by the total number of individuals in the population. This measure helps to provide a snapshot of the overall burden of a disease, making it valuable for understanding chronic conditions that persist over time. Prevalence data is essential for health care planning, resource allocation, and evaluating the long-term impact of public health policies and interventions. Unlike incidence, prevalence does not distinguish between new and old cases, making it a more comprehensive measure of the disease's impact on a community. To better understand the difference between incidence and prevalence, observe Eqs. (1) and (2).

$$Incidence = \frac{New\ cases}{Population\ at\ risk} \tag{1}$$

$$Prevalence = \frac{All\ cases}{Population\ at\ risk} \tag{2}$$

2. Relative Risk (RR)

Risk Ratio (RR), also known as Relative Risk, is a measure used in epidemiology to compare the risk of a certain event, such as developing a disease, between two groups. It is particularly useful in cohort studies where individuals are followed over time to observe the occurrence of an event. The RR is calculated by dividing the risk of the event in the exposed group by the risk of the event in the unexposed group. The formula for RR is:

$$Risk\ Ratio(RR) = \frac{Risk\ in\ exposed\ group}{Risk\ in\ unexposed\ group} = \frac{Incidence\ in\ exposed\ group}{Incidence\ in\ unexposed\ group} \quad (3)$$

To illustrate, if we have a group of individuals exposed to a certain factor (e.g., a chemical or a behavior) and another group not exposed, we can track how many in each group develop the condition of interest. Take a look back at Figure 1 for example. Suppose 20 out of 30 exposed individuals develop the disease (a risk of 2/3) and 10 out of 30 unexposed individuals develop the disease (a risk of 1/3). Using Eq. (3), the RR would be:

$$Risk\ Ratio(RR) = \frac{2/3}{1/3} = 2.$$

The interpretation of RR is straightforward: an RR of 1 indicates no difference in risk between the two groups, an RR greater than 1 indicates a higher risk in the exposed group, and an RR less than 1 indicates a lower risk in the exposed group. This RR of 2 suggests that the exposed group has twice the risk of developing the disease compared to the unexposed group.

Risk Ratio (RR) is primarily used in cohort studies because these studies involve following groups of individuals over time to observe the occurrence of new cases of a disease or condition. In a cohort study, researchers can directly measure the incidence of disease in both the exposed and unexposed groups, allowing them to calculate the risk of disease and subsequently the RR. RR is not applicable in case-control or cross-sectional studies due to the nature of these study designs. For case-control and cross-sectional studies, another measure called Odds Ratio (OR) are used.

3. Odds Ratio (OR)

In case-control studies, participants are selected based on the presence (cases) or absence (controls) of a disease, and past exposure status is then assessed. In cross-sectional studies, data on exposure and disease status are collected simultaneously at a single point in time. Since the incidence of disease cannot be directly measured in this design, the odds ratio (OR) is used instead to estimate the association between exposure and disease. The OR compares the odds (not risk) of the outcome occurring in the exposed group to the odds of the outcome occurring in the unexposed group.

To understand the OR, it's essential to first grasp the concept of "odds." The odds of an event is the ratio of the probability of the event occurring to the probability of it

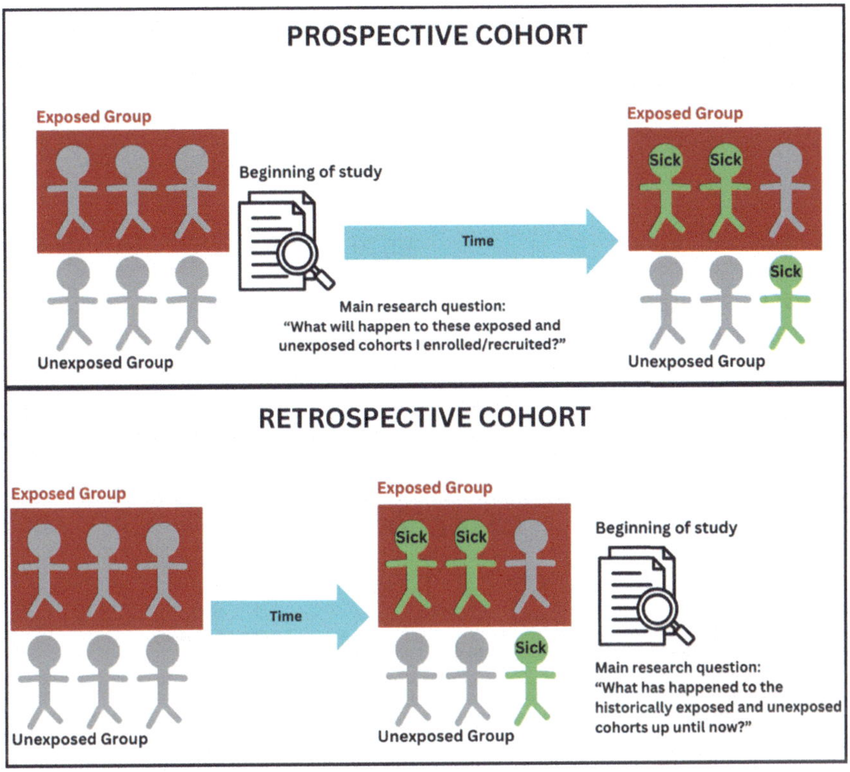

Fig. 1 Cohort study design. Red background illustrates environmental exposure

not occurring. For instance, if 20 people out of 100 develop a disease, the probability (risk) is 0.2, and the odds are 0.2/(1−0.2) = 0.25.

The formula for calculating the OR is:

$$Odds\ Ratio = \frac{Odds\ of\ exposure\ in\ cases}{Odds\ of\ exposure\ in\ controls} \tag{4a}$$

Or equivalently,

$$Odds\ Ratio = \frac{\left(\begin{array}{l}Number\ of\ cases\ with\ exposure\\ /Number\ of\ cases\ without\ exposure\end{array}\right)}{\left(\begin{array}{l}Number\ of\ controls\ with\ exposure\\ /Number\ of\ controls\ without\ exposure\end{array}\right)} \tag{4b}$$

To illustrate, in case-control studies, we have two groups of individuals: case group and control group, where we trace back their exposure status. We can then quantify the odds ratio by counting the number of cases with and without exposure, and also number of controls with and without exposure. Consider Fig. 2. Suppose we have 30 individuals with asthma (cases) and 30 healthy individuals (controls). 20 of 30 cases and 10 of 30 controls have a history of living near industrial area. Using Eq. (4b), the OR would be:

$$Odds\,Ratio = \frac{20/10}{10/20} = 4.$$

This OR of 4 suggests that the odds of living near industrial area are 4 times higher among the sick individuals compared to the healthy individuals. The interpretation of the OR is similar to that of the Risk Ratio (RR). An OR of 1 indicates no association between exposure and outcome. An OR greater than 1 indicates a positive association between exposure and outcome, while an OR less than 1 suggests a negative

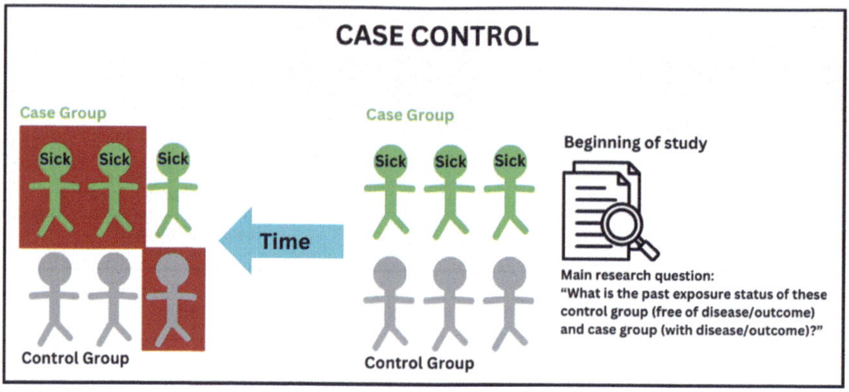

Fig. 2 Case–control study design. Red background illustrates environmental exposure

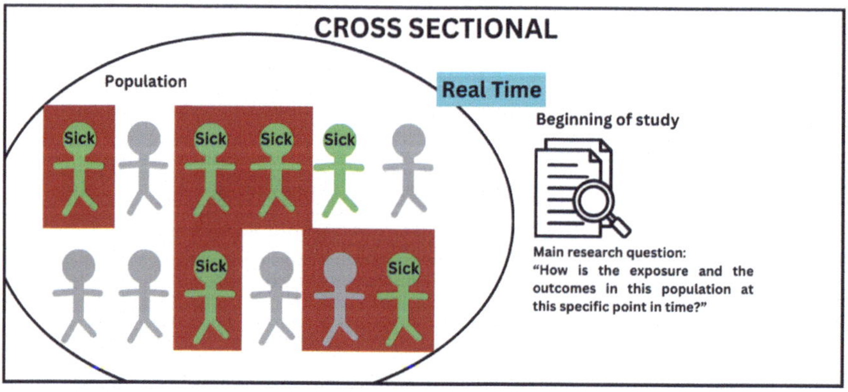

Fig. 3 Cross-sectional study design. Red background illustrates environmental exposure

association. In this context, the OR helps researchers understand the potential link between the exposure and the outcome, even though it does not establish a causal relationship. One of the advantages of using OR in case-control studies is that it allows researchers to study rare diseases, where it might be impractical to follow a large cohort over time. However, it's important to note that while OR can approximate RR in rare diseases, it tends to overestimate the risk when the outcome is common. Thus, careful interpretation is required, especially when applying findings to broader populations.

4. Confidence Interval (CI)

Confidence intervals (CIs) are a crucial component of statistical analysis in epidemiological studies, often reported alongside measures like relative risk (RR) and odds ratio (OR). They provide a range of values within which the true effect size is likely to fall, offering a measure of precision and reliability of the estimate. A 95% confidence interval, for example, suggests that there is a 95% probability that the interval contains the true population parameter. This means if the study were repeated numerous times, 95% of the intervals calculated from those studies would contain the true effect size. When interpreting CIs in the context of RR and OR, several key points should be considered.

Firstly, range and precision. A narrower confidence interval (CI) indicates greater precision in the estimate, suggesting less variability and higher certainty about the true effect size. This is because a narrower range means that the data points are closer together, leading to a more consistent and reliable estimate. Conversely, a wider CI suggests more variability and less precision, indicating greater uncertainty about the true effect size. For instance, an RR of 2.0 with a 95% CI of 1.8–2.2 is more precise than an RR of 2.0 with a 95% CI of 1.2–3.5, as the former shows a tighter clustering of data points, making the estimate more reliable.

Secondly, statistical significance. If the CI for an RR or OR does not include the value 1.0, the result is considered statistically significant because a value of 1.0 indicates no association between exposure and outcome. For example, an OR of 1.5 with a 95% CI of 1.1–2.0 is statistically significant, suggesting a true association between the exposure and the outcome, as the interval does not cross 1. Conversely, an OR of 1.5 with a 95% CI of 0.8–2.5 is not statistically significant, as the interval includes 1.0, indicating that the observed association could be due to chance.

Thirdly, direction and strength of association. The position of the CI relative to 1.0 helps determine the direction (positive or negative) and the strength of the association. For example, an RR of 0.8 with a 95% CI of 0.7–0.9 indicates a statistically significant protective effect of the exposure, reducing the risk of the outcome, as the entire interval is below 1.0. Conversely, an RR of 1.5 with a 95% CI of 1.2–1.8 indicates a statistically significant increased risk of the outcome due to the exposure, as the interval is entirely above 1.0.

2.6 Statistics and Confounding Challenges in Environmental Epidemiology

Environmental epidemiology investigates how environmental factors influence health outcomes. The outcomes and diseases studied often have multiple contributing factors and causes. This complexity necessitates precise measurement and analysis to discern the true impact of environmental exposures. For instance, biomarker levels or clinical measures can vary widely between individuals and populations, influenced by genetics, lifestyle, and other environmental factors.

Those factors are usually called confounding factors (Pearce and Greenland 2024). Confounding factors are variables that can obscure the true relationship between the exposure and the outcome of interest. They are particularly challenging in environmental epidemiology because they can introduce bias, leading to incorrect conclusions about causal relationships. For instance, if a study finds a correlation between air pollution and respiratory disease but fails to account for potential confounding factors such as age, socioeconomic factors, and lifestyles such as occupation and smoking, the results may overestimate the effect of air pollution (Hajat et al. 2021).

To address confounding, researchers employ various strategies. Randomization in experimental studies helps to evenly distribute confounders across study groups. In observational studies, statistical adjustments, such as stratification and multivariable regression, are used to control for confounding factors. Propensity score matching and sensitivity analysis are additional methods to mitigate confounding effects (D'Onofrio et al. 2020). To gain the basic understanding of addressing confounders, consider this simple example.

Suppose a study aims to investigate the relationship between air pollution and the incidence of asthma in urban populations. Researchers collect data on air pollution levels and asthma cases across several cities over a period of time. Using the dataset given in Table 1, a correlation analysis between exposure concentration and asthma incidence yields a correlation value of 0.473 with a p-value of 0.035, indicating a significant association between air pollution exposure and asthma. However, these values are considered "unadjusted" because potential confounding factors are not taken into account.

Next, using the same dataset but accounting for smoking status as a confounder, we run a partial correlation and regression analysis. In this analysis, asthma is the outcome (dependent variable), and exposure concentration and smoking status are the independent variables (predictors). The results show a correlation value of 0.189 and a p-value of 0.438, indicating no significant association between air pollution exposure and asthma when adjusting for smoking status.

This example illustrates how a single confounder can significantly alter the results and interpretation of an environmental epidemiology study. In real-world epidemiology studies, the measures of exposure and effects are often more complex than in this simplified example. In some environmental epidemiology studies examining multiple environmental factors or pollutants, one factor or pollutant can even be considered a confounder for another. Therefore, robust statistical methods and

Table 1 Example dataset

Number	Exposure concentration ($\mu g/m^3$)	Asthma (0 = No, 1 = Yes)	Smoking (0 = No, 1 = Yes)
1	80	1	1
2	70	0	0
3	60	0	0
4	50	0	0
5	40	0	0
6	50	1	1
7	55	1	1
8	45	0	0
9	65	1	1
10	75	0	0
11	35	0	0
12	80	1	1
13	45	0	0
14	55	1	0
15	60	1	1
16	50	1	1
17	40	0	0
18	60	1	0
19	70	1	1
20	35	0	0

strategies are essential to uncover the true relationships between environmental factors or exposures and their effects. Unfortunately, some environmental epidemiology research papers do not adequately account for confounding factors, potentially leading to misleading conclusions. A research paper highlighted that many studies examining the relationship between air quality and acute mortality suffer from serious statistical problems. Without adjusting for multiple comparisons and modeling, unadjusted p-values can be misleading, resulting in unreliable claims (Young 2017). This underscores the importance of rigorous statistical methodologies, including adjustments for multiple testing, to ensure the reliability of findings in environmental epidemiology.

A careful interpretation of environmental epidemiology studies is often needed to avoid being misled by unreliable studies. Readers, including environmental engineers, scientists, policymakers, and other multidisciplinary practitioners, should critically evaluate the methodology of environmental epidemiology studies. They should check whether the study has considered potential confounders and use appropriate statistical adjustments. It is also important to assess the sample size, study design, and data collection methods to ensure they are robust and appropriate for the research question. Reviewing the discussion and conclusion sections for any acknowledgment of limitations and confounding factors can also provide insights into the study's reliability. By taking these steps, readers can better interpret the findings and apply them appropriately in their work.

2.7 Environmental Epidemiology in Supporting Pollution Control

We discussed that in environmental pollution control, there are four basic concepts covering a wide range of control measures or actions: (1) pollution prevention, (2) pollution reduction, (3) waste treatment and management, and (4) remediation. However, where and how can the results or information from environmental epidemiology studies contribute to more effective pollution control measures? After successfully learning the fundamentals of environmental epidemiological studies and how to interpret them, it is essential to discuss how these results have historically helped in pollution control and how they can continue to do so in the future. This subchapter explores how environmental epidemiology has historically supported pollution control, focusing on pollution prevention through policymaking and regulations, and guiding remediation efforts.

Pollution Prevention

In 1997, the United States Environmental Protection Agency (EPA) added PM2.5 to the National Ambient Air Quality Standards (NAAQS), a parameter that was not regulated in the previous standards set in 1987. The new standards set the acceptable annual average concentration of PM2.5 at 15 μg per cubic meter and the 24-h standard at 65 μg per cubic meter. A substantial contribution to these new standards was the Harvard Six Cities Study, a landmark environmental epidemiology study.

The researchers at that time realized that people living in cities with higher levels of air pollution were dying sooner than those in cleaner cities. Through cohort design, the Harvard Six Cities Study revealed a strong link between mortality risk and concentrations of fine particulate matter (PM2.5) (Dockery et al. 1993), with people in the more polluted cities having shorter life expectancies than those in less polluted areas. These findings were surprising, especially since the "dirty" cities met the Environmental Protection Agency (EPA) air quality standards at the time, which did not classify them as unhealthy (Feldscher 2014). The study's findings, which showed a significant link between fine particulate matter (PM2.5) and increased mortality rates, necessitated immediate action. The EPA swiftly reviewed the data and recognized the need to address the previously unregulated PM2.5.

Without environmental epidemiology studies like the Harvard Six Cities Study, policymaking and pollution control might not be as effective, as the impact of pollution on life expectancy differences was not apparent under the previous standards. The concentration of PM10 did not exceed the standards, masking the health risks posed by PM2.5. Thus, environmental epidemiology has substantially contributed to effective pollution control through policy and regulations, raising awareness among both the public and authorities. This demonstrates the critical role of environmental epidemiology in guiding evidence-based decisions for public health protection and environmental pollution management.

Remediation

The Harvard Six Cities Study was not the first environmental epidemiology study that played a crucial role in pollution control and public health protection. In the late 1970s, residents of the Love Canal neighborhood in New York discovered that their homes and a school had been built on a toxic waste dump. The site, previously used by the Hooker Chemical Company to bury over 21,000 tons of hazardous chemical waste, began to leach toxins into the surrounding environment, causing serious health issues among the local population.

Environmental epidemiology studies were instrumental in linking the contamination to adverse health effects in the community. Reports of increased rates of birth defects, miscarriages, and other health problems led to widespread public concern and media attention (National Research Council 1991). The data gathered through these epidemiological investigations provided compelling evidence of the health risks posed by the buried toxic waste, catalyzing both local and national action.

In response to the Love Canal crisis, Congress enacted the Comprehensive Environmental Response, Compensation, and Liability Act (CERCLA) in 1980, commonly known as the Superfund program. CERCLA established a federal program to clean up hazardous waste sites and respond to environmental emergencies involving hazardous substances. The program's creation marked a significant shift in the federal government's approach to environmental contamination, emphasizing the importance of identifying and mitigating health risks associated with toxic sites. The EPA's Superfund program is dedicated to addressing some of the country's most severely contaminated sites and managing environmental crises, including oil spills and natural disasters. This initiative primarily focuses on remediation efforts, which often begin with identifying and understanding contamination through robust assessment and methods, which includes environmental epidemiology, as was the case with the Love Canal incident.

EPA's Superfund program exemplifies how environmental health assessments and methods, such as environmental epidemiology, support pollution control through site remediation. The Superfund cleanup process begins with three key steps before deciding on a remedy and taking action, according to the EPA (US Environmental Protection Agency 2023): (1) Preliminary assessment/site inspection, which involves testing the air, water, and soil at the site to determine the presence of hazardous substances and their potential threat to human health; (2) National priorities site listing process, which uses a numerical scoring system to assess the relative potential of sites to pose a threat to human health; and (3) Remedial feasibility study, which determines the extent of contamination, available technology options, and evaluates cost and performance.

Environmental epidemiology can be highly supportive in the Superfund program's cleanup process by providing valuable insights and data. In the preliminary assessment and site inspection phase, epidemiological studies help detect hazardous substances and assess their potential threat to human health. This ensures that testing efforts focus on the most harmful contaminants. During the national priorities site listing process, data from epidemiological research feed into the numerical scoring system, helping to rank sites based on their potential threat to human health through the mapping of impacted populations. This prioritization ensures that the most dangerous sites receive the necessary attention and resources.

Furthermore, in the remedial feasibility study phase, environmental epidemiology contributes valuable information on the extent of contamination and the associated health risks. This aids in evaluating the effectiveness and cost of different remediation technologies. Additionally, strong epidemiological evidence linking polluters to health problems in affected communities supports legal actions and regulatory measures, ensuring that responsible parties contribute to cleanup efforts.

3 Conclusions, Future Challenges, and Potential

Environmental epidemiology (EE) has proven to be an invaluable tool in understanding the health impacts of environmental pollutants and guiding pollution control measures. Historical cases like the Love Canal tragedy and landmark studies such as the Harvard Six Cities Study have demonstrated EE's crucial role in identifying health risks, informing policy decisions, and guiding remediation efforts. However, as we look to the future, EE must continue to evolve and address new challenges and identify future potentials to remain effective in protecting public health.

One of the most pressing challenges for EE is the rise of emerging contaminants and the complexity of chemical mixtures. Industrial advancements and consumer products introduce new chemicals into the environment, many of which have unknown health impacts. These emerging contaminants include endocrine-disrupting chemicals, pharmaceuticals, and nanomaterials. Traditional EE methods, which often focus on single pollutants, may not suffice to understand the combined effects of these complex mixtures. Thus, EE must develop innovative methodologies to study these multi-pollutant exposures and their interactions, which is critical for accurately assessing health risks and informing regulatory actions.

Additionally, the field of EE has the potential to explore emerging and re-emerging diseases, such as COVID-19, as environmental factors or pollutants may play a role in exacerbating or controlling these diseases (Oginawati et al. 2022; Weaver et al. 2022; Zeka et al. 2020). Understanding how environmental exposures influence disease dynamics can provide valuable insights for public health interventions and disease prevention strategies. The findings of such studies can inform policymakers about the environmental determinants of disease spread and severity, enabling them to implement targeted measures to control diseases and enhance public health. For instance, identifying pollution hotspots that correlate with higher disease incidence can lead to stricter regulations on emissions and better urban planning. Ultimately, EE can guide the development of policies that not only aim to reduce pollution but also mitigate the impact of environmental factors on disease outcomes, leading to healthier communities.

Thirdly, the rapid advancements in technology and data analysis present significant potential for evolving EE. Big data analytics, biomonitoring, remote sensing, and advanced statistical methods offer unprecedented opportunities to enhance the precision and scope of EE studies. For instance, wearable sensors and mobile health applications can provide real-time exposure data, while satellite imagery can track pollution levels over large areas and extended periods. These technological innovations can improve exposure assessment, enable more detailed and personalized health risk evaluations, and facilitate large-scale epidemiological studies.

Lastly, interdisciplinary collaborations are essential for the future of EE. Addressing complex environmental health issues and effectively controlling pollution requires integrating knowledge and methods from various scientific disciplines, including toxicology, environmental science, public health, data science, environmental engineering, and chemical engineering. Collaborative efforts can enhance the comprehensiveness and robustness of EE studies. For example, combining toxicological data with epidemiological findings can provide a more detailed understanding of dose–response relationships. Partnerships with data scientists can help manage and analyze large datasets more effectively. Working with environmental and chemical engineers can guide the development of targeted technologies and methods for treating and managing waste, wastewater, and flue gas, ensuring these technologies effectively reduce targeted, prioritized pollutants and mitigate health risks. Additionally, collaborating with policymakers and community organizations can ensure that research findings are translated into actionable policies and public health interventions. These interdisciplinary collaborations can maximize the impact of EE on pollution control and health protection. This chapter is written in the hope of initiating and sparking such collaborations by sharing the fundamentals of EE with scientists and practitioners in these disciplines.

References

Bloom MS (2019) Environmental epidemiology. In: Encyclopedia of environmental health. Elsevier, pp 419–427. https://doi.org/10.1016/B978-0-12-409548-9.10635-9

Dockery DW, Pope CA, Xu X, Spengler JD, Ware JH, Fay ME, Ferris BG, Speizer FE (1993) An association between air pollution and mortality in Six U.S cities. New England J Med 329:1753–1759. https://doi.org/10.1056/NEJM199312093292401

D'Onofrio BM, Sjölander A, Lahey BB, Lichtenstein P, Öberg AS (2020) Accounting for confounding in observational studies. Annu Rev Clin Psychol 16:25–48. https://doi.org/10.1146/annurev-clinpsy-032816-045030

Edo GI, Itoje-akpokiniovo LO, Obasohan P, Ikpekoro VO, Samuel PO, Jikah AN, Nosu LC, Ekokotu HA, Ugbune U, Oghroro EEA, Emakpor OL, Ainyanbhor IE, Mohammed WA-S, Akpoghelie PO, Owheruo JO, Agbo JJ (2024) Impact of environmental pollution from human activities on water, air quality and climate change. Ecol Front. https://doi.org/10.1016/j.ecofro.2024.02.014

Epstein L (2014) Fifty years since silent spring. Annu Rev Phytopathol 52:377–402. https://doi.org/10.1146/annurev-phyto-102313-045900

Feldscher K (2014) Landmark air pollution study turns 20 [WWW Document]. https://www.hsph.harvard.edu/news/features/six-cities-air-pollution-study-turns-20/

Hajat A, MacLehose RF, Rosofsky A, Walker KD, Clougherty JE (2021) Confounding by socioeconomic status in epidemiological studies of air pollution and health: challenges and opportunities. Environ Health Perspect 129. https://doi.org/10.1289/EHP7980

National Research Council (1991) Environmental epidemiology, vol 1. National Academies Press, Washington, D.C. https://doi.org/10.17226/1802

Oginawati K, Nathanael RJ, Pasaribu US, Mukhaiyar U, Humam A, Ilmi NFF, Susetyo SH (2022) Analysis of the effect and role of indoor environmental quality in the COVID-19 transmission. Aerosol Air Qual Res 22:210339. https://doi.org/10.4209/aaqr.210339

Pearce N, Greenland S (2024) Confounding and interaction. In: Handbook of epidemiology. Springer New York, New York, NY, pp 1–31. https://doi.org/10.1007/978-1-4614-6625-3_10-1

Semionov A (2018) Minamata disease—Review. World J Neurosci 08:178–184. https://doi.org/10.4236/wjns.2018.82016

Tulchinsky TH, Varavikova EA (2014a) A history of public health. In: The new public health. Elsevier, pp 1–42. https://doi.org/10.1016/B978-0-12-415766-8.00001-X

Tulchinsky TH, Varavikova EA (2014b) Measuring, monitoring, and evaluating the health of a population. In: The new public health. Elsevier, pp 91–147. https://doi.org/10.1016/B978-0-12-415766-8.00003-3

US Environmental Protection Agency (2023) About the superfund cleanup process [WWW Document]. https://www.epa.gov/superfund/about-superfund-cleanup-process. Accessed 30 June 24

Vineis P (2018) From John Snow to omics: the long journey of environmental epidemiology. Eur J Epidemiol 33:355–363. https://doi.org/10.1007/s10654-018-0398-4

Voisin EM, Ruthsatz M, Collins JM, Hoyle PC (1990) Extrapolation of animal toxicity to humans: interspecies comparisons in drug development. Regul Toxicol Pharmacol 12:107–116. https://doi.org/10.1016/S0273-2300(05)80052-2

Weaver AK, Head JR, Gould CF, Carlton EJ, Remais JV (2022) Environmental factors influencing COVID-19 incidence and severity. Annu Rev Public Health 43:271–291. https://doi.org/10.1146/annurev-publhealth-052120-101420

Yadav D, Rangabhashiyam S, Verma P, Singh P, Devi P, Kumar P, Hussain CM, Gaurav GK, Kumar KS (2021) Environmental and health impacts of contaminants of emerging concerns: recent treatment challenges and approaches. Chemosphere 272:129492. https://doi.org/10.1016/j.chemosphere.2020.129492

Young SS (2017) Air quality environmental epidemiology studies are unreliable. Regul Toxicol Pharmacol 86:177–180. https://doi.org/10.1016/j.yrtph.2017.03.009

Zeka A, Tobias A, Leonardi G, Bianchi F, Lauriola P, Crabbe H, Vardoulakis S, Guo Y, Honda Y, Gasparrini A, Hashizume M, Vicedo AM, Knudsen LE, Sera F, Ashworth M (2020) Responding to COVID-19 requires strong epidemiological evidence of environmental and societal determining factors. Lancet Planet Health 4:e375–e376. https://doi.org/10.1016/S2542-5196(20)30169-8

Sustainable Approaches Mitigate Environmental Pollution

Mohd Taufiq Mat Jalil

Abstract Issues with global environmental degradation do not only impact large numbers of people. Rather, they are those enormous issues that impact the entire Earth. Not everyone will be impacted by these issues at the same time or to the same extent. The causes of today's more complicated environmental pollution issues may have started decades or even centuries ago and are the result of cumulative, linked occurrences. Pollution develops when dangerous substances are released into the environment. Pollutants are a collective term for these dangerous substances. Natural pollutants include things like ash from volcanoes, the result of human activities, such as factory run-off or waste. Pollution is a result of many human-useful things. Pollutants stream out of automobile exhaust pipes. Air pollution results from burning coal to produce energy. Garbage and sewage from homes and businesses can contaminate the land and water. Pesticides are chemical toxins that damage animals by seeping into streams and killing vegetation and insects. In addition to seriously contributing to climate change, pollution hinders economic growth and exacerbates poverty and inequality in both urban and rural areas. Those with less money, who are unable to shield themselves from the damaging effects of pollution, bear the brunt of the consequences. The main environmental factor that contributes to illness and early mortality is pollution. In fact, pollution on a worldwide scale is the result of natural climate variations paired with human activity. Global issues also contain intricate political, social, and economic elements and may not have instantaneous solutions. This review aims to search for sustainable approaches to mitigate environmental pollution. Several sustainable approaches were discussed in this review article including the use of nanotechnology, bioremediation, and green technology. These sustainable approaches are doable, able to maintain a safe environment with reduced side effects, and have a low impact on the environment.

Keywords Environmental pollution · Pollutants · Sustainable approaches · Nanotechnology · Bioremediation · Green technology

---

M. T. M. Jalil (✉)
School of Biology, Faculty of Applied Sciences, Universiti Teknologi MARA, 40450 Shah Alam, Selangor, Malaysia
e-mail: taufiqjalil@uitm.edu.my

A. Z. Yaser et al. (eds.), *Controlling Environmental Pollution*,
https://doi.org/10.1007/978-981-97-8931-3_9

1 Introduction

Any material (solid, liquid, or gas) or energy (heat, sound, or radioactivity) added to the environment at a rate greater than its ability to be distributed, diluted, decomposed, recycled, or stored in a harmless form is referred to as pollution or environmental pollution. The three main categories of pollution, which are typically categorized by environmental factors, are air, water, and land contamination. Particular pollution kinds, such as plastic, light, and noise pollution, are also an issue for modern culture. All forms of pollution frequently have an adverse influence on human health and well-being as well as on the environment and wildlife (Nathanson 2024). Although environmental pollution is not a recent occurrence, it continues to be the biggest issue facing humanity and the primary cause of disease and mortality in the environment. Human activity is the primary cause of environmental contamination worldwide, including mining, industry, urbanization, and exploration. This burden is shared by developed and developing countries equally, while environmental protection in richer countries has benefited more from increased awareness and stronger legislation (Ukaogo et al. 2020). The absence of creative and practical ways to lessen the problem of environmental pollution is a significant and rising worry in society. One enormous issue of worldwide concern is the uncontrolled discharge of toxins into the environment as a result of industrialization and urbanization. The sustainable use of natural resources to maintain a safe environment for both the current and future generations is the ethical and responsible relationship between humans and the environment.

One of the main problems facing industrial civilizations, both developing and developed, is soil degradation brought on by an increase in human activity. Petroleum is one of the main soil pollutants; during production, operation, use, and transportation, it releases poisonous hydrocarbons that harm both terrestrial and aquatic ecosystems. Microorganisms that are contaminated with petroleum have different metabolic processes. Numerous oil wells dispersed around the area have caused serious petroleum contamination in China's northern Shaanxi Province (Gao et al. 2022). Hydrocarbons derived from petroleum are the most common type of environmental contamination. An unspoiled environment's characteristics are quickly altered by the presence of petroleum hydrocarbons, which lowers ecosystem functionality. In addition to direct toxicity, hydrocarbon exposure causes plants to become unable to receive light, nutrients, and water because the oil obstructs their capacity to pass through the soil matrix. All of these effects severely reduce plant productivity. Weathering is the process that takes place when petroleum hydrocarbons get into the environment. Physical (dispersion), physiochemical (evaporation, dissolution, sorption), chemical (photo-oxidation, auto-oxidation), and biological (plant and microbial hydrocarbon catabolism) factors all play a role in the weathering of petroleum hydrocarbons in the environment (Truskewycz et al. 2019).

Heavy metal (HM) contamination of soil and water is a global concern since it harms human health. It is also one of the most serious environmental problems that lowers crop yield and either directly or indirectly endangers the survival of nearly all living things on Earth. HMs are well recognized as environmental pollutants because of their toxicity, enduring nature in the environment, and capacity for bioaccumulation. As a result, heavy metals including cadmium, chromium, arsenic, mercury, lead, nickel, copper, zinc, and so forth are released in excess into natural resources like soil and marine environments (Mohammad Ali et al. 2021). Toxic metals that plants absorb leave chemical residues on produce that is sold, which can trigger mutagenesis reactions and cause cancer in humans. Since wildlife depends on plants, heavy metal pollution also affects them, upsetting mother nature's delicate balance and lowering biodiversity (Das et al. 2023). To remove heavy metals (HMs) from contaminated soil and water, remediation strategies based on physical, chemical, and biological approaches are currently widely used (Liu and Tran 2021). Even at low concentrations, heavy metals are dangerous and pose serious dangers of exposure to both humans and animals.

Antibiotics have a high potential for bioaccumulation in the environment, are nonbiodegradable, and can endure for extended periods in aquatic environments. The excessive use of antibiotics may lead to antibiotic-resistant disease. The treatment of antibiotic-resistant diseases is predicted to cost the US economy $35 billion annually, placing a heavy financial strain on nations such as the USA. Worse, due to technological and financial obstacles, the rate of antibiotic discovery has decreased over the past few decades, creating an "antibiotic crisis." World leaders were prompted by this prediction to demand an immediate cessation of antibiotic use. Over the past few decades, there has been a steady growth in the usage of antibiotics worldwide, which can be attributed to growing use in both human medicine and other economic sectors. For instance, the amount of antibiotics used on cattle reached 63,151 tons in 2010 and is expected to rise by an additional 67% by 2030. Due to intensive farming, aquaculture—the food industry with the fastest global growth—is likewise using more antibiotics. Because of this, pharmaceutical-grade antibiotics are now widely present in man-made ecosystems including sewage and wastewater treatment plants (WWTPs). Furthermore, the presence of antibiotic compounds in freshwater, marine, and terrestrial habitats is growing due to inadequate regulation of antibiotic pollution at both local and global levels (Kraemer et al. 2019).

The presence of dyes in water streams has ordinary impacts on organisms. Since 1,000,000 tons of dyes are produced worldwide, they are used by small- and large-scale enterprises such as tanneries, food, cosmetic, textile, and pharmaceutical industries. The textile industry is mostly responsible for dye emissions into the environment. Only industries that are fading out release roughly 7.5 metric tons a year. The dye's complex structures, which are made up of aromatic rings bound to various functional groups with an electron, can absorb light in the 380–700 nm range. The presence of chromogens and chromophores gives them color. Several synthetic and natural dyes are very carcinogenic, although azo group proliferation has been linked to amine and benzidine emissions. In addition, the dye molecules' inability to biodegrade prolongs their half-life in the environment, posing risks (Maheshwari et al. 2021). The area's residents and other species are exposed to the possibly dangerous effects of wastewater discharge from textile and industrial manufacturing due to the growth in textile manufacturing. Numerous dye varieties released from textile effluent negatively impact the aquatic ecosystem. Previous studies reported the various dye types and their uses in the textile business, including azo, sulfur, acidic, basic, and direct dyes. Nonetheless, a variety of industries, such as those that make food, medications, paper, cosmetics, textiles, leather, and other goods, commonly use azo dyes (Tripathi et al. 2023).

To mitigate environmental pollution, several sustainable approaches can be implemented including phytoremediation, phycoremediation, mycoremediation, nanotechnology, and green technology. Bioremediation is a novel technique that can be used in a variety of soil and water settings because of the adaptability of microorganisms to remove harmful pollutants. Microbial bioremediation is a potential technique to reduce these pollutants from the contaminated water since several methods have been tried to remove the contaminants and maintain water quality (Saravanan et al. 2023). According to studies, the most effective way to change pollutants into less hazardous and mobile forms or to mineralize hydrocarbon contaminants into H_2O, CO_2, and microbial biomass is by microbial bioremediation, which uses bacteria, fungi, and microalgae (Kebede et al. 2021).

Microbial remediation can be categorized into three types based on the use of microbes namely bacterial bioremediation and mycoremediation (fungi and yeast), and phycoremediation (microalgae). Bacterial remediation is the use of bacteria either aerobic or anaerobic in the bioremediation process. Bacteria have a great potential for eliminating organic pollutants since they are highly suited for biological processes, easy to cultivate, and multiply more quickly. Extensive studies have been conducted on the capacity of bacteria in anaerobic settings to decolorize azo dyes. On the other hand, the azoreductase enzyme breaks azo bonds as a result of the activity of redox mediators. In aerobic settings, however, certain species of bacteria can decolorize wastewater. There are several benefits of using microbial consortia over pure cultures. Numerous strains can target the dye chemicals in various locations. Moreover, the breakdown products generated by the metabolic activity of one strain might serve as a substrate for other strains in the culture (Saeed et al. 2022). Bacteria such as *Nitrosomonas*, *Xanthobacter*, *Arthrobacter*, *Pseudomonas*, *Bacillus*, *Mycobacterium*, *Corynebacterium*, and *Flavobacterium* have been reported to perform bioremediation under aerobic conditions. Nevertheless, several microorganisms have been employed in the anaerobic bioremediation process, including sulfate-reducing bacteria, *Aeromonas*, and *Pseudomonas* (Bala et al. 2022).

One economical and environmentally friendly way to remediate wastewater is through bioremediation. Fungi are rarely employed in biological wastewater treatment systems, whereas bacteria predominate. As the most efficient waste decomposers on Earth, fungi secrete a variety of extracellular enzymes, such as laccase, catalase, peroxidase, and others, to support the soil food web. Polycyclic aromatic hydrocarbons, pesticides, effluent from hospitals, breweries, and the food sector are all treated with them. Since fungi have functional groups in their cell walls that may bind metal, such as hydroxyl, amine, and carboxyl, they are superior metal sequesterers. The filamentous fungi that create pellets are characterized by their huge specific area, porosity (which facilitates mass and oxygen transport), and net-like gaps on their surface (Negi and Das 2023). For the removal of heavy metals and dyes, a wide variety of molds and filamentous microorganisms are employed, including *Saccharomyces cerevisiae*, *Aspergillus niger*, *Penicillium simplicissimum*, *Aspergillus fumigatus*, *Termitomyces clypeatus*, *Penicillium brevicompactum*, and *Trichoderma* (Koul et al. 2021).

Intending to detoxify and decolorize the dangerous dye-laden effluents, phycoremediation is becoming a new era technique. The fundamental processes involved in the phycoremediation of dyes are biodegradation, biosorption, and biocoagulation. These processes are impacted by several physical, chemical, and biological variables, including pH, temperature, contact time, surface features, algal biomass concentration and particle size, and dye concentration and composition (Sarkar and Dey 2021). The microalgae with great promise for wastewater treatment include Scenedesmus, Spirulina, and Chlorella. The ability of these microscopic organisms to absorb and accumulate nutrients, organic pollutants, heavy metals, and other contaminants from wastewater results in an environmentally friendly and cleaner effluent. Microalgae's high photosynthetic efficiency makes it possible for carbon dioxide (CO_2) to be converted into biomass, providing the combined benefits of carbon sequestration and wastewater treatment (WWT). Algae effectively lower the carbon footprint (the amount of carbon dioxide in the atmosphere that causes pollution, global warming, and greenhouse effects) and are a great carbon dioxide sink. They are generally categorized as macroalgae, microalgae, and marine algae—often referred to as seaweeds—because they are abundant in nature and well-suited to a variety of settings. With an estimated 200,000–800,000 species globally, microalgae are rich in biodiversity. Because of their remarkable adaptogenic capacity, which enables them to thrive in various conditions, they are excellent candidates for both biofuel production and WWT (Koul et al. 2022).

Soil pollution has been the main application for phytoremediation. The use of plants in bioremediate water has increased in availability due to the development of hydroponic techniques and a mechanical approach. The morphological and physiological characteristics linked to the pollution uptake, compartmentalization, volatilization, filtration, and numerous other processes, nevertheless, need to be comprehended to create a replicated water phytoremediation model. Numerous studies have demonstrated the value of plants in the treatment of both organic and inorganic contaminants, including those that may be resistant to treatment or difficult to metabolize. Heavy metals are significant pollutants that several approaches attempt to address and that may persist in water and sediments. On the other hand, phytoremediation is a new method that can aid in the removal of HMs (Delgado-González et al. 2021). Many phytoremediation techniques can be used to clean up heavy metal-contaminated soils. These techniques include (i) phytostabilization, which uses plants to lower the bioavailability of heavy metals in the soil, (ii) phytoextraction, which uses plants to extract and remove heavy metals from the soil; (iii) phytovolatilization, which uses plants to absorb heavy metal from the soil and release it into the atmosphere as volatile compounds; and (iv) phytofiltration, which uses hydroponically cultured plants to absorb or adsorb heavy metal ions from groundwater and aqueous waste (Yan et al. 2020). Native plant-based phytoremediation can be used for both ecological restoration and site cleanup. Native plants' rhizospheres serve as the perfect home for a variety of inorganic and organic pollutants' enzymatic accumulation, stabilization, biodegradation, or volatilization. In a case study comparing *Medicago sativa* as the reference plant and *Chromolaena odorata*, a native plant from crude oil-polluted land in Nigeria, for polycyclic aromatic hydrocarbons (PAHs) remediation, it was found that the native plant degraded, thrived, and could withstand PAHs better than the reference plant, but there was no discernible difference in PAH degradation. Phytoremediation with native species is easier to cultivate, less expensive, safer for the environment, and socially and ecologically acceptable than non-native alternatives. It may also be more effective and efficient than non-native alternatives (Futughe et al. 2020).

Owing to the special qualities of nanoparticles, nanotechnology has many uses in the food, agriculture, energy, and environmental sectors. Environmental nanotechnology is not only capable of addressing many environmental issues, but its products and processes are also regarded as the most inventive and efficient means of achieving sustainability objectives. To produce biofuel, treat wastewater, reduce greenhouse gas emissions, and remediate the environment sustainably, nanostructured materials such as nanocomposites, functionalized nanomaterials, metal–organic frameworks, nanocatalysts, carbonaceous materials, nano zeolites, nano silica, nanolubricants, and nanocoatings, among others, offer a plethora of opportunities (Chausali et al. 2023).

Since Brawn and Wield initially proposed the idea of "green technology" in 1994, as environmental issues have gotten worse, the value of innovation in this sector has been acknowledged more and more. The use of creative and environmentally friendly solutions to solve problems with the environment and advance sustainable development is known as "green technology." It covers a broad range of industries, such as waste reduction, energy-efficient transportation, green building materials, and renewable energy (Zhou et al. 2023). Green technology is friendly to the environment and works to preserve it. It can also be used to describe the generation of clean energy, which entails the use of environmentally friendly fuels and alternative technologies. Infrastructure technology is used to conserve natural resources, recycle trash, filter water, and produce clean energy, among other things (Mohammed 2021). Reducing air pollution, lessening soil contamination, and slowing down global warming are benefits of green technology. This strategy also reduces plastic pollution and water contamination. Green technology might reduce waste through efficient recycling, preserve natural resources, and enable us to feed more people worldwide.

2 Types of Environmental Pollution

2.1 *Air Pollution*

Air pollution is the term used to describe any chemical, physical, or biological factor that tampers with the natural properties of the atmosphere to contaminate the indoor or outdoor environment. Common causes of air pollution include motor vehicles, industrial operations, household combustion appliances, and forest fires. Particulate matter, ozone, nitrogen dioxide, sulfur dioxide, and carbon monoxide are among the pollutants that pose a serious threat to public health. Air pollution, both indoors and outdoors, is a major cause of morbidity and mortality and causes diseases such as respiratory disorders.

2.2 *Water Pollution*

Water pollution is a worldwide problem, and to address it, scientists and decision-makers are developing new strategies and mitigating techniques. When dangerous materials typically chemicals or microbes contaminate a stream, river, lake, ocean, aquifer, or other body of water, the water quality decreases, and the water becomes hazardous to people or the environment.

2.3 *Land Pollution*

Land pollution is the deposition of solid or liquid waste materials on land or underground in a manner that can contaminate the soil and groundwater, threaten public health, and cause unsightly conditions and nuisances. It is caused by human activities such as littering and waste washed ashore from boats, oil rigs, and sewage treatment plants. The waste materials that cause land pollution are broadly classified as municipal solid waste (MSW), construction and demolition (C&D) waste or debris, and hazardous waste. MSW includes nonhazardous garbage, rubbish, and trash from homes, institutions, commercial establishments, and industrial facilities. C&D waste (or debris) includes wood and metal objects, wallboard, concrete rubble, asphalt, and other inert materials produced when structures are built, renovated, or demolished. Hazardous wastes include harmful and dangerous substances generated primarily as liquids but also as solids, sludges, or gases by various chemical manufacturing companies, petroleum refineries, paper mills, smelters, machine shops, dry cleaners, automobile repair shops, and many other industries or commercial facilities.

2.4 Thermal Pollution

Any process that modifies the ambient water temperature can lead to thermal pollution, which is the deterioration of water quality. When heat is produced by elevated industrial effluents or by anthropogenic (human) modifications to stream bank vegetation that boosts the water system's temperature as a consequence of solar radiation, it is regarded as a water pollutant. The use of water as a coolant by power plants and industrial enterprises is a common source of thermal pollution. The abrupt temperature shift that occurs when coolant-treated water is released into the natural environment lowers oxygen availability and has an impact on the ecosystem (Speight 2020).

2.5 Industrial Pollution

Industrial pollution refers to the massive amounts of garbage that are released from several types of industrial plants (Red, Orange-A, Orange-B). Manufacturing and industrial processes produce pollutants or industrial waste. Water bodies are receiving an increasing amount of wastewater due to the quick production growth. The primary sources of toxins that disturb marine biota and deteriorate water quality are thought to be solid and liquid industrial wastes that are seeping into watercourses either physical, chemical, or biological (Salam and Sultana 2022).

2.6 Soil Pollution

Soil pollution is the contamination of soil with anomalous concentrations of toxic substances. It can be caused by various processes such as agriculture, industry, or poor waste disposal. Soil pollution can have serious health and environmental consequences, such as cancer, leukemia, and water pollution. Some of the most hazardous soil pollutants are xenobiotics—substances that are not naturally found in nature and are synthesized by human beings. Heavy metals such as lead, mercury, cadmium, and polycyclic aromatic hydrocarbons (PAHs) are some of the common pollutants that contaminate soil.

3 Sustainable Approaches—Nanotechnology

Numerous applications in the fields of agriculture, food, energy, and the environment are possible thanks to nanotechnology. Nanotechnology is not only capable of addressing many environmental issues but its products and processes are also regarded as the most inventive and efficient means of achieving sustainability objectives. Nanostructured materials such as nanocomposites, functionalized nanomaterials, metal-organic frameworks, nanocatalysts, carbonaceous materials, nano zeolites, nano silica, nano lubricants, and nano coatings, among others, offer a plethora of opportunities to produce biofuel, treat wastewater, reduce greenhouse gas emissions, and remediate the environment sustainably (Chausali et al. 2023).

3.1 Nanopowder

Nanoparticles offer hitherto unheard-of prospects to address global warming issues in water purification, renewable energy technologies, greenhouse gas management, materials supply and utilization, green manufacturing, and chemistry. Nanoparticles can range in size from 1 to 100 nm and are made up of several atoms and molecules bonded together. Their physical and chemical properties differ because of their small size and high surface area-to-volume ratio. Hence researchers are focusing on searching for new sustainable eco-friendly products (Dinesh et al. 2023). Nanopowder is a powdery substance made up of minuscule particles ranging in size from one to one hundred nanometers (Wang et al. 2019). It is also known as an ultrafine particle or nanoparticle. They are invisible because the particles in nanopowders are smaller than visible light's wavelength. Nanopowders offer remarkable qualities including tremendous strength because of their incredibly high surface area-to-volume ratio.

3.2 Nanocatalyst

Catalytic technologies are essential to the chemical process, energy, and environmental industries both now and in the future. Catalytic technologies are used in the manufacturing of various petrochemical and chemical products, the conversion of crude oil, coal, and natural gas into fuels and chemical feedstock, and the control of carbon monoxide (CO), hydrocarbons, and nitric oxide (NO) emissions. Nanocatalysis has proven to be an excellent and efficient way of purifying water areas, and it plays a significant part in the mineralization of harmful organic compounds at room temperature. Additionally, the majority of nanocatalysts have certain benefits, such as being inexpensive, less hazardous, chemically stable, easily accessible, and possessing outstanding photoactive characteristics at the nanoscale, or within the range of 1–100 nm. Broadly speaking, nanocatalysts have several advantages, including low cost, environmental issues, and stability (both chemical and physical).

For instance, the most popular nanocatalyst, TiO_2, has strong photostability. However, due to their photo-corrosion property, many other nanocatalysts, including zinc oxide, metal sulfide, copper-based materials, etc., have relatively low chemical stability. Depending on the materials, they either oxidize or decrease when exposed to light. The oxidation levels of these materials fluctuate as a result of the generation of holes or electrons, which ultimately causes the photocatalysts to degrade and lose efficiency. As a result, the synthesis of nanocomposites is desperately needed in order to provide robust photocatalysts for extended use. The effect of quantum size, which expands the energy range and contracts particle size, is the primary benefit of the nanoscale (Wassel et al. 2020).

Numerous benefits of the photodegradation process include its low cost, reusability, and generally full decomposition. Despite all of the advancements, there are still issues with the nanocatalyst, including toxicity and catalyst recovery from the mixture. This kind of issue restricts the nanocatalyst's use and range to a greater degree. As a result, alternative nanocomposites composed of various components that can lessen toxicity in water treatment are currently the focus of scientific attention. It is therefore recommended that to achieve stable results, the academic community synthesize novel photocatalysts that can work in visible areas and encourage the use of photocatalysts doped with other materials, such as graphene and its derivatives, to lessen their hazardous effects (Wassel et al. 2020).

3.3 Nanotubes

Nanotubes are a special kind of material because their shape affects their qualities in addition to their composition. The diameter, van der Waals forces, chirality, number of walls, length, and quality all influence the characteristics and functionality of nanotubes (Schulz et al. 2014). Researchers are interested in nano-sized materials because of their exceptional qualities, which might be effectively applied to a wide range of tasks. There are two types of sp2 hybridized carbon nanotubes (CNTs), which are intriguing forms of elemental carbon. Single-walled carbon nanotubes (SWCNTs) have a single-layer cylinder, whereas multi-walled carbon nanotubes (MWCNTs) have concentric rings with a distinct space between the layers. Significant progress has been made in the utilization of CNTs for the efficient removal of metal ions and pollutants as they possess large surface area, high mechanical strength, and excellent thermal, and electrical properties and are small, hollow, and layered structures (Parlayici et al. 2015). Several MWCNTs adsorbents were developed for the treatment of aqueous solutions contaminated with Cr(VI). A previous study revealed that the developed magnetic Fe_2O_3 nanoparticle-MWCNTs composite was checked for the efficient removal of Cr(VI) at different temperatures with a qmax of 42.02 mg g^{-1} at 35 °C (Lua et al. 2017). Fe-Ag/f-MWCNT/PES Nanostructured-Hybrid Membranes developed by (Masheane et al. 2017) could remove 94.8% of Cr(VI) from an aqueous solution. Chitosan was immobilized in nanoparticles and carbon nanotubes by forming a nanocomposite that could efficiently remove Cr(VI) up to 84% within a short contact time (Salam 2017).

3.4 Nanocomposites

Polymers containing nanoparticles less than 100 nm are called nanocomposites. There are three types of nanocomposites: metal matrix, ceramic matrix, and polymer-matrix. Electrically conductive polymer-matrix nanocomposites can be synthesized to create nanocomposite films for application in nanoelectronics devices (Khalifeh 2020). Due to their heterogeneous nature, nanocomposites' characteristics are influenced by the same elements that affect traditional composites: component characteristics, composition, structure, and interfacial interactions. However, compared to microcomposites, their structures are typically more complex; this is particularly true for polymer/layered silicate nanocomposites. In addition to the commonly recognized individual silicate platelets and tactoids, layered silicate nanocomposites have the potential to include larger particles and the formation of a silicate network at high exfoliation levels (Hári and Pukánszky 2011).

3.5 Nanofilms

Nanofilms are extremely thin layers of material that range in thickness from a few nanometers to several micrometers. They stand in for an atomic-thick boundary between the surrounding environment where the bulk of physicochemical processes occur. As a result, a thin layer of a certain material can affect how it behaves and give surface and bulk engineering leverage (Ranzoni and Cooper 2017). Nanofilm created by the electrospinning technique is more susceptible to an alkaline environment due to the increased porosity. A gentle and flexible technique that makes it simple and flexible to create nanoscale fiber films is electrospinning. The large specific surface area and high porosity of the resulting nanofilm are advantageous since they increase the material's ability to respond to volatile ammonia. Moreover, the encapsulation efficiency of both pH indicators and antimicrobial agents is high, allowing for the integration of the intelligent packaging system with active packaging (Wu et al. 2024). A smart electrospun bilayer made of poly(lactic acid)-Viola odorata extract and carboxymethyl cellulose-cellulose was found to be effective in monitoring the freshness of rainbow trout filets, chicken fillets, minced lamb meat, and Pacific white prawns in real-time in a prior study (Rezaei et al. 2023). Biodegradable and sustainable biopolymers are chosen over synthetic polymers due to the growing emphasis on environmental conservation. Ethyl cellulose/gelatin-based multifunctional nanofilm, which serves as an environmentally benign and biodegradable packaging material obtained from natural sources, offers a wide range of potential uses in active and intelligent food packaging (Wu et al. 2024).

4 Sustainable Approaches—Bioremediation

The development of bioremediation offered a sustainable, economical, and environmentally acceptable alternative for the treatment of hazardous waste. To remove or break down inorganic and organic pollutants, some bacteria and other microorganisms are applied. Certain bacteria can release enzymes from contaminated locations that can remove dangerous contaminants. Depending on the kind and concentration of contaminants, the cost, site features, and other factors, bioremediation can be carried out either *in-situ* or *ex-situ*. Site characterization, which determines the most appropriate and effective bioremediation strategies, is the most important phase in the successful use of bioremediation (Xiong et al. 2022).

4.1 Phycoremediation (Microalgae)

There is a lack of water resources that are fit for direct human consumption as a result of the world economy and population growing so quickly. Water cleanup will hence unavoidably take the front stage on a worldwide basis. Wastewater of several kinds can support the growth of microalgae (WW). They have a great potential to eliminate pollutants from urban and industrial effluents (Abdelfattah et al. 2023). Phycoremediation is the use of macroalgae or microalgae for the removal or biotransformation of pollutants, including heavy metals (HMs), nutrients, and other toxins such as xenobiotics from wastewater and CO_2 from waste air. Algae effectively reduces the carbon footprint (the amount of carbon dioxide in the atmosphere that causes pollution, global warming, and greenhouse effects) and is a great carbon dioxide sink. They are generally categorized as macroalgae, microalgae, and marine algae often referred to as seaweeds because they are abundant in nature and well-suited to a variety of settings (Koul et al. 2022). The exceptional ability of some microalgal species, such as Scenedesmus, *Chlorella, Botryococcus, Phormidium, Limnospira* (previously *Arthrospira, Spirulina*), and *Chlamydomonas*, to bioremediate nutrients, HMs, emerging contaminants, and pathogens from WW has been confirmed. In WW, several microalgae species have grown effectively, including *Scenedesmus, Chlorella,* Euglena, *Oscillatoria, Chlamydomonas,* and *Ankistrodesmus* (Abdelfattah et al. 2023). They also show excellent resistance to WW toxins. Thus, for the treatment of livestock, agro-industrial, municipal, and industrial wastewater, microalgae-based methods have attracted a lot of attention lately. Microalgae can reduce the chance of eutrophication by taking out P and N (Priyadharshini et al. 2021). These are regarded as multipurpose substitutes for biological processing, converting unwanted organic and inorganic components into valuable biomass (Mastropetros et al. 2022). Additionally, microalgae harvested from specific treatment ponds can serve a variety of purposes in a wide range of sectors and be utilized as a source of food and feedstock for numerous products (Abdelfattah et al. 2023).

4.1.1 Wastewater Management

Wastewater management, which is the process of gathering, treating, and reusing wastewater, is a crucial strategy for safeguarding water resources. Unwanted performance in the wastewater acquiring network, one of the crucial infrastructures, can result in a variety of issues. Furthermore, a variety of physical and chemical problems affect the wastewater treatment process, which is the reason factor prediction in wastewater treatment plants is crucial. On the other hand, municipal treated wastewater is a suitable resource to allocate to different uses, like industry and agriculture; nevertheless, inappropriate use of these resources can result in threats to the environment and human health, which in turn can cause social unrest and occasionally economic difficulties (Roozbahani 2021).

4.1.2 Biohydrogen Production

As a potential solution to the majority of the issues that fossil fuels cause, biohydrogen generation is regarded as a development that is essential to a sustainable, clean energy source for the entire world. The main benefit of using biohydrogen for electricity is that it doesn't produce greenhouse gas emissions because pure water is the sole byproduct of burning hydrogen or using fuel cells to convert it into energy. The intended clean result of the microbial process would be biohydrogen with the application of suitable technologies (Show et al. 2011).

4.1.3 Biodiesel

Biofuel is described as a liquid, solid, or gaseous fuel created by converting biomass, such as biogas from anaerobic waste decomposition, bioethanol from sugar cane or corn, and charcoal or woodchips. Fuels made from organic materials, such as plants and animals, are referred to as biofuels. These fuels derived from plants and animals include, among others, liquid fuels like acetone, methanol, ethanol, and biodiesel, as well as gaseous fuels like hydrogen and methane (Muiruri et al. 2023).

Biodiesel production using microalgae is one of the sustainable methods to mitigate air pollution. Fatty acid alkyl esters, or "biodiesel," are an environmentally friendly substitute fuel for diesel engines that have gained a lot of attention. Biodiesel is a renewable, biodegradable, and nontoxic fuel with several benefits (Tamalampudi and Fukuda 2011). Proteins, carbohydrates, and lipids are all present in significant proportions in the microalgae biomass that is produced, and because microalgae are photosynthetic microorganisms, they can also produce a wide range of other valuable compounds that serve as a precursor to the generation of biofuels. Notably, for CO_2 biofixation and biofuel production, microalgae and cyanobacteria have been discovered including *Anacystis*, *Botryococcus*, *Coelastrum* sp. SM, *Chlamydomonas*, *C. vulgaris*, *Emiliania*, *Gracilariopsis*, *Monoraphidium*, *Rhodobacter*, *Synechococcus*, and *Sc. Obliquus* (Li et al. 2022).

There are two major steps involved in biodiesel production which are the extraction of microalgae lipid components and the transesterification of microalgae lipids. It is possible to further convert the lipid that was isolated from microalgae into biodiesel. Microalgae biomass must be dehydrated before extraction in order to prolong both its own shelf life and that of the finished product. Certain microalgae species, like *Chlorella*, *Spirulina*, and *Scenedesmus*, have been dried using a variety of methods; the most popular ones include freeze-drying, spray-drying, drum-drying, and sun-drying (Mata et al. 2010). Biodiesel is unable to be produced without further conversion following the extraction of microalgae oil. The most common process for producing biodiesel from microalgae lipids is transesterification (Zhou et al. 2012). Triglycerides are typically combined with methanol (methanolysis) to make monoglycerides and diglycerides, which ultimately produce fatty acid methyl ester (FAME, also known as biodiesel) and corresponding glycerol (byproducts) in a multi-step continuous reaction termed transesterification (Gong and Jiang 2011).

4.2 Phytoremediation

The approach of phytoremediation employs plants to clean up contaminated media, such as soil and water. Because it uses green plants to retain, sequester, or detoxify toxins from contaminated soil and water, phytoremediation is a technology that benefits both the environment and the economy (Ashraf et al. 2019). To degrade, remove, or incapacitate the contaminants, phytoremediation employs a variety of mechanisms, such as degradation (rhizodegradation, phytodegradation), accumulation (phytoextraction, rhizofiltration), dissipation (phytovolatilization), and immobilization such as hydraulic control and phytostabilization (Kafle et al. 2022). Moreover, Kafle et al. (2022) reported several processes involved in the phytoremediation process including phytoextraction or phytoaccumulation, phytostabilization/phytoimmobilization, rhizofiltration, phytovolatilization, phytodegradation and rhizodegradation, Phytodesalination.

Phytostabilization reduces the number of inorganic contaminants that are transported by water or dust by immobilizing them in the soil, such as heavy metal(loid)s. By utilizing microbial activity connected to plant roots, this method has the potential to speed the breakdown of organic pollutants like pesticides and hydrocarbons into harmless forms (Bolan et al. 2011). A promising method for remediating contaminated soil is phytoextraction or phytoaccumulation, which can quickly absorb heavy metals and remove them from the soil. Plants possess an innate ability to absorb and retain nutrients based on their soil bioavailability and the needs of the plant (Ghori et al. 2016).

Similar to phytoaccumulation, rhizofiltration is a phytoremediation method that differs mostly in the kind of polluted environment that is preferentially remediated. Rhizofiltration is used to clean up surface or groundwater, particularly when it has been contaminated by metals, whereas phytoaccumulation is primarily used to clean up soil. Rhizofiltrating plants can store pollutants in their roots, which allows them to absorb them. This method can also be used for *ex-situ* remediation, negating the requirement to undertake toxicity tests on shoots (Tonelli et al. 2022). Numerous organic molecules can interact with plants, influencing the movement and destiny of various environmental pollutants. Direct phytovolatilization of volatile organic compounds occurs when they are released from stems or leaves, while indirect phytovolatilization occurs when plant roots are involved. From localized contaminant leaks to global methane fluxes originating from ecosystems that biochemically reduce organic carbon, the volatilization of pollutants from plants is significant on a variety of scales. Numerous pollutants, both inorganic and organic, have been shown to undergo phytovolatilization. Plants can release volatile versions of several inorganic chemicals, such as selenium, arsenic, and mercury (Limmer and Burken 2016).

The microbial decomposition of organic pollutants in the soil and sediment of the root zone (rhizosphere) is referred to as rhizodegradation. This method lowers the concentrations of organic contaminants in soil and sediment by using plants' innate ability to control the biological, chemical, and physical properties of the rhizosphere. In plant-assisted remediation efforts, rhizoremediation was proposed as the main mechanism accountable for PAH breakdown in sediment (Lu et al. 2011). Organic substances can undergo phytodegradation either inside the plant or in its rhizosphere. This technique can be used to remove a wide range of different substances and classes of compounds from the environment, including volatile compounds in the air, petroleum and aromatic compounds in soils, and solvents in groundwater (Newman and Reynolds 2004). Since salinity is one of the abiotic stressors that reduce plant output, efforts are being made to develop more tolerant varieties of plants. Using extremely productive halophytes for desalination is one of the practical and affordable reuse solutions. Halophytes possess an intrinsic capacity to adjust to the salinized environment via various methods of removing or accumulating salt. Through traditional breeding programs, halophytic cash crops can be domesticated, increasing their productivity and tolerance to salt. Several salt-tolerant crops and halophytes have been effectively used in the desalination process,also, a small number of commercial halophytes have been grown on degraded land, indicating that these plants would make excellent bioprospecting candidates (Saddhe et al. 2020).

Generally, phytoremediation uses plants to absorb and eliminate elemental pollutants or reduce their bioavailability in soil heavy metals through bioabsorption (biosorption), bioaccumulation, bioleaching, and bioprecipitation. Through their roots, plants may take up ionic substances from the soil, even in very little amounts. In order to recover contaminated soil and stabilize soil fertility, plants grow their roots into the soil matrix and create a rhizosphere ecosystem, which collects heavy metals and modifies their bioavailability. Using phytoremediation has the following benefits: (a) It is economically possible since phytoremediation is an autotrophic system that runs on solar energy, making it easy to maintain and requires little upkeep. Its benefits include: (b) being environmentally friendly and less harmful to the environment—it can lessen the amount of pollutants that are exposed to the environment and ecosystem; (c) being easily applied over a large field and disposed of; (d) preventing erosion and metal leaching by stabilizing heavy metals, which lowers the risk of contaminants spreading; and (e) improving soil fertility by releasing various organic matters into the soil (Yan et al. 2020). Four key traits identify plants that are good candidates for phytoremediation: (1) large biomass and quick growth; (2) a deep root system; (3) easy harvesting; and (4) accumulation of high concentrations of heavy metals in shoots (Nedjimi 2021).

The phytoremediation effect is more impacted by ornamental plants. For instance, flowering plants with lots of potential to accumulate heavy metals like Cd, Cr, and Pb include African marigold, French marigold, Celosia, Tubose, Chrysanthemum, and more. The importance of industrial landscaping has grown recently as a means of preserving a sustainable and clean environment. Since most plants are hazardous to heavy metals, growing plants becomes a laborious task. Several hyper accumulator plants, including marigold, celosia, tuberose, and others, as well as some trees, like neem and willow, can be grown to get around the restriction and improve the soil. Because flowering plants are non-edible and have the potential to be a profitable endeavor, using them offers the advantage of preventing heavy metals from entering the food chain (Sadasivam and Jawaharlal 2022).

4.3 Mycoremediation

Mycoremediation is an affordable, environmentally responsible, and successful method of addressing the growing issue of soil and water contamination. Fungi are a great option for the remediation of a variety of pollutants due to their robust growth, extensive hyphal network, production of versatile extracellular ligninolytic enzymes, high surface area-to-volume ratio, resistance to heavy metals, flexibility in responding to changes in pH and temperature, and presence of metal-binding proteins (Akhtar and Mannan 2020). Mycoremediation is more successful than bacterial remediation procedures due to several factors such as greater fungal biomass than bacterial biomass, a network of hyphae, a high surface-to-cell ratio, tolerance to changes in pH and temperature, rapid development, and the presence of metal-binding proteins and polysaccharides of fungus (Akhtar and Mannan 2020). Fungi remediate organic and inorganic pollutants, including heavy metals, by biosorption or bioreduction. Through the mechanism of biosorption, which allows microbial biomass to collect organic and inorganic species actively or passively in aqueous solutions, these harmful substances can be safely and responsibly removed while maintaining a clean environment. Fungal cell walls are rich in functional groups including carboxy, amine, and hydroxyl that aid in biosorption. Fungal biomass is a potentially useful biosorbent due to its high cell wall composition. This is because fungi's capacity to sequester metals is enhanced when the amount of metal-binding functional groups on their cell wall surface increases (Negi and Das 2023).

Fungal enzymes can biodegrade a broad spectrum of contaminants found in industrial effluent. Additionally, they can transform harmful anions, such as selenite, into harmless nanoparticle form. The type of pollutants and extracellular or intracellular enzymes present in the growing media determine whether nanoparticles develop.

Because of their great efficiency, selectivity, and ecologically benign reactions, enzymes are being used more and more to minimize or remove industrial and environmental pollutants. Amylases, catalases, cellulases, laccases, lipases, peroxidases, proteases, and xylanases are examples of fungal enzymes used in bioremediation that have industrial uses and clean up organic waste. Numerous polymeric compounds found in organic wastes, such as food scraps, produce market leftovers, leaves, litter, etc., are hydrolyzed by enzymes derived from fungus. Additionally, the hydrolyzed materials can be composted or used to produce useful products like volatile fatty acids and biogas (Negi and Das 2023).

Fungi can live in a wide range of environments; the main site of fungal colonization is the complex soil matrix. Freshwater and marine environments also exhibit sustained fungal colonization. The majority of fungi can grow in soil found in a variety of climates, including those with extreme temperatures. They can spread by releasing spores into the atmosphere, and they can also aid in preserving the ecosystem's equilibrium. Additionally, it has been documented that they thrive in various waste fluids being treated by effluent treatment plants (ETPs). Fungal variety in habitats and their capacity to secrete a wide range of enzymes make them attractive options for bioremediation at different sites (Deshmukh et al. 2016).

The primary biodegradation of ligneous material in nature, which aids in the global recycling of carbon, is facilitated by white-rot fungus. The breakdown of pharmaceuticals and personal care products (PPCPs) that contain trihalomethanes (TrOCs) and endocrine-disrupting chemicals (EDCs) has attracted a lot of attention due to the potential for bioaccumulation, acute and chronic toxicity to aquatic organisms, and possibly harmful effects on human health. The majority of research has shown that white-rot fungi, including *Phanerochaete chysosporium*, *Trametes versicolor*, *Bjerkandera adjusta*, and *Pleurotus* sp., have the capacity for bioremediation due to their production of several ligninolytic enzymes, including laccases and peroxidases (Santos Bazanella et al. 2013).

With about 5 million species—only 5% of which have been described—fungi are theoretically the most diverse group of unicellular eukaryotes on Earth. Therefore, maritime environments may include a wealth of unidentified fungi with potential uses in biotechnology. It is commonly known that marine fungi can produce novel enzymes, biosurfactants, polysaccharides, polyunsaturated fatty acids, and secondary metabolites. They can also be used in the bioremediation of hydrocarbons and heavy metals. The biological advantage these mushrooms have over terrestrial fungi is their capacity to adapt to high salinities and extreme pH values (Damare et al. 2012). The ability of marine microbes to remove metal ions efficiently suggests that extremophilic organisms hold great promise for both bioremediation and nanotechnology. Thatoi et al. (2013) reviewed the role of marine fungi from mangrove areas with an eye toward their various potential applications. They focused particularly on the diversity, significant ecological role, and biotechnological potential of these fungi as a source of novel drugs, enzymes, biodiesel, biopesticides, and bioremediation. It has been reported that several fungal species, including *Cladosporium*, *Aspergillus*, *Cunninghamella*, *Penicillium*, *Fusarium*, and *Mucor*, are involved in the breakdown of more resistant aromatic hydrocarbons in addition to aliphatic hydrocarbons. Fungal species that are useful for cleaning up contaminated areas include *Pleurotus ostreatus*, *Agaricus bisporus*, *Trametes versicolor*, and *Phanerochaete chrysosporium*. Few studies have investigated the possibility of fungal bioremediation of petroleum hydrocarbons in marine environments, even though several terrestrial fungal taxa have been reported as having a significant role in hydrocarbon biodegradation. However, many studies have shown that fungi found in various maritime environments may effectively break down petroleum hydrocarbons (Dell'Anno et al. 2021).

Microorganisms known as extremophiles are those that can grow and survive in a variety of harsh conditions, such as high or low temperatures, an acidic or alkaline pH, high pollutant concentrations, and salts, among others. These organisms have special physiological and enzymatic traits that enable them to thrive in hostile conditions, making them ideal candidates for environmental biotechnology. These bacteria can be employed for bioremediation of areas contaminated with highly resistant contaminants because of their stability and endurance in harsh environments (Giovanella et al. 2020). At a temperature of 55 °C, Taha et al. (2014) investigated the adsorption of the thermophilic fungus *Thermomucor indicae-seudatica* in a mixture of several azo and anthraquinone dyes at a high concentration (1000 mg L^{-1}). It has been demonstrated that the inactive biomass decolorizes more quickly than the active fungus. An alkali-, thermo-, and halo-tolerant strain of *Aspergillus lentulus* was studied by Kaushik et al. (2014) at various pH levels. They found that the strain preferred to adsorb textile dyes in an acidic environment, eliminating about 94% of the dye in under five hours.

4.3.1 Polycyclic Aromatic Hydrocarbons

Chemical substances with a cyclic arrangement of carbon and hydrogen molecules are known as polycyclic aromatic hydrocarbons or PAHs. Risks to human health, particularly carcinogenesis, are linked to PAHs. The route of exposure to these PAHs is via the consumption of tainted food can happen during high-temperature food preparation and processing (such as grilling, smoking, toasting, roasting, and frying), as well as through exposure to PAHs found in the environmental pollution such as soil, air, and water (Sampaio et al. 2021).

4.3.2 Heavy Metal

Over the past century, industrialization has rapidly expanded. As a result, it has raised demand for the reckless exploitation of Earth's natural resources, aggravating the global environmental pollution issue. Heavy metal pollution is one of the emerging environmental pollution issues. The substances are defined as heavy metals because of their high density or large atomic weight. Heavy metals are metallic chemical compounds and metalloids that are toxic to both humans and the environment. Certain metalloids and lighter metals like aluminum, arsenic, and selenium are hazardous (Briffa et al. 2020).

4.3.3 Agricultural Effluents

Any solid, liquid, or gas that is released into the environment as a byproduct of farming is referred to as an agricultural effluent. The process of pesticides and other chemicals used in agriculture flowing with water is one example of an agricultural effluent. Furthermore, agriculture effluent also involves soil erosion brought on by inadequate farming practices, tank washing operations' slurry, liquid manure, silage juices, brown waters, green waters, wine-making effluents, sediment, metals, organic pollutants, nutrients, and pesticides particularly herbicide and agricultural fertilizer run-off, pesticide use, and livestock effluents.

4.3.4 Pharmaceutical Wastes

The extremely awful surroundings are being rendered worsened by pharmaceutical pollutants that are discharged from the effluents of medication manufacturing facilities. These substances are released into the environment, endangering the health of people and animals as well as significantly reducing the amounts of readily available drinking water. One of the main drug classes that is used extensively worldwide is antibiotics; regrettably, a lot of antibiotics are also released into the environment through hospital waste, human waste, expired medication disposal, pharmaceutical industry effluents, and careless use in the aquaculture industry and livestock husbandry (Akhtar and Mannan 2020).

4.3.5 Phthalates, Dyes, and Detergents

A class of synthetic esters of phthalic acid compounds known as phthalates (PAEs) are widely employed in most products and industries, primarily as plasticizers in plastic materials. They readily leach out of the polymeric matrix since they are plasticizers in plastic materials and are not chemically bonded to it. Although PAEs should naturally be present in the environment, until recently, little was understood about their prevalence, movement, destiny, and consequences. PAEs are classified as contaminants of emergent concern (CECs) because they may have detrimental effects on human and ecological health. They are also known to be carcinogenic, endocrine-disrupting compounds, and agents that cause liver and kidney failure (Baneshi et al. 2024). The chemicals called dyes are used to provide color to leather, textiles, and other materials. These brightly colored materials form chemical bonds with any surface they are applied. After dyeing, the colored materials may seem appealing, but the waste that they produce is harmful to the environment. Because so much industrial waste ends up in rivers, lakes, etc., dyes harm water bodies and contribute to pollution (Garg and Chopra 2022). Among the various pollutants, detergent poses a significant harm to natural ecosystems as a pollutant. Detergents can also enter wastewater treatment facilities and negatively impact their operation. They are a necessary component of human life and are consumed for a variety of reasons, most notably hygiene. Detergent components can therefore find their way into soil and water bodies through a variety of channels. Detergents have both direct and indirect effects on ecosystems, affecting both flora and fauna. The consequences of eutrophication, foaming, and changing temperature, salinity, turbidity, and pH must be monitored and regulated since they are increasingly significant. The majority of detergents can be broken down by aerobic processes, according to research, but anaerobic breakdown is impossible due to the toxicity and limited metabolic routes of detergents (Mousavi and Khodadoost 2019).

4.4 Microbial Bioremediation

The technique known as "microbial remediation" uses microorganisms to reduce the bioavailability of pollutants, particularly organic toxins, therefore reducing their ecological toxicity. It is well acknowledged that microbial bioremediation is an economical, adaptable, quick, and environmentally responsible way to eliminate pollutants from the environment. Using live or dead biomass, bioremediation is the process of transforming, degrading, and detoxifying contaminants to an innoxious state and then mineralizing them into substances like carbon dioxide, nitrogen, and water (Yadav et al. 2018). For heavy metal bioremediation, cyanobacteria, algae, bacteria, plants, and fungi are studied and used extensively. The best candidates for the job are microorganisms since they are simpler to work with, cultivate, and use. The capacity of microorganisms, particularly bacteria, to adsorb, solubilize, and precipitate heavy metals through the processes of bioabsorption (biosorption), bioaccumulation, bioleaching, and bioprecipitation has gained attention in recent years. Bacteria are the ideal choice for bioremediation because of their omnipresence, abundance, diversity, tiny size, and special ability to grow and propagate under both controlled and uncontrolled circumstances with environmental resilience (Kapahi and Sachdeva 2019). Aerobic bacteria commonly used in bioremediation include *Mycobacterium*, *Pseudomonas*, *Acinetobacter*, *Flavobacterium*, *Rhodococcus*, and *Nocardia* (Saeed et al. 2023). Numerous factors, such as biological parameters, environmental circumstances, nutrition availability, heat, oxygen and moisture levels, pH, site assessment and selection, metal ions, and hazardous compounds, influence ex-situ or in-situ microbial bioremediation.

In recent years, heavy metal-containing wastewater has been widely released into the environment. Freshwater reservoirs contain the most hazardous pollutants, such as lead (Pb), nickel (Ni), chromium (Cr), copper (Cu), and cadmium (Cd). Heavy metals pose serious health hazards due to their carcinogenic and teratogenic properties, which can accumulate in human bodies through the food chain (Lee et al. 2012). Therefore, the heavy metal treatment of contaminated sewage becomes necessary. Rapid urbanization and industrialization are releasing large volumes of wastewater, which is gradually being used as an important irrigation resource by the agricultural sector. It creates a lot of economic activity, especially for poor farmers, supports a lot of livelihoods, and drastically changes the characteristics of the water in natural bodies of water (Singh et al. 2021).

The application of bacterial bioremediation is common for the treatment of industrial wastewater. Numerous bacteria, including *B. subtilis*, *Pseudomonas*, endophytes, and some bacterial strains, are employed in this type of treatment. The pharmaceutical industry's samples are collected in various forms from nine different locations, including MEE (multiple-effect evaporator), ETP (effluent treatment plant), cooling tower water, condensate online water, S.T.P. (sewage treatment plant), condensate water, processed salts, MEE feed water, and boiler blow down water, according to a study done in Tamil Nadu, India (Das et al. 2022). The bacterial consortium was produced using *Rhodococcus*, *Aspergillus niger*, *Nitrosomonas*, *Bacillus licheniformis*, *Pseudomonas putida*, *Pseudomonas fluorescens*, *Bacillus megaterium*, *Nitrobacter*, *Bacillus pumilus*, *Bacillus subtilis*, and bacteria that solubilize phosphate. Prior to treatment, several parameters were determined, such as sulfates, COD, TDS (total dissolved solids), and TSS (total suspended solids). Next, wastewater was treated using the bacterial consortium that was previously mentioned. When compared to data collected prior to treatment, it was evident that all samples had much lower levels of TDS, sulfates, and TSS following the treatment. Conversely, COD exhibited few alterations during therapy and tended to stay stable across all samples. According to the study, bacterial strains with different bacterial consortia are very good at breaking down sulfates and TSS but are unable to increase the COD of water (Saeed et al. 2022).

4.5 *Challenges and Recent Advancements in Bioremediation*

4.5.1 Challenges in Bioremediation

While bioremediation has great promise in the treatment of medium contaminated by inorganic pollutants, it is not without its difficulties. Among the difficulties with bioremediation is cleanup after bioremediation takes a long period. The majority of hyperaccumulator plants have moderate growth rates and low biomass, which generally limits their phytoextraction efficacy. The effectiveness of phytoremediation is greatly influenced by environmental factors, which may not always be sufficient for the majority of species. The adoption of particular species that are suited or tolerant to the environmental circumstances and contaminants present is necessary when there are various pollution sources present. After phytoextraction, the contaminated biomass must be properly disposed of (as hazardous waste). Because of the restricted bioavailability of the pollutants in the soil, it is difficult to mobilize the more firmly bound fraction of metal ions from the soil. It is necessary to prevent the introduction of invasive plant species as hyperaccumulators, as they have the potential to reduce the diversity of native flowers. Sites with low to moderate pollution levels resulting from unsustainable plant or microbe development in highly contaminated substrates can benefit from the application of bioremediation. Microorganisms applied externally have the potential to upset biological treatment systems' equilibrium (Wang et al. 2021).

4.5.2 Recent Advancements in Bioremediation

One effective method for removing trash from contaminated areas and locations is bioremediation. Nowadays, several strategies have been established for bioremediation of emerging pollutants. Nanotechnologies have been introduced in bioremediation whereby many toxic substances can be removed with their help because of their unique abilities against various recalcitrant contaminants.

Nanotechnology has opened our eyes to new possibilities in technology, including water purification. Environmentally friendly methods are now classified as nanofiltration (Sanghvi et al. 2020). Water that has been treated with effective microorganisms through the application of effective microbes (EM) technology can be utilized for irrigation. Electromagnetics (EM) and nanotechnology can be useful for purifying water. Recalcitrant organic pollutants such as polycyclic aromatic hydrocarbons (PAHs) with numerous benzene rings cause a plethora of persistent environmental concerns. Polycyclic aromatic hydrocarbons, or PAHs, are not biodegradable and cause mutations (Shukla 2020; Vázquez-Núñez et al. 2020).

Furthermore, hydrophobic contaminant bioremediation using engineered polymeric nanoparticles has been introduced. In this strategy, Organic contaminants, such PAHs and petroleum hydrocarbons, are soluble and mobile in soil and lose some of their environmental influence as a result. Polymeric nano-network particles enhance the solubility of phenanthrene and its release from contaminated aquifer material. Polymeric nanoparticles are made from precursor chains of poly-(ethylene) glycol-modified urethane acrylate (PMUA). The qualities of PMUA nanoparticles are intended to last in the presence of a wide variety of bacterial populations (Bala et al. 2022).

Besides that, enzyme technology can be used for the bioremediation of pollutants. Lead, chromium, mercury, polymers, azo dyes, PAHs, and other organic and inorganic contaminants can all be remedied by enzymes. Pollutant bioremediation has made use of a variety of enzymes that have been extracted from different species. Finding novel enzymes and subtypes with particular physicochemical properties could be a viable approach to developing more effective and affordable pollution cleanup instruments. Several enzymes can be utilized to bioremediate organic pollutants, including hydrolase, esterase, nitrilase, organophosphorus hydrolase, and cytochrome p450 monooxygenase. These enzymes have been reported to have the ability to degrade organic pollutants such as herbicides, pesticides, dyes, drugs, plastics, polymers, aromatic molecules, polycyclic aromatic hydrocarbons (PAHs), chlorinated hydrocarbons, steroids, and organocyanides (Mousavi et al. 2021).

4.5.3 Advantages and Disadvantages of Bioremediation

Because bioremediation is a natural process, the general public views it as a suitable waste treatment method for contaminated materials like soil. When a contaminant is present, the number of microbes that can degrade it increases; when the contamination is degraded, the population of biodegradative microbes decreases. The treatment's leftovers, which typically consist of carbon dioxide, water, and cell biomass, are safe byproducts. In theory, a wide range of pollutants can be completely destroyed using bioremediation. Numerous substances that are deemed dangerous by law can be converted into safe goods. By doing this, the possibility of future liability for handling and getting rid of tainted material is eliminated. It is feasible to destroy target pollutants rather than transmit them from one environmental medium to another, such as from land to water or air. Bioremediation can frequently be completed on-site without significantly interfering with regular operations. This also removes the need to move large amounts of waste off-site and the hazards that transportation-related hazards may pose to the environment and public health. Comparing bioremediation to other approaches for hazardous waste cleanup, it may end up being less expensive.

Exclusively biodegradable chemicals can be used in bioremediation and not every substance can break down completely and quickly. There are worries that the byproducts of biodegradation could be poisonous or more persistent than the original substance and biological processes are frequently quite specialized. The existence of metabolically competent microbial populations, ideal environmental growth conditions, and acceptable concentrations of pollutants and nutrients are important site requirements needed for success. Extrapolating results from bench and pilot-scale research to large-scale field operations is challenging. Bioremediation systems that are suitable for locations with complex combinations of pollutants that are not evenly distributed in the environment require research and development. It is possible for contaminants to exist as gases, liquids, or solids. Compared to alternative treatment methods like soil excavation and removal or burning, bioremediation frequently requires more time. Regarding acceptable performance parameters for bioremediation, there is still confusion in the regulations. There being no agreed-upon definition of "clean," it is challenging to assess bioremediation's effectiveness, and there are no recognized benchmarks for bioremediation procedures.

4.5.4 Sustainability and Long-Term Efficacy of Bioremediation

A new remediation technique called bioremediation breaks down or transforms harmful contaminants into more appropriate or environmentally friendly products through natural processes. It also restores the damaged area and prevents further contamination without the need for chemical reagents. To degrade waste to undetectable, nontoxic, or acceptable levels below regulatory agencies' permissible environmental limits—or, ideally, to completely mineralize organopollutants to CO_2 under controlled conditions—this method uses biological agents such as algae, fungi, bacteria, viruses, and some specialized human cells. One of the most important problems facing the globe today is environmental pollution, which is harming the environment permanently with every year that goes by. The industrial sector's explosive growth has produced massive amounts of hazardous new pollutants, including hospital effluents, polycyclic aromatic hydrocarbons, bisphenol A, and polychlorinated biphenyls. These pollutants ultimately contribute to serious ecological issues. Both humans and other living things are at risk from the buildup of toxic toxins in the environment. The advancement of technology has influenced the creation of methods capable of efficiently getting rid of harmful pollutants and breaking down resistant ones. By incorporating bioremediation techniques like the use of biosurfactants, bioinformatics, immobilization, adsorption, and electrostatic binding, application of genetically engineered microorganisms, entrapment in porous matrix and encapsulation, electrokinetic remediation, metagenomics, nanotechnology, and cometabolic, remediation strategies have recently been focused toward improved biological remediation technologies. These techniques have been proven to offer several advantages, including on-site cleanup, affordability, environmental friendliness, minimal energy use, and broad acceptance. For this reason, one of the most practical solutions for getting rid of pollutants or contaminants from the environment is bioremediation. Compared to other traditional remediation procedures, the bioremediation process is thought to be more economical, safe, dependable, reasonably efficient, and environmentally benign because it uses bacteria and their metabolic activity to break down toxins. There are two methods for carrying out the bioremediation process: *ex-situ* and *in-situ* bioremediation (Tyagi and Kumar 2021). Based on the previous explanation, bioremediation is believed to be an alternative sustainable method and has long-term efficacy in addressing pollution issues.

4.6 Green Technology

Green technologies are methods and remedies that transcend beyond what is considered environmental technology. According to definitions, "green" refers to the combination of sustainable and friendly concerning the environment, economy, societies, and technologies (Wu and Strezov 2023). Several green technologies have been reported including implementing solar panels, vertical farming and hydroponics, plant-based protection for fresh produce, and biogas.

4.6.1 Solar Panels

The most well-known energy-gathering transducer is the solar panel. They are mostly used in conjunction with storage batteries to control the variable energy supplied by a light source, such as the sun or interior lights, and are made up of arrays of photovoltaic cells. The surface area of the photovoltaic cell array, its orientation concerning the light source, the light's spectrum makeup, the time of day, and latitude in the case of sunlight are all factors that affect solar cell power (Bensky 2019).

The capacity to use the sun, a clean and limitless source of energy, is the main advantage of using a solar power system to generate electricity. Solar panels are independent of limited resources, in contrast to nuclear energy and fossil fuels. PV cell electricity generation doesn't result in any hazardous byproducts. The energy produced by solar panels during their lifespan—usually a few years of operation—balances the energy required to build and install them. Due to its ability to reduce carbon emissions that conventional power plants would otherwise emit; solar panels greatly contribute to green energy sources. This reduction is more than just an abstract idea. It is significant and quantifiable. Solar power reduces a building's need for grid-supplied electricity, which is still produced using carbon-intensive methods, in both residential and commercial settings. Furthermore, solar firms never stop innovating. In this approach, they advance the development of more cost-effective and efficient solar technologies, advancing global energy sustainability.

Fossil fuels have a greater effect on water sources than solar energy. Many households overlook the fact that solar power generation does not require water, which is one advantage of switching to solar electricity. Large volumes of water are used in the power generation process by power plants that produce electricity using natural gas and crude oil, usually for cooling equipment. On the other hand, water is not needed for the energy production or management of rooftop solar panels. Not only is water necessary for life support, but it is also needed for almost all global production, including food production, manufacturing, and other activities. Water conservation is more important than ever because droughts are growing more frequent and severe in many sections of the country. Water is the most valuable resource needed for life, thus taking it away from plants and animals always stresses the ecosystem and may cause certain species to go extinct or become endangered. Furthermore, by lowering the quantity of carbon dioxide emitted into the atmosphere during the combustion of fossil fuels, solar energy helps prevent global warming. The gradual rise in global temperature caused by the accumulation of carbon dioxide can be mitigated by switching to solar energy (Pruette 2022).

Finally, switching to solar energy contributes to better overall air quality. The amount of pollutants released into the air decreases as we burn fewer fossil fuels for energy. As was already said, pollution can have a negative impact on the environment as a whole, but it also exacerbates lung problems in both people and other animals. The EPA reports that since the 1990s, there has been an improvement in air quality, most likely as a result of growing solar energy usage and more ecologically conscious corporate operations. In general, switching to solar energy keeps the planet safe and livable for coming generations (Pruette 2022).

4.6.2 Vertical Farming and Hydroponics

Vertical farming (VF) is a technique of producing crops in layers that are piled vertically. To maximize plant growth, it frequently combines soilless farming methods like hydroponics, aquaponics, and aeroponics with controlled-environment agriculture. VF system is an indoor farm located in a high-rise building equipped with sophisticated agricultural systems and climate control technologies. Lights, ventilation or warmth, irrigation supply units, fertilizer solution, CO_2, a soilless medium, the farming structure to support the growing units, and irrigation are the main components of a vertical farming system. The automation system and sensor technologies are two other essential components of VF. These technologies are currently being used throughout the world in numerous indoor farms and greenhouses to provide steady, year-round growth conditions. Furthermore, automated methods are used to reduce operating expenses because labor costs in farming account for half of the output costs (Sengodan 2022).

4.6.3 Biogas

One byproduct of the breakdown of organic matter is biogas. A variety of materials can be utilized as raw materials to produce biogas, including coal, agro-industrial waste, algal biomass, and petroleum coke. Anaerobic digestion and gasification are two methods for producing biogas. Gasification requires a number of stages to produce biogas, including feeding the system with microalgae biomass and raising the temperature to 1200 °C. In this process, biogas—also known as synthesis gas—is finally produced as the temperature rises due to the evaporation of water, which causes partial oxidation of the biomass. Anaerobic digestion can produce biogas from a variety of substrates, including biomass and organic materials. Chemical processes such as hydrolysis, acidification, acetate generation, and methane production occur at different phases of the biogas extraction process. Methane and carbon dioxide are the byproducts of this chemical reaction that generates biogas (Nunes et al. 2021).

Lignocellulosic waste is one of the materials that can be used as a source for biogas production. Pretreatment is required to help increase the yield of biogas by allowing enzymes to break down the complicated structure of lignocellulosic biomass more easily. Following the pretreatment phase, the biowastes undergo a four-stage anaerobic breakdown process. Enzymes including cellulase, protease, xylanase, and lipase are secreted by hydrolytic bacteria during the first stage of food digestion. These enzymes convert carbohydrates, proteins, and fats into small-chain sugars, amino acids, and fatty acids. In the second stage, organic acids (acetic, propionic, and butyric acid), alcohols, H_2, and CO_2 are created by fermentative bacteria using the fatty acids and amino acids that were produced during the hydrolysis stage. By converting byproducts from earlier stages, methanogenesis-specific microbes aid in the creation of biogas components in the fourth and final step (Pocha et al. 2022).

4.6.4 Plant-Based Protection for Fresh Produce

Edible coatings can be made from a variety of materials with the ability to form films. These substances ought to be soluble in substances such as alcohol, water, etc. The commercially available coating material can be categorized according to its structural ingredients, such as polysaccharides, proteins, lipids, etc., and is typically derived from plants and other environmentally benign resources. This material has many benefits, including non-toxicity, accessibility, biodegradability, colorlessness, odorlessness, and transparency. The way the structural chains of the coating material are arranged has a big impact on how the coatings behave in terms of things like water permeability, tensile strength, solubility, etc. (Bajaj et al. 2023).

5 Conclusion

Worldwide environmental contamination is regarded as a multifaceted worldwide public health concern. This significant issue is linked to social, economic, and legal issues as well as lifestyle choices. Undoubtedly, global urbanization and industrialization are escalating to previously unheard-of and unsettling levels in our day and age. The introduction of pollutants into the environment that are hazardous to humans and other living things is known as pollution. Pollutants are dangerous substances that are created in more than normal amounts and lower the quality of our environment. They can be solids, liquids, or gases. In this review, several sustainable approaches to mitigate environmental pollution have been discussed. Methods such as bioremediation have been utilized to deal with environmental pollution due to several advantages. The advantages of bioremediation include its capacity to handle a wide range of pollutants, reduced costs, enhanced efficiency, and less environmental impact. Additionally, the complete removal of dangerous toxins found in the environment rather than their transfer from one location to another is another benefit of bioremediation. This method is an economical technique using the least amount of sophisticated instruments and equipment and using microorganisms in place of hazardous chemicals is an environmentally friendly method. Most of the time, on-site work can be done to save money on transportation. Minimal disturbance and damage to the site. Reduce the degree of liability. Minimal use of energy. As a result of its great effectiveness, dependability, and ease of use, this strategy is recognized by the general public and regulatory bodies. Nevertheless, this approach necessitates close monitoring and is restricted to pollutants and biodegradable trash. Specificity, as a biological process, has substantial drawbacks when it comes to elements like microbe species, environmental growth circumstances, nutrient requirements, and pollutants.

Additionally, bioremediation is a time-consuming procedure that may produce unidentified and perhaps hazardous byproducts. On the other hand, the use of nanotechnology has great potential in mitigating environmental pollution. Through the direct application of nanoparticles for the detection, prevention, and removal of pollutants as well as the indirect application of them through improved industrial design processes and the creation of environmentally friendly products, nanotechnology has the potential to improve the environment. The small size and high surface area of nanoparticles contribute to their increased reactivity. Although this feature has many uses and advantages, there are also potential hazards to worker and environmental safety, including the potential to remain suspended in the air for extended periods, the potential for environmental buildup, ease of absorption, and harm to different bodily organs. Besides, green technology has been reported as an efficient way to mitigate environmental pollution. Green technology lessens the negative effects of human activity on the environment by emphasizing resource efficiency, pollution control, and the use of renewable energy sources. Smart infrastructure and energy-efficient techniques encourage responsible energy use, reduce waste, and preserve natural resources. However, the quest for improved techniques and materials that significantly and effectively reduce environmental pollution is an ongoing endeavor. With time, sustainable mitigation techniques have improved and now have fewer restrictions while yet offering higher performance.

References

Abdelfattah A, Ali SS, Ramadan H, El-Aswar EI, Eltawab R, Ho S-H, Elsamahy T, Li S, El-Sheekh MM, Schagerl M, Kornaros M, Sun J (2023) Microalgae-based wastewater treatment: mechanisms, challenges, recent advances, and future prospects. Environ Sci Ecotechnol 13:100205. https://doi.org/10.1016/j.ese.2022.100205

Akhtar A, Mannan MA (2020) Mycoremediation: expunging environmental pollutants. Biotechnol Reports 26(2020):e00452

Ashraf S, Ali Q, Zahir ZA, Ashraf S, Asghar HN (2019) Phytoremediation: environmentally sustainable way for reclamation of heavy metal polluted soils. Ecotoxicol Environ Saf 174:714–727

Bajaj K, Adhikary T, Gill PPS, Kumar A (2023) Edible coatings enriched with plant-based extracts preserve postharvest quality of fruits: a review. Prog Org Coat 182:107669. https://doi.org/10.1016/j.porgcoat.2023.107669

Bala S, Garg D, Thirumalesh BV, Sharma M, Sridhar K, Inbaraj BS, Tripathi M (2022) Recent strategies for bioremediation of emerging pollutants: a review for a green and sustainable environment. Toxics 10(8):484. https://doi.org/10.3390/toxics10080484

Baneshi M, Tonney-Gagne J, Halilu F, Pilavangan K, Abraham BS, Prosser A, Marimuthu NK, Kaliaperumal R, Britten AJ, Mkandawire M (2024) Unpacking phthalates from obscurity in the environment. Molecules 29(1):106. https://doi.org/10.3390/molecules29010106

Bensky A (2019) Chapter 14—Technologies and applications. In Bensky A (ed) Short-range wireless communication, 3rd edn. Newnes, pp 387–430. https://doi.org/10.1016/B978-0-12-815405-2.00014-2

Bolan NS, Park JH, Robinson B, Naidu R, Keun Young Huh KY (2011) Chapter four—Phytostabilization: a green approach to contaminant containment. In: Sparks DL (ed) Advances in agronomy, vol 112, pp 145-204. Academic Press. https://doi.org/10.1016/B978-0-12-385538-1.00004-4

Briffa J, Sinagra E, Blundell R (2020) Heavy metal pollution in the environment and their toxicological effects on humans. Heliyon 6(9):e04691. https://doi.org/10.1016/j.heliyon.2020.e04691

Chausali N, Saxena J, Prasad R (2023) Nanotechnology as a sustainable approach for combating the environmental effects of climate change. J Agric Food Res 12:100541. https://doi.org/10.1016/j.jafr.2023.100541

Damare S, Singh P, Raghukumar S (2012) Biotechnology of marine fungi. Prog Mol Subcell Biol 53:277–297. https://doi.org/10.1007/978-3-642-23342-5_14

Das R, Lindström T, Sharma PR, Kai Chi K, Benjamin S, Hsiao BS (2022) Nanocellulose for sustainable water purification. Chem Rev 122(9):8936–9031. https://doi.org/10.1021/acs.chemrev.1c00683

Das S, Sultana KW, Ndhlala AR, Mondal M, Chandra I (2023) Heavy metal pollution in the environment and its impact on health: exploring green technology for remediation. Environ Health Insights 17:11786302231201260. https://doi.org/10.1177/11786302231201259

Delgado-González CR, Madariaga-Navarrete A, Fernández-Cortés JM, Islas-Pelcastre M, Oza G, Iqbal HMN, Sharma A (2021) Advances and applications of water phytoremediation: a potential biotechnological approach for the treatment of heavy metals from contaminated water. Int J Environ Res Public Health 18(10):5215. https://doi.org/10.3390/ijerph18105215

Dell'Anno F, Rastelli E, Sansone C, Brunet C, Ianora A, Dell'Anno A (2021) Bacteria, fungi, and microalgae for the bioremediation of marine sediments contaminated by petroleum hydrocarbons in the omics era. Microorganisms 9(8):1695. https://doi.org/10.3390/microorganisms9081695

Deshmukh R, Khardenavis AA, Purohit HJ (2016) Diverse metabolic capacities of fungi for bioremediation. Indian J Microbiol 56(3):247–264. https://doi.org/10.1007/s12088-016-0584-6

Dinesh A, Yuvaraj S, Abinaya S, Bhanushri S (2023) Nanopowders as an additive for strength and durability enhancement of cement composite: review and prospects, Mater Today: Proc. https://doi.org/10.1016/j.matpr.2023.03.651

dos Santos Bazanella GC, Araujo AV, Castoldi R, Maciel GM, Inacio FD, de Souza CGM, Bracht A, Peralta RM (2013) Ligninolytic enzymes from white-rot fungi and application in the removal of synthetic dyes. In: Polizeli TM, Rai M, De Lourdes M (eds) Fungal enzymes. CRC Press, Boca Raton, pp 258–279

Futughe AE, Purchase D, Jones H (2020) Phytoremediation using native plants. In: Shmaefsky B (eds) Phytoremediation. Concepts and strategies in plant sciences. Springer, Cham, pp 285–327. https://doi.org/10.1007/978-3-030-00099-8_9

Gao H, Wu M, Liu H, Xu Y, Liu Z (2022) Effect of petroleum hydrocarbon pollution levels on the soil microecosystem and ecological function. Environ Pollut 293:118511. https://doi.org/10.1016/j.envpol.2021.118511

Garg A, Chopra L (2022) Dye waste: a significant environmental hazard. Mater Today: Proc 48(5):1310–1315. https://doi.org/10.1016/j.matpr.2021.09.003

Giovanella P, Vieira GAL, Otero IVR, Pellizzer EP, Fontes B-J, Sette LD (2020) Metal and organic pollutants bioremediation by extremophile microorganisms. J Hazard Mater 382:121024. https://doi.org/10.1016/j.jhazmat.2019.121024

Ghori Z, Iftikhar H, Bhatti MF, Nasar-um-Minullah, Sharma I, Kazi AG, Ahmad P (2016) Chapter 15—Phytoextraction: the use of plants to remove heavy metals from soil. In: Ahmad P (ed) Plant metal interaction. Elsevier, pp 385–409. https://doi.org/10.1016/B978-0-12-803158-2.00015-1

Gong Y, Jiang M (2011) Biodiesel production with microalgae as feedstock: from strains to biodiesel. Biotech Lett 33:1269–1284. https://doi.org/10.1007/s10529-011-0574-z

Hári J, Pukánszky B (2011) 8—Nanocomposites: preparation, structure, and properties. In: Kutz M (ed) Plastics design library, applied plastics engineering handbook. William Andrew Publishing, pp 109–142. https://doi.org/10.1016/B978-1-4377-3514-7.10008-X

Kafle A, Timilsina A, Gautam A, Adhikari K, Bhattarai A, Aryal N (2022) Phytoremediation: mechanisms, plant selection and enhancement by natural and synthetic agents. Environ Adv 8:100203. https://doi.org/10.1016/j.envadv.2022.100203

Kapahi M, Sachdeva S (2019) Bioremediation options for heavy metal pollution. J Health Pollut 9:191203. https://doi.org/10.5696/2156-9614-9.24.191203

Kaushik P, Mishra A, Malik A, Pant KK (2014) Biosorption of textile dye by *Aspergillus lentulus* pellets: process optimization and cyclic removal in aerated bioreactor. Water Air Soil Pollut 225. https://doi.org/10.1007/s11270-014-1978-x

Kebede G, TafeseT, Abda EM, Kamaraj M, Assefa F (2021) Factors influencing the bacterial bioremediation of hydrocarbon contaminants in the soil: mechanisms and impacts. Hindawi J Chem 2021:17. Article ID 9823362. https://doi.org/10.1155/2021/9823362

Khalifeh S (2020) 1—Introduction to polymers for electronic engineers. In: Khalifeh S (ed) Polymers in organic electronics. ChemTec Publishing, pp 1–31. https://doi.org/10.1016/B978-1-927885-67-3.50006-7

Koul B, Ahmad W, Singh J (2021) Chapter 30—Mycoremediation: a novel approach for sustainable development. In: Kumar A, Singh VK, Singh P, Mishra VK (eds) Woodhead publishing series in food science, technology and nutrition, microbe mediated remediation of environmental contaminants. Woodhead Publishing, pp 409–420. https://doi.org/10.1016/B978-0-12-821199-1.00030-4

Koul B, Sharma K, Shah MP (2022) Phycoremediation: a sustainable alternative in wastewater treatment (WWT) regime. Environ Technol Innov 25:102040. https://doi.org/10.1016/j.eti.2021.102040

Kraemer SA, Ramachandran A, Perron GG (2019) Antibiotic pollution in the environment: from microbial ecology to public policy. Microorganisms 7(6):180. https://doi.org/10.3390/microorganisms7060180

Lee JC, Son YO, Pratheeshkumar P, Shi X (2012) Oxidative stress and metal carcinogenesis. Free Radic Biol Med 53(4):742–757

Li S, Li X, Ho S-H (2022) Microalgae as a solution of third world energy crisis for biofuels production from wastewater toward carbon neutrality: an updated review. Chemosphere 291:132863. https://doi.org/10.1016/j.chemosphere.2021.132863

Liu Z, Tran K-Q (2021) A review on disposal and utilization of phytoremediation plants containing heavy metals. Ecotoxicol Environ Saf 226:112821. https://doi.org/10.1016/j.ecoenv.2021.112821

Limmer M, Burken J (2016) Phytovolatilization of organic contaminants. Environ Sci Technol 50(13):6632–6643. https://doi.org/10.1021/acs.est.5b04113

Lu H, Zhang Y, Liu B, Liu J, Ye J, Yan C (2011) Rhizodegradation gradients of phenanthrene and pyrene in sediment of mangrove (*Kandelia candel* (L.) Druce). J Hazard Mater 196:263–269. https://doi.org/10.1016/j.jhazmat.2011.09.031

Lua W et al (2017) One-pot synthesis of magnetic iron oxide nanoparticle-multiwalled carbon nanotube composites for enhanced removal of Cr(VI) from aqueous solution. J Colloid Interf Sci 505:1134–1146

Maheshwari K, Agrawal M, Gupta AB (2021) Dye pollution in water and wastewater. In Muthu SS, Khadir A (eds) Novel materials for dye-containing wastewater treatment. sustainable textiles: production, processing, manufacturing & chemistry. Springer, Singapore, pp 1–25. https://doi.org/10.1007/978-981-16-2892-4_1

Masheane ML et al (2017) Synthesis of Fe-Ag/f-MWCNT/PES nanostructured-hybrid membranes for removal of Cr(VI) from water. Sep Purif Technol 184:79–87

Mastropetros SG, Pispas K, Zagklis D, Ali SS, Kornaros M (2022) Biopolymers production from microalgae and cyanobacteria cultivated in wastewater: recent advances. Biotechnol Adv 60:107999

Mata TM, Martins AA, Nidia S, Caetano NS (2010) Microalgae for biodiesel production and other applications: a review. Renew Sustain Energy Rev 14(1):217–232. https://doi.org/10.1016/j.rser.2009.07.020

Mohammad Ali M, Hossain D, Al-Imran, Suzan Khan Md., Begum M, Hasan Osman M (2021) Environmental pollution with heavy metals: a public health concern. IntechOpen. https://doi.org/10.5772/intechopen.96805

Mohammed SI (2021) Advantages of green technology to mitigate the environment problems. IOP Conf Ser: Earth Environ Sci 779:012133. https://doi.org/10.1088/1755-1315/779/1/012133

Mousavi SA, Khodadoost F (2019) Effects of detergents on natural ecosystems and wastewater treatment processes: a review. Environ Sci Pollut Res 26:26439–26448. https://doi.org/10.1007/s11356-019-05802-x

Mousavi SM, Hashemi SA, Iman Moezzi SM, Ravan N, Gholami A, Lai CW, Chiang WH, Omidifar N, Yousefi K, Behbudi G (2021) Recent advances in enzymes for the bioremediation of pollutants. Biochem Res Int 2021:5599204. https://doi.org/10.1155/2021/5599204

Muiruri JK, Yeo JCC, Loh XJ, Chen G-Q, He C, Li Z (2023) Chapter 9—Poly(hydroxyalkanoates) (PHAs) based circular materials for a sustainable future. In: Li Z, Lim JYC, Wang C-G (eds) Circularity of plastics. Elsevier, pp 273–303. https://doi.org/10.1016/B978-0-323-91198-6.00002-4

Nedjimi B (2021) Phytoremediation: a sustainable environmental technology for heavy metals decontamination. SN Appl Sci 3:286. https://doi.org/10.1007/s42452-021-04301-4

Nathanson JA (2024) Pollution. Encyclopedia Britannica. https://www.britannica.com/science/pollution-environment

Negi BB, Das C (2023) Mycoremediation of wastewater, challenges, and current status: a review. Bioresour Technol Reports 22:101409. https://doi.org/10.1016/j.biteb.2023.101409

Newman LA, Reynolds CM (2004) Phytodegradation of organic compounds. Curr Opin Biotechnol 15(3):225–230. https://doi.org/10.1016/j.copbio.2004.04.006

Nunes NSP, Ansilago M, de Oliveira NN, Leite RSR, Marcelo Fossa da Paz MF, Fonseca GG (2021) Chapter 6—Biofuel production. In: Galanakis CM, Microalgae. Academic Press, pp 145–171. https://doi.org/10.1016/B978-0-12-821218-9.00006-2

Parlayici S, Eskizeybek V, Avcı A, Pehlivan E (2015) Removal of chromium (VI) using activated carbon-supported functionalized carbon nanotubes. J Nanostruct Chem 5:255–263

Pocha CKR, Chia SR, Chia WY, Koyande AK, Nomanbhay S, Chew KW (2022) Utilization of agricultural lignocellulosic wastes for biofuels and green diesel production. Chemosphere 290:133246. https://doi.org/10.1016/j.chemosphere.2021.133246

Priyadharshini SD, Babu PS, Manikandan S, Subbaiya R, Govarthanan M, Karmegam N (2021) Phycoremediation of wastewater for pollutant removal: a green approach to environmental protection and long-term remediation. Environ Pollut 290:117989

Pruette R (2022) Environment benefits of solar panels. https://eponline.com/articles/2022/11/18/environment-benefits-of-solar-panels.aspx. Accessed 25 June 2024

Ranzoni A, Cooper MA (2017) Chapter One—The growing influence of nanotechnology in our lives. In: Skwarczynski M, Toth I (eds) Micro and nano technologies, micro and nanotechnology in vaccine development. William Andrew Publishing, pp 1–20. https://doi.org/10.1016/B978-0-323-39981-4.00001-4

Rezaei F, Tajik H, Shahbazi Y (2023) Intelligent double-layer polymers based on carboxymethyl cellulose-cellulose nanocrystals film and poly(lactic acid)-Viola odorata petal anthocyanins nanofibers to monitor food freshness. Int J Biol Macromol 252. https://doi.org/10.1016/j.ijbiomac.2023.126512

Roozbahani A (2021) Chapter 6—Application of Bayesian networks modelling in wastewater management. In: Karri RR, Ravindran G, Mohammad Hadi Dehghani MH (eds) Soft computing techniques in solid waste and wastewater management. Elsevier, pp 111–130. https://doi.org/10.1016/B978-0-12-824463-0.00002-1

Sadasivam S, Jawaharlal M (2022) Role of ornamental plants in phytoremediation—An overview. Chem Sci Rev Lett 11(44):478–488. https://doi.org/10.37273/chesci.cs205309530

Saddhe AA, Manuka R, Nikalje GC, Penna S (2020) Halophytes as a potential resource for phytodesalination. In: Grigore MN (eds) Handbook of halophytes. Springer, Cham. https://doi.org/10.1007/978-3-030-17854-3_92-1

Saeed MU, Hussain N, Sumrin A, Shahbaz A, Noor S, Bilal M, Aleya L, Iqbal HMN (2022) Microbial bioremediation strategies with wastewater treatment potentialities—A review. Sci Total Environ 818:151754. https://doi.org/10.1016/j.scitotenv.2021.151754

Saeed MU, Hussain N, Javaid M, Zaman H (2023) Chapter 14—Microbial remediation for environmental cleanup. In: Gangola S, Kumar S, Joshi S, Bhatt P (eds) Developments in applied microbiology and biotechnology, advanced microbial technology for sustainable agriculture and environment. Academic Press, pp 247–274. https://doi.org/10.1016/B978-0-323-95090-9.00010-8

Salam MA (2017) Preparation and characterization of chitin/magnetite/multiwalledcarbon nanotubes magnetic nanocomposite for toxic hexavalent chromium removal from solution. J Mol Liq 233:197–202

Salam MA, Sultana N (2022) Chapter 18—Pollution of water resources. In: Tiwari AK, Kumar A, Singh AK, Singh TN, Suozzi E, Gagan Matta G, Russo SL (eds) Current directions in water scarcity research. Elsevier, vol 5, pp 355–378. https://doi.org/10.1016/B978-0-323-85378-1.00018-0

Sampaio GR, Guizellini GM, da Silva SA, de Almeida AP, Pinaffi-Langley ACC, Rogero MM, de Camargo AC, Torres EAFS (2021) Polycyclic aromatic hydrocarbons in foods: biological effects, legislation, occurrence, analytical methods, and strategies to reduce their formation. Int J Mol Sci 22(11):6010. https://doi.org/10.3390/ijms22116010

Sanghvi G, Thanki A, Pandey S, Singh NK (2020) Bioremediation of pollutants. In: Engineered bacteria for bioremediation. Elsevier, Amsterdam, The Netherlands, pp 359–374

Saravanan A, Kumar PS, Duc PA, Rangasamy G (2023) Strategies for microbial bioremediation of environmental pollutants from industrial wastewater: a sustainable approach. Chemosphere 313:137323. https://doi.org/10.1016/j.chemosphere.2022.137323

Sarkar P, Dey A (2021) Phycoremediation—An emerging technique for dye abatement: an overview. Process Saf Environ Prot 147:214–225. https://doi.org/10.1016/j.psep.2020.09.031

Schulz MJ, Ruff B, Johnson A, Vemaganti K, Li W, Sundaram MM, Hou G, Krishnaswamy A, Li G, Fialkova S, Yarmolenko S, Wang A, Liu Y, Sullivan J, Alvarez N, Shanov V, Pixley S (2014) Chapter 2—New applications and techniques for nanotube superfiber development. In: Schulz MJ, Shanov VN, Zhangzhang Yin Z (eds) Nanotube superfiber materials. William Andrew Publishing, pp 33–59. https://doi.org/10.1016/B978-1-4557-7863-8.00002-5

Sengodan P (2022) An overview of vertical farming: highlighting the potential in malaysian high-rise buildings. Pertanika J Sci Technol 30(2):949–981

Show K-Y, Lee D-J, Zhang Z-P (2011) Chapter 20—Production of biohydrogen: current perspectives and future prospects. In: Pandey A, Larroche C, Ricke SC, Dussap C-G, Gnansounou E (eds) Biofuels. Academic Press, pp 467–479. https://doi.org/10.1016/B978-0-12-385099-7.00021-8

Shukla P (2020) Microbial nanotechnology for bioremediation of industrial wastewater. Front Microbiol 11:590631. https://doi.org/10.3389/fmicb.2020.590631

Singh AK, Bilal M, Iqbal HM, Raj A (2021) Trends in predictive biodegradation for sustainable mitigation of environmental pollutants: recent progress and future outlook. Sci Total Environ 770:144561. https://doi.org/10.1016/j.scitotenv.2020.144561

Speight JG (2020) 5—Sources of water pollution. In: Speight JG (ed) Natural water remediation. Butterworth-Heinemann, pp 165–198. https://doi.org/10.1016/B978-0-12-803810-9.00005-X

Taha M, Adetutu EM, Shahsavari E, Smith AT, Ball AS (2014) Azo and anthraquinone dye mixture decolourization at elevated temperature and concentration by a newly isolated thermophilic fungus, *Thermomucor indicae-seudaticae*. J Environ Chem Eng 2:415–423. https://doi.org/10.1016/j.jece.2014.01.015

Tamalampudi S, Fukuda H (2011) 3.07—Biodiesel. In: Moo-Young M (ed) Comprehensive biotechnology, 2nd edn. Academic Press, pp 63–70. https://doi.org/10.1016/B978-0-08-088504-9.00163-X

Thatoi H, Behera BC, Mishra RR (2013) Ecological role and biotechnological potential of mangrove fungi: a review. Mycology 4:54–71. https://doi.org/10.1080/2150123.2013.785448

Tonelli FCP, Tonelli FMP, Lemos MS, Nunes NAd-M (2022) Chapter 3—Mechanisms of phytoremediation. In: Rouf Ahmad Bhat RA, Tonelli FMP, Gowhar Hamid Dar GH, Khalid Hakeem K (eds) Phytoremediation. Academic Press, pp 37–64. https://doi.org/10.1016/B978-0-323-89874-4.00023-6

Tripathi M, Singh S, Pathak S, Kasaudhan J, Mishra A, Bala S, Garg D, Singh R, Singh P, Singh PK, Shukla AK, Pathak N (2023) Recent strategies for the remediation of textile dyes from wastewater: a systematic review. Toxics 11(11):940. https://doi.org/10.3390/toxics11110940

Truskewycz A, Gundry TD, Khudur LS, Kolobaric A, Taha M, Aburto-Medina A, Ball AS, Shahsavari E (2019) Petroleum hydrocarbon contamination in terrestrial ecosystems-fate and microbial responses. Molecules (Basel, Switzerland) 24(18):3400. https://doi.org/10.3390/molecules24183400

Tyagi B, Kumar N (2021) Chapter 1—Bioremediation: principles and applications in environmental management. In: Saxena G, Kumar V, Shah MP (eds) Bioremediation for environmental sustainability. Elsevier, pp 3–28. https://doi.org/10.1016/B978-0-12-820524-2.00001-8

Ukaogo PO, Ewuzie U, Onwuka CV (2020) Environmental pollution: causes, effects, and the remedies. Microorgan Sustain Environ Health. https://doi.org/10.1016/b978-0-12-819001-2.00021-8

Vázquez-Núñez E, Molina-Guerrero CE, Peña-Castro JM, Fernández-Luqueño F, de la Rosa-Álvarez M (2020) Use of nano-technology for the bioremediation of contaminants: a review. Processes 8:826. https://doi.org/10.3390/pr8070826

Wang J, Wu S, Suo X, Liao H (2019) Chapter 2—The processes for fabricating nanopowders. In: Yang G, Suo X (eds) Micro and nano technologies, advanced nanomaterials and coatings by thermal spray. Elsevier, pp 13–25. https://doi.org/10.1016/B978-0-12-813870-0.00002-4

Wang M, Chen S, Jia X, Chen L (2021) Chapter 1—Concept and types of bioremediation. In: Hasanuzzaman M, Prasad MNVP (eds) Handbook of bioremediation. Academic Press, pp 3–8. https://doi.org/10.1016/B978-0-12-819382-2.00001-6

Wassel AR, El-Naggar ME, Shoueir K (2020) Recent advances in polymer/metal/metal oxide hybrid nanostructures for catalytic applications: a review. J Environ Chem Eng 8(5):104175. https://doi.org/10.1016/j.jece.2020.104175

Wu J, Strezov V (2023) Green technologies and sustainability: a new trend. Green Technologies and Sustainability 1(1):100008. https://doi.org/10.1016/j.grets.2023.100008

Wu J, Liao J, Hu T, Zong M, Wen P, Wu H (2024) Fabrication of multifunctional ethyl cellulose/gelatin-based composite nanofilm for the pork preservation and freshness monitoring. Int J Biol Macromol 265:130813. https://doi.org/10.1016/j.ijbiomac.2024.130813

Xiong X, Zhang Y, Wang L, Tsang DCW (2022) Chapter 1—Overview of hazardous waste treatment and stabilization/solidification technology. In: Tsang DCW, Wang L (eds) Low carbon stabilization and solidification of hazardous wastes. Elsevier, pp 1–14. https://doi.org/10.1016/B978-0-12-824004-5.00031-1

Yadav M, Gupta R, Sharma RK (2018) Green and sustainable pathways for wastewater purification. Adv Water Purif Technol 2018:355–383. https://doi.org/10.1016/B978-0-12-814790-0.00014-4

Yan A, Wang Y, Tan SN, Mohd Yusof ML, Ghosh S, Chen Z (2020) Phytoremediation: a promising approach for revegetation of heavy metal-polluted land. Front Plant Sci 11:359. https://doi.org/10.3389/fpls.2020.00359

Zhou S, Liu L, Wang B, Xu F, Sun RC (2012) Biodiesel preparation from transesterification of glycerol trioleate catalyzed by basic ionic liquids. Chin Chem Lett 23(4):379–382. https://doi.org/10.1016/j.cclet.2012.01.034

Zhou J, Zhou Y, Bai X (2023) Can green-technology innovation reduce atmospheric environmental pollution? Toxics 11(5):403. https://doi.org/10.3390/toxics11050403

Sustainable Electronic Waste Management Among Households in Kelantan, Malaysia

Wan Nor Azilawanie Tun Ismail, Mohd Khairul Amri Kamarudin, and Mohd Armi Abu Samah

Abstract Electronic waste management involves the proper handling and managing of electronic waste, including devices like phones, computers, tablets, and other such items. Electronic waste (e-waste) is currently seeing the most rapid growth among all waste streams on a global scale. Therefore, it is crucial to develop strategies for waste management to address this issue before it becomes too severe. Therefore, this study aims to identify the awareness hazards of e-waste, recycling e-waste behavior, and sustainable waste management levels among households in Kelantan, Malaysia. A total of 300 responses were collected using an online survey distributed via Google Forms and data was analyzed through SPSS. The results indicated that the three variables are at a high level and very satisfactory. These findings emphasize the significance of continuous endeavors to educate communities about the dangers of e-waste, advocate for responsible recycling methods, and promote the implementation of sustainable waste management strategies. Through the ongoing efforts of raising awareness and implementing efficient waste management policies, Kelantan and other areas can strive to decrease the environmental impact of e-waste and secure a healthier and more sustainable future.

Keywords Electronic waste management · Sustainable · Households · Malaysia

W. N. A. T. Ismail (✉) · M. K. A. Kamarudin
Faculty of Applied Social Sciences, Universiti Sultan Zainal Abidin, Gong Badak Campus, 21300 Kuala Nerus, Terengganu, Malaysia
e-mail: azilawanieismail@unisza.edu.my

M. K. A. Kamarudin
e-mail: mkhairulamri@unisza.edu.my

M. K. A. Kamarudin
Faculty of Applied Social Sciences, Universiti Sultan Zainal Abidin, Gong Badak Campus, 21300 Kuala Nerus, Terengganu, Malaysia

M. A. A. Samah
Kuantan Campus, International Islamic University Malaysia (IIUM), Jalan Sultan Ahmad Shah, 25200 Bandar Indera Mahkota, Kuantan, Pahang, Malaysia
e-mail: marmi@iium.edu.my

A. Z. Yaser et al. (eds.), *Controlling Environmental Pollution*,
https://doi.org/10.1007/978-981-97-8931-3_10

1 Introduction

Electronic waste, sometimes referred to as Waste of Electronic and Electrical Equipment (WEEE) or e-waste, encompasses electronic and electrical components that have become obsolete or are no longer required by users. E-waste refers to electrical and electronic gadgets that are no longer functional, broken, outdated, or considered obsolete. Examples of such devices include laptops, computers, mobile phones, TVs, DVD players, refrigerators, washing machines, medical devices, and air conditioners (Thukral et al. 2023). The problem of e-waste has long been a concern in the field of solid waste management and is currently having a growing impact on the environment. The proliferation of electronic equipment in the market, resulting from advancements in the electronics sector, has led to a surplus of products. This surplus has caused a significant decrease in the price of electronic equipment, making it more affordable for a majority of individuals. Consequently, many people now possess many electronic devices. The widespread ownership of electronic devices by the majority of folks is not a concern. Nevertheless, the accumulation of electronic waste might be a problem when the equipment becomes obsolete or reaches its lifespan.

The United Nations Environment Program report indicates an annual estimate of electronic waste ranging from 20 to 50 tons. The Global E-waste Monitor 2020 documented that in the year 2020, a total of 53.6 million metric tons of e-waste were generated globally, with an average of 7.3 kg per person. The analysis forecasts that global e-waste will increase to 74 metric tons by 2030, representing an annual growth of 2 metric tons (Adeel et al. 2023; Thukral et al. 2023). Based on the Department of Environment (DOE), Malaysia produced 12.344 million units of electronic garbage (e-waste) in 2016, which subsequently increased by 38% to 17.065 million units in 2020. Furthermore, it is expected that the volume will continue to rise, reaching 24.504 million units by the year 2025 (Department of Environment 2020). The surge in e-waste is primarily fueled by the rapid adoption of electrical and electronic devices, their limited lifespan, and the scarcity of repair alternatives.

As the production of electrical and electronic items increases, there is a corresponding increase in the amount of e-waste that must be disposed of or controlled. Previous research conducted by Fadhullah et al. (2022) revealed that the community residing on the East Coast of Malaysia exhibits a low level of awareness as well as practice about waste management. Furthermore, it was observed that a significant portion of the waste is disposed of without proper segregation. Improper disposal of e-waste, such as dumping it in rivers, and landfills, or burning it indiscriminately, can pose risks to life, impact health, and degrade the environment's quality (Islam et al. 2021). Poisonous and hazardous compounds in e-waste necessitate environmentally responsible disposal to prevent their infiltration into the soil and water sources, posing a threat to aquatic life and human health. Furthermore, the electronic components found in e-waste also contain valuable metals that possess a significant recycling worth. Therefore, it must be effectively controlled from manufacturing until final disposal. Malaysian society, as a whole, continues to overlook this issue due to its inclination toward individualism and lack of concern for environmental sustainability.

2 Demographic Profile of Respondents

Out of the 300 participants in this survey, there was an equal distribution of 50% male and 50% female respondents. The most populous age group consists of those aged 18–24, accounting for 42% of the population. This is followed by individuals aged 25–34, making up 25% of the population. The age group of 35–49 represents 17.3%, while those aged 50–64 account for 8.7%. Lastly, individuals aged 65 and above comprise 7.0% of the population. In terms of education level, 37.3% of the respondents held a degree, 21.7% had completed secondary school, 17% had a diploma, 10.3% had a certificate, 5.0% had completed STPM, 4.3% had only completed primary school, 2.3% had other qualifications, and 2.0% held a master's or Ph.D. degree. The respondents were primarily self-employed (40.3%), followed by those in the private sector (25.7%), the government sector (16.3%), and other occupational categories (17.7%). Finally, a majority of 46% of the respondents reported an income ranging from RM1001 to RM3000. This was followed by 42% reporting an income below RM1000, 8.7% reporting an income between RM3001 and RM5000, and 3.3% reporting an income of RM5001 and above. The demographic characteristics of the household are displayed in Table 1.

3 Awareness of Hazards of E-Waste

The primary catalyst for effective recycling and e-waste management is the promotion of public awareness (Kadir et al. 2022). Educating households about the detrimental impact of e-waste on the environment and human health is crucial. The objective of this study is to ascertain the extent of households' awareness in Kelantan regarding the hazards associated with e-waste. This study had a total of 300 participants. A survey was employed to gather data regarding household awareness of the hazards associated with electronic waste. The mean scores derived from the Likert scale can be interpreted based on the following criteria in Table 2. Table 2 presents the mean score analysis, indicating whether it falls into the categories of very low, low, moderate, high, or very high.

The interpretation of the mean score for items 1–3 (refer to Table 3) is very high with values of 4.38, 4.41, and 4.44, respectively. Based on this mean score, the interpretation that can be formulated is that respondents strongly agree that e-waste might affect human health with the highest mean score of 4.44. In addition, respondents strongly agree that e-waste is a serious problem for the environment (4.41). Next, respondents strongly agree on the hazardousness of e-waste (4.38).

Table 1 Demographic profile (n = 300)

Demographic profile		Frequency (n)	Percentage (%)
Gender	Male	150	50.0
	Female	150	50.0
Age	18–24	126	42.0
	25–34	75	25.0
	35–49	52	17.3
	50–64	26	8.7
	65 and above	21	7.0
Level of education	Primary school	13	4.3
	Secondary school	65	21.7
	STPM	15	5.0
	Certificate	31	10.3
	Diploma	51	17.0
	Degree	112	37.3
	Master/Ph.D.	6	2.0
	Others	7	2.3
Job categories	Government sector	49	16.3
	Private sector	77	25.7
	Self-employment	121	40.3
	Others	53	17.7
Incomes	Below RM1000	126	42.0
	RM1001 to RM3000	138	46.0
	RM3001 to RM5000	26	8.7
	RM5001 and above	10	3.3

Table 2 The criteria of the interpretation of mean score (5 Likert scales)

Range	Interpretation
1.00–1.80	Very low
1.81–2.60	Low
2.61–3.40	Moderate
3.41–4.20	High
4.21–5.00	Very high

Source Bahri et al. (2016)

Table 3 Mean score awareness of hazards of e-waste

No	Statement	Mean	Sd
1	I am aware of the hazardousness of e-waste	4.38	0.646
2	I am aware that e-waste is a serious problem for the environment	4.41	0.640
3	I am aware that e-waste might affect human health	4.44	0.634

Awareness plays a crucial role in fostering sustainable electronic waste management practices. Educating households about the environmental and health implications of improper e-waste disposal and the benefits of sustainable management is important. By raising awareness and providing guidance on proper disposal methods and the importance of recycling, organizations can contribute to a culture of responsible electronic waste management. Besides, proficient comprehension of environmental awareness is vital for environmentalists, educators, stakeholders, and policymakers to strategize sustainable development for household socioeconomic circumstances (Fadhullah et al. 2022).

4 Recycling E-Waste Behavior

An individual's practice or behavior significantly influences e-waste management (Borthakur and Govind 2019). Recycling methods are highly efficient in managing electronic waste, offering a sustainable approach to handling and repurposing gadgets and reducing environmental impact. The objective of this particular section is to determine the recycling behavior of the household. Consequently, respondents were requested to express their degree of agreement with a total of four assertions. The items were assessed using a Likert scale ranging from strongly agree, agree, neutral, disagree to strongly disagree. Table 4 shows that the mean scores indicate that two items demonstrate a very high level of e-waste recycling behavior. These items are 'I am willing to implement recycling behavior for my family' (4.34) and 'I am willing to recycle e-waste to protect the environment' (4.29). Next, two items showing the recycling e-waste behavior of respondents were high, namely 'I am practicing recycling' (4.18) and 'It is my responsibility to encourage my neighbors to recycle e-waste' (4.13).

According to Afroz et al. (2020), the recycling initiative in Malaysia is currently in its early phase, and there is a conspicuous dearth of involvement among the populace. The recycling rate in the country is significantly lower than that of other developed countries. Therefore, effective e-waste management in the community yields several advantages, including waste reduction, cost savings, pollution reduction, job creation, promotion of green technology, and environmental conservation. Each person has a crucial responsibility in effectively managing e-waste and must possess information

Table 4 Mean score recycling e-waste behavior

No	Statement	Mean	Sd
1	I am practicing recycling	4.18	0.929
2	I am willing to recycle e-waste to protect the environment	4.29	0.727
3	It is my responsibility to encourage my neighbors to recycle e-waste	4.13	1.005
4	I am willing to implement recycling behavior for my family	4.34	0.701

Table 5 Mean score sustainable waste management

No	Statement	Mean	Sd
1	I consider the equipment that is easy to repair and reuse before purchase	4.49	0.696
2	I separate my electronic waste from ordinary garbage so I can sell them	4.43	0.730
3	I only buy electronic equipment when there is an immediate need	4.49	0.641
4	I should not waste anything if it can be used again	4.54	0.586

about e-waste to be able to differentiate it and refrain from disposing of it thoughtlessly. This is essential to prevent contamination and minimize the quantity of waste being sent to landfills.

5 Sustainable Waste Management

Sustainable waste management is a strategy that aims to reduce the adverse effects on the environment caused by the creation and disposal of waste, while also encouraging the conservation of resources and the long-term well-being of the environment. It entails executing strategies and practices that reduce, reuse, recycle, and recover waste, aiming to minimize the quantity of waste disposed of in landfills and incinerators.

As seen in Table 5, the interpretation of the mean score for all items is very high. Specifically, items 1 and 3 have a mean score of 4.49, item 2 has a mean score of 4.43, and item 4 has a mean score of 4.54. The respondents strongly agreed that they should use something wisely if it can be used again with a high mean score of 4.54. In addition, respondents strongly agree that they consider the equipment easy to repair and reuse before purchase and that they only buy electronic equipment when there is an immediate need (4.49). Furthermore, respondents strongly agree that they separate their electronic waste from ordinary garbage so they can sell it (4.43).

Overall, the high mean scores across all items suggest a positive and environmentally conscious mindset among the respondents. They prioritize waste reduction, reuse, and responsible consumption practices, particularly in the context of electronic waste. These findings could be valuable for understanding and promoting sustainable behaviors among the surveyed population.

6 Conclusion

In conclusion, the study of household awareness regarding e-waste management practice is crucial in assessing their comprehension and knowledge of e-waste management. The concerns and difficulties addressed can serve as a point of reference for the government, non-governmental organizations, authorities, and the community in their efforts to cultivate a more sustainable community life. By integrating green procurement practices, collaborating with stakeholders, and prioritizing education and awareness, organizations can work toward achieving sustainable electronic waste management. This holistic approach addresses the various dimensions of electronic waste management and helps minimize the environmental and health hazards associated with electronic waste. The community can contribute to a more sustainable and environmentally conscious future by embracing these practices.

References

Adeel S, Nayab A, Qureshi MU, Channa KA (2023) University students' awareness of e-waste and its disposal practices in Pakistan: a construction of the conceptual framework. J Mater Cycles Waste Manag 1–14

Afroz R, Muhibbullah M, Farhana P, Morshed MN (2020) Analyzing the intention of the households to drop off mobile phones to the collection boxes: empirical study in Malaysia. Ecofeminism Clim Change 1(1):3–20

Borthakur A, Govind M (2019) Computer and mobile phone waste in urban India: an analysis from the perspectives of public perception, consumption and disposal behaviour. J Environ Planning Manag 62(4):717–740

Department of Environment (2020) What is e-waste? Department of Environment, Malaysia. https://ewaste.doe.gov.my/index.php/what-is-e-waste/

Fadhullah W, Imran NIN, Ismail SNS, Jaafar MH, Abdullah H (2022) Household solid waste management practices and perceptions among residents in the East Coast of Malaysia. BMC Public Health 22(1):1–20

Islam MT, Dias P, Huda N (2021) Young consumers' e-waste awareness, consumption, disposal, and recycling behavior: a case study of university students in Sydney, Australia. J Clean Prod 282:124490

Kadir WHWA, Hassin NH, Noor ANM, Abas MA, Muhamad M, Dauda SA (2022) Public knowledge, perception and attitude on e-waste recycling management in Kulim, Kedah. In: AIP conference proceedings (2454), 1. AIP Publishing

Thukral S, Shree D, Singhal S (2023) Consumer behaviour towards storage, disposal and recycling of e-waste: a systematic review and future research prospects. Benchmark Int J 30(3):1021–1072

Qualitative Contents Analysis on Current Practices Clinical Waste Management During Pandemic in Healthcare Establishment, Kelantan

F. N. Mansor, A. A. Suhaimi, Mohd Armi Abu Samah, S. S. Nurud, and A. Hidayah

Abstract Clinical waste management (CWM) has become a greater concern since the COVID-19 pandemic. Malaysia has reported a 27% increase in the generation of clinical waste, mostly attributed to COVID-19-related materials. Current clinical waste management practices must be handled carefully in accordance with existing CWM policies and laws, such as the Schedule Waste Regulations, the Environmental Quality Act, and the standard operating procedures provided by the Ministry of Health Malaysia. This research employs a qualitative design, conducting in-depth interviews with healthcare establishment (HCE) staff to explore CWM practices before and during the pandemic and to compare these practices over time. The study revealed that clinical waste management in government hospitals adhered to existing guidelines for COVID-19 waste, with some additional precautions and regulations implemented by the involved concession companies. It was concluded that a few additional standard operating procedures (SOPs) need to be followed, but no new policies or regulations regarding CWM have been implemented.

Keywords Clinical waste management · Healthcare establishments · Pandemic COVID-19 · COVID-19 · Hospitals

F. N. Mansor · A. A. Suhaimi (✉) · S. S. Nurud · A. Hidayah
Occupational Safety and Health Research Group, Faculty of Industrial Sciences and Technology, Universiti Malaysia Pahang Al-Sultan Abdullah, Kuantan, Pahang, Malaysia
e-mail: afiza@umpsa.edu.my

M. A. A. Samah
Kulliyyah of Science, International Islamic University Malaysia, Kuantan, Pahang, Malaysia

A. Z. Yaser et al. (eds.), *Controlling Environmental Pollution*,
https://doi.org/10.1007/978-981-97-8931-3_11

1 Introduction

COVID-19 was declared a pandemic by the World Health Organization (WHO) on March 11, 2020. Most countries worldwide reported an increasing trend in COVID-19 cases. Similarly, healthcare establishments in Malaysia received many COVID-19 admissions daily. While the greatest concern was the potential for the virus to spread across the country and strain the healthcare system, another significant concern emerged: the increase in clinical waste generation due to COVID-19.

During the pandemic, the generation of clinical waste, such as personal protective equipment (PPE), gloves, and hand sanitizers, became commonplace. This waste contributed significantly to the load on Clinical Waste Management (CWM) systems (WHO 2020). The Ministry of Health (MOH) Malaysia reported a 27% increase in weight in CW generation, primarily due to COVID-19-related waste (Berita Awani 2021; Agamuthu and Barasarathi 2020). CW is produced in public health labs, blood donation centers or blood banks, vaccination or immunization clinics, including COVID-19 Quarantine and Treatment Centres (PKRC), and programs associated with mental health and cognitive disorders (Dehghani et al. 2019; Bendjoudi et al. 2008). Sharps, human tissues or body parts, and other infectious materials in CW pose potential health and environmental risks (DOE 2009; Mustaffa et al. 2023; Saad et al. 2023).

Clinical waste can be dangerous because it contains infectious agents, including bacteria, viruses, fungi, and protozoans. According to the WHO (2022), CW involves potentially infectious microorganisms that can quickly affect other patients, healthcare staff, or the general population if not treated and disposed of properly. This includes COVID-19-related waste. Frontline medical workers, such as doctors, nurses, caregivers, and others handling COVID-19 patients, must wear appropriate personal protective equipment (PPE). This includes face masks, medical gloves, aprons, and disposable protective clothing such as boots, long-sleeved gowns, thick gloves, masks, goggles, or face shields. They must also practice hand hygiene after removing PPE (WHO 2020), especially when treating COVID-19-related waste.

Since COVID-19 can remain on the surface of objects for a few hours to several days, a large volume of used PPE is considered infectious waste and must be handled as clinical waste (Hantoko et al. 2021; Wiktorczyk-Kapischke et al. 2021). Other infectious agents, such as HIV, skin infections, and infectious Hepatitis B, from healthcare establishments, can also spread to the environment if proper handling and CW management are not implemented. Therefore, careful attention to the handling and disposal of CW is crucial to reduce the risk of public health threats or contamination (Agamuthu and Barasarathi 2020). This study aims to explore the current practices of CWM before and during the pandemic in healthcare establishments (HCEs) and to compare these practices over time.

2 Research Methodology

A qualitative exploratory research design was employed to allow participants to share their thoughts, opinions, and experiences regarding how healthcare establishments (HCEs) handle clinical waste. This design was chosen because it can address a greater number of elements that emerge in the data, yielding abundant information about participants' behaviors, views, experiences, and thoughts, which can be used as evidence to interpret observed changes (Tong et al. 2012). Four participants were purposely selected for interviews to provide sufficient information on clinical waste management. To be eligible, participants had to have worked in HCEs for at least one year to ensure they had sufficient knowledge about clinical waste management.

Each in-depth interview lasted about one hour and was conducted in Malay. The interviews took place in a private room at the HCEs to encourage naturalistic responses (Article 2009). Participants were asked to answer semi-structured questions about current clinical waste management practices before and during the COVID-19 pandemic, CWM policy and safety practices before and during the pandemic, and the challenges of CWM during the pandemic. The responses were recorded with the participants' permission for further analysis. The collected data were kept confidential and securely stored. Each recorded discussion was transcribed and translated into English.

3 Result and Discussion

3.1 The Explores of Clinical Waste Management Before and During Pandemic

Generally, the participants reported that clinical waste has been managed similarly to how it was before the COVID-19 pandemic, with only slight differences and minimal changes to the main standard operating procedures. COVID-19 CW is collected and labeled as COVID-19 CW in a yellow bag with a biohazard symbol, just as it was before the pandemic.

The yellow bag is then placed in a COVID-19 collection bin that is properly labeled and locked. Access keys are given only to designated cleaners who handle COVID-19 clinical waste. Every day, the collection bins are changed, and the COVID-19 CW collection bin is stored in cold storage while awaiting transportation to the incineration facility outside the hospital (Teluk Panglima Garang).

According to the DOE (2009), clinical waste is defined as waste that includes human or animal tissue, blood or body fluids, excretions, drugs, pharmaceutical products, soiled swabs or dressings, syringes, needles, and sharps—essentially any waste that has come into contact with or been mixed with any of these types of clinical waste. Unless rendered safe, such waste may prove hazardous to anyone who comes into contact with it. This definition and categorization were also discussed by the participants.

From the in-depth interviews, three respondents mentioned that food waste from COVID-19 patients was initially categorized as clinical waste. However, at the end of August 2021, the Ministry of Health (MOH) announced that it should be classified as general waste, as it was before the pandemic:

> In the COVID-19 pandemic, clinical waste also includes PPE that we wore while working. The latest amendment in August 2021 reclassified food waste from COVID-19 patients as domestic or general waste. Before this, we treated food waste from COVID-19 patients as clinical waste. We disinfected the food waste about five times before disposing of it as general waste, which would be collected by the city council to ensure the bacteria and viruses were inactive and fully dead. (Staff 2)

> Okay, for your information, clinical waste includes human tissues, blood, drugs, sharps, vaccine bottles, needles, and PPE. Before last August 2021, there was a lot of clinical waste, including PPE and food waste from COVID-19 patients. When the COVID-19 pandemic first began, food waste from COVID-19 patients had to be treated as clinical waste. (Staff 3)

> Previously, we handled food waste from COVID-19 patients as clinical waste, placing it inside yellow plastic bags and disposing of it in yellow bins. At the end of August 2021, MOH said the food waste from COVID-19 patients could be treated as general waste. Because of that, we disinfected the food waste multiple times to ensure the virus and bacteria were all dead before disposing of it in normal trash to be collected by the city council. (Staff 4)

Following the categorization of clinical waste, a systematic management process is implemented, which includes waste segregation, storage, transportation, disposal, and addressing spills or accidents that may occur at the hospital. Hospitals routinely practice segregating clinical waste from general waste. All respondents indicated the need to segregate clinical waste into two categories: sharp and non-sharp. Sharps are disposed of in puncture-proof sharps bins, while non-sharp items are placed in standard yellow bins.

> Before COVID-19 and during this COVID-19, the waste segregation of clinical waste remains the same, we will separate into 2 types only, sharp and non-sharp. (Staff 1)

> We will segregate clinical waste and general waste. Whichever is included in the clinical waste category will be put in the yellow bin. There are two types of clinical waste of yellow bins "sharp and non-sharp". The sharp ones will be put in a sharp bin. If the waste is not sharp, we must put it in a yellow bin can lined with yellow plastic. General or domestic waste we will put it on black plastic. Sharp are not only the needles, if a glass bottle that has the potential to break or that is already broken, we will categorize it as sharp waste. As well as with the vaccine bottle after used we put it on sharp bin. (Staff 2)

> Also, the same before and during COVID-19 until now, the segregation there will be 2 types, one sharp and not sharp, the sharp ones are all thrown in the sharp bin, if not sharp in the regular yellow bin. (Staff 4)

After the clinical waste has been segregated, it will undergo the process of storage. According to one of the experienced respondents, the central storage in healthcare establishment in Tanah Merah Kelantan was stored in the hospital.

> All the clinical waste will be stored at the hospital, this is included from PKRC, district health clinic and mortuary because the destination from the hospital is not far, it will only take about 5 to 10 minutes only. (Staff 3)

Other participants also added, that "Clinical waste here (PKRC), the porter will collect to store them at the hospital." (Staff 4)

Further exploration on clinical waste storage has been conducted and most of the participants stated that all clinical waste storage bins will be kept in cold storage until they are transported to Teluk Panglima Garang for incineration. COVID-19 clinical waste collection bins will be labeled and secured:

Storage will be done less than 24 hours in the hospital store and will go to the disposal place according to the set schedule. (Staff 2)

Usually, the porter will make a collection three times a day and it will be stored in the hospital store. as well as clinical waste from PKRC will be taken to the hospital store because PKRC with the hospital is not far, just approximately five minutes only. (Staff 3)

In the hospital store we call it the cabin. All clinical waste storage is located there and the storage only will be taken in less than 24 hours. If the transportation comes late, the clinical waste will be placed in a cold store or we call it a freezer. (Staff 4)

Next, the clinical waste will be transported from central storage locations to designated facilities using specialized vehicles that to ensure secure from the infected bacteria and viruses to people and environment.

We have an airtight truck so it doesn't spread if it gets infected. Transportation must be made by a competent licensed officer. (Staff 2)

Transportation uses airtight trucks by licensed officers. Before the transportation process, there is a weighing process that will be done by us witnessed by people from the MOH. (Staff 3)

Transportation from PKRC to the hospital store is made by the porter staff and there will be a licensed officer who will transport used truck clinical waste to Teluk Panglima Garang. (Staff 4)

The documentation and records of clinical waste processed or disposed of at the landfill must be prepared and verified by hospital staff (Yi and Jusoh 2023). For disposal activities, all the responses said that this disposal will be done at Teluk Panglima Garang:

All the clinical waste will be sent to their final destination to dispose at Teluk Panglima Garang, where we used our new incinerator that used green technology. (Staff 1)

3.2 *The Differences of Clinical Waste Management Practices in the Healthcare Establishments Before and During the Pandemic of COVID-19*

The differences in CWM before and during pandemic have been shown in Fig. 1: Flowchart A (CWM before pandemic) and Fig. 2: Flowchart B (CWM during pandemic). Flowchart A focused on the guidelines of all activities in CWM before pandemic meanwhile Flowchart B shown the differences in activities before pandemic and more focuses on the additional SOPs that need to be considered start from waste collection and transportation of the CWM during pandemic.

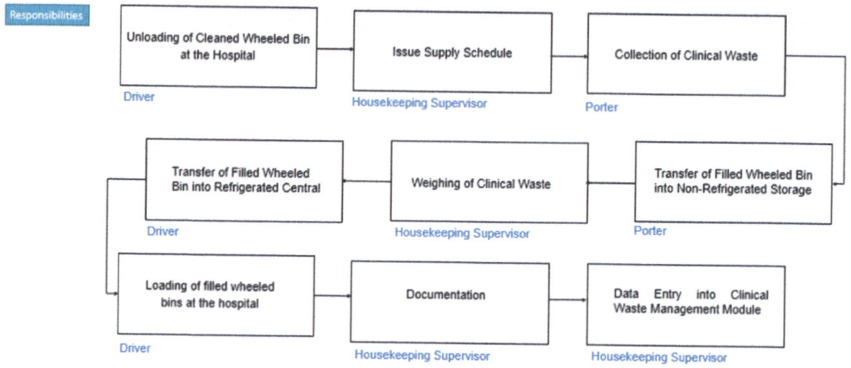

Fig. 1 Flowchart A CWM before pandemic

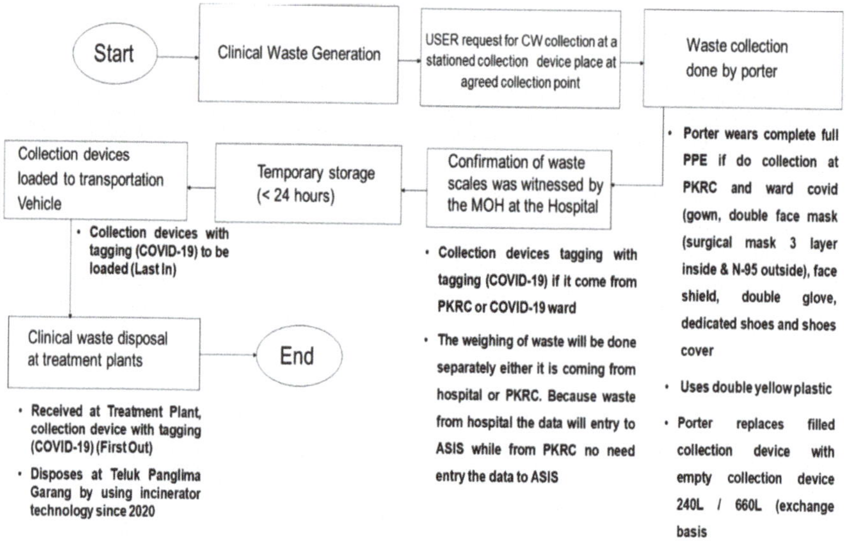

Fig. 2 Flowchart B CWM during pandemic

4 Conclusion

This study explored clinical waste management (CWM) at healthcare establishments (HCEs) in Kelantan. There were slight differences in CWM practices before and during the COVID-19 pandemic. Only a few additional standard operating procedures (SOPs) needed to be adhered to, and no new policies or regulations were implemented regarding CWM. The Environmental Quality Act 1974 has been applied with recent modifications appropriate to the COVID-19 situation. CWM became more challenging during the pandemic due to the increase in clinical waste produced and the critical use of personal protective equipment (PPE), which must be handled appropriately. The CWM practices remain the same even when dealing with COVID-19-related waste. With proper management and disposal systems, any unwanted virus infection or spread can be avoided by adhering to the appropriate SOPs.

Acknowledgements A very thanks to all the participants for their cooperation and University Malaysia Pahang Al-Sultan Abdullah.

References

Agamuthu P, Barasarathi J (2020) Clinical waste management under COVID-19 scenario in Malaysia. Waste Manag Res 5:1–12. https://doi.org/10.1177/0734242X20959701

Article J (2009) Open Research Online. https://doi.org/10.1111/j.16541103.2009.01102.x

Bendjoudi Z, Taleb F, Abdelmalek F, Addou A (2008) Healthcare waste management in Algeria and Mostaganem department. Waste Manag 29(4):1383–1387. https://doi.org/10.1016/j.wasman.2008.10.008

Berita Awani (2021) COVID-19: clinical waste management still under control—Dr. Noor Hisham | Astro Awani, 2020. https://www.astroawani.com/berita-malaysia/COVID-1919-clinical-waste-management-still-under-control-dr-noor-hisham-244319. Accessed 03 May 2021

Dehghani MH, Ahrami HD, Nabizadeh R, Heidarinejad Z, Zarei A (2019) Medical waste generation and management in medical clinics in South of Iran. MethodsX 6:727–733. https://doi.org/10.1016/j.mex.2019.03.029

DOE (2009) Guidelines on the handling and management of clinical wastes in Malaysia. J Exp Psychol Gen 136(1):7–9

Hantoko D, Li X, Pariatamby A, Yoshikawa K, Horttanainen M, Yan M (2021) Challenges and practices on waste management and disposal during COVID-19 pandemic. J Environ Manag 286:112140. https://doi.org/10.1016/j.jenvman.2021.112140

Mustaffa H, Kamarudin MKA, Toriman ME, Rosli MH, Sunardi S (2023) Impact of suspended sediment on Pahang river development using geographic information system. Plan Malaysia 21(1):116–133

Saad MHM, Kamarudin MKA, Toriman ME, Wahab NA, Ata FM, Samah MAA, Manoktong SN (2023) Analysis of the flash flood event and rainfall distribution pattern on Relau river basin development, Penang, Malaysia. Plan Malaysia 21(1):58–71

Tong A, Flemming K, McInnes, Oliver ES, Craig J (2012) Enhancing transparency in reporting the synthesis of qualitative research: ENTREQ. BMC Med Res Methodol 12:181

WHO (WHO Malaysia | World Health Organization) (2020) https://www.who.int/malaysia. Accessed 02 May 2023

WHO (World Health Organisation) (2022) https://www.who.int/news/item/20-05-2022-world-health-statistics-2022

Wiktorczyk-Kapischke N, Grudlewska-Buda K, Wałecka-Zacharska E, Kwiecińska-Piróg J, Radtke L, Gospodarek-Komkowska E, Skowron K (2021) SARS-CoV-2 in the environment—non-droplet spreading routes. Sci Total Environ 770:145260

Yi TC, Jusoh MNH (2023) Overview of clinical waste management in Malaysia. Front Water Environ 1(1):47–57

Designed Landscape for Environmental Sustainability

Mohd Ramzi Mohd Hussain and Mohd Armi Abu Samah

Abstract The chapter elaborated on the keyword of designed landscape for environmental sustainability. As it is well defined, a designed landscape for environmental sustainability is known to build, plan, and manage landscapes that consider the ecology of a site and produce landscapes that benefit both people and the rest of the ecosystem. Consequently, landscape design improves land management and reduces the environmental impact of urban living. It improves air quality, and plants may purify the air and soil. It will be combined in landscape design planning to provide holistic answers to complex issues. Understanding ecological wisdom is a systematic land development approach that avoids environmentally sensitive areas and works with natural dynamics. Ecological wisdom reflects a socio-ecological intelligence that embodies a cultural conscientiousness oriented toward species protection and a built environment integrated with nature and natural systems. In landscape ecological restoration, it is also necessary to use nutrient-rich resources to effectively promote the virtuous cycle of restoration work, reduce the cost of restoration work, and contribute to the construction of conservation-oriented ecologically designed landscapes. However, many traditional landscapes already contain some components of sustainability. The Islamic perspectives highlight the critical roles humans play in sustaining the ecological landscape and natural resources for the benefit of human beings and other living creatures.

Keywords Designed landscape · Environmental sustainability · Ecological wisdom · Landscape ecological restoration · Islamic perspective

---

M. R. M. Hussain (✉) · M. A. A. Samah
International Islamic University Malaysia (IIUM), Kuala Lumpur, Malaysia
e-mail: ramzi@iium.edu.my

M. A. A. Samah
e-mail: marmi@iium.edu.my

A. Z. Yaser et al. (eds.), *Controlling Environmental Pollution*,
https://doi.org/10.1007/978-981-97-8931-3_12

1 Introduction

The word landscape, which derives in English from German and Dutch origins in the Middle Ages as "Landschaft" and "landschap," has been given several definitions and interpretations. Landscape, which includes topography, vegetation and associated plants and soil, water bodies, and their spatial configuration, is one of people's most visual needs. These human–nature interactions lead human beings to have a contrasting preference for the surrounding landscape and environment because a pleasing landscape can bring mental and physical benefits to people. The understanding and preference of people for their surrounding landscape provide a significant contribution to human development. That is, individuals acknowledge the intrinsic value of nature and its subsequence right to exist irrespective of its function for humanity.

Landscapes deliver goods, functions, and values on which human welfare and quality of life are built (Angelstam et al. 2013a). Every landscape is a function of its abiotic and biotic template combined with its unique history of human intervention (O'Farrell and Anderson 2010). On the other part, landscapes are complex entities, dynamic and further complicated in the human dimension of how they are perceived. Landscapes from the ecological perspective focus on the spatial arrangement, size, shape, connectivity, and ecological function of landscapes.

Angelstam et al. (2013b) distinguish four interpretations and aspects of landscapes: (i) ecological—the interactions between biophysical components such as soil, water, biota, and vegetation and resultant patterns and processes; (ii) anthropogenic—landscapes as human phenomena with constructs such as infrastructure, land uses, electoral and political constituencies; (iii) intangible—as cognitive representations of space, organization, and systems; and (iv) coupled socio-ecological systems with a combination of all three above and the interfaces between them.

The design has two main landscape elements: the hard landscape (hardscape) and the soft landscape (softscape) elements. The hardscape elements are the man-made features, which include spaces, patios, benches, proportions, and others. Meanwhile, the softscape elements are soft materials known as plants (trees and vegetation), plantation patterns, and flowers. The hardscape elements are the man-made structure that functions as a frame of all decoration and spaces of landscape design. The softscape is a living element in landscape designs which can be permanently used, such as trees and shrubs. They can also be temporary, such as seasonal plants that are appreciated for their characteristics, such as fruits and flowers, which give color to the surrounding environment (Angelstam et al. 2013b).

2 Sustainability and Ecological Concept of Designed Landscape

A sustainable landscape is a new concept of designing parks or outdoor spaces in an urban environment that considers environmental sustainability (Kawther and Hassan 2021). It includes designing, constructing, and managing residential and commercial parks surrounding various types of buildings and scales. Further, Kawther and Hassan (2021) define the designed landscape as a territory of land altered principally by individuals for aesthetic impact. This definition includes the natural spaces in which they are used, management, preservation, and restoration by people in an environment. It has the importance of increasing the natural capital, preserving the ecosystem, and preserving the cultural features of the urban population. The priority areas in landscape planning and environmental management include water resources management, transport, waste management, maintenance, and the overall design for sustainable development (Kawther and Hassan 2021).

Over time, the potential of design for sustainable development has been noticed; an integration from planning to design has started to be realized in a process in which environmental problems bring ecological approaches to the agenda (Celik 2017). Celik (2017) adds that sustainable landscape architecture is a category of sustainable design concerned with the planning and design of outdoor space. Sustainable landscape architecture considers economic, social, and environmental aspects. It begins with appropriate systems which address function, cost, energy efficiency, beauty, and environment. Sustainable landscape architecture integrates ecological, social, cultural, and economic factors in designing landscapes to help protect habitat, contribute to stormwater management, and conserve water, among other objectives. The current trend in landscape architecture is to find the balance of "aesthetics and function" required for successful sustainable design. Creating a sustainable landscape means working toward a thoughtful balance between resources used (both in construction and maintenance) and results gained.

Sustainable landscapes require as much, if not more, planning as traditional landscapes. For example, plants that are an integral part of landscape design help reduce pollution and mitigate some of the effects of heat, sound, wind, and air (Chen 2023). People and other living things are increasingly influenced by landscape design to create conducive living conditions. Designed landscapes significantly impact a community's social, aesthetic, and environmental qualities. They also consider the aesthetics of innovation to provide a natural atmosphere conducive to our health and the health of our loved ones. Trees and vegetation contribute strongly to desirable ecological, social, and economic conditions in any city, which enhances the city's sustainability. Trees also have sounds, such as when leaves rustle in the wind or a carpeted floor of fallen leaves. Trees give people joy through the senses. Trees provide shade. Shade is at the root of many of the social, economic, and environmental benefits provided by trees in the city. It is vital in promoting health, as shade trees block harmful ultraviolet radiation. Consequently, Chen (2023) adds that designed landscapes improve land management and reduce the environmental impact of urban living.

In 1969, a Scottish landscape architect, Ian McHarg, introduced a comprehensive method of utilizing the ecological wisdom of landscape design and planning. According to Liao and Chan (2016), ecological wisdom principles is to have clear directions that a practitioner follows to develop an ecologically wise resolution to a practical problem. They add that, in ecological planning and design, the "design with nature" approach, which Ian McHarg championed, is a systematic land development approach that avoids environmentally sensitive areas and works with natural dynamics. The beauty of ecological wisdom lies in its endurance over time, efficacy in practice, and ability to predict project performance decades, if not centuries, in advance (Yang and Li 2016). Wang et al. (2016) add that ecological wisdom reflects a socio-ecological intelligence that embodies a cultural conscientiousness oriented toward species protection as well as a built environment integrated with nature and natural systems. There are three responsibilities of ecological wisdom ethical values that designers, planners, and stakeholders need to be conscious of (a) the duty to guard the well-being of all species and their habitats; (b) the obligation to preserve resources for generations to come; and (c) the obligation to perform competency in making, selecting, and implementing plans and designs (Wang et al. 2016).

Landscape ecology believes that at the landscape scale, each independent ecosystem (or landscape unit) can be regarded as a broad patch, narrow corridor, or background matrix, and energy, species, and nutrients are in the landscape (Zhou 2021). The flow between the two forms a landscape function, and the play of the landscape function mainly involves the functional characteristics of corridors, substrates, and patches. The landscape and the other three functions integrate technologies, policies, and actions to realize ecosystem utilization's productivity, safety, protection, feasibility, and acceptability. Figure 1 shows the theoretical framework of landscape ecology.

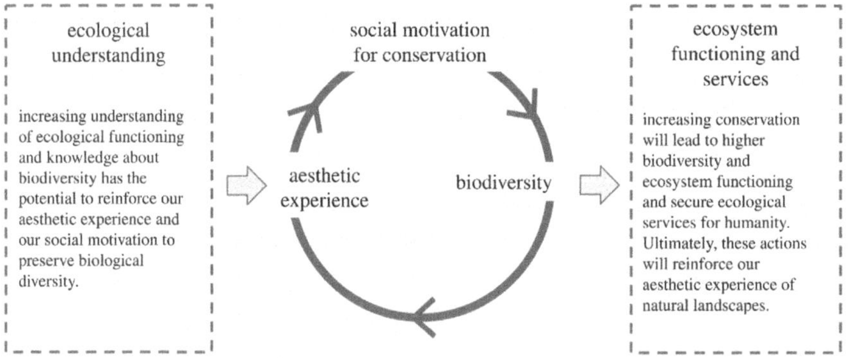

Fig. 1 The theoretical framework of landscape ecology (adapted from Zhou 2021, p. 2)

Further, Zhou (2021) adds that the process at the landscape level includes both natural and human aspects. They play an essential role in forming the landscape spatial structure, and the landscape structure also has a primary control effect on the process. Therefore, only a specific landscape structure can realize a particular landscape function by "the pattern counter-evolving process, the process counter-evolving mechanism, and the mechanism revealing the law."

Ecological planning offers a theoretical framework and method that can help encourage the efficient use of natural resources and the protection of the environment (Fu et al. 2016). The contribution of ecological planning, such as the inland suitability approach, apply bio-ecology and human ecology concepts in finding ways to reduce the environmental impact of urban development; the participation of ecological planning and design professionals, which include natural areas and green space protection, low impact development, transfer of development rights, green infrastructure, brownfield/derelict redevelopment, urban ecological design, and sustainable urbanism (Wang et al. 2016). Green (or environmental) infrastructure is one strategy to enhance and expand ecosystem services (Steiner 2016; Meerow and Newell 2017). Green infrastructure can also reduce and mitigate the human consequences of natural disasters. Human beings should be treated as a component of an ecosystem which interacts between organisms and the environment to find the fittest environment and ecological adaptation (Fu et al. 2016).

In his book, "Design with Nature," McHarg presents an ecological design based on ecosystem services and layering techniques as the foundation for using ecological science for design and decision-making (Steiner 2016). As the book title suggests, following nature's lead in planning and design is the wisdom of achieving sustainability. At the same time, anthropogenic uses or interventions shall become an integral part of the natural processes (Yang and Li 2016). Steiner (2016) adds two of the most significant changes in ecological science relevant to designers and planners since the publication of Design with Nature: The Emergence of Landscape and Urban Ecology. As McHarg is concerned, the landscape and urban ecology will be the most significant challenges for planners and urban designers where the urban regions are growing, and many cities are located in places increasingly vulnerable to climate change. It will affect society, which includes the poor, the elderly, the young, and the disabled, who are at risk in urban regions due to the danger of flooding, hurricanes, wildfires, and earthquakes (Steiner 2016). Thus, planning and designing safe places for people to live is fundamental.

Furthermore, apart from protecting people from natural disasters, landscape architects, and planners can enhance the ecosystem services produced in urban areas (Steiner 2016). These include those natural processes that provide food and water; that regulate, such as control of the climate and disease; that support, such as nutrient cycles and crop production; and that contribute to culture, such as spiritual and recreational benefits. To make human actions in tune with natural processes, ecological planning and design should pay particular attention to the human–environment relationships in making plans (Wang et al. 2016).

Ian McHarg formalized a design practice that prioritized incorporating nature throughout the urban landscape and promoted landscape architecture as the instrument of environmentalism. He insisted that landscape architecture should find its "rules" in nature, and those rules emerged from the scientific study of nature where ecologists provide "not only explanation but also a command." The expansion of urban ecology is a step toward reconceiving urban regions that provide many environmental benefits, such as the efficient use of energy and water (Steiner 2016). Apart from the complex ecosystem, urban ecology comprises flows, networks, corridors, matrices, patterns, nodes, habitats, communities, and various interactions. Steiner (2016) highlights the challenges of making cities fairer, more sustainable, more resilient, and more productive ecosystems to create a regenerative city in which sources of energy and materials are restored, renewed, and revitalized.

McHarg focuses on the natural, social, and cultural processes as an iterative design process primarily shaped by human and ecosystem interactions (Yang et al. 2013). An iterative process would address the dynamic nature of landscapes in pursuing, understanding, and developing multifunctional landscapes (O'Farrell and Anderson 2010). The book (Design with Nature) stresses that design should let nature perform the maximum amount of work. According to Yang et al. (2013), there was increasing evidence that design can achieve multifunctional benefits should nature's role be considered; that is, design with nature warrants multifunctional landscapes. Some examples—such as decentralized and naturalized ways of managing the stormwater—suggest the benefits of runoff reduction and water quality enhancement, and increasing tree canopy coverage mitigates the urban heat island (UHI) effect and may reduce the incidence of heat-related diseases. In the context of a design process, McHarg called upon interdisciplinary professionals and started with land-carrying capacity, which integrates and facilitates the procedures, functions, and services of multiple ecosystems.

Ecological integrity and spatial heterogeneity are the concentrated expressions of the core of landscape ecology theory (Zhou 2021). The landscape is a complex system composed of the organic connections of landscape elements, with a hierarchical structure, independent functional and prominent visual characteristics. A healthy landscape ecosystem has structural integrity and must achieve functional continuity. Furthermore, Zhou (2021) adds that the landscape is a mosaic composed of heterogeneous landscape elements. The spatial heterogeneity at the landscape scale includes three aspects: spatial composition, spatial configuration, and spatial correlation. Heterogeneity determines the diversity and biodiversity of landscape patterns and is the core of landscape ecology. It is generally believed that high heterogeneity is conducive to the symbiosis of species but not conducive to the survival of rare species in the landscape. The increase in heterogeneity within a specific range can help improve the stability of the landscape. Consciously increasing and maintaining the heterogeneity of the landscape is necessary. Figure 2 shows the relationship between ecological integrity and spatial heterogeneity.

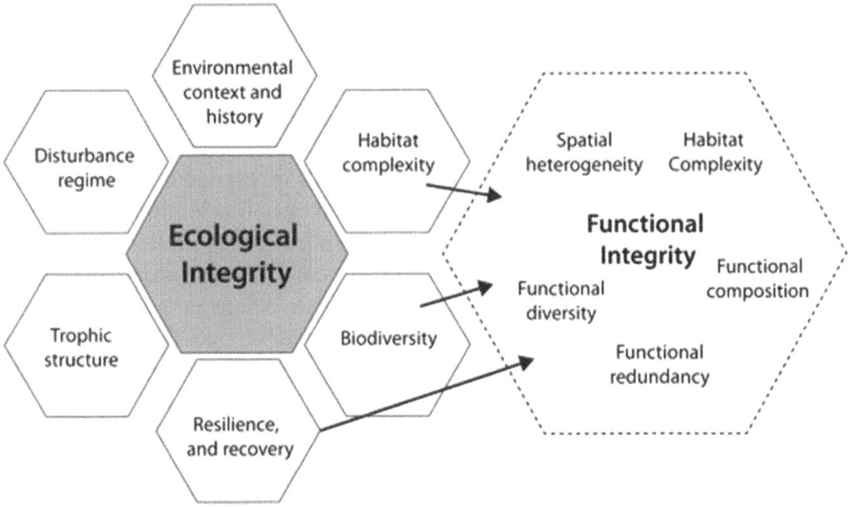

Fig. 2 Relationship between ecological integrity and spatial heterogeneity (adapted from Zhou 2021, p. 3)

3 Multifunctional Approach of Designed Landscape

The multifunctional landscape is defined as multifunctional benefits (Yang et al. 2013). Brandt and Vejre (2001) describe three types of landscape functionality which reflect different approaches to the landscape. They are: (a) functionality of ecosystems and landscape ecosystems—the interaction between living and dead biomass and the abiotic frames; (b) functionality about land use—human-ecological aspect of the landscape; and (c) social functionality—social interaction within the ecosystem. Thus, the planning and realization of multifunctional landscapes imply that each land use must be managed within certain limits and modified to suit competing land uses and their claims on the landscape resources (Brandt and Vejre 2001). In response to the challenge of climate change, a *multifunctional landscape* is a metaphor for the holistic approach and a strategic form of art.

> Multifunctional Landscapes can provide food, energy, water storage and flood mitigation as well as a valuable resource for biodiversity and promote health and well-being. The holistic approach traditionally taken by the landscape architect to master planning, landscape design and management means that the profession is responsible for environmental stewardship. The profession also has a wider role to plan, create and design sustainable communities, bringing a balanced environmental, social, and economic approach to tackling climate change.
>
> Sustainable landscape planning, design, and management are essential to adapt our environments to a changing climate and mitigate future change. In many instances, landscape responses incorporate various mitigation and adaptation principles, with many interlinked and mutually reinforcing, while also providing wider socio-economic and environmental benefits. For example, provision of urban green space will mitigate through carbon storage and reduce surface run-off, an important aspect of adaptation.

> Landscape architects provide a holistic approach to the protection, conservation, and enhancement of urban and rural landscapes, which take account of environmental, social and economic conditions. This approach will assist in mitigating and adapting to future climate change and ensure the future sustainability of our landscapes and their communities (Williamson and Lovell 2008, p. 3).

Furthermore, we need landscapes that assist species in responding to "increasing climate pressure, facilitating movement, and establishing new emerging ecosystems" (O'Farrell and Anderson 2010, p. 59). According to Yang et al. (2013), multifunctional landscapes are attributed to integrating ecosystems' metabolism into the design process, which creates and maintains reasonable human settlements and ecosystems. Yang et al. (2013) outlined four main aspects of ecosystem services that can be solved by applying multifunctional landscapes. They include (a) supporting and biophysical functions (e.g., protecting and enhancing biodiversity and water quantity and quality); (b) provisioning services (e.g., production of energy and other utilitarian resources); (c) regulating services (e.g., waste reduction and reuse); (d) cultural and social services (e.g., visual quality, beauty, human health, and recreational opportunity). These will allow for increased socio-ecological resilience and improved service provision under the scenario of change (O'Farrell and Anderson 2010).

Sustainable multifunctional landscapes are created and managed to integrate human production and landscape use into a landscape's ecological fabric, maintaining critical ecosystem function, service flow and biodiversity retention (O'Farrell and Anderson 2010). Multifunctional landscapes are perceived as mixed natural and cultural interacting systems (O'Farrel and Anderson 2010). According to O'Farrell and Anderson (2010), creating and designing sustainable multifunctional landscapes require transdisciplinary approaches using available science and technology. As discussed above, McHarg's design with nature approach warrants multifunctionality because of its natural interdisciplinary approach.

4 Designed Landscape: An Ecology Approach for the Conservation of Biodiversity and Ecosystem

The contribution of ecological planning—such as inland suitability approach; applying bio-ecology and human ecology concepts in finding ways to reduce the environmental impact of urban development; the participation of ecological planning and design professionals, which include natural areas and green space protection, low impact development, transfer of development rights, green infrastructure, brownfield/derelict redevelopment, urban ecological design, and sustainable urbanism. Green (or environmental) infrastructure is one strategy to enhance and expand ecosystem services. Green infrastructure can also reduce and mitigate the human consequences of natural disasters. Human beings should be treated as a component of an ecosystem which interacts between organisms and the environment to find the fittest environment and ecological adaptation.

In addition, the landscape approach is defined as a strategy for the integrated management of land, water, and living resources that equitably promotes conservation and sustainable use. Applying the ecosystem approach will help reach a balanced objective of landscape conservation, sustainable use, and the fair and impartial sharing of the benefits arising from the utilization of genetic resources. This statement integrates the economic, social, and ecological aspects of development. How people, organizations, or institutions view a landscape approach depends on their objectives and considerations for initiating it.

There is a set of ten principles for landscape approaches which include: (a) Continual learning and adaptive management—The adaptive collaborative management have emerged as practical approaches to continual learning; (b) Common concern entry point—This process will build the confidence and the trust needed to address further issues; (c) Multiple scales—the high- and low-level processes can improve local interventions, policy and good governance, and coordinate administrative entities; (d) Multiple stakeholders—developing a landscape approach requires a patient iterative process of identifying stakeholders and recognizing theirs concerns and aspirations; (e) Negotiated and transparent change logic—all members need to understand and accept the general logic, legitimacy, and justification for a course of action and to be aware of the risks and uncertainties; (f) Clarification of rights and responsibilities—rules on resource access and land use shape social and conservation outcomes and need to be clear as a basis for good management; (g) Participatory and user-friendly monitoring—all members should be able to generate, gather, and integrate the information they require to interpret activities, progress, and threats; (h) Resilience—the challenge in landscape approach is often to bring about transformational change while maintaining the attributes of the landscape that provide resilience to undesirable changes; (i) Strengthened stakeholder capacity—effective participation by members in accepting various roles and responsibilities.

In these approaches, the interaction between the stakeholders and users of biodiversity would be made visible to enable spatial planning. Ideally, all stakeholders would have access to the same information, knowledge, and technology for making optimal decisions on land use and be able to monitor the effects and progress. Therefore, the ten principles of the landscape approach are innovations that should address the challenges in planning and designing a sustainable and smart landscape to restore and conserve the environment. It will further help to prevent risk and epidemic spread to the public health.

5 Islamic Perspective of Sustainability-Designed Landscape

The relationship between man and the environment has been recognized as the trust for the man to carry out his duty as a "caliph" on earth. A man should conserve and preserve the resources for the use of present and future generations, which is in conformity with the concept of sustainable development. In the Quran, nature is an estate belonging to Allah which constitutes a testing ground for man's morality and behaviors, and whatever right man possesses to have dominion over nature. This was formed from the three central concepts of Islam: Tawhid (unity), Khilafa (stewardship), and amana (trust).

For example, the concept of khilafa (stewardship) and amana (trust) emerges from the principle of tawhid (Zen, 2008). The Quran explains that mankind holds a privileged position among God's creation on earth as stewardship and carries the responsibility of caring for God's earthly creation. Each on earth is given this task and privilege in the form of God's trust. The environmental crisis is, in effect a failure of the trusteeship. Thus, nature becomes an index of how well a particular society has performed responsible toward Allah s.w.t. However, the human being was the only one in Allah's creation (i.e., nature) that was prepared to undertake this task of trusteeship (amana) willingly, according to the verse:

> We did intend to offer the trust (amana) to the heaven and the earth and the mountains: but they refused to undertake it, being afraid thereof: but man undertook it, he was indeed unjust and foolish. (Al-Ahzab, 72)

Plants, together with every creation, are important and essential living beings, worshipping and glorifying Allah in a manner that humankind cannot perceive. Plants have also been placed on earth to bring benefits not only for human but also for animal life. As human is part of nature, therefore humans live by its rule. Furthermore, man has a very strong relationship with natural marvels where peaceful co-existence with the natural environment remains unchanged in providing basic needs, and goals to the purpose of men's creation in this valuable life. In the Quran mentions,

> And within the land are neighbouring plots and gardens of grapevines and crops and palm trees, [growing] several from a root or otherwise, watered with one water; but We make some of them exceed others in [quality of] fruit. Indeed, in that are signs for a people who reason (Al-Ra'd, 13:4)

The above *ayat* declares about the existence of living environment and their purposes for human life. The earth has perhaps more than 30,000 plant species. The existence of these numerous and divergent plant species has been very beautifully mentioned in the Quran:

> For it is He who has brought into being gardens – [both] the cultivated ones and those growing wild – and the date-palm, and fields bearing multiform produce, and the olive tree, and the pomegranate: [all] resembling one another and yet so different! (Al-Anaam, 6:141)

In Surah Ar-Rahman, it is mentioned that:

> And the earth has He spread out for all living beings, with fruit thereon, and palm trees with sheathed clusters [of dates], and grain growing tall on its stalks, and sweet-smelling plants. (Ar-Rahman, 55:10–12)

Islam encourages people to take care of plants from any damage or death and prohibits cutting or destruction of trees or plants without any meaningful reason. Where the life of one tree is appreciated, one can see what the Islamic position is toward destroying millions of trees as a result of humans directly acting upon nature (e.g., deforestation) or indirectly (e.g., acid rain). Hence, Islam encourages its followers to plant, maintain, and conserve plants since its importance to mankind is undeniable.

6 Conclusion

The chapter discusses and interprets the ecological knowledge of designed landscapes, in particular, for urban design. The specific reference to Design with Nature (Ian McHarg 1969) proved the essential need for ecological wisdom and knowledge in conserving and preserving the landscape's natural resources. Designed landscapes can be a solution part of sustainable development goals, and landscape architects can help the world achieve sustainable development goals. A landscape architect interested in realizing sustainable development goals with landscape design might focus on innovative design approaches. For landscape architects, it is critical to focus on the local culture and ecology and be sure that they not only use it in designs but let it guide design. The Landscape Architect must incorporate ecological planning and design concepts into the design. Under the premise of respecting nature and regionality, find a more favorable design method for urban development and apply local characteristic culture to the design to make landscape gardens, functionality, and aesthetics coexist. Adhering to the concept of ecological landscape design, a Landscape Architect should also pursue the art of landscape design instead of adhering to the ecological concept and abandoning garden art, let alone sacrificing the natural ecological environment while pursuing the artistic landscape.

Creating a sustainable landscape means working toward a thoughtful balance between resources used (both in construction and maintenance) and results gained. Sustainable landscapes require as much, if not more, planning as traditional landscapes. In landscape ecological restoration, it is also necessary to use nutrient-rich resources to effectively promote the virtuous cycle of restoration work, reduce the cost of restoration work, and contribute to the construction of conservation-oriented ecologically designed landscapes. However, many traditional landscapes already contain some components of sustainability. The Islamic perspectives highlight the critical roles humans play in sustaining the ecological landscape and natural resources for the benefit of human beings and other living creatures.

References

Al-Quran

Angelstam P, Elbakidze M, Axelson R, Koch NE, Tyupenko TI, Mariev AN, Myhrman L (2013a) Knowledge production and learning for sustainable landscapes: forewords by the researchers and stakeholders. AMBIO 42:111–115. Springer

Angelstam P, Grodzynskyi M, Anderson K, Axelsson R, Elbakidze M, Khoroshev A, Kruhlov I, Naumov V (2013b) Measurement, collaborative learning and research for sustainable use of ecosystem services: landscape concepts and Europe as laboratory. AMBIO 42:129–145. Springer

Brandt J, Vejre H (2001) Multifunctional landscapes, definitions and applications. Roskilde: Roskilde Universitet. (Publikationer fra Institut for Geografi og international udviklingsstudier. Arbejdspapir; No. 172)

Celik F (2017) The role and the importance of landscape design to the realization of sustainable development goals. Int J Landsc Plan Archit 3(2):52–61

Chen X (2023) Environmental landscape design and planning system based on computer vision and deep learning. J Intell Syst 32

Fu X, Wang X, Schock C, Stuckert T (2016) Ecological wisdom as benchmark in planning and design. Landsc Urban Plan 155:79–90. Elsevier

Kawther KK, Hassan RH (2021) The sustainable design rules of landscape. In: E3S web of conference, ICGE 2021

Liao K-H, Chan JKH (2016) What is ecological wisdom and how does it relate to ecological knowledge? Landsc Urban Plan 155:111–113. Elsevier

McHarg IL (1969) Design with nature. Natural History Press, NY, USA

Meerow S, Newell JP (2017) Spatial planning for multifunctional green infrastructure: growing resilience in Detroit. Landsc Urban Plan 159:62–75. Elsevier

O'Farrel PJ, Anderson PML (2010) Sustainable multifunctional landscape: a review to implementation. Current Opin Environ Sustain 2:59–65. Elsevier

Steiner F (2016) The application of ecological knowledge requires a pursuit of wisdom. Landsc Urban Plan 155:108–110. Elsevier

Wang X, Palazzo D, Carper M (2016) Ecological wisdom as an emerging field of scholarly inquiry in urban planning and design. Landsc Urban Plan 155:100–107. Elsevier

Williamson N, Lovell J (2008) Position statement: landscape architecture and the challenge of climate change. Landscape Institute, London, UK

Yang B, Li MH, Li S (2013) Design-with-nature for multifunctional landscapes: environmental benefits and social barriers in community development. Int Environ Res Public Health 10:5433–5458

Yang B, Li S (2016) Design with nature: Ian McHarg's ecological wisdom as actionable and practical knowledge. Landsc Urban Plan 155:21–32. Elsevier

Zen I (2008) Vision of an Islamic city. In: Sarkawi AA, Abdullah A (eds) Urban planning: an Islamic perspective. Arah Publications, Kuala Lumpur

Zhou L (2021) Research on landscape architecture design based on ecological restoration and sustainable utilization. In: EMCEME 2020. IOP conference series: earth and environmental science, vol 692

Advancements in Metal Recovery from Industrial Wastes: A Comprehensive Overview

Mohd Aizudin Abd Aziz, Muhammad Auni Hairunnaja, Mohd Azmir Arifin, and Khairuddin Md. Isa

Abstract The rapid pace of industrialization has driven the demand for heavy metals, while the depletion of high-grade natural ore reserves necessitates the exploration of alternative metal sources. Simultaneously, industrial activities generate vast quantities of waste, often relegated to landfills or incineration, culminating in environmental hazards and health risks. These industrial waste streams contain a spectrum of hazardous metals (such as Au, Ag, Ni, Mo, Co, Cu, Zn, and Cr), compounding their potential harm to both the environment and human health. In response, strategies for industrial waste management, including reuse, remanufacturing, and recycling, have gained prominence due to their long-term cost-effectiveness and potential for profitable metal recovery. This study presents a comprehensive review of state-of-the-art technologies for recovering precious metals from industrial waste streams with a focus on sustainability. Diverse industrial wastes, including spent petroleum catalysts, medical waste, electronic scraps, battery residues, metal finishing industry waste, and fly ash, constitute some of the largest industrial waste categories. Various metal recovery processes, categorized as physical, chemical, and thermal methods, leverage the unique characteristics of waste streams and target metals for separation and extraction. Challenges in conventional pyrometallurgy, advancements in hydrometallurgical methods, physical and chemical techniques, and other emerging technologies are extensively discussed. The hydrometallurgical approach, involving dissolution and leaching, stands out as a proven and successful method for recovering metals from diverse raw materials. Although alternative recovery methods exist, many are specialized for specific metal properties within industrial waste. Following metal dissolution, recovery solutions are further concentrated and purified through adsorption, cementation, chemical precipitation, ion exchange, membrane filtration, and ion flotation techniques, which can be applied to various liquid waste streams. The feasibility of a recovery method hinges on its cost-effectiveness in comparison

M. A. Abd Aziz (✉) · M. A. Hairunnaja · M. A. Arifin
Faculty of Chemical and Process Engineering Technology, Universiti Malaysia Pahang Al-Sultan Abdullah (UMPSA), 26300 Gambang, Pahang, Malaysia
e-mail: maizudin@umpsa.edu.my

K. Md. Isa
Faculty of Chemical Engineering Technology, Universiti Malaysia Perlis (UniMAP), 02600 Arau, Perlis, Malaysia

A. Z. Yaser et al. (eds.), *Controlling Environmental Pollution*,
https://doi.org/10.1007/978-981-97-8931-3_13

to the value of the precious metals. Stringent waste disposal regulations and environmental legislation mandate environmentally friendly metal recovery technologies. This review offers critical insights to assist researchers in selecting appropriate metal recovery methods for distinct industrial waste types. Moreover, it aids stakeholders and decision-makers in shaping waste management strategies that align with sustainability goals.

Keywords Metal recovery · Industrial wastes · Sustainability

1 Introduction

The challenge of industrial waste management and the potential for metal recovery from these waste streams have garnered significant attention in recent years. This introduction sets the stage for our comprehensive review by highlighting the pressing issues at the intersection of waste management, resource depletion, and environmental impact. In 2019, the world generated a staggering 53.6 million metric tons (Mt) of electronic waste (e-waste), with only 17.4% officially documented as properly collected and recycled (Forti et al. 2020; Rahman and Aziz 2022). This glaring disparity underscores the global waste management challenge. Improper disposal practices, whether for e-waste or other solid waste types, have wide-reaching consequences. Waste from workplaces and households often ends up in landfills, posing significant environmental and human health risks. This situation is particularly acute in countries like China, India, and various African nations, where substantial waste accumulations demand effective treatment solutions (Chauhan and Punia 2022; Ismail et al. 2021; Xia et al. 2023).

Compounding these waste management challenges is the rapid growth of the global population and the finite reserves of natural wealth. Many countries continue to exert heavy demands on the planet's natural resources, while others confront critical issues such as food and water shortages. In response to these mounting concerns, the world has increasingly turned its focus toward mitigating pollution, minimizing waste, and enhancing recycling efforts, particularly regarding end-of-life products (ELP) (Abd Aziz et al. 2017; Salai and Kelambakkam 2020). Further complicating the environmental landscape is the generation of industrial solid and liquid waste, notable for its non-biodegradable properties. Approximately 10–15% of total industrial waste falls into the category of non-biodegradable and hazardous materials, with this proportion steadily increasing each year. Industries such as iron and steel plants, fertilizer industries, chemical, drug, and dye industries contribute significantly to this hazardous waste stream, perpetuating environmental harm (Chandraseagar et al. 2019; Razali et al. 2018; Saw et al. 2021). Currently, many industries resort to incinerating waste or incorporating it into construction materials like cement. However, this approach carries the risk of toxic compounds leaching into the environment and the loss of valuable metals within the waste stream (He and Kappler 2017; Japar et al. 2018). In response to these challenges, industries are increasingly focusing on waste recovery, subjecting waste materials to rigorous testing and regulations before their incorporation into construction materials. For example, in Sweden, boiler companies and the Swedish Energy Agency have initiated research programs to promote the use of non-coal ash (Ortiz-Ribbing et al. 2007).

Considering these intertwined challenges, scientists, engineers, and researchers have proposed a compelling alternative solution for enhanced solid waste management: the recovery of metals from these waste streams. Recovering metals from waste has proven to be more energy-efficient and environmentally friendly than extracting primary metals from mineral ores (Fan et al. 2020; Ismail et al. 2019a). Notably, the extraction of metals from e-waste consumes only a fraction (between 10 and 15%) of the energy required for smelting and refining metal ore concentrates, as demonstrated by Boliden Rönnskär in Sweden (Mostaghel et al. 2011). Over the centuries, numerous studies have contributed to the development of various metal extraction methods. Nonferrous metals such as aluminium, copper, chromium, nickel, lead, titanium, zinc, and precious metals like silver, gold, and platinum are extensively recycled as scrap (Aziz et al. 2019; Chauhan et al. 2020). Metallurgical processes, known for their practicality, play a pivotal role in extracting metals from ores and solid wastes. Additionally, microbial action has emerged as a viable technique for metallurgical processes and the treatment of metal-containing wastewater (Chauhan et al. 2020).

In the subsequent sections of this review, we will delve into various physiochemical techniques, including adsorption, cementation, chemical precipitation, membrane filtration, ion exchange, and ion flotation, which are utilized for metal recovery from industrial wastewater. A survey conducted by C. Pathiarana reveals that ion exchange and membrane filtration are the most frequently employed treatments for metal-containing wastewater. However, there remains a pressing need to develop cost-effective and environmentally friendly technologies for recovering precious metals from industrial waste (Pathirana et al. 2019; Suhaila et al. 2018).

This paper aims to provide a state-of-the-art review of the technologies currently applied to the recovery of precious metals from industrial wastes. It offers a comprehensive analysis of recent developments in metal recovery technologies and their potential to contribute to sustainable and energy-efficient metal product recovery from both solid and liquid waste streams. Moreover, it addresses the economic and environmental safety aspects related to the recovery of valuable metals from dispersed toxic compounds in industrial wastewater, emphasizing the commercial value of these metal compounds (Razali et al. 2020; Zhang and Xu 2016).

2 Metallurgical Extraction of Solid Waste

In the contemporary digital landscape, the rapid evolution and accessibility of electronic products have driven an upsurge in demand, resulting in a considerable increase in electronic waste, commonly known as e-waste. This section delves into the intricacies of extracting valuable metals from e-waste and other forms of solid waste, meticulously considering both the challenges and the environmental consequences associated with these extraction processes.

The escalating demand for electronic gadgets, coupled with the shortened lifecycle of these products, has led to a surge in discarded electronic items. This burgeoning e-waste poses a dual challenge—the need to manage electronic waste responsibly and the opportunity to recover precious metals embedded within these discarded items. Understanding the complexities of metallurgical extraction from solid waste is pivotal in addressing environmental concerns and harnessing the potential for resource recovery.

As we explore metallurgical extraction from solid waste, it is imperative to recognize the environmental implications intertwined with the disposal and treatment of e-waste. E-waste contains a myriad of metals, some of which are scarce and valuable. The conventional methods employed in metallurgical extraction from solid waste involve intricate processes to separate and reclaim these metals. However, the environmental impact of these extraction processes, including the release of hazardous substances and energy consumption, must be scrutinized.

This section aims to provide a comprehensive narrative of the challenges and advancements in the metallurgical extraction of solid waste. From the environmental repercussions of electronic waste to the evolving techniques for metal reclamation, this exploration seeks to shed light on a critical aspect of modern waste management, where technological innovation and environmental responsibility intersect.

2.1 Understanding Global Waste Generation Statistics

To address the challenges posed by e-waste, it is crucial to understand the global waste generation landscape. In 2016, the world produced an estimated 44.7 million tons of e-waste, with projections indicating a growth rate that could reach 52.2 million tons by 2021. (Hairunnaja et al. 2023; Ma et al. 2016). Shockingly, only 1% of this waste is properly collected and recycled, while a staggering 80% remains undocumented, often ending up in illegal landfills. Developed countries tend to generate more e-waste per capita than their developing counterparts (Jagannath et al. 2017).

Asia leads the world in e-waste generation, with China, Japan, and India being the primary contributors. Europe, despite a 35% collection rate, produces 12.3 million tons of e-waste, and the Americas generate 11.3 million tons, primarily due to the United States' contribution. Africa and Oceania also play a role in e-waste generation, albeit on a smaller scale (Awasthi et al. 2016). E-waste legislation has been enacted in many countries, with China and India implementing stringent regulations. Recycling programs, such as Australia's National Television and Computer Recycling Program, have demonstrated the potential for metal recovery from e-waste (Thakur and Kumar 2020).

2.2 *Pyrometallurgical Techniques: Harnessing the Power of Heat*

Pyrometallurgical techniques offer a comprehensive solution for extracting valuable metals from various industrial wastes. This approach is highly efficient for recovering metals like Titanium (Ti), Zirconium (Zr), Niobium (Nb), Tantalum (Ta), and Molybdenum (Mo) (Ismail et al. 2019b; Ojeda et al. 2009). Typically, valuable metals in solid waste exist in oxide form. The pyrometallurgical process involves a series of steps, including incineration, sintering, and melting at high temperatures. Furnaces like smelting furnaces, thermal reactors, and plasma processes are commonly used in industrial practices (Dahawi et al. 2019; Rao 2006). While pyrometallurgy is effective in metal recovery, it is energy-intensive, emits pollutants, and may result in metal loss during combustion. Researchers are actively working to minimize the environmental impact and enhance efficiency. Innovative methods, including automated disassembly systems and reducing dioxin emissions through lower furnace temperatures, are being explored (Ma et al. 2016). Moreover, pyrometallurgical processes have successfully facilitated the recovery of metals such as gold, silver, copper, and zinc from various waste sources. Methods like chlorination have been employed for metal extraction (Ismail et al. 2020; Jena and Brocchi 1997).

2.3 *Hydrometallurgical Techniques: The Power of Aqueous Solutions*

Hydrometallurgy, a metal recovery process conducted in aqueous or organic solutions, offers several advantages, including impurity control, low investment costs (ideal for small-scale applications), reduced environmental impact, and high metal recovery potential (Chmielewski et al. 1997; Jha et al. 2001). This process comprises leaching, concentration/purification, and recovery stages. Numerous treatment routes are available to recycle waste, with a primary focus on efficient metal recovery (Abdu Rahman et al. 2019; de Buzin et al. 2017). The selective dissolution of desired metals using robust chemical reagents is followed by various extraction protocols such as cementation, solvent extraction, filtration, and ion exchange, which play pivotal roles in the concentration and purification of metals (Zhang and Chen 2012; Zhang et al. 2013). Hydrometallurgical methods, in comparison to pyrometallurgy, offer a more environmentally friendly alternative and are suitable for extracting metals from different aqueous leach and waste effluents. Ongoing efforts aim to minimize liquid waste generation in hydrometallurgical processes (Li et al. 2020; Love et al. 2019).

3 Recent Advances in Metal Recovery Techniques from Wastewater

In recent years, the field of metal recovery has undergone a significant transformation with the introduction of cutting-edge techniques, revolutionizing both metallurgical extraction and the recuperation of metals from wastewater. This section is dedicated to exploring the forefront of research, highlighting innovative methodologies that have emerged to address prevailing challenges and shape the future of metal recovery practices. Traditional pyrometallurgical methods, long employed for metal extraction, have recently undergone refinement and enhancement. Pioneering studies led by Moskalyk and Alfantazi (2003) have pushed the boundaries of efficiency and sustainability in high-temperature processes. These advancements signify a crucial shift toward energy-efficient and environmentally conscious pyrometallurgical practices, heralding a new era in metal extraction technologies. Within the realm of hydrometallurgical techniques, the works of Kinoshita et al. (2005) have introduced groundbreaking methods utilizing aqueous solutions for metal recovery. These studies delve into the intricate interplay between chemistry and metallurgy, yielding promising outcomes that lay the groundwork for more sustainable hydrometallurgical practices. The reduction of environmental impact is a key focus, demonstrating a commitment to eco-friendly approaches in metal recovery. Recognizing the imperative to mitigate the environmental impact of metallurgical extraction, recent investigations have concentrated on understanding repercussions and implementing strategies for sustainable waste management. The research efforts of Zheng and Kozinsk (1996), particularly in evaluating the environmental impact of metallurgical extraction, including the transformation of slags, contribute valuable insights to the broader discourse on eco-friendly practices. This section delves into recent advancements in metal recovery from wastewater, exploring both chemical and physical methods. Notable contributions from Byambaa et al. (2018) have leveraged chemical methods such as cementation, chemical precipitation, ion exchange, and ion flotation. These studies present innovative principles, applications, and advantages, redefining the landscape of metal recovery from industrial effluents and paving the way for sustainable practices. Breakthroughs in physical methods, particularly adsorption and membrane technologies, are elucidated through the work of Hosseini et al. (2016). These studies showcase the potential of these methods in enhancing efficiency and selectivity, contributing to the evolution of sustainable practices in metal recovery. The synthesis of cutting-edge physical techniques underscores their role in shaping the future of efficient and environmentally conscious metal recovery processes.

By immersing ourselves in these recent research studies, this section endeavors to encapsulate the dynamic developments in metal recovery, offering readers a comprehensive perspective on state-of-the-art techniques that hold the promise of shaping the future of the field.

3.1 The Significance of Metal Recovery from Wastewater

Metals play a vital role in modern industry, from electronics manufacturing to chemical processing, but their presence in industrial wastewater poses a significant environmental and economic challenge. Discharging metal-contaminated effluents into natural water bodies can have detrimental effects on ecosystems and human health. However, this challenge also presents a valuable opportunity—the recovery of metals from industrial wastewater. This section delves into the importance of metal recovery and provides a comprehensive exploration of various methods to achieve this critical goal.

Understanding the significance of metal recovery from wastewater is essential in recognizing the multifaceted benefits it offers:

- **Environmental Stewardship**: Efficient metal recovery mitigates the environmental impact of metal discharges, reducing the contamination of natural water bodies. This is vital for preserving aquatic ecosystems, safeguarding aquatic life, and ensuring safe drinking water sources (Oteng et al. 2023).
- **Resource Conservation**: Many metals present in industrial wastewater are finite and valuable resources. By recovering these metals, industries can reduce their dependence on virgin raw materials, thus conserving natural resources and contributing to sustainable practices (Hemdan et al. 2022; Nikfar and Mozaffari 2024).
- **Economic Gains**: Metal recovery can yield significant economic benefits. Valuable metals, once reclaimed, can either be reused within the industry or sold as valuable commodities. This presents an opportunity for industries to generate additional revenue streams while reducing disposal costs (Al-Sharrah and Marafi 2023; Vijayan and Bhattacharya 2023).
- **Regulatory Compliance**: Strict environmental regulations necessitate the limitation of metal discharges from industrial processes. Effective metal recovery ensures compliance with these regulations, helping industries avoid legal and financial penalties while enhancing their environmental performance (Goronovski et al. 2021).

3.2 Methods for Metal Recovery

Recovering metals from wastewater is a complex process, involving various techniques that can be broadly categorized into chemical and physical methods. The selection of the most appropriate method depends on factors such as the type of metal ions present, their concentration, and the specific requirements of the industry. The following sub-sections provide detailed insights into these methods.

3.2.1 Chemical Methods

Chemical methods for metal recovery harness specific chemical reactions to separate and reclaim metal ions from wastewater. The primary chemical methods include.

Cementation: Principles, Applications, and Advantages

Cementation stands out as a precipitation-based technique crucial for extracting desired metals from waste streams. This method operates on the principle of selective metal ion reduction, where a metal-bearing solution is introduced to the appropriate metal powder or scrap, causing certain selected metals to precipitate as a metal sponge. This precipitation occurs due to the variance in reduction potentials among different metals (Wallace 1978). The process of cementation has proven highly effective in recovering valuable metals, notably precious metals like gold (Au), palladium (Pd), rhodium (Rd), platinum (Pt), copper (Cu), and zinc (Zn), from diverse waste solutions. These solutions include stop baths, stabilizers, and solution tailings (Zhang et al. 2023). One of the distinctive features of cementation is its ability to selectively recover specific metals, showcasing its versatility and adaptability in treating various metal-containing waste streams.

Copper recovery through cementation, particularly in an acid solution, leads to the dissolution of iron, which is subsequently extracted through lime precipitation along with any remaining copper. This cascade effect underscores the complexity and interconnected nature of the cementation process in metal recovery (Ahmed et al. 2021). Extensive research has explored the application of different materials for metal cementation, including aluminium, iron, copper, and zinc (Nhan Hau Nguyen and Seung Lee 2023). The versatility, efficiency, and relatively straightforward operation of cementation make it a valuable technique in the treatment of metal-containing waste solutions. Its effectiveness, especially in copper recovery, has been well-documented in various studies, emphasizing its potential for high metal removal efficiencies (Amrane and Bouhidel 2018). Overall, cementation emerges as a vital precipitation method, contributing significantly to the selective extraction of valuable metals from complex waste matrices.

Chemical Precipitation: Principles, Applications, and Advantages

Chemical precipitation emerges as a cornerstone in the realm of metal recovery, celebrated for its cost-effectiveness and versatile applications. This method orchestrates a strategic transformation of the solution's composition by adjusting the pH to basic conditions (pH 9–10). The fundamental principle at play is the induction of insoluble metal complexes, primarily in the form of metal hydroxides, marking the efficiency of chemical precipitation in facilitating the removal of metals from solution (Kim et al. 2022). One of the stellar applications of chemical precipitation lies in the removal of heavy metals from wastewater. Variants such as hydroxide precipitation (OH^-), sulphide precipitation (S^{2-}), and carbonate precipitation (CO_3^{2-}) offer tailored solutions to the challenges posed by heavy metal-laden wastewater (Zhang et al. 2024). Among these, hydroxide precipitation stands out as a preferred technique, cherished for its affordability, simplicity, and effective pH control capabilities, making it ideal for scenarios demanding a straightforward and cost-efficient approach (Güven and Ateş 2024).

Delving into the realm of sulphide precipitation unveils a method notable for its lower metal solubility, selectivity at lower pH values, and higher kinetic reaction rates. This nuanced approach to metal removal, highlighted by the advantages of sulphide precipitation, finds its application in situations requiring a more intricate strategy for effective metal recovery (Güven and Ateş 2024). While not explicitly detailed, carbonate precipitation within chemical precipitation methods offers a realm of potential advantages for specific metal removal scenarios. This underscores the adaptability and diverse applications of chemical precipitation, tailoring its approach based on the unique characteristics of the target metals and wastewater.

In the symphony of metal recovery methods, chemical precipitation takes center stage. Its fundamental principle, inducing insoluble metal complexes through pH adjustments, establishes it as a cost-effective and robust method. The versatile applications, including hydroxide, sulphide, and potentially carbonate precipitation, underline its significance in addressing the multifaceted challenges posed by heavy metal-laden wastewater. As the story unfolds, chemical precipitation emerges not just as a method but as a dynamic and adaptable solution, seamlessly navigating through the complexities of metal recovery.

Ion Exchange: Principles, Applications, and Advantages

Ion exchange, a sophisticated and reversible chemical reaction, plays a pivotal role in metal recovery by facilitating the exchange of ions in a wastewater solution with similarly charged ions attached to an immobile solid particle (Abeywickrama et al. 2023). This method's versatility is exemplified in its widespread application, particularly in the chemical process industries, where it serves as a go-to solution for removing metals from diverse waste streams. The fundamental principle of ion exchange revolves around a reversible chemical reaction. In this dynamic process, ions from the wastewater solution undergo an exchange with similarly charged ions immobilized on synthetic organic ion exchange resins (Zhang et al. 2022). This reversible exchange mechanism forms the essence of ion exchange, allowing for a controlled and efficient removal of targeted metal ions. Ion exchange takes center stage in metal recovery efforts, with a notable focus on chemical process industries. Synthetic organic ion exchange resins, renowned for their efficacy, find common use in recovering metals like uranium from ores (Vecino et al. 2021). The method's versatility extends to the removal of various metals, including copper, lead, and cadmium, from diverse waste streams, showcasing its adaptability across different industrial contexts (Zhang et al. 2022).

The advantages of ion exchange are manifold. This method boasts high rates of metal recovery and selectivity, making it a preferred choice in situations where precision and efficiency are paramount. Its versatility is further emphasized by its capability to remove various metal ions from wastewater, offering a tailored approach based on factors such as pH and the specific metal ions present in the solution (Vecino et al. 2021). In essence, ion exchange stands as a dynamic and efficient player in the metal recovery arena, leveraging its reversible chemical reaction for precise and controlled removal of metals from wastewater. The strategic use of synthetic organic ion exchange resins and the method's adaptability across different industrial scenarios underscore its significance in shaping modern approaches to metal recovery.

Ion Flotation: Principles, Applications, and Advantages

Ion flotation emerges as a sophisticated separation process, orchestrating a dance of surfactants or collectors with opposite charges to target ions, ultimately creating surfactant complexes. The extraction of these complexes is achieved by introducing moving gas bubbles through the solution, effectively and selectively removing metal ions from the liquid phase or aqueous solution. (Hoseinian et al. 2023). At its core, ion flotation hinges on the creation of surfactant complexes through the strategic introduction of surfactants or collectors with charges opposing those of the target ions. The ensuing formation of these complexes sets the stage for their removal through the passage of gas bubbles, forming the essence of ion flotation as a precise and controlled separation process (Chakankar et al. 2024).

Ion flotation shines in its applicability to the removal of heavy metal ions from wastewater, showcasing its versatility and effectiveness. The method's suitability for a broad spectrum of target ions underlines its adaptability in addressing diverse metal-containing wastewater scenarios. Notably, dissolved-air flotation (DAF) emerges as a prominent technique within ion flotation, specifically tailored for treating wastewater laden with metal ions (Pooja et al. 2022). The allure of ion flotation lies in its simplicity, low energy requirements, and rapid operational capabilities. This method stands as a cost-effective solution, offering an array of advantages that make it a preferred choice in metal removal efforts. Its effectiveness in treating metal-containing wastewater, coupled with the benefits of simplicity and low operating costs, positions ion flotation as a versatile and efficient player in the realm of separation processes (Taseidifar et al. 2019; Zhang et al. 2012). In the symphony of metal removal techniques, ion flotation takes center stage, harnessing the power of surfactants, gas bubbles, and strategic charges to choreograph a precise and efficient dance of separation. The method's adaptability and efficiency, coupled with its inherent advantages, mark it as a valuable tool in the quest for treating metal-containing wastewater.

3.2.2 Physical Methods

Physical methods for metal recovery rely on physical processes rather than chemical reactions. They are often favored for their simplicity and efficiency. Key physical methods include adsorption and membranes.

Ion Adsorption: Principles, Applications, and Advantages

Adsorption emerges as a promising and versatile approach for the treatment of wastewater laden with heavy metals, orchestrating a physicochemical interaction on the surface of an adsorbent with the adsorbate. This interaction triggers the transfer of the adsorbate from the liquid phase to the solid surface, creating a molecular or atomic film that encapsulates the heavy metal particles (Zhao et al. 2024). At the heart of adsorption lies the intricate dance of molecular affinity, where the surface of an adsorbent engages in a physical and/or chemical interaction with the adsorbate. This dynamic interplay results in the formation of a molecular or atomic film on the adsorbent surface, effectively capturing heavy metal ions from the wastewater effluent (Zhou et al. 2022). Adsorption stands as a versatile method capable of removing a spectrum of heavy metals, including Ni (II), Cr (VI), Cd (II), Cu (II), and Zn (II), from wastewater effluents. The method's flexibility is exemplified by the diverse range of adsorbents employed, such as activated carbon, clay minerals, zeolites, and industrial solid waste. Activated carbon, with its expansive surface area and specific reactivity, takes a prominent role in heavy metal removal. Natural zeolites, particularly clinoptilolite, showcase significant potential in eliminating cationic heavy metal particles (Singh and Gupta 2016).

Adsorption offers high removal efficiency, particularly for low concentrations and trace quantities of heavy metals. The widespread use of activated carbon attests to its efficacy, driven by its remarkable surface area and reactivity. To further enhance adsorption selectivity and performance, researchers have ventured into developing composites by combining adsorbents with other materials, presenting a pathway for improved heavy metal capture (Zhou et al. 2022). In the realm of heavy metal removal, adsorption takes center stage, utilizing the principles of molecular interaction to create a protective film on the adsorbent surface. Its adaptability to a range of heavy metals and the strategic use of diverse adsorbents underscores its significance as an efficient and versatile player in the treatment of heavy metal-containing wastewater.

Membranes: Principles, Applications, and Advantages

Membrane-based methods stand as stalwarts in the realm of wastewater treatment, wielding their semipermeable prowess to eliminate a spectrum of contaminants, from organic compounds to inorganic pollutants like heavy metals and suspended solids (Zhao et al. 2024). The crux of membrane-based methods lies in their utilization of semipermeable membranes, strategically allowing the passage of water while retaining metal ions and other contaminants based on their size and charge (Zhou et al. 2022). These membranes, including ultrafiltration (UF), nanofiltration (NF), and reverse osmosis (RO), function as selective barriers, offering a nuanced approach to heavy metal recovery. UF membranes, with pore sizes ranging from 5 to 20 nm, emerge as selective isolators, effectively segregating metals, molecules, and suspended solids from inorganic solutions (Lin 2018). Nanofiltration, blending sieving and Donnan effects, specializes in separating charged solutes and larger neutral solutes. RO, propelled by pressure, masterfully retains metal ions while permitting water to traverse the membrane (Wang et al. 2023). Each of these membrane-based methods serves distinctive roles in heavy metal removal, forming a comprehensive arsenal against environmental contamination.

Membrane technology unfolds an array of advantages, encompassing simplicity, rapid operation, minimal space requirements, and reduced solid waste generation (Azmi et al. 2018). RO boasts high effectiveness, with rejection rates surpassing 96% for metal concentrations ranging from 21 to 200 mg/L. However, it grapples with fouling risks, primarily from suspended solids and oxidized compounds, showcasing the need for meticulous management strategies (Suffia and Dutta 2024). In the subsequent sections, the multifaceted world of membrane-based methods will be dissected, delving deeper into their specific applications, unrivaled advantages, and intrinsic limitations. This exploration equips industries with the insights needed to make informed decisions, aligning their metal recovery strategies with sustainability goals and regulatory imperatives.

4 Challenges and Considerations in Metal Recovery from Wastewater

While the methods discussed in the previous section offer effective ways to recover metals from wastewater, there are various challenges and considerations that must be addressed to ensure successful implementation. This section explores these challenges and provides insights into how to overcome them.

4.1 Environmental Impact and Management

Effective environmental management is vital in metal recovery processes to mitigate potential risks. Residues like spent ion exchange resins, metal-laden sludge, and waste chemicals must be managed properly to reduce their environmental impact. Implementing waste treatment strategies is essential for minimizing the adverse effects (Gümüşsoy et al. 2023; Guo et al. 2023).

4.2 Process Optimization and Efficiency

Selecting the right recovery method is critical, considering factors like metal specificity and wastewater composition. While some methods are highly selective, others offer versatility but with reduced selectivity. Balancing energy consumption, as some methods can be energy-intensive, is crucial. Exploring energy-efficient alternatives is a means to reduce operational costs and environmental footprints. The development of sustainable alternatives to chemical usage addresses cost and environmental concerns (Al-Rajabi et al. 2024).

4.3 Regulatory Compliance and Scalability

To avoid legal consequences, adherence to environmental regulations and discharge limits is paramount. Ensuring that metal recovery processes meet regulatory standards is essential. Scalability poses challenges when transitioning from laboratory-scale processes to large-scale industrial applications. Evaluating economic viability through cost–benefit analyses is crucial for informed decision-making (Arun et al. 2024).

4.4 Technological Advancements and Monitoring

Staying informed about the latest advancements in metal recovery technologies is essential for optimization. These continual advancements offer opportunities to improve efficiency, selectivity, and sustainability in metal recovery processes. Implementing real-time monitoring and control systems is critical to ensure efficiency and reliability. Such systems help in detecting and addressing issues promptly, minimizing downtime, and maximizing metal yield (Chandnani et al. 2023; Jadhao et al. 2022).

5 Environmental Impact and Sustainability in Metal Recovery

Metal recovery from waste materials not only reclaims valuable resources but also contributes significantly to sustainable practices, fostering positive environmental outcomes. This section meticulously explores these dimensions, shedding light on the favorable environmental impact and showcasing successful initiatives in sustainable metal recovery:

In the realm of metallurgical extraction, particularly in the pyrometallurgical smelting process, the generation of melting slag is a consequential aspect of metal extraction from raw ore and other materials. This process leads to the formation of slags that play a dual role, impacting both smelting furnace operation and the quality of the extracted metal. The discharge and cooling of these slags result in substantial quantities of waste material, primarily composed of oxides such as CaO, SiO_2, and FeO, originating from the pyrometallurgical industry. Despite their high-temperature exposure, these slags exhibit stable physical and chemical properties and undergo treatments, including air cooling, water cooling, granulation, and extension (Isa et al. 2017; Zheng and Kozinsk 1996).

The challenge lies in mitigating the environmental impact associated with slag disposal. Efforts are ongoing and multifaceted, focusing on reducing emissions, implementing recycling strategies, and overall efficiency improvements (Japar et al. 2019; Ghodrat et al. 2018). This comprehensive approach underscores the commitment to environmentally conscious practices within the metallurgical industry, aligning with broader sustainability goals.

Metal recovery from waste materials offers various environmental benefits and plays a crucial role in sustainable practices. This section delves into these aspects, showcasing the positive environmental impact and highlighting successful sustainable metal recovery initiatives.

5.1 Environmental Benefits

Metal recovery from waste materials offers substantial environmental advantages. Firstly, it conserves natural resources by reducing the need for fresh mining and extraction activities, thereby preserving the environment. Secondly, metal recovery contributes to waste reduction, minimizing the volume of metal-containing waste materials and reducing the demand for landfill space. Additionally, recycling metals generally consumes less energy compared to primary metal production, leading to lower greenhouse gas emissions, which is crucial in the fight against climate change. Lastly, the recovery processes can effectively remove heavy metals and pollutants from wastewater, promoting cleaner water sources and enhancing the overall quality of the environment (Al Hassan et al. 2023; Iqbal et al. 2023).

5.2 Sustainable Practices

Sustainability plays a central role in metal recovery endeavors. One way in which it aligns with sustainability principles is through promoting a circular economy. Metal recovery encourages the reuse and recycling of materials, significantly reducing waste generation and lessening the environmental burden associated with waste disposal. Moreover, sustainable practices in metal recovery focus on reducing emissions, enhancing energy efficiency, and minimizing environmental impact, which are essential aspects of responsible resource management. Furthermore, life cycle assessment is employed to assess the environmental impact throughout the entire life cycle of metal recovery processes, helping identify areas for improvement and sustainability enhancements (Barbhuiya and Das 2023; Rachid et al. 2023).

5.3 Research and Development Case Studies

Examining successful sustainable metal recovery initiatives through case studies provides real-world examples of environmentally responsible practices. For instance, urban mining initiatives recover valuable metals from electronic waste, diminishing the environmental impact of electronic disposal and reducing the need for new resource extraction (Erdiaw-Kwasie et al. 2024; Xavier et al. 2023). Battery recycling, specifically for lithium-ion batteries, effectively reduces resource depletion and pollution while extending the lifespan of valuable metals. Furthermore, industries implement sustainable metal recovery in wastewater treatment to meet environmental regulations and reduce pollution levels, contributing to cleaner water sources and overall environmental quality (Barbhuiya and Das 2023; Rachid et al. 2023). These case studies exemplify how metal recovery contributes to a more eco-friendly and responsible approach to resource management, showcasing its practical benefits and applications. By focusing on environmental benefits, sustainable practices, and case studies of successful initiatives, this section demonstrates how metal recovery contributes to a more eco-friendly and responsible approach to resource management.

6 Conclusion

Metal recovery from wastewater stands at the intersection of environmental responsibility, technological innovation, and industrial sustainability. This comprehensive exploration of metal recovery methods, environmental considerations, and sustainable practices underscores the multifaceted nature of this critical endeavor. Metal recovery processes offer an array of benefits, including resource conservation, waste reduction, energy savings, and reduced pollution. By reclaiming metals from waste materials, industries contribute to the preservation of natural resources and the reduction of greenhouse gas emissions. Moreover, hydrometallurgical processing emerges as an effective and flexible method for recovering valuable metals, especially when compared to the energy-intensive pyrometallurgy method. The challenges and considerations in metal recovery, ranging from environmental concerns to economic viability, necessitate a holistic approach. Industries must not only select the most suitable recovery method but also ensure the proper management of byproducts and residues. The level of impurities in each waste stream varies considerably, making it challenging to separate metals from each other and from non-metal components. While chemical methods are more rapid and efficient, they require higher energy and chemical consumption. In contrast, physical methods involve zero sludge production and fewer chemicals, emphasizing their importance in sustainable metal recovery. The obstacles posed by the heterogeneous and complex nature of solid wastes necessitate innovative approaches. Embracing circular economy principles, reducing emissions, and conducting life cycle assessments are pivotal steps toward achieving sustainable metal recovery. Additionally, addressing the organic content in wastewater by biological, physical, and chemical means is essential, but it remains a challenge to remove or separate inorganic metals due to their intrinsic properties such as solubility and the tendency to form metal complexes. Real-world case studies exemplify the successful implementation of sustainable metal recovery initiatives. From urban mining to battery recycling and industrial effluent treatment, these examples demonstrate the tangible benefits of environmentally responsible practices. In conclusion, metal recovery from wastewater is an essential component of a more sustainable and eco-conscious industrial landscape. By addressing environmental challenges, embracing sustainable practices, and drawing inspiration from successful initiatives, industries can optimize their metal recovery processes. This not only leads to economic advantages but also fosters a greener and more sustainable future for all. The journey toward responsible metal recovery is ongoing, driven by technological advancements, regulatory requirements, and a shared commitment to environmental stewardship. As industries continue to innovate and refine their practices, the prospect of a more sustainable and resource-efficient world becomes increasingly attainable.

References

Abd Aziz MA, Md Isa K, Rashid RA (2017) Pneumatic jigging: influence of operating parameters on separation efficiency of solid waste materials [Article]. Waste Manage Res 35(6):647–655. https://doi.org/10.1177/0734242X17697815

Abdu Rahman NW, Japar NSA, Aziz MAA, Razik AHA, Yunus MYM (2019) Sodium grease formulation from waste engine oil. IOP Conf Ser Earth Environ Sci 257(1):012018. https://doi.org/10.1088/1755-1315/257/1/012018

Abeywickrama J, Hoth N, Grimmer M, Haubrich F, Drebenstedt C (2023) Optimizing metal recovery from slag leaching solutions: advanced ion exchange techniques for sustainable resource extraction. J Water Process Eng 56:104482. https://doi.org/10.1016/j.jwpe.2023.104482

Ahmed MS, Zewail TM, El-Ashtoukhy ESZ, Farag HA, El Azab IH, Albatati FA, Abdel-Aziz MH et al (2021) Enhancement of heavy metals recovery from aqueous solutions by cementation on a rotating cylinder using a stationary wiper. J Ind Eng Chem 97:460–465. https://doi.org/10.1016/j.jiec.2021.03.001

Al-Rajabi MM, Abumadi FA, Laoui T, Atieh MA, Khalil KA (2024) Capacitive deionization for water desalination: cost analysis, recent advances, and process optimization. J Water Process Eng 58:104816. https://doi.org/10.1016/j.jwpe.2024.104816

Al-Sharrah G, Marafi M (2023) Process simulation and techno-economic evaluation of recovery of metals and alumina from spent hydroprocessing catalysts using leaching with EDTA. J Eng Res. https://doi.org/10.1016/j.jer.2023.10.019

Al Hassan MK, Alfarsi A, Nasser MS, Hussein IA, Khan I (2023) Ionic liquids and NADES for removal of organic pollutants and heavy metals in wastewater: a comprehensive review. J Mol Liq 391:123163. https://doi.org/10.1016/j.molliq.2023.123163

Amrane C, Bouhidel K-E (2018) Integrated diffusion dialysis precipitation—cementation for selective recovery of leaching chemicals and metal values from electroplating sludge. Hydrometallurgy 177:34–40. https://doi.org/10.1016/j.hydromet.2018.02.011

Arun M, Barik D, Chandran SSR (2024) Exploration of material recovery framework from waste—a revolutionary move towards clean environment. Chem Eng J Adv 18:100589. https://doi.org/10.1016/j.ceja.2024.100589

Awasthi AK, Zeng X, Li J (2016) Comparative examining and analysis of e-waste recycling in typical developing and developed countries. Procedia Environ Sci 35:676–680. https://doi.org/10.1016/j.proenv.2016.07.065

Aziz MAA, Isa KM, Miles NJ, Rashid RA (2019) Pneumatic jig: effect of airflow, time and pulse rates on solid particle separation [Article]. Int J Environ Sci Technol 16(1):11–22. https://doi.org/10.1007/s13762-018-1648-4

Azmi AA, Jai J, Zamanhuri NA, Yahya A (2018) Precious metals recovery from electroplating wastewater: a review. IOP Conf Ser Mater Sci Eng 358:012024. https://doi.org/10.1088/1757-899x/358/1/012024

Barbhuiya S, Das BB (2023) Life cycle assessment of construction materials: methodologies, applications and future directions for sustainable decision-making. Case Stud Constr Mater 19:e02326. https://doi.org/10.1016/j.cscm.2023.e02326

Byambaa M, Dolgor E, Shiomori K, Suzuki Y (2018) Removal and recovery of heavy metals from industrial wastewater by precipitation and foam separation using lime and casein. J Environ Sci Technol 11:1–9

Chakankar M, Pollmann K, Rudolph M (2024) Selective removal of Gallium from mixed metal solutions with Arsenic by ion flotation using the biosurfactant rhamnolipid. J Water Process Eng 58:104879. https://doi.org/10.1016/j.jwpe.2024.104879

Chandnani K, Rajput N, Jadav T, Pillai M, Dhakne P, Tekade RK, Sengupta P (2023) Technological advancement and current standing of microfluidic chip based devices for targeted analysis of biomarkers. Microchem J 195:109532. https://doi.org/10.1016/j.microc.2023.109532

Chandraseagar S, Abdulrazik AH, Abdulrahman SN, Abdaziz MA (2019) Aspen Plus simulation and optimization of industrial spent caustic wastewater treatment by wet oxidation method. IOP Conf Ser Mater Sci Eng 702(1):012011. https://doi.org/10.1088/1757-899X/702/1/012011

Chauhan G, Kaur PJ, Pant K, Nigam K (2020) Sustainable metal extraction from waste streams. Wiley. https://doi.org/10.1002/9783527826704

Chauhan NS, Punia A (2022) Role of education and society in dealing plastic pollution in the future. In: Plastic and microplastic in the environment: management and health risks, pp 267–281. https://doi.org/10.1016/S0304-386X(96)00090-4

Chmielewski A, Urbański T, Migdał W (1997) Separation technologies for metals recovery from industrial wastes. Hydrometallurgy 45(3):333–344. https://doi.org/10.1016/S0304-386X(96)00090-4

Dahawi YA, Abdulrazik A, Abu Seman MN, Abd Aziz MA, Mohd Yunus MY (2019) Aspen plus simulation of bio-char production from a biomass-based slow pyrolysis process. In: Key engineering materials, vol 797, pp 336–341

de Buzin PJWK, Heck NC, Vilela ACF (2017) EAF dust: an overview on the influences of physical, chemical and mineral features in its recycling and waste incorporation routes. J Mater Res Technol 6(2):194–202. https://doi.org/10.1016/j.jmrt.2016.10.002

Erdiaw-Kwasie MO, Abunyewah M, Baah C (2024) A systematic review of the factors—barriers, drivers, and technologies—affecting e-waste urban mining: on the circular economy future of developing countries. J Clean Prod 436:140645. https://doi.org/10.1016/j.jclepro.2024.140645

Fan E, Li L, Wang Z, Lin J, Huang Y, Yao Y, Wu F et al (2020) Sustainable recycling technology for Li-ion batteries and beyond: challenges and future prospects. Chem Rev 120(14):7020–7063. https://doi.org/10.1021/acs.chemrev.9b00535

Forti V, Baldé CP, Kuehr R, Bel G (2020) The global e-waste monitor 2020. In: United Nations University (UNU), International Telecommunication Union (ITU) & International Solid Waste Association (ISWA), Bonn/Geneva/Rotterdam, 120 pp. https://ewastemonitor.info/wp-content/uploads/2020/11/GEM_2020_def_july1_low.pdf

Ghodrat M, Rhamdhani MA, Khaliq A, Brooks G, Samali B (2018) Thermodynamic analysis of metals recycling out of waste printed circuit board through secondary copper smelting. J Mater Cycles Waste Manag 20(1):386–401. https://doi.org/10.1007/s10163-017-0590-8

Goronovski A, Rivera RM, Van Gerven T, Tkaczyk AH (2021) Radiological assessment of bauxite residue processing to enable zero-waste valorisation and regulatory compliance. J Clean Prod 294:125165. https://doi.org/10.1016/j.jclepro.2020.125165

Gümüşsoy A, Başyiğit M, Uzun Kart E (2023) Economic potential and environmental impact of metal recovery from copper slag flotation tailings. Resour Policy 80:103232. https://doi.org/10.1016/j.resourpol.2022.103232

Guo F, Guo Y, Chen L, Jia W, Zhu Y, Li Y, Wu J et al (2023) Multitudinous components recovery, heavy metals evolution and environmental impact of coal gasification slag: a review. Chemosphere 338:139473. https://doi.org/10.1016/j.chemosphere.2023.139473

Güven HM, Ateş H (2024) A holistic approach to the recovery of valuable substances from the treatment sludge formed from chemical precipitation of fruit processing industry wastewater. Sci Total Environ 917:170372. https://doi.org/10.1016/j.scitotenv.2024.170372

Hairunnaja MA, Aziz MAA, Bashari NAF, Arifin MA, Nedumaran N, Isa KM, Ali UFM (2023) The grease formulation using waste substances from palm oil refinery and other industrial wastes: a review [review]. Pertanika J Sci Technol 31(5):2505–2529. https://doi.org/10.47836/pjst.31.5.25

He J, Kappler A (2017) Recovery of precious metals from waste streams. Microbial Biotechnol 10(5):1194. https://doi.org/10.1111/1751-7915.12759

Hemdan B, Garlapati VK, Sharma S, Bhadra S, Maddirala S, Varsha KM, Aminabhavi TM (2022) Bioelectrochemical systems-based metal recovery: Resource, conservation and recycling of metallic industrial effluents. Environ Res 204:112346. https://doi.org/10.1016/j.envres.2021.112346

Hosseini SS, Bringas E, Tan NR, Ortiz I, Ghahramani M, Shahmirzadi MAA (2016) Recent progress in development of high performance polymeric membranes and materials for metal plating wastewater treatment: A review. J Water Process Eng 9:78–110

Hoseinian FS, Ramshini S, Rezai B, Kowsari E, Safari M (2023) Toxic heavy metal ions removal from wastewater by ion flotation using a nano collector. Miner Eng 204:108380. https://doi.org/10.1016/j.mineng.2023.108380

Iqbal MS, Aslam AA, Iftikhar R, Junaid M, Imran SM, Nazir MS, Ahmadipour M (2023) The potential of functionalized graphene-based composites for removing heavy metals and organic pollutants. J Water Process Eng 53:103809. https://doi.org/10.1016/j.jwpe.2023.103809

Isa KM, Kasim FH, Saad SA, Rahim MAA, Aziz MAA, Ali UFM (2017) Influence of operating parameters on biomass conversion under sub- and supercritical water conditions [Article]. Chem Eng Technol 40(3):537–545. https://doi.org/10.1002/ceat.201600343

Ismail NA, Abd Aziz MA, Hisyam A, Abidin MA (2021) Separation of samarium from medium rare earth mixture using multi-stage counter-current extraction [Article]. Chem Eng Commun 208(5):764–774. https://doi.org/10.1080/00986445.2020.1746654

Ismail NA, Abdul Aziz MA, Hisyam A (2020) Experimental and theoretical study on Sm/Eu-Gd extraction by P204 and [A336][P204]. IOP Conf Ser Mater Sci Eng

Ismail NA, Aziz MAA, Yunus MYM, Hisyam A (2019a) Selection of extractant in rare earth solvent extraction system: a review [Article]. Int J Recent Technol Eng 8(1):728–743. https://www.scopus.com/inward/record.uri?eid=2-s2.0-85067864920&partnerID=40&md5=728173400fad6eb9a51f1f0452c478081

Ismail NA, Mohd Yunus MY, Abdul Aziz MA, Abidin MA (2019b) Comparison of optimal solvent extraction stages between P204 and [A336][P204] for the separation of europium and gadolinium. IOP Conf Ser Mater Sci Eng

Jadhao PR, Ahmad E, Pant KK, Nigam KDP (2022) Advancements in the field of electronic waste recycling: critical assessment of chemical route for generation of energy and valuable products coupled with metal recovery. Sep Purif Technol 289:120773. https://doi.org/10.1016/j.seppur.2022.120773

Jagannath A, Shetty V, Saidutta M (2017) Bioleaching of copper from electronic waste using Acinetobacter sp. Cr B2 in a pulsed plate column operated in batch and sequential batch mode. J Environ Chem Eng 5(2):1599–1607. https://doi.org/10.1016/j.jece.2017.02.023

Japar NSA, Aziz MAA, Razali MN, Rahman NWA (2018) Grease and its application on electrical equipment: a review [Article]. Int J Eng Technol (UAE) 7(3):23–29. https://doi.org/10.14419/ijet.v7i3.26.17455

Japar NSA, Aziz MAA, Razali MN (2019) Formulation of fumed silica grease from waste transformer oil as base oil. Egypt J Pet 28(1):91–96. https://doi.org/10.1016/j.ejpe.2018.12.001

Jena P, Brocchi E (1997) Metal extraction through chlorine metallurgy. Miner Process Extr Metull Rev 16(4):211–237. https://doi.org/10.1016/j.rser.2016.04.020

Jha MK, Kumar V, Singh R (2001) Review of hydrometallurgical recovery of zinc from industrial wastes. Resour Conserv Recycl 33(1):1–22. https://doi.org/10.1016/S0921-3449(00)00095-1

Kinoshita T, Yamaguchi K, Akita S, Nii S, Kawaizumi K, (2005) Hydrometallurgical recovery of zinc from ashes of automobile tire wastes. Chemosphere 59(8):1105–1111

Kim J, Yoon S, Choi M, Min KJ, Park KY, Chon K, Bae S (2022) Metal ion recovery from electrodialysis-concentrated plating wastewater via pilot-scale sequential electrowinning/chemical precipitation. J Clean Prod 330:129879. https://doi.org/10.1016/j.jclepro.2021.129879

Li X, Monnens W, Li Z, Fransaer J, Binnemans K (2020) Solvometallurgical process for extraction of copper from chalcopyrite and other sulfidic ore minerals. Green Chem 22(2):417–426. https://doi.org/10.1039/C9GC02983D

Lin P (2018) Adsorption of cadmium and nickel by modified agricultural byproducts [The University of Arizona]

Love J, Miguirditchian M, Chagnes A (2019) New insights into the recovery of strategic and critical metals by solvent extraction. In: Moyer BA (ed) Ion exchange and solvent extraction, 1st ed, pp 1–44. https://doi.org/10.1201/9781315114378-1

Ma C, Yu J, Wang B, Song Z, Xiang J, Hu S, Sun L et al (2016) Chemical recycling of brominated flame retarded plastics from e-waste for clean fuels production: a review. Renew Sustain Energy Rev 61:433–450. https://doi.org/10.1016/j.rser.2016.04.020

Moskalyk RR, Alfantazi AM (2003) Review of copper pyrometallurgical practice: today and tomorrow. Miner Eng 16(10):893–919

Mostaghel S, Yang Q, Samuelsson C (2011) Recycling of cathode ray tube in metallurgical processes: influence on environmental properties of the slag. Glob J Environ Sci Technol 1(Artikelnr: 19)

Nhan Hau Nguyen T, Seung Lee M (2023) Recovery of Au and Pd from the etching solution of printed circuit boards by cementation, solvent extraction, reduction, and precipitation. J Ind Eng Chem 126:214–223. https://doi.org/10.1016/j.jiec.2023.06.011

Nikfar S, Mozaffari S (2024) Resource conservation and recovery Act (USA). In: Wexler P (ed) Encyclopedia of toxicology, 4th ed. Academic Press, pp 233–237. https://doi.org/10.1016/B978-0-12-824315-2.00894-0

Ojeda M, Perino E, Ruiz MdC (2009) Gold extraction by chlorination using a pyrometallurgical process. Miner Eng 22(4):409–411. https://doi.org/10.1016/j.mineng.2008.09.002

Ortiz-Ribbing L, Roskamp G, Roegge M (2007) Preventative foliar fungicide and insecticide applications and their impact on soybean yield components. Phytopathology

Oteng D, Zuo J, Sharifi E (2023) An evaluation of the impact framework for product stewardship on end-of-life solar photovoltaic modules: an environmental lifecycle assessment. J Clean Prod 411:137357. https://doi.org/10.1016/j.jclepro.2023.137357

Pathirana C, ZiyathAM, Jinadasa K, Egodawatta P, Sarina S, Goonetilleke A (2019) Quantifying the influence of surface physico-chemical properties of biosorbents on heavy metal adsorption. Chemosphere 234:488–495. https://doi.org/10.1016/j.chemosphere.2019.06.074

Pooja G, Kumar PS, Indraganti S (2022) Recent advancements in the removal/recovery of toxic metals from aquatic system using flotation techniques. Chemosphere 287:132231. https://doi.org/10.1016/j.chemosphere.2021.132231

Rachid S, Taha Y, Benzaazoua M (2023) Environmental evaluation of metals and minerals production based on a life cycle assessment approach: a systematic review. Miner Eng 198:108076. https://doi.org/10.1016/j.mineng.2023.108076

Rahman NWBA, Aziz MABA (2022) The effects of additives on anti-wear properties of lubricating grease formulated from waste engine oil [Article]. Egypt J Pet 31(3):71–76. https://doi.org/10.1016/j.ejpe.2022.07.002

Rao SR (2006) Pyrometallurgical processing. In: Waste management series, vol 7. Elsevier, pp 127–165

Razali MN, Aziz MAA, Jamin NFM, Salehan NAM (2018) Modification of bitumen using polyacrylic wig waste. In: AIP conference proceedings

Razali MN, Mohd Isa SNE, Md Salehan NA, Musa M, Abd Aziz MA, Nour AH, Yunus RM (2020) Formulation of emulsified modification bitumen from industrial wastes [Article]. Indonesian J Chem 20(1):96–104. https://doi.org/10.22146/ijc.40888

Salai RG, Kelambakkam KD (2020) A descriptive study to assess the knowledge, practice and attitude on iron deficiency Anemia among Adolescent girls at selected private school, Kanchipuram Dt, Tamilnadu, India. Executive Editor 11(03):3409. https://doi.org/10.37506/ijphrd.v11i3.1176

Saw HS, bin Azmi AA, Chew KW, Show PL (2021) 4.2 sustainability and development of industry 5.0. https://doi.org/10.1016/j.spc.2022.08.003

Singh N, Gupta S (2016) Adsorption of heavy metals: a review. Int J Innov Res Sci Eng Technol 5(2):2267–2281

Suffia S, Dutta D (2024) Applications of deep eutectic solvents in metal recovery from E-wastes in a sustainable way. J Mol Liq 394:123738. https://doi.org/10.1016/j.molliq.2023.123738

Suhaila N, Japar A, Aizudin M, Aziz A, Razali MN (2018) Formulation of lubricating grease using Beeswax thickener. IOP Conf Ser Mater Sci Eng 342(1):012007. https://doi.org/10.1088/1757-899X/342/1/012007

Taseidifar M, Ziaee M, Pashley R, Ninham B (2019) Ion flotation removal of a range of contaminant ions from drinking water. J Environ Chem Eng 7:103263. https://doi.org/10.1016/j.jece.2019.103263

Thakur P, Kumar S (2020) Metallurgical processes unveil the unexplored "sleeping mines" e-waste: a review. Environ Sci Pollut Res 27(26):32359–32370. https://doi.org/10.1007/s11356-020-09405-9

Vecino X, Reig M, López J, Valderrama C, Cortina JL (2021) Valorisation options for Zn and Cu recovery from metal influenced acid mine waters through selective precipitation and ion-exchange processes: promotion of on-site/off-site management options. J Environ Manag 283:112004. https://doi.org/10.1016/j.jenvman.2021.112004

Vijayan SK, Bhattacharya S (2023) Processing of automotive shredder residues: economic evaluation of a process for energy and high-value metals recovery. Clean Chem Eng 6:100103. https://doi.org/10.1016/j.clce.2023.100103

Wallace RA (1978) Recovery of waste heavy metals from solutions by cementation with aluminum. In: Google patents

Wang H, Wang W, Zhu G, Cao Y, Zhang L (2023) A perspective of microemulsions in critical metal separation and recovery: implications for potential application of CO_2-responsive microemulsions. Chemosphere 338:139494. https://doi.org/10.1016/j.chemosphere.2023.139494

Xavier LH, Ottoni M, Abreu LPP (2023) A comprehensive review of urban mining and the value recovery from e-waste materials. Resour Conserv Recycl 190:106840. https://doi.org/10.1016/j.resconrec.2022.106840

Xia H, Han J, Milisavljevic-Syed J (2023) Predictive modeling for the quantity of recycled end-of-life products using optimized ensemble learners. Resour Conserv Recycl 197:107073. https://doi.org/10.1016/j.resconrec.2023.107073

Zhang K, Chen J (2012) The regulation of integrin function by divalent cations. Cell Adh Migr 6(1):20–29. https://doi.org/10.4161/cam.18702

Zhang L, Xu Z (2016) A review of current progress of recycling technologies for metals from waste electrical and electronic equipment. J Clean Prod 127:19–36. https://doi.org/10.1016/j.jclepro.2016.04.004

Zhang M, Zhu G, Zhao Y, Feng X (2012) A study of recovery of copper and cobalt from copper–cobalt oxide ores by ammonium salt roasting. Hydrometallurgy s 129–130:140–144. https://doi.org/10.1016/j.hydromet.2012.06.014

Zhang T, Liu J-M, Huang X-F, Xia B, Su C-Y, Luo G-F, Qiu R-L et al (2013) Chelant extraction of heavy metals from contaminated soils using new selective EDTA derivatives. J Hazard Mater 262:464–471. https://doi.org/10.1016/j.jhazmat.2013.08.069

Zhang X, Ge T, Xu L, Sun Y, Yang C, Yao Y, Zhao Z et al (2023) Oxidative acid leaching of indium phosphide waste and recovery of indium metal by cementation with aluminum. Hydrometallurgy 221:106109. https://doi.org/10.1016/j.hydromet.2023.106109

Zhang X, Yang S, Deng C, Liu W, Xiang D, Liang L, Pan K et al (2024) Selective recovery of metals in spent batteries by electrochemical precipitation to cathode material for sodium-ion batteries. Heliyon e27127. https://doi.org/10.1016/j.heliyon.2024.e27127

Zhang Y, You C, Ren M, Liu M, Xu X, Zhang Y, Cui J et al (2022) Ion exchange membrane optimized light-driven photoelectrochemical unit for efficiency simultaneous organic degradation and metal recovery from the mine wastewater. J Hazard Mater 429:128352. https://doi.org/10.1016/j.jhazmat.2022.128352

Zhao S, Samadi A, Wang Z, Pringle JM, Zhang Y, Kolev SD (2024) Ionic liquid-based polymer inclusion membranes for metal ions extraction and recovery: fundamentals, considerations, and prospects. Chem Eng J 481:148792. https://doi.org/10.1016/j.cej.2024.148792

Zheng GH, Kozinsk JA (1996) Solid waste remediation in the metallurgical industry: application and environmental impact. Environ Prog 15(4):283–292. https://doi.org/10.1002/ep.670150419

Zhou X, Wang B, Wang R (2022) Insights into ion-imprinted materials for the recovery of metal ions: preparation, evaluation and application. Sep Purif Technol 298:121469. https://doi.org/10.1016/j.seppur.2022.121469

Harvesting Wealth: Phytomining as a Sustainable Solution for Mine Waste Management and Economic Prosperity

Mohd. Azrul Naim Mohamad and Zakuan Azizi Shamsul Harumain

Abstract Phytomining is a tool that utilizes plants to take up economic metals and metalloids such as nickel, cobalt, manganese, and gold into its aerial tissues from sub-economic ore bodies such as mine wastes. The output from phytomining produces highly valuable bio-ores that can be extracted from its biomass enabling a dual role as a practical sustainable cleanup with economic gain. Mine wastes generated from mining activity around the world typically comprise crushed rock, process effluent, sand, clay, and water, alongside notable quantities of metal residues that are economically impractical to extract using traditional methods. These wastes often lack sufficient organic matter or macronutrients, hindering the growth and survival of various organisms, including bacteria, resulting in un-vegetated areas. Environmental concerns regarding mine waste management also have escalated due to high levels of toxic metals and metalloids and metalloids such as arsenic that can be chemically leached out together with the mined metals and metalloids and metalloids due to its presence in the same ore deposits. This review discusses potential solutions for managing mine waste through phytomining.

1 Introduction

Mining activity for economic metals and metalloids such as gold, palladium, platinum, and nickel increased annually to meet the rising demand from various sectors including jewelry, medical, electronics, and automotive industries. This led to the production of mine wastes, also known as tailings, which are the residual materials

Mohd. A. N. Mohamad · Z. A. S. Harumain (✉)
Department of Biotechnology, Kulliyyah of Science, International Islamic University Malaysia, 25200 Pahang, Malaysia
e-mail: zak@iium.edu.my

Research Unit in Bioinformatics and Computational Biology, Kulliyyah of Science, International Islamic University Malaysia, 25200 Pahang, Malaysia

Environmental Biology and Biotechnology Research Unit, Kulliyyah of Science, International Islamic University Malaysia, 25200 Pahang, Malaysia

A. Z. Yaser et al. (eds.), *Controlling Environmental Pollution*,
https://doi.org/10.1007/978-981-97-8931-3_14

left behind following the processing of ore, ore concentrate, or other mined materials to extract valuable components like metals and metalloids. Typically, these wastes consist of ground rock, process effluent, sand, clay, and water, along with significant quantities of metal residues that are not economically feasible to be extracted using conventional methods [1]. The rise in demand for economic metals and metalloids across industries has led to an increase in mining activities, subsequently resulting in a higher volume of mine wastes generated annually. Latest survey of 107 mining companies estimated at least 44.5 billion cubic meters of tailings are currently stored at mine sites and have estimated a global annual tailings production of about 13 billion tons [2]. Norilsk Nickel, the world's largest palladium producer, reported disposing of 13 million tons of waste material, including soils, hard rocks, and concentrated mine wastes, in 2013. In Poland, mining plants had accumulated over 26 million metric tons of waste by the end of 2016 [3]. Throughout the mining process, these mine wastes are stored in designated tailings storage facilities. The water from these mine wastes is often recirculated for reuse in the mining process, while the solid materials remain within the storage facility.

Environmental concerns regarding the management of mine wastes have emerged due to the presence of elevated levels of toxic metals and metalloids, such as arsenic, mercury, cadmium, and nickel, as well as polycyclic aromatic hydrocarbons like phenanthrene [4]. Tailings can remain unvegetated for extended periods, leading to environmental degradation and the spread of tailings over vast areas through erosion. Additionally, mine wastes often contain relatively low levels of organic matter or macronutrients, which hinder the growth and survival of many living organisms, including bacteria, resulting in barren landscapes [5, 6]. Old mining areas, in particular, pose significant risks of metal contamination due to inadequate waste management practices and lack of environmental awareness at the time. These sites now present threats to human health due to the dispersion of toxic metal-containing particles in the air, soil, and water [7].

Proper management of mine wastes is essential to accommodate the substantial volume generated and prevent the direct discharge of toxic metals, metalloids, and organics into surface water. Inadequate management practices can also lead to contamination of groundwater and open water sources, further exacerbating environmental concerns associated with mining activities. This review describes how phytomining technology can be used as a sustainable solution for mine waste management and economic prosperity.

2 Current Practice on Mine Waste Management

There exist various methods for long-term waste storage management, including techniques such as "conventional impoundment," "in-pit," or "backfill" methods. Conventional impoundment serves to retain mine waste for extended durations while also facilitating the separation of undesirable solids prior to discharge [8]. Within this process, embankments are engineered to hold water and mine wastes. Nonetheless, this strategy heightens the risk of flooding and may induce significant environmental repercussions if the embankment fails to accommodate the stored mine waste adequately.

In-pit disposal refers to the placement of waste materials directly into the excavated areas of the mine, which can help reduce the need for additional land for waste storage [9]. Backfill processing entails mixing mine waste with cement, subsequently utilizing it as a supporting pillar within the underground mining area. While this technique can mitigate problems related to dust generation and surface water contamination, it entails substantial operational and maintenance costs, coupled with CO_2 emissions from cement usage [10].

Alternatively, submarine disposal of mine waste involves the deep-sea discharge of waste material. This approach might be viable in cases where onshore processes are unfeasible due to terrain challenges, high seismic activity, excessive rainfall, and/or limited land availability [10]. Nevertheless, environmental apprehensions persist regarding the unpredictable dispersion of mine waste within seawater, potentially augmenting contamination migration.

Other than physical methods for managing mine waste, sustainable waste management technologies like bioremediation and phytoremediation can also be adopted by many industries to manage mine waste constituents, such as acid mine drainage and toxic heavy metals and metalloids.

Acid mine drainage is caused by the oxidation of sulfidic constituents that can be easily found in ores, especially when exposed to water and air, forming sulfuric acid. The use of sulfate-reducing bacteria for acid mine bioremediation helps mitigate the harmful effects of acid mine tailings by reducing it into sulfide, which is more insoluble and less bioavailable, thus reducing its mobility to contaminate the environment [11]. Several bacterial strains such as Bacillus [12] and Acinetobacter [13] can also transform thiocyanate into sulfate, ammonia, and carbon dioxide, providing a bio-based alternative in managing cyanide contamination in mine wastes. Additionally, the mixture of organic waste such as crab shell and shrimp shell was found to promote micro-environments for active growth and metabolism of functional microorganisms leading to increased sulfate and heavy metal removal in mine waste (Chai et al. 14).

In addition to acid mine drainage, gold mine tailings often contain high levels of toxic heavy metals and metalloids such as arsenic due to its presence in the same ore deposits as gold. The use of cyanide as a chemical lixiviant in the gold leaching process can increase the bioavailability of arsenic, which can contaminate the groundwater system. To overcome this, bacteria that possesses the ability to produce arsenite oxidase can be used to oxidize highly toxic arsenite to arsenate, which is much less toxic and more easily immobilized in the environment [15]. Alternatively, researchers have also identified arsenic-tolerant plant species such as *Pteris vittata* [16] that are not only able to grow on arsenic-rich tailings but also able to hyperaccumulate arsenic into their aerial tissues. [17] suggested that *Acacia mangium* can be a good candidate to phytostabilize arsenic in mine waste considering its high tolerance to high levels of arsenic (790 mg kg^{-1}). Phytostabilization strongly retains targeted contaminants such as arsenic in the root area, avoiding from potential leaching of toxic metals and metalloids into groundwater system [7].

Considering all the alternatives mentioned above, less attention is given to recovering significant amounts of economic metals and metalloids from the tailings. Therefore, phytomining offers not only the opportunity to remediate mine wastes but also serves as a sustainable and environmentally friendly tool for extracting economic metals and metalloids to meet industry demand.

3 Elemental Sustainability of Economic Metals and Metalloids

The concept of "elemental sustainability" has been elucidated by Hunt et al. [18] as the principle that ensures the availability of every element in the periodic table for the utilization of present and future generations. There is a growing global recognition of the significance of maintaining the production of valuable metals and metalloids due to an escalating demand in contrast to the limited supply. These precious elements are widely dispersed and diluted throughout the technosphere, rendering them challenging to recapture and reuse. To address this challenge, it is imperative to develop multiple efficient methods for the recovery and reuse of these metals and metalloids.

The life cycle of a metal, as proposed by Reck et al. [19], typically involves stages such as mining and milling, smelting, refining, fabricating, manufacturing, utilization, and waste management. According to the metal life cycle depicted by Hunt et al. [18], one of the key stages where opportunities for recovery may arise is during the initial stage where it involves the extraction of metals and metalloids from mine waste materials generated during mining and processing operations. These mine wastes are usually voluminous and contain relatively low metal concentrations. Nevertheless, this method often entails high consumption of acidic leaching solutions and substantial operational and maintenance costs [1]. Hence, there is a pressing need for more sustainable approaches, such as phytomining, to recycle economic metals and metalloids in mine tailings in an environmentally friendly manner to meet the demands of various industries.

4 Phytomining at a Glance

To understand phytomining, one must be able to grasp the main phytotechnology mechanism behind it which is phytoextraction. Phytoextraction is a technology that involves the use of plants to uptake, transport, and accumulate metals and metalloids from the soil into their above-ground tissues [20]. To enhance the efficiency of phytoextraction, specific plant attributes are necessary, including a rapid growth rate, a high metal transport capacity from roots to shoots, a substantial biomass, a deep root system, and a high tolerance to the target metal [21, 22]. The process of phytoextraction begins with the absorption of metals and metalloids by the plant roots, followed by their transportation through the xylem to the aerial parts of the plant. Subsequently, the accumulated metals and metalloids can be stored in various plant compartments such as the cell wall, vacuole, or they may undergo reduction through cytoplasmic chelation [23].

Phytomining, on the other hand, is a specialized form of phytoextraction, where it can be described as the process of extracting metals and metalloids from sub-economic ore bodies or contaminated mine sites in situ, with the additional objective of recuperating economically viable amounts of metal from the mine waste [24]. Traditional mining practices often fail to extract 100% of the desired metal from the surrounding minerals, leading to the retention of some metals and metalloids in the mine wastes [25]. The recovery of valuable metals, metalloids, and platinum group metals (PGMs) from mine wastes may not be economically feasible using conventional methods. It has been observed that mine wastes from historical, now-defunct mines contain notably high concentrations of these metals and metalloids due to inefficient extraction techniques employed in the past, and many of these abandoned sites have not undergone sufficient remediation efforts [25].

Although many literatures related to phytomining are focusing on the phytoextraction of nickel from nickel-rich ultramafic soil, phytomining is also suggested to be used to recover remaining economic metals left in the mine wastes or tailings. Field trial on phytomining of mine wastes was initially done on gold where Anderson et al. [26] had initially grown *Brassica juncea* (Indian mustard) on waste-rock dump at the Fazenda Brasileiro mine site containing 0.6 g/ton of gold for 6 week. It was revealed that an average of 39 mg/kg of gold was extracted from *Brassica juncea's* aerial tissues after being dosed with sodium cyanide to increase the bioavailability of gold for plant uptake. *Sorghum bicolor* grown on a plot of a former uranium mining site in Ronneburg, Germany, also showed a high concentration of manganese accumulation in its aerial tissues (798 mg/kg) when the soil was amended with mycorrhiza and streptomyces [27]. Recent work by Medina-Díaz et al. [28] identified *Spergularia rubra* as a new hyperaccumulator species for zinc. It was observed that 17,800 mg/kg of Zn was accumulated in the aerial tissues when the plants were electrokinetically assisted with a voltage gradient of 1/2 V cm^{-1} for 6–12 h per day when grown on mine tailings taken from an abandoned mine site in Spain.

Phytomining often relies on hyperaccumulator species, which can efficiently grow and concentrate metals and metalloids in their above-ground tissues. Previous research on phytomining of nickel and gold has identified promising hyperaccumulator candidates such as Brassica spp., *B. coddii*, and Alyssum spp. [29]. Progress toward the commercialization of nickel phytoextraction has been notable, with advancements in breeding techniques to develop improved agronomic traits in Alyssum sp. cultivars [25]. However, the search for suitable hyperaccumulator species for other economic metals and metalloids like gold, palladium, and other platinum group metals (PGMs) remains ongoing.

Mine wastes often exhibit detrimental characteristics such as high concentrations of heavy metals and metalloids, suboptimal pH levels, low organic matter content, and limited nutrient availability, which can impede plant growth and reduce biomass production. Consequently, there is a quest for plant species that can thrive in the challenging conditions prevalent in metal-rich soils and waste materials. Research has focused on identifying plant candidates suitable for field applications, including non-hyperaccumulator species like willow (Salix sp.) and miscanthus known for its exceptional tolerance to toxic metals and metalloids concentrations (e.g., copper, cadmium, nickel, zinc) and ability to achieve substantial biomass in nutrient-deficient soils [24].

Unpublished work done by Harumain et al. [20] reveals that willow (*Salix viminalis* & *S. purpurea*) and miscanthus (*Miscanthus x giganteus*) are strong candidates for revegetating, phytomining, and remediating palladium-containing mine wastes. The work discovered that although the amount of palladium accumulated in the leaf is not sufficient enough to be commercially equivalent for catalytic activity, the ability of the plants to withstand low nutrient conditions and high toxic metals and metalloids concentrations is something that needs to be considered especially in real field applications of phytomining.

A significant challenge in implementing phytomining is the limited uptake of metals and metalloids from mine waste and soil due to their low bioavailability. These elements typically exist in chemically inert, zero-valent forms, or are bound to minerals. The plant's ability to access metals and metalloids is significantly influenced by the geochemical characteristics of the specific metal present in the soil. Studies have shown that incorporating metal chelators like EDTA [30], N-hydroxyethyl-EDTA (HEDTA) [31], ethylenediamine-N, N'-disuccinic acid (EDDS) [32], nitrilotriacetic acid (NTA), citric acid [33], and thiocyanate [34, 35] can significantly increase the uptake and accumulation of metals and metalloids such as copper, nickel, and cadmium [36]. A greater challenge lies in metals and metalloids such as gold and palladium, which have very low bioavailability in nature compared to other economically significant metals and metalloids like nickel, manganese, and cobalt. Phytomining of gold was first documented by Anderson et al. [37], who explored the use of ammonium thiocyanate to induce the accumulation of gold in *Brassica juncea*. A maximum accumulation of 57 $\mu g\ g^{-1}$ of gold (Au) was observed in the leaves of *B. juncea* when 0.64 $g\ kg^{-1}$ of ammonium thiocyanate was added to the substrate. Harumain et al. [20] further observed that the addition of 1 $\mu g\ mL^{-1}$ of potassium cyanide had increased the accumulation of palladium (Pd) in the leaves of *Arabidopsis thaliana*, *Brassica alba* L., *Miscanthus x giganteus*, and *Salix viminalis* grown on synthetic ore materials containing various concentrations of palladium. Interestingly, bio-ore produced from the pyrolysis of *Arabidopsis thaliana* biomass also exhibited good catalytic activity based on the production of trans-methyl cinnamate through the Heck reaction, further highlighting the advantage of using cyanide as a chemical lixiviant in phytomining of palladium. Nevertheless, the downside of using chemical lixiviant lies in the costs incurred in purchasing the chemicals, especially in large-scale applications. The excessive use of chemical lixiviant may also lead to the leaching of targeted metals, metalloids, and other contaminants in mine wastes into the groundwater system or nearby ecosystems. EDTA was found to increase the uptake of other heavy metals and metalloids such as cadmium [30], lead [31], zinc [31], and chromium [38] together with the targeted metal, limiting the effectiveness of phytomining. The utilization of natural, low molecular weight organic acids (NLMWOAs), like citric acid, to enhance metal solubility has emerged as a substitute for EDTA [33]. However, the effectiveness of NLMWOAs is uncertain, given that these substances can undergo rapid biodegradation by soil microorganisms [39], leading to the disintegration of complexes potentially prior to metal uptake by plants [36].

Considering both the advantages and disadvantages of using induced phytomining techniques, a thorough investigation is needed to consider for the best chemical lixiviant that can maximize the uptake and accumulation of targeted metals and metalloids into plants without causing any detrimental damage to the environment.

5 Suggested Pathways in Implementing Phytomining in Mine Waste Management

5.1 Mine Waste Location Selection and Characterization

Phytomining begins with the selection of potential mine waste areas, followed by a thorough inspection of various chemical and physical characteristics. In rehabilitating ex-mining land, afforestation is identified as the most suitable method due to low macronutrient levels, high sulfide minerals, acidic conditions, and elevated concentrations of toxic heavy metals and metalloids like zinc and aluminum that can hinder plant growth [40]. Afforestation involves the establishment of forests on land that has not been forested for a significant period or has never been forested before. This process typically involves planting trees on barren land or converting non-forest land into forested areas. In this case, phytomining can be a good additional benefit for afforestation initiatives on ex-mining land as it offers the use of metal hyperaccumulator species that can address soil contamination issues and at the same time, recovering valuable metals and metalloids during the restoration process (Fig. 1).

In the afforestation initiative, a detailed soil survey will be conducted to evaluate initial soil properties to fit with the restoration efforts. Mapping the mine waste area using a global positioning system (GPS) and conducting soil profiling up to 100 cm depth on the mine waste area facilitate a more accurate delineation of soil types. According to the successful afforestation project of an ex-mining land in Klian Intan, Perak, Malaysia, the soil types were categorized into five distinct categories which are sandy loam to sandy with gravel, sandy overlying silty clay, sandy to sandy loam, sandy over clay atop sandy clay loam, and sandy with gravel over silty clay [41]. This is crucial to understand the soil composition and characteristics, which can aid in effective soil preparation for various purposes. Additionally, further description of the area can also be grouped as plantable, steep slope, and gully, depending on the depth of the area as different plant species can be introduced to fit with the specific conditions of the area.

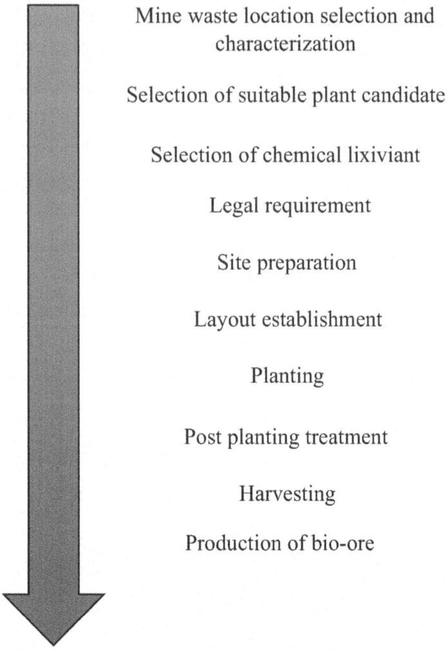

Fig. 1 Suggested pathways for application of phytomining on mine waste area

Heavy metal quantification analysis allows researchers to quantify the targeted metal to be extracted using plants, as well as the presence of other heavy metals and metalloids that could hinder the uptake ability of the targeted metal [42]. Since different plant species may have varying tolerance levels toward concentrations of heavy metals and metalloids, this analysis also enables researchers to make informed decisions when selecting suitable candidates for optimizing uptake capacity. Adjusting the soil pH using calcium carbonate or hydrated lime helps neutralize acids in mine soils. This can enhance soil microbial activity, thus improving the uptake of targeted metals and metalloids into plant roots [43, 44, 45]. The nutrient and organic matter content present in the mine waste soil facilitates the quantification of the requirements for amending the mine waste with fertilizers, thereby preventing plants from dying due to insufficient nutrients [43, 44, 45].

5.2 Selection of Suitable Plant Candidate

The search for suitable metal hyperaccumulators can commence once soil characterization profiling is completed. According to the latest definition by van der Ent and Rylott [46], the widely accepted description of metal hyperaccumulators pertains to plants that, when growing naturally rather than in artificially enriched substrates, exhibit elemental concentrations exceeding specific thresholds: 100 $\mu g\ g^{-1}$ for cadmium (Cd), thallium (Tl), or selenium (Se); 300 $\mu g\ g^{-1}$ for cobalt (Co) or copper (Cu); 1000 $\mu g\ g^{-1}$ for nickel (Ni), arsenic (As), or Rare Earth Elements (REEs); 3000 $\mu g\ g^{-1}$ for zinc (Zn); and 10,000 $\mu g\ g^{-1}$ for manganese (Mn) dry weight in shoot tissue. Previous definitions also include precious metals and metalloids like 1 $\mu g\ g^{-1}$ for Palladium (Pd) [47] and Gold (Au) [34]. To date, only 721 plant species have been identified as metal hyperaccumulators for various metals and metalloids worldwide. It is crucial to select the right plant candidate to be grown on selected mine waste to maximize the yield extracted from the mine waste and minimize total losses due to ineffective phytomining plant candidates. Using local metal hyperaccumulators is advisable, as they are suitable for the local geographical and climatic conditions of specific areas. Scattered mine wastes around the world present a challenge in finding suitable plant candidates that can withstand various environmental and climatic conditions. *Phyllanthus rufuschaneyi*, which can accumulate as high as 23,300 $\mu g\ g^{-1}$ of nickel in its aerial tissues, is among the best candidates for tropical regions, especially mine wastes found in ASEAN countries. *Rinorea cf. bengalensis*, which is also located in Sabah, was also found to be suitable for phytomining of cobalt considering its ability to accumulate up to 1200 $\mu g\ g^{-1}$. *Alyssum murale* was found to be suitable for phytomining activities of nickel in the Balkans region based on a 7-year field trial of agromining of nickel on Mg-rich vertisols soil in Albania [48]. As for manganese, a field trial conducted on manganese tailings in China using *Phytolacca americana* showed a maximum concentration of 8000 $\mu g\ g^{-1}$ of Mn in its leaf dry matter, suggesting its suitability for phytomining manganese in an Asian climate. Reeves et al. [49] documented a total of 721 metal hyperaccumulator species worldwide, with nickel being the most prevalent metal hyperaccumulator, comprising 523 species. In terms of nations, Cuba leads with 128 species of documented hyperaccumulator plants [50], followed by New Caledonia with 65 species [51], Turkey with 59 species [52], Brazil with at least 30 species [53], and Malaysia with 24 species [54].

5.3 *Selection of Chemical Lixiviant*

After certain species have been identified, it is also suggested to consider the use of chemical organic lixiviant if needed due to the limitations in terms of the bioavailability of certain metals and metalloids, which hinder their uptake into root cells. Various organic lixiviants have been proven to significantly increase the bioavailability of metals and metalloids, thus increasing their adsorption and accumulation into plant cells. Metals and metalloids such as palladium and gold are known for their low bioavailability in nature compared to nickel. To address this, Anderson et al. [34] and Harumain et al. [20] investigated the use of low concentrations of cyanide, which were found to boost the accumulation of gold and palladium in the aerial tissues of plants. A recent field trial on gold phytomining was conducted in Central Lombok, Indonesia using Tobacco plants grown on cyanidation tailings containing around 1 mg kg^{-1} of gold and further induced with sodium cyanide [55]. The ash was then treated with borax but results showed that it only significantly induces the uptake of Ag into plants instead of Au suggesting a considerable work must still be done to optimize valuable metal recovery by plants at the field scale. Harumain et al. [20] found that the inclusion of 1 $\mu g\ mL^{-1}$ of potassium cyanide increased palladium (Pd) accumulation in the leaves of *Arabidopsis thaliana, Brassica alba L.*, *Miscanthus x giganteus*, and *Salix viminalis* grown on synthetic ore materials with varying palladium concentrations although higher yield of palladium is required to show a good catalytic activity. However, the application of chemical lixiviant should only be considered when there are no alternative options utilizing metal hyperaccumulators, as it could potentially result in harmful effects on the environment without proper monitoring, and also lead to increased costs, thus making phytomining not a favorable option.

5.4 Legal Requirement

As biosecurity regulations have been tightened in many countries to control illegal trading, it is also very important to comply with all legal requirements such as access, import, or export licenses provided by the local authorities. Since phytomining provides significant commercial value from the output of the project, many countries are now contemplating the protection of their metal hyperaccumulator species, most of which are endemic plant species, through various legislative acts. The import and export of biological resources generally are protected by various acts and policies such as Convention of Biological Diversity (CBD), Convention for International Trade of Endangered Species of Wild Fauna and Flora (CITES), and as well as internal legislation specific to each country. For instance, in Sabah, Malaysia, any natural resources intended for extraction or exportation outside of Sabah must procure access and import licenses from the Sabah Biodiversity Centre (SaBC), failure to do so may result in a fine of not more than MYR 100,000 or imprisonment for up to 5 years. The importation of metal hyperaccumulator plants into Malaysia is further safeguarded by the Plant Quarantine Act 1976, which stipulates the necessary requirements and regulations to prevent potential economic, environmental, or health-related risks associated with alien species. These regulations are crucial as alien species have the potential to negatively impact biodiversity by competing with, preying upon, or transmitting pathogens to native species, leading to their decline or elimination and the disruption of local ecosystems and ecosystem functions.

5.5 Site Preparation

Once approval is obtained from local authorities, site preparation can commence, ensuring that the soil in the designated area is adequately prepared for planting metal hyperaccumulator plants for phytomining purposes. Usually for ex-mining areas, the preparation involves the mechanical introduction of mineral soils from nearby areas, sealing of existing acidic layer, and reducing the level of acidity for tree plantings. The introduction of mineral soil from nearby areas may encourage the foundation of the seedling's roots as ex-mining soil contains deleterious elements which might hinder the growth of the seedlings. Prior to doing that, a layer of magnesium limestone powder can be broadcasted on the surface of the tailings before adding the mineral soils for approximately around 45 cm thickness. The addition of another layer of magnesium limestone powder on top of the mineral soil allows for a better pH environment for the plant to grow. Smith et al. [56] propose that incorporating an extra layer of magnesium limestone powder above the mineral soil can improve the pH conditions for plant growth, creating a buffered and less acidic substrate that enhances the overall success of seedling establishment on ex-mining sites. Finally, another 15 cm of forest topsoil can be added on top of the 2nd magnesium powder layer to produce a 60 cm thick mechanically introduced soil that acts as a barrier, expanding room for the roots and separating the roots of planted trees from the acidic layer [41].

5.6 Layout Establishment

There are few ways in establishing a planting layout on ex-mining land. Following to successful model of afforestation of ex-mining land by Ahmad-Zuhaidi et al. [41], any phytomining project should come out with a layout that must include the exact location and shape for the ex-mining land, road network plan, waterways for fire-fighting purposes, buffer lines and access, spacing, and mapping. As countries like Indonesia are often susceptible to open-fire burning during periods of extreme heat, the establishment of proper waterways for firefighting and direct access to water can significantly mitigate the impact of such incidents. For areas smaller than 10,000 m, the proposed scale of 1:5000 is recommended for use [41]. Three types of spacing can be used depending on the suitability of the location which are square (4 × 4 m, 3 × 3 m), rectangular (2 × 8 m), or triangular (4.3 × 3.72 m). The use of the spacing will determine the total number of seedlings to be planted within the area which relates to the maintenance and the income generation based on the total biomass obtained from the leaf. In phytomining, various types of metal hyperaccumulators can be used which include herbaceous non-woody plants like *Noccaea caerulescens* as well as small trees such as *Phyllanthus rufuschanyei.* The types of metal hyperaccumulators to be used will determine the appropriate number of plants to be cultivated on the land, optimizing the overall yield achievable through phytomining. Previous field trials conducted on *Noccaea caerulescens* in extracting cadmium indicated that 4–12 weeks-old seedlings were grown at densities of 16–200 plants m^{-2} [57]. Bani et al. [58] found that cultivating *Alyssum murale* at a spacing of 1.5 plants per square meter resulted in the highest biomass yield of 10.8 tons per hectare, with a maximum nickel yield of 76.6 kg per hectare, compared to growing 6 plants per square meter. The lower biomass obtained from densely populated *Alyssum murale* suggests that competition within the same species, especially for nutrient uptake, may impact both biomass output and subsequent nickel content. Currently, field trials on nickel uptake in *Phyllanthus rufuschaneyi* are being conducted in Ranau, Sabah. These trials may offer valuable insights into the impact of plant spacing on nickel uptake, particularly considering the species' relatively large biomass compared to other nickel hyper-accumulators. Another crucial measure to consider is the use of barriers to prevent ruminants such as cows, goats, or other wild animals from consuming the plant materials. Since metal hyperaccumulators may contain high concentrations of metals and metalloids, this can be toxic and potentially lethal to the animals.

5.7 Planting

After the layout of the site has been established, the seedlings can be planted into the dedicated hole prepared earlier. It is suggested that planting should be conducted during the rainy season because the soil is still wet which facilitates the establishment of roots of the seedlings before the dry season starts. Planting the seedlings in the morning also reduces the evapotranspiration from the seedlings, prolonging the chance for seedlings to survive. The addition of fertilizers is recommended, as ex-mining land typically lacks sufficient nutrients. Inorganic fertilization has been shown to promote biomass, increase phytoavailability of targeted metals and metalloids, and subsequently influence the uptake of various metals and metalloids in plants. The addition of NPK fertilizer was found to increase the accumulation of nickel in *Alyssum bertolonii*, *Berkheya coddii*, and *Streptanthus polygaloides* by increasing the biomass of the plants when grown on ultramafic soil [59]. Sulfur added to ultramafic soil was also found to boost the uptake of nickel in *Berkheyya coddii* by increasing its phytoavailability [60]. Nevertheless, the addition of fertilizer must be based on the soil analysis as the excess amount of inorganic fertilizer might also hinder the growth of the plants. Muarif et al. [61] highlighted that continuous use of inorganic fertilizers can degrade soil fertility, leading to soil acidity and environmental pollution. Increased soil acidity may increase the phytoavailability of other contaminants beyond the tolerance level of the plants. Ahmad-Zuhaidi et al. [41] suggested that the use of fertilizer is essential on poor sites only. Other than using inorganic fertilizer, the use of bacterial inoculation was also found to increase nickel yield by increasing nickel uptake and biomass production of the metal hyperaccumulator plant based on a pot study of *Bornmuellera tymphaea* and *Noccaea tymphaea* treated with plant growth promoting rhizobacteria on ultramafic soil [62].

5.8 Post-planting Treatment

Throughout the planting period, several maintenance procedures should be conducted to maintain the quality of the plants. Activities such as replanting dead seedlings with a new one should be done immediately. Therefore, it is very important to have enough stock of seedlings before any phytomining effort to be executed. Soil amendment using lixiviants and fertilizers should be monitored regularly to maximize the growth biomass and subsequently the accumulation of the targeted metals and metalloids into the aerial tissues. Watering is not necessary during the wet season but should be well monitored especially when it gets too hot. Weed control was found to be beneficial in reducing competition for essential nutrients between metal crops and other weeds resulting in a higher yield of nickel accumulated in the aerial tissues [63]. The use of pesticides can also be considered if necessary, as pests such as *Pteroma pendula* (bagworm) are quite common and can lay eggs between young leaves, causing them to turn brown when the larvae feed on them [41]. However, the use of common pesticides such as Furadan 3G and Cyperin 550 should be carefully regulated to prevent environmental pollution.

5.9 Harvesting

Harvesting should be properly planned and conducted to prevent the reintroduction of metals and metalloids onto the ground, especially into the topsoil, to avoid economic losses and, most importantly, potential pollution. A 6-year field trial study on the use of poplar clones grown on cadmium and zinc-contaminated land showed increased concentrations of metals and metalloids in the topsoil due to litterfall caused by improper harvesting management [64].

The duration for the harvesting depends on the type of metal hyperaccumulator species used and the concentration of targeted metals and metalloids in the soil. Different plant species may have different growth periods depending on various factors such as available nutrients and climate conditions. In phytomining, the goal is to maximize the biomass of leaf or aerial tissues that can be harvested every cycle while maintaining the accumulation capacity [43a, b, c]. Herbaceous metal hyperaccumulator such as *Alyssum murale* grows faster than small trees hyperaccumulator like *Phyllanthus rufuschaneyi*. But in terms of harvested biomass per cycle, the latter showed a much higher value considering its relatively bigger size of leaf materials [65]. This distinction highlights the importance of considering both growth rate and biomass yield when evaluating the effectiveness of different hyperaccumulator plant species in phytomining operations. *Blepharidium guatemalense (Standl.) Standl*, a woody nickel hyperaccumulator plant in Mexico, was grown in soil containing various concentrations of nickel for a one-year duration. It was observed that harvesting twice a year resulted in more dry leaf biomass and subsequently higher nickel yields compared to harvesting once a year [66].

In another interesting approach, the intercropping of *Sedum alfredii* with oilseed rape on soil containing 0.82 mg kg^{-1} of cadmium demonstrated a higher yield of cadmium extracted from the shoot of *Sedum alfredii* after 7 months of harvest [67].

Generally, phytomining can be harvested once a year to maximize biomass production from the plants used and allow sufficient time for the harvested plants to regenerate [68]. Field trials have indicated that an annual harvesting period is ideal for phytomining, ensuring a balance between biomass yield and metal uptake efficiency [69]. The success of phytomining is contingent on achieving high biomass yield and metal hyperaccumulation in the plants, making it a commercially viable proposition [43, 44, 45]. The harvesting period also determines the total expected income generated through phytomining. Taking nickel as an example, an economic feasibility study conducted by Dr. Antony van der Ent estimated that approximately 80 kg of nickel could be harvested using phytomining technology on a one-acre plot within a year, with the harvested yield valued at RM 18,154.00/acre based on the actual date of nickel prices. In comparison, one harvest of rice, which also occurs annually, can yield RM 8,500.00/acre. This presents an opportunity for implementing phytomining technology in agriculture, to sustainably absorb metals and metalloids and bring economic benefits to the people.

For some species like *Alyssum murale*, harvesting can be conducted when the biomass is considered high which can be determined during the mid-flowering stage, allowing for maximum recovery of nickel [48]. Another method that can be used is regular quantification of the metal concentration in the leaf using portable devices such as an XRF analyzer. Each metal hyperaccumulator species may have a specific range of maximum concentration that they can accumulate. However, this range must also be accompanied by the physical conditions of the plant as some plants might die due to the presence of other contaminants rather than the targeted metal in the mine waste. Therefore, a proper field trial study on each metal hyperaccumulator species on various potential economic metals and metalloids can give us reliable data that can be used to estimate the suitable harvesting period in real phytomining scenario.

5.10 Production of Bio-ore

Once the biomass is harvested, several physical treatments can be done to enhance the process of extracting targeted metal from the material. Drying the material will allow the removal of water remaining in the leaf, thus increasing the concentration of the targeted metal to be extracted. It can be done either using a dedicated oven or under direct sunlight which might take a few days to be fully dry. Once it is dried, the material can be crushed to enhance the exposure of metals and metalloid-rich components, facilitating better contact with extracting agents and improving the overall extraction efficiency. The increased surface area resulting from crushing promotes a more effective reaction during subsequent processing steps. Sieving allows the removal of unwanted impurities, therefore ensuring the purity of the extracted metal. Weighing the total dried biomass in the end will allow for an estimation of the yield output of the targeted metal.

The dried biomass can either be processed through combustion or a non-combustion approach. The non-combustion approach involves direct leaching using acids and the targeted metal can be further extracted using water at room temperature [70, 71]. Washing the ash with water allows for the removal of water-soluble salts such as potassium, which can then be utilized as fertilizer. This process subsequently reduces the amount of acid required to leach out the targeted metals and metalloids [70]. However, Simonnot et al. [70] proposed that the water-leached metal contains organic matter derived from biomass which may complicate the product recovery. The majority of post-harvesting treatment utilizes combustion technique as this can help to remove organic compounds such as rich carbon sources, chlorine, and phosphate [70]. Combustion also allows the disposal of contaminated biomass, generates heat as energy, and chemically enhances the concentration of the targeted metal. Established work on nickel extraction through combustion on various nickel hyperaccumulator plants showed that *Phyllanthus securinegioides* [72] contains lower potassium, calcium, and magnesium in the form of oxides and carbonates compared to *Alyssum murale* [71] and *Rinorea bengalensis* (Vaughan et al. 2016a). The ashing of the biomass can be conducted in an electrical furnace at temperatures more than 550 °C but higher temperatures may also lead to the removal of certain metals and metalloids through volatilization. Harumain et al. [20] used 300 °C as temperature level to produce bio-ore from Arabidopsis biomass treated with various concentrations of potassium tetrachloropalladate. This temperature was chosen based on previous observations by Parker et al. [73] where a higher temperature (800 °C) leads to the production of larger nanoparticle size which subsequently failed to show any catalytic functions. However, the temperature used for this experiment may not be suitable for other plant materials as Arabidopsis is known for its softer physical structure compared to other metal hyperaccumulator plants.

Most electric furnaces are not suitable for large-scale ashing production. Although the use of open burning allows large biomass to be combusted, this method is not suggested as it may be difficult to control the quality of the bio-ore produced [70]. The selection of suitable furnace is important to control the temperature and time duration required for the combustion process. Higher temperature utilizes a shorter time to pyrolyze the biomass material but comes with a bigger running cost and energy used whereas burning the biomass at temperatures around 200–300 °C may require a longer combustion period. The most suitable furnace for burning large-scale biomass from phytomining, while maintaining temperatures suitable for efficient combustion, would be a MILD combustion furnace. MILD (Moderate or Intense Low-oxygen Dilution) combustion furnaces are known for their ability to operate at lower temperatures while ensuring high efficiency. Riaza et al. [74] indicate that MILD combustion furnaces can effectively handle biomass particles with reduced soot formation, making them suitable for burning biomass from phytomining. The ability of MILD combustion to provide uniform temperature distribution and efficient combustion makes it an ideal choice for large-scale biomass combustion processes.

The bio-ore product, although can already be sold on the market, can still be further refined to produce high purity salt. Take nickel, for example, the bio-ore can undergo an acid leaching process using sulfuric acid to produce nickel leachate. The nickel leachate can also then be selectively precipitated using calcium hydroxide to remove iron by precipitation of iron hydroxide. Magnesium remaining in the leachate can be removed by producing magnesium fluoride precipitate using sodium fluoride before the solution is evaporated to produce high purity ammonium nickel sulfate hexahydrate (ANSH) salt [70]. Nevertheless, several articles showed that the bio-ore produced can already be used as a catalyst in many catalytic reactions. Harumain et al. [20] pyrolyzed Arabidopsis dosed potassium tetrachloropalladate to produce bio-ore containing palladium nanoparticles with high catalytic activity based on the production of trans-methyl cinnamate through Heck reaction. Dube et al. [75] observed a strong catalytic property of ruthenium-bio-ore produced from *Manihot esculenta* biomass grown on ruthenium-rich solutions to catalyze the hydrogenation of furfural to furfural alcohol using triethylamine as a base and formic acid as a hydrogen source. Akinbile et al. [76] also recently published the use of nickel-bio-ore produced from *Manihot esculenta* as a good catalyst candidate for the conversion of carbon dioxide to formate.

6 Concluding Remarks

In the sphere of environmental reclamation and resource management, phytomining emerges as a pioneering solution, addressing the intricate complexities inherent in mine waste management. This nascent methodology capitalizes on plants' exceptional capacity to absorb and amass economically significant metals and metalloids from sub-economic ore deposits, thereby presenting a sustainable avenue for both environmental remediation and economic advancement.

Fundamentally, phytomining signifies a fundamental shift in our approach to confronting the enduring legacies of mining activities. Traditionally, the disposal of mine residues, encompassing crushed rock, process effluent, sand, clay, and aqueous solutions, has constituted a substantial environmental challenge. These residues frequently harbor substantial quantities of metal remnants, including nickel, cobalt, manganese, and gold, among others, rendering them economically impractical to extract using conventional means. Consequently, extensive swathes of land persist in a state of contamination, posing risks to ecosystems and human health.

Furthermore, the environmental ramifications of mine waste management are exacerbated by the presence of toxic metals and metalloids such as arsenic, which can leach out alongside mined metals and metalloids, amplifying contamination concerns and perpetuating long-term threats to water sources and biodiversity. Within this context, the imperative for innovative and sustainable strategies to mitigate the environmental impacts of mining operations has never been more pressing.

Enter phytomining, a pioneering strategy that harnesses natural plant physiological processes to address these challenges. Through the mechanism of phytoextraction, certain plant species have evolved mechanisms to sequester and concentrate metals and metalloids from the soil via their roots, subsequently accumulating these metals and metalloids within their aerial tissues. As illustrated in this chapter, by judiciously selecting and cultivating hyperaccumulating plant species, such as *Phyllanthus rufuschaneyi* for nickel or *Pteris vittata* for arsenic, phytomining offers a cost-effective and environmentally benign means of extracting metals and metalloids from mine residues.

The allure of phytomining resides in its adaptability and scalability. In contrast to conventional mining methodologies, which necessitate extensive infrastructural investments and energy-intensive operations, phytomining can be deployed across various scales, ranging from localized remediation initiatives to expansive commercial endeavors. Furthermore, phytomining obviates the necessity for hazardous chemicals or the generation of toxic by-products, rendering it inherently safer and more sustainable than traditional mining practices.

A pivotal advantage of phytomining lies in its potential to transmute mine wastes into high-value bio-ores. Once metals and metalloids are assimilated by the plants, they can be extracted from the biomass via phytoextraction, tailored to specific metals and metalloids and plant species, facilitating the recovery of metals and metalloids in a format conducive to downstream processing. By effecting this conversion of mine residues into bio-ores, phytomining not only facilitates the remediation of contaminated sites but also engenders a valuable resource with tangible economic utility.

Furthermore, phytomining proffers a distinctive solution to the predicament of rehabilitating un-vegetated areas affected by mining activities. Many mine residues exhibit a paucity of organic matter and nutrients requisite for sustaining plant growth, resulting in barren landscapes devoid of vegetation. By introducing hyperaccumulating plant species into these environments, phytomining not only facilitates metal extraction but also fosters the establishment of vegetative cover, thereby reinstating ecosystem functionality and biodiversity.

In addition to its environmental and economic merits, phytomining holds promise as a mechanism for addressing the social and regulatory challenges inherent in mine waste management. By furnishing a sustainable alternative to conventional remediation approaches, phytomining can assist mining enterprises in navigating increasingly stringent environmental regulations and fulfilling their obligations to local communities and stakeholders. Moreover, the economic incentives associated with phytomining may engender enhanced collaboration between industry, governmental, and environmental entities, fostering innovation and yielding positive social and economic outcomes.

Nevertheless, notwithstanding its considerable promise, phytomining is not devoid of challenges and constraints. Chief among these is the identification and optimization of hyperaccumulating plant species for specific metals and metalloids and environmental contexts. While substantial strides have been made in this regard, further research is imperative to enhance the efficiency and efficacy of phytomining techniques.

Additionally, the scalability of phytomining may be constrained by factors such as land availability, soil quality, and climatic conditions. In locales characterized by limited land resources or unfavorable growing conditions, alternative approaches such as hydroponic or aeroponic phytomining may represent viable alternatives.

Moreover, the economic viability of phytomining hinges upon factors such as fluctuating metal prices, extraction costs, and market demand, necessitating vigilance and adaptability.

In spite of these challenges, the transformative potential of phytomining in revolutionizing mine waste management and resource extraction is undeniable. As we confront the exigencies of environmental degradation and resource depletion, innovative methodologies like phytomining offer glimpses of a more sustainable and resilient future. By harnessing the inherent capabilities of nature, we can unlock the latent value of mine residues, transmuting environmental liabilities into economic assets and charting a course toward a cleaner, more prosperous world.

References

1. Sarker SK, Pownceby MI, Bruckard W, Haque N, Bhuiyan M, Pramanik BK (2023) Unlocking the potential of sulphide tailings: a comprehensive characterization study for critical mineral recovery. Chemosphere 328:138582
2. Franks DM, Stringer M, Torres-Cruz LA, Baker E, Valenta R, Thygesen K, Barrie S et al (2021) Tailings facility disclosures reveal stability risks. Sci Rep 11(1):5353
3. Kaźmierczak U, Blachowski J, Górniak-Zimroz J, Wirth H (2018) Quantitative and qualitative research on the waste from the mining of rock raw materials in lower Silesia. Minerals 8(9):375. https://doi.org/10.3390/min8090375
4. Marques JE, Martins V, Santos P, Ribeiro J, Mansilha C, Melo A, Flores D et al (2021) Changes induced by self-burning in technosols from a coal mine waste pile: a hydropedological approach. Geosciences 11(5):195. https://doi.org/10.3390/geosciences11050195
5. Krzaklewski W, Pietrzykowski M (2002) Selected physico-chemical properties of zinc and lead ore tailings and their biological stabilisation. Water Air Soil Pollut 141(1):125–141
6. Mendez MO, Glenn EP, Maier RM (2007) Phytostabilization potential of quailbush for mine tailings. J Environ Qual 36(1):245–253
7. Gil-Loaiza J, White SA, Root R, Solís-Dominguez FA, Hammond CM, Chorover J, Maier RM et al (2016) Phytostabilization of mine tailings using compost-assisted direct planting: translating greenhouse results to the field. Sci Total Environ 565:451–461. https://doi.org/10.1016/j.scitotenv.2016.04.172
8. Wei Z, Yin G, Wang J, Wan L, Li G (2012) Design, construction and management of tailings storage facilities for surface disposal in China: case studies of failures. Waste Manag Res J Sustain Circ Econ 31(1):106–112. https://doi.org/10.1177/0734242X11424404
9. Badiozamani MM, Ben-Awuah E, Askari-Nasab H (2019) Mixed integer linear programming for oil sands production planning and tailings management. J Environ Inf
10. Asif Z, Chen Z (2016) Environmental management in North American mining sector. Environ Sci Pollut Res 23(1):167–179
11. Ayangbenro A, Olanrewaju O, Babalola O (2018) Sulfate-reducing bacteria as an effective tool for sustainable acid mine bioremediation. Front Microbiol 9
12. Rai A, Gowrishetty KK, Singh S, Chakrabarty J, Bhattacharya P, Dutta S (2021) Simultaneous bioremediation of cyanide, phenol, and ammoniacal-N from synthetic coke-oven wastewater using Bacillus sp. NITD 19. J Environ Eng 147(1):04020143
13. Welman-Purchase MD, Castillo J, Gomez-Arias A, Matu A, Hansen RN (2024) First insight into the natural biodegradation of cyanide in a gold tailings environment enriched in cyanide compounds. Sci Total Environ 906(number_4):1–461
14. Chai G, Wang D, Zhang Y, Wang H, Li J, Jing X, Lin Y et al (2023) Effects of organic substrates on sulfate-reducing microcosms treating acid mine drainage: performance dynamics and microbial community comparison. J Environ Manag 330

15. Basu S, Dutta A, Mukherjee SK, Hossain ST (2023) Isolation and characterization of an As (III) oxidizing bacterium, Acinetobacter sp. TMKU4 from paddy field for possible arsenic decontamination. J Hazard Mater Adv 10
16. Wang W, Meng D, Tan X, Zheng M, Xiao J, Li S, Li H et al (2024) Nitrogen addition accelerates litter decomposition and arsenic release of *Pteris vittata* in arsenic-contaminated soil from mine. Ecotoxicol Environ Safety 271(number_1):1–461
17. Rosli RA, Harumain ZA, Zulkalam MF, Hamid AA, Sharif MF, Mohamad MA, Shahari R et al (2021) Phytoremediation of arsenic in mine wastes by *Acacia mangium*. Remediat J 31(3):49–59
18. Hunt AJ, Farmer TJ, Clark JH (2013) Elemental sustainability and the importance of scarce element recovery. In: Element recovery and sustainability. The Royal Society of Chemistry, pp 1–28. https://doi.org/10.1039/9781849737828-00001
19. Reck BK et al (2008) Anthropogenic nickel cycle: insights into use, trade, and recycling. Environ Sci Technol 42(9):3394–3400
20. Harumain ZAS, Parker HL, Munoz-García A, Austin MJ, McElroy CR, Hunt AJ, Clark JH, Meech JA, Anderson CWN, Ciacci L, Graedel TE, Bruce NC, Rylott EL (2017) Toward financially viable phytoextraction and production of plant-based palladium catalysts. Environ Sci Technol 51(5):2992–3000. https://doi.org/10.1021/acs.est.6b05079
21. Smith A, Johnson B, Brown C (2018) Enhancing phytoextraction efficiency through plant attribute optimization. J Environ Sci 15(2):45–58
22. Johnson E, Brown L (2019) The role of plant attributes in phytoextraction processes. Plant Physiol Rev 25(4):321–335. https://doi.org/10.1007/s12298-019-00700-w
23. Jones D (2017) Phytoextraction: plant traits and metal uptake mechanisms. Environ Biotechnol 8(3):112–125. https://doi.org/10.12691/env-8-3-3
24. Smith A, Johnson B (2021) Advances in phytomining: extracting metals from sub-economic ore bodies. J Environ Sci 18(3):112–125
25. Brown C (2020) Challenges in metal recovery from mine wastes: a review of conventional methods. Mining Eng J 25(4):321
26. Anderson C, Moreno F, Meech J (2005) A field demonstration of gold phytoextraction technology. Miner Eng 18(4):385–392
27. Phieler R, Merten D, Roth M, Büchel G, Kothe E (2015) Phytoremediation using microbially mediated metal accumulation in Sorghum bicolor. Environ Sci Pollut Res 22(24):19408–19416. https://doi.org/10.1007/s11356-015-4471-1
28. Medina-Díaz HL, López-Bellido FJ, Alonso-Azcárate J, Fernández-Morales FJ, Rodríguez L (2024) A new hyperaccumulator plant (*Spergularia rubra*) for the decontamination of mine tailings through electrokinetic-assisted phytoextraction. Sci Total Environ 912:169543
29. Deng T, CloquetC, Tang Y, Sterckeman T, Echevarria G, Estrade N, Qiu R et al (2014) Nickel and zinc isotope fractionation in hyperaccumulating and nonaccumulating plants. Environ Sci Technol 48(20)
30. Blaylock MJ, Salt DE, Dushenkov S, Zakharova O, Gussman C, Kapulnik Y (1997) Enhanced accumulation of Pb in Indian mustard by soil-applied chelating agents. Environ Sci Technol 31(3):860–865
31. Huang JW, Blaylock MJ, Kapulnik Y, Ensley BD (1997) Phytoremediation of lead-contaminated soils: role of synthetic chelates in lead phytoextraction. Environ Sci Technol 31(3):800–805. https://doi.org/10.1021/es960559e
32. Wilson-Corral V, Anderson CWN, Rodriguez-Lopez M (2012) Gold phytomining. A review of the relevance of this technology to mineral extraction in the 21st century. J Environ Manag 111:249–257
33. do Nascimento CWA, Amarasiriwardena D, Xing B (2006) Comparison of natural organic acids and synthetic chelates at enhancing phytoextraction of metals from a multi-metal contaminated soil. Environ Pollut 140(1):114–123
34. Anderson CWN, Brooks RR, Chiarucci A, LaCoste CJ, Leblanc M, Robinson BH, Meech JA et al (1999) Phytomining for nickel, thallium and gold. J Geochem Explor 67(1–3):407–415

35. Anderson CWN, Stewart RB (2003) Gold phytomining. Novel developments in a plant-based mining system. In: Gold 2003 Conference, World Gold Council, Vancouver, pp 35–45
36. Hunt AJ, Anderson CWN, Bruce NC, Garcia AM, Graedel TE, Hodson ME, Meech JA, Nassar NT, Parker HL, Rylott EL, Clark JH (2014) Phytoextraction as a tool for green chemistry. Green Process Synth 3(1):3–22. https://doi.org/10.1515/gps-2013-0071
37. Anderson CWN, Stewart RB, Moreno FN, Wreesmann CT, Gardea-Torresday JL, Robinson B, Meech JA (1998) Harvesting a crop of gold in plants. Nature 395(6702):553–554
38. Duo L-A, Gao Y-B, Zhao S-L (2005) Heavy metal accumulation and ecological responses of turfgrass to rubbish compost with EDTA addition. J Integr Plant Biol 47(9):1047–1054
39. Meers E et al (2004) Enhanced phytoextraction: in search of EDTA alternatives. Int J Phytorem 6(2):95–109
40. Ahirwal J, Kumar A, Maiti S (2020) Effect of fast-growing trees on soil properties and carbon storage in an afforested coal mine land (india). Miner 10(10):840
41. Ahmad-Zuhaidi Y, Jeyanny V, Hashim MN, Mohamad-Fakhri I (2018) Model afforestation of ex-mining lands: establishment of new forests in Klian Intan, Perak. FRIM Res Pam 140:66
42. Chaney RL, Li YM, Brown SL (2007) Phytoremediation of contaminated soil and water. Phytoremediation: Methods Rev (pp. 1–22). Human Press
43. van der Ent A, Baker AJM, Reeves RD, Chaney RL, Anderson C, Meech J, Mulligan DR et al (2015a) Agromining: farming for metals in the future? Environ Sci Technol 49(8):4773–4780. https://doi.org/10.1021/es506031u
44. van der Ent A, Erskine P, Sumail S (2015b) Ecology of nickel hyperaccumulator plants from ultramafic soils in Sabah (Malaysia). Chemoecology 25(5):243–259
45. van der Ent A, Repin R, Sugau J, Wong KM (2015c) Plant diversity and ecology of ultramafic outcrops in Sabah (Malaysia). Aust J Bot 63(4):204–215
46. Van der Ent A, Rylott EL (2024) Phytomining: a sustainable approach to metal recovery. Environ Sci Technol 58(2):1234–1245
47. Masinire MM, Moyo P, Moyo N (2022) Phytoremediation of heavy metals: a review of the mechanisms and applications. Environ Adv (7)100150
48. Bani A, Ghosh M (2018) Phytoremediation of heavy metal contaminated soils: a review. Environ Sci Pollut Res 25(14):13456–13470
49. Reeves RD, Baker AJM, Jaffré T, Erskine PD, Echevarria G, van der Ent A (2018) A global database for plants that hyperaccumulate metal and metalloid trace elements. New Phytol 218(2):407–411
50. Reeves RD, Baker AJM (1999) Metal-accumulating plants. Phytoremediation of Toxic Metals: Using Plants to Clean Up the Environment (pp. 1–18). John Wiley & Sons
51. Jaffre T, Reeves RD (2013) The role of hyperaccumulators in phytoremediation. Front Plant Sci 4:279
52. Reeves RD, Adıguzel A (2008) Hyperaccumulators of metals and metalloids: a global overview. New Phytol 179(3):763–782
53. Reeves RD, Baker AJM, Jaffré T (2007) The role of hyperaccumulators in the biogeochemical cycling of metals. New Phytol 174(1):1–10
54. Van der Ent A (2016) Phytomining: a sustainable technology for metal recovery. Nat Sustain 1(1):1–3
55. Krisnayanti BD, Anderson CWN, Sukartono S, Afandi Y, Suheri H, Ekawanti A (2016) Phytomining for artisanal gold mine tailings management. Fortschritte Auf Dem Gebiet Mineral 6:1–9. https://doi.org/10.3390/min6030055
56. Smith A, Johnson B, White C (2019) Ecological restoration techniques in mining sites. Environ Sci J 15(2):45–58
57. Sterckeman T, Puschenreiter M (2018) Phytoremediation of heavy metal contaminated soils: a review of the current state of knowledge. Environ Sci Pollut Res 25(14):13456–13470
58. Bani A, Echevarria G, Zhang X, Benizri E, Laubie B, Morel JL (2015b) The effect of plant density in nickel-phytomining field experiments with *Alyssum murale* in Albania. Aust J Bot 63:72

59. Álvarez-López V, Prieto-Fernández Á, Cabello-Conejo MI, Kidd PS (2016) Organic amendments for improving biomass production and metal yield of Ni-hyperaccumulating plants. Sci Total Environ 548:370–379
60. Robinson BH, Brooks RR, Clothier BE (1999) Soil amendments affecting nickel and cobalt uptake by *Berkheya coddii*: potential use for phytomining and phytoremediation. Ann Bot 84(6):689–694
61. Muarif S, Sulistyaningsih E, Handayani VDS, Isnansetyo A (2022) Substituting sargassum sp. compost for inorganic fertilizer improves the growth and yield of shallot (*allium cepa l.* aggregatum group). Pertanika J Trop Agric Sci 45(4):867–880. https://doi.org/10.47836/pjtas.45.4.02
62. Durand C, Jaffré T (2015) Nickel hyperaccumulation in plants: a review of the mechanisms and applications. Front Plant Sci 6:1–12
63. Bani A, Echevarria G, Sulçe S, Morel JL (2015a) Improving the agronomy of *Alyssum murale* for extensive phytomining: a five-year field study. Int J Phytoremed 17:117–127
64. Michels E, Annicaerta B, De Moor S, Van Nevel L, De Fraeye M, Meiresonne L, Meers E et al (2018) Limitations for phytoextraction management on metal-polluted soils with poplar short rotation coppice—evidence from a 6-year field trial. Int J Phytorem 20(1):8–15
65. Bouman RW, Welzen Pv, Sumail S, Echevarria G, Erskine PD, van der Ent A (2018) *Phyllanthus rufuschaneyi*: a new nickel hyperaccumulator from Sabah (Borneo Island) with potential for tropical agromining. Bot Stud 59(1)
66. Gutiérrez DMN, Nkrumah PN, van der Ent A, Pollard J, Baker AJM, Torralba FN, Pons M-N, Sánchez JAC, Hernández TG, Echevarria G (2021) The potential of *Blepharidium guatemalense* for nickel agromining in Mexico and Central America. Int J Phytorem. https://doi.org/10.1080/15226514.2021.1881039
67. Cao X, Wang X, Lu M, Hamid Y, Lin Q, Liu X, Yang X et al (2021) The Cd phytoextraction potential of hyperaccumulator *Sedum alfredii*-oilseed rape intercropping system under different soil types and comprehensive benefits evaluation under field conditions. Environ Pollut 285
68. Sheoran V, Sheoran AS, Poonia P (2009) Phytomining: a review. Miner Eng 22(12):1007–1019
69. van der Ent A, Callahan DL, Noller BN, Mesjasz-Przybylowicz J, Przybylowicz WJ, Barnabas A et al (2017) Nickel biopathways in tropical nickel hyperaccumulating trees from Sabah (Malaysia). Sci Rep 7(41861)
70. Simonnot MO, Vaughan J, Laubie B (2018) Processing of bio-ore to products. In: Agromining: farming for metals: extracting unconventional resources using plants, pp 39–51
71. Zhang H (2014) Advances in phytoremediation technologies for heavy metal-contaminated soils. Environ Sci Pollut Res 21(10):6468–6480
72. Vaughan JG, McCarthy J (2016a) The role of plants in the bioremediation of heavy metals. Environ Sci Pollut Res 23(12):12234–12245
73. Parker JR, Baker AJM (2014) The role of hyperaccumulators in phytoremediation. Environ Sci Pollut Res 21(10):6468–6480
74. Riaza J, Khatami R, Levendis Y, Álvarez L, Gil M, Pevida C, Pís J et al (2014) Combustion of single biomass particles in air and in oxy-fuel conditions. Biomass Bioenerg 64:162–174. https://doi.org/10.1016/j.biombioe.2014.03.018
75. Dube S, Matsinha LC, Makhubela BC, Ambushe AA (2022) Investigating cyanogen rich Manihot esculenta efficacy for Ru phytomining and application in catalytic reactions. RSC Adv 12(2):1165–1176
76. Akinbile BJ, Matsinha LC, Ambushe AA, Makhubela BC (2021) Catalytic conversion of CO_2 to formate promoted by a biochar-supported nickel catalyst sourced from nickel phytoextraction using cyanogen-rich cassava. ACS Earth Space Chem 5(10):2846–2854
77. Chaney RL, Angle JS, Baker AJM, Li YM (1998) Method for phytomining of nickel, cobalt and other metals from soil. US Patent No. 5,711,784
78. Clemens S (2001) Molecular mechanisms of plant metal tolerance and homeostasis. Planta 212(4):475–486

79. Clemens S, Palmgren MG, Krämer U (2002) A long way ahead: understanding and engineering plant metal accumulation. Trends Plant Sci 7(7):309–315
80. Deng F, Yamaji N, Xia J, Ma JF, Chen Z (2013) A member of the heavy metal P-type ATPase OsHMA5 is involved in xylem loading of copper in rice. Plant Physiol 163(3):1353–1362
81. Harris AT, Bali R (2007) On the formation and extent of uptake of silver nanoparticles by live plants. J Nanopart Res 10(4):691–695. https://doi.org/10.1007/s11051-007-9229-2
82. Li YM, Chaney R, Brewer E, Roseberg R, Angle JS, Baker A et al (2003) Development of a technology for commercial phytoextraction of nickel: economic and technical considerations. Plant Soil 249(1):107–115
83. Mari S et al (2006) Root-to-shoot long-distance circulation of nicotianamine and nicotianamine–nickel chelates in the metal hyperaccumulator *Thlaspi caerulescens*. J Exp Bot 57(15):4111–4122
84. Nemutandani T, Dutertre D, Chimuka L, Cukrowska E, Tutu H (2006) The potential of *Berkheya coddii* for phytoextraction of nickel, platinum, and palladium contaminated sites. Toxicol Environ Chem 88(2):175–185. https://doi.org/10.1080/02772240600585842
85. Nkrumah PN, Tisserand R, Chaney RL, Baker AJM, Morel JL, Goudon R et al (2019) The first tropical ‘metal farm’: some perspectives from field and pot experiments. J Geochem Explor 198:114–122
86. Paul ALD, Nkrumah PN, Echevarria G, Erskine PD, Chaney RL, Spiers KM et al (2020) Cobalt hyperaccumulation in *Rinorea cf. bengalensis* (Violaceae) from Sabah: accumulation potential and tissue and cellular-level distribution of cobalt. Plant Soil 455(1):289–303
87. Shah K, Nongkynrih JM (2007) Metal hyperaccumulation and bioremediation. Biol Plant 51(4):618–634
88. van der Ent A, Ocenar A, Tisserand R, Sugau JB, Echevarria G, Erskine PD (2019) Herbarium X-ray fluorescence screening for nickel, cobalt and manganese hyperaccumulator plants in the flora of Sabah (Malaysia, Borneo Island). J Geochem Explor 202:49–58
89. van der Pas L, Ingle RA (2019) Towards an understanding of the molecular basis of nickel hyperaccumulation in plants. Plants 8(1):11
90. Akinbile CO, Yusoff MK (2022) Phytoremediation of heavy metals: a review of the mechanisms and applications. Environ Sci Pollut Res 29(1):1–15
91. Chaney RL, Malik M, Li YM, Brown SL, Brewer EP (1998) Phytoremediation of soil metals. Curr Opi Biotechnol 9(3):222–226
92. Clemens S, Palmgren MG, Krämer U (2002) A long way ahead: Understanding metal transport in plants. Nat Revi Mol Cell Biol 3(7):552–563
93. Deng Y, Wang Y, Zhang J (2013) Phytoremediation of heavy metal contaminated soil by using hyperaccumulator plants. Environ Sci Pollut Res 20(5):2901–2909
94. Harris JA, Bali K (2007) Phytoremediation: A sustainable approach to environmental cleanup. Environmental Science & Technology, 41(9), 3271-3276
95. Li, Y. M., Chaney, R. L., & Brown, S. L. (2003). Phytoremediation of contaminated soils: a review. Environ Pollut 122(3):305–314
96. Mari M, Mazzarino MJ, Rojas C (2006) Phytoremediation of heavy metals in contaminated soils: A review. J Environ Manage 78(1):1–10
97. Nemutandani S, Mulaudzi RB, Makhado E (2006) Phytoremediation of heavy metals: a review. African J Biotechnol 5(25):2421–2425
98. Nkrumah P, Osei EK, Osei M (2019) Phytoremediation potential of some indigenous plant species in heavy metal contaminated soils. Environ Monit Assess 191(3):1–12
99. Paul S, Saha S, Ghosh S (2020) Phytoremediation of heavy metals: a review of the current status and future prospects. Environ Sci Pollut Res 27(12):13967–13982
100. Shah AA, Nongkynrih B (2007) Phytoremediation: a green technology for the removal of heavy metals from contaminated soils. Indian J Environ Health 49(1):1–6
101. Van der Ent A, Baker AJM, Reeves RD (2019) Phytomining: A review of the potential for metal recovery from contaminated soils. Environ Sci Technol 53(12):6887–6896
130. Van der Pas LJ, Ingle RA (2019) Advances in phytoremediation: a review of the current state of research. Environ Sci Pollut Res 26(10):9275–9286

Recent Advances in Processed Water Hyacinth (Eichhornia Crassipes) for the Removal of Organic Contaminants from Water

Mohd Aizudin Abd Aziz, Muhammad Auni Hairunnaja, Nur Amira Fatihah Bashari, and Khairuddin Md. Isa

Abstract Water hyacinth (WH), a globally recognized invasive species, poses a substantial threat to aquatic biodiversity and ecosystems. Traditional control methods involving chemical herbicides contribute to secondary water pollution and generate organic waste. This study introduces a sustainable approach by repurposing waste from Processed Water Hyacinth (PWH) into adsorbents, showcasing its potential for water resource remediation. The critical global need for clean water underscores the importance of this research. Investigating various remediation approaches, the study emphasizes PWH's role in phytoremediation and the removal of diverse organic contaminants, with a specific focus on organic dyes. This underscores the versatile applications of PWH-based adsorbents. The study delves into the intricate mechanisms governing adsorption processes, providing depth to our understanding of PWH-based water remediation. The imperative for future research to harness PWH's potential in remediating water resources is underscored, offering a pathway for effective and environmentally responsible water resource management.

Keywords Processed water hyacinth · Phytoremediation · Organic contaminant removal

1 Introduction

Water pollution caused by the presence of organic contaminants is an increasingly worrisome issue within the fields of environmental science and public health. Organic contaminants, encompassing a wide range of pollutants like industrial

M. A. Abd Aziz (✉) · M. A. Hairunnaja · N. A. F. Bashari
Faculty of Chemical and Process Engineering Technology, Universiti Malaysia Pahang Al-Sultan Abdullah (UMPSA), 26300 Gambang, Kuantan, Pahang, Malaysia
e-mail: maizudin@umpsa.edu.my

K. Md. Isa
Faculty of Chemical Engineering Technology, Universiti Malaysia Perlis (UniMAP), 02600 Arau, Perlis, Kangar, Malaysia

A. Z. Yaser et al. (eds.), *Controlling Environmental Pollution*,
https://doi.org/10.1007/978-981-97-8931-3_15

chemicals, pesticides, pharmaceuticals, and personal care products, pose significant threats to both aquatic ecosystems and the quality of drinking water (Chandraseagar et al. 2019). These contaminants typically enter water bodies through industrial discharges, agricultural runoff, and domestic sewage, ultimately impacting the health of humans and the environment. Consequently, the development of effective methods for removing organic contaminants from water has become imperative in the quest for clean, safe, and sustainable water resources.

The rapid expansion and proliferation of Water Hyacinth (Eichhornia crassipes), formally referred to as WH, in nutrient-rich aquatic environments have raised global concerns. Based on published data, this free-floating perennial hydrophyte swiftly spreads across continents and the globe (Hashem et al. 2020). In Thailand, the growth of WH in several major water bodies surged to 5.6 million tons per month (Jirawattanasomkul et al. 2021). WH plants grow to dense volumes exceeding 60 kg m^{-2}, often severely obstructing waterways (Gaurav et al. 2020). While these plants can reach a maximum height of 1 m, a more common height is around 40 cm (Liu et al. 2018). Due to its substantial size, WH intercepts sunlight, thereby depriving aquatic life of vital resources, such as oxygen from the atmosphere. Furthermore, the presence of WH in natural ecosystems negatively impacts tourism industries in some countries, where visitor's frequent reservoirs and dams for recreational and leisure purposes (Ayanda et al. 2020; Abd Aziz et al. 2017).

Recognizing the ecological impact of water hyacinths and the demand for sustainable solutions, recent research has witnessed a paradigm shift. Beyond the conventional application of water hyacinth for organic contaminant removal, there is a pronounced focus on processed water hyacinth—a novel approach that holds promise for improved remediation practices. This transition marks a pivotal juncture in water hyacinth utilization, where the emphasis now lies on exploring innovative processing methods, such as the production of biochar, to enhance its adsorption capacities and overall effectiveness in removing organic contaminants from water. This paper delves into the nuances of this transformative shift, offering insights into the potential of processed water hyacinth, specifically biochar, for advancing the field of water remediation.

Water hyacinth's rapid nutrient uptake and vigorous growth make it an appealing choice for phytoremediation—the application of plants to eliminate contaminants from water or soil. Since the 1980s, WH has been employed to combat water pollution (Liu et al. 2020). WH biomass can remove both metals and organic substances from water (Madikizela 2021; Nyamunda et al. 2019). Recently, there has been increased interest in its potential for effectively eliminating organic dyes (Liu et al. 2020; Salahuddin et al. 2021; Sharma et al. 2021). Some researchers have favored the use of WH in its native form for adsorption studies, while others have preferred modified WH-based adsorbents (Chen et al. 2019; Sahoo et al. 2019; Aziz et al. 2019).

This review paper is dedicated to exploring recent advancements in the use of processed water hyacinth, particularly biochar as an efficient agent for removing organic contaminants from water. It is a timely and necessary endeavor to synthesize the latest research findings and methodologies, given the increasing demand for sustainable, cost-effective, and eco-friendly solutions to water contamination.

Numerous reports in the field critically evaluate the potential of utilizing biochar derived from WH for phytoremediation (Hashem et al. 2020; Nidheesh et al. 2021; Gwenzi et al. 2017; Japar et al. 2019). This research aims to critically assess the mechanisms that influence pollution removal, which are essential considerations in planning future research initiatives. The paper also delves into the characteristics of WH and biochar derived from its biomass.

Furthermore, the paper highlights the potential of processed WH, particularly biochar, in supporting sustainable waste management. As with previously published studies, the primary goal here is to offer a comprehensive assessment of the use of processed WH, specifically biochar, for remediating organic contaminants from water bodies. The literature covers strategies for removing organic contaminants, including adsorption technologies and plant-based adsorption from contaminated freshwater.

In summary, this review paper's primary objective is to present a comprehensive overview of the recent developments in utilizing processed water hyacinth, particularly biochar, for the removal of organic contaminants from water. It explores the ecological significance of water hyacinth, its diverse applications, and the challenges it poses. This information is critical for comprehending the current state of water treatment and guiding future research in this crucial field.

2 Water Hyacinth (Eichhornia Crassipes) Characteristics and Processed Forms

Water hyacinth (Eichhornia crassipes), often abbreviated as WH, has received substantial attention for its potential as a biomass in recent research investigations (Gaurav et al. 2020). This exceptional plant possesses distinct qualities that distinguish it from other aquatic species, rendering it a valuable resource in the field of water purification. WH predominantly comprises structural carbohydrates, including lignin, crystalline cellulose, and hemicellulose polymers (Zhang et al. 2020). These structural elements impart distinctive surface characteristics to WH, marked by essential functional groups like carboxyl, hydroxyl, and carbonyl. These functional groups serve as catalysts for the adsorption of waterborne contaminants onto plant-based adsorbents (Brown et al. 2020). Additionally, WH's roots are notably rich in functional groups, featuring entities such as PO4, C=O, and C–H (Milke et al. 2020).

The composition of WH fibers is another noteworthy aspect. It comprises a significant proportion of cellulose in the form of hemicellulose (33%), cellulose (25%), and lignin (10%) (Salahuddin et al. 2021). This composition has drawn the attention of scientists who have conducted research-based experiments utilizing cellulose derived from WH to eliminate water contaminants (Emam et al. 2020; Salahuddin et al. 2021; Ismail et al. 2021). The abundance of hydroxyl groups within WH's cellulose framework plays a pivotal role in the adsorption process, as these groups are chemically tailored to enhance adsorption (Singh and Chandra 2019).

WH is particularly renowned for its porous structure, prompting investigations into the carbonization of this material to produce mesoporous carbon, thereby amplifying its adsorption capabilities (Zhang et al. 2020). It's important to note that WH-based adsorbents possess a comparatively smaller surface area compared to synthetic counterparts designed for adsorption purposes. Research suggests that the surface areas of WH's roots range between 4.5 and 5.8 $m^2 g^{-1}$, while its petioles exhibit surface areas between 2.5 and 3.3 $m^2 g^{-1}$ (Madikizela 2021). Intriguingly, a significant proportion of WH-based adsorbents is sourced from its roots.

In specific instances, chemical treatment is applied to WH to expand its surface area, consequently enhancing the adsorption process. This approach is of critical significance as the adsorption capacity exhibits a direct correlation with the specific surface area (Gwenzi et al. 2017). Moreover, recent advancements have focused on the transformation of WH into biochar through the process of pyrolysis. Biochar, derived from water hyacinth, presents a carbon-rich and porous structure, further enhancing its adsorption capacities. This transformative step not only stabilizes WH but also significantly contributes to its effectiveness in sustainable water remediation practices (Awe et al. 2020; Talaiekhozani et al. 2021; Tkaczyk et al. 2020; Isa et al. 2017).

This biochar, endowed with enhanced adsorption capacities due to its increased surface area and porous nature, becomes a powerful tool for efficiently removing organic contaminants from water. The pyrolysis process not only stabilizes water hyacinth but also introduces a valuable, sustainable component to the realm of water remediation (Hashem et al. 2020). The resulting biochar, with its heightened adsorption capabilities, becomes a formidable agent in sustainable water treatment practices. In the context of water hyacinth's characteristics, it's imperative to note that this transformative approach is a pivotal juncture in maximizing the potential of water hyacinth for environmental benefit. The porous structure of biochar, coupled with its carbon-rich composition, makes it an excellent candidate for adsorbing various contaminants, including organic dyes and other pollutants present in water bodies (Sharma et al. 2021). The introduction of biochar derived from water hyacinth represents a significant stride toward more effective, eco-friendly, and sustainable water reclamation strategies. This transformative shift from raw water hyacinth to biochar is seamlessly integrated into the broader discussion of water hyacinth's characteristics. It represents not just a modification in form but a strategic enhancement that aligns with the growing need for innovative and sustainable solutions in the field of water remediation (Madikizela 2021). These characteristics underscore the compelling potential of water hyacinth, both in its raw and processed forms, as a natural and eco-friendly tool for water reclamation, particularly in the context of eliminating organic contaminants.

3 Processed Water Hyacinth (PWH) for Organic Contaminant and Dye Removal

Water hyacinth (Eichhornia crassipes), in its processed form known as Processed Water Hyacinth (PWH), has emerged as a promising solution for tackling organic contaminants and dyes in water systems. Unlike its raw counterpart, PWH undergoes a transformative process, often through pyrolysis, resulting in the production of biochar. This biochar is characterized by a carbon-rich and porous structure, making it an effective adsorbent for a wide range of organic pollutants (Bapat and Jaspal 2020; Cuerda-Correa et al. 2019; Mamane et al. 2020; Razali et al. 2018).

One of the key characteristics that distinguish PWH from its raw form is its enhanced adsorption capacity. The pyrolysis process leads to the creation of a highly porous material with an increased surface area, providing more active sites for the adsorption of organic contaminants (Gaurav et al. 2020). The carbon-rich nature of biochar plays a pivotal role in attracting and immobilizing organic molecules, contributing to the overall removal efficiency (Chen et al. 2019; Pan et al. 2021; Zhu et al. 2022).

PWH's effectiveness in organic contaminant removal is particularly notable when considering its application as an adsorbent. The porous structure of biochar allows for the physical trapping of contaminants, while the surface functional groups, such as hydroxyl and carboxyl, chemically interact with organic molecules, facilitating their adsorption. This dual mechanism enhances the overall performance of PWH as a natural and eco-friendly adsorbent for organic contaminants in water (Bapat and Jaspal 2020). In addition to its prowess in handling organic pollutants, PWH exhibits significant capabilities in the removal of synthetic dyes from water. The porous nature of biochar proves advantageous in adsorbing dye molecules, leading to the purification of water resources. The adsorption process is influenced by various factors, including the type of dye, initial concentration, and contact time, all of which contribute to the versatility of PWH in dye removal applications (Enaime et al. 2020; Samsami et al. 2020; Sonu et al. 2020; Razali et al. 2020).

As researchers continue to explore the potential of PWH, advancements in regeneration techniques are also being investigated. Strategies for regenerating PWH adsorbents aim to prolong their lifespan and enhance their reusability, contributing to sustainable water remediation practices. The utilization of PWH in organic contaminant and dye removal represents a significant step forward in harnessing the potential of water hyacinth in its processed form for environmentally friendly and effective water purification.

3.1 Characteristics of PWH for Organic Contaminant Removal

Water hyacinth, when transformed into its processed form known as Processed Water Hyacinth (PWH), exhibits distinctive characteristics that position it as an exceptional solution for the removal of organic contaminants from water systems. The transformative process, often involving pyrolysis, results in the creation of biochar, a carbon-rich and porous material. This chapter explores the key characteristics that distinguish PWH and highlights its effectiveness in tackling organic contaminants.

3.1.1 PWH as an Adsorbent

Water hyacinth, renowned for its carbon-based materials, emerges as a potent solution for combatting diverse organic contaminants in water bodies, spanning phenol, heavy metals, polycyclic aromatic hydrocarbons (PAHs), and challenging dyes (Talaiekhozani et al. 2021). Its adsorption prowess hinges on a symphony of factors, from the specific contaminant type to pH, contact time, and the inherent properties of the adsorbent (Feng et al. 2021; Pan et al. 2021; Azhar et al. 2021). The distinctive ability of water hyacinth lies in its high surface area, porous structure, and endowed functional groups like hydroxyl and carboxyl on its surface, fostering chemical interactions with organic pollutants and orchestrating their extraction from the water matrix (Zhu et al. 2022).

In the quest for an even more potent adsorbent, the narrative shifts to Processed Water Hyacinth (PWH), unveiling a transformative journey through the pyrolysis process (Sahoo et al. 2019). This metamorphosis begets PWH biochar, a dynamic material with an augmented surface area, establishing an arsenal of active sites for the adsorption ballet with a myriad of organic pollutants in water (Gwenzi et al. 2017). The carbon-rich essence of PWH biochar becomes a linchpin, entrapping and immobilizing organic molecules through a dual mechanism—an intricate dance within the porous structure and chemical liaisons with surface functional groups. This tandem act propels PWH into the limelight, a stalwart ally in the efficient removal of organic contaminants from water sources.

Delving into the specifics, PWH's efficacy as an adsorbent unravels in its ability to physically trap contaminants within its porous structure, while surface functional groups like hydroxyl and carboxyl engage in a chemical tango with organic molecules, ushering them into the biochar domain (Gaurav et al. 2020). The versatility of PWH extends beyond organic contaminant removal, finding success in adsorbing heavy metals like cadmium and lead (Gaurav et al. 2020). However, this performance is no solo act, as factors such as pH, initial contaminant concentration, temperature, and contact time take center stage in influencing the orchestrated adsorption process. The optimization of these parameters becomes the conductor's wand, orchestrating a harmonious removal efficiency (Mamane et al. 2020).

As the curtain falls on the remarkable adsorption tale of water hyacinth and its processed counterpart, the stage is set for a new chapter. This narrative stands as a testament to nature's ingenuity, where an aquatic plant and its processed avatar become pivotal players in the relentless pursuit of purifying our water sources. The saga continues, with researchers pushing the boundaries of knowledge, exploring the depths of regeneration techniques and innovative applications. Water hyacinth and its processed iterations stand not just as solutions but as symbols—symbols of sustainability, efficacy, and the unwavering commitment to a cleaner, healthier aquatic world.

3.1.2 Regeneration of PWH Adsorbents

Regeneration of Processed Water Hyacinth (PWH) adsorbents is a critical aspect in ensuring their sustained efficacy and promoting environmentally friendly water remediation practices. After successive adsorption cycles, the biochar derived from PWH may become saturated with adsorbed contaminants, necessitating regeneration to restore its adsorption capacity. The regeneration process involves several techniques designed to desorb and remove adsorbed contaminants from the PWH biochar, allowing it to be reused. One of the notable advantages of using water hyacinth-based adsorbents, including PWH, is their potential for regeneration and reusability (Cuerda-Correa et al. 2019). This aspect contributes to cost savings and reduced waste generation, aligning with the principles of sustainable water treatment.

Several methods employed for regenerating water hyacinth-based adsorbents can be considered for PWH as well. Thermal treatment, standing out as a prominent regeneration technique for PWH adsorbents, involves subjecting the biochar to elevated temperatures through processes like pyrolysis or calcination. This thermal regeneration process not only removes the adsorbed pollutants but also rejuvenates the porous structure of the biochar, enhancing its surface area and adsorption capabilities. However, careful control of temperature is crucial to avoid any damage to the biochar structure during regeneration (Pan et al. 2021).

Chemical regeneration, another common method, involves the use of desorbing agents such as acids, bases, and organic solvents to break the chemical bonds between the PWH biochar and the adsorbed contaminants. This approach offers the advantage of selectively desorbing specific contaminants, providing a tailored and effective regeneration strategy. Biological regeneration, harnessing the capabilities of microorganisms, represents an environmentally friendly method for breaking down organic pollutants on the PWH biochar. While considered sustainable, biological regeneration may require longer durations compared to thermal or chemical methods (Zhu et al. 2022; Ismail et al. 2019).

Successful regeneration not only extends the lifespan of PWH adsorbents but also contributes to the overall sustainability of water treatment practices. Researchers are continually exploring innovative regeneration techniques to optimize the performance and reusability of PWH biochar for organic contaminant removal. As we delve into the intricacies of regeneration, it becomes evident that the choice of regeneration method depends on factors such as the type of adsorbent, the nature of contaminants, and the desired degree of regeneration. The development of efficient and sustainable regeneration strategies plays a pivotal role in maximizing the benefits of Processed Water Hyacinth (PWH) adsorbents for long-term and eco-friendly water remediation applications (Tkaczyk et al. 2020).

3.2 Dye Removal Using PWH

Water hyacinth (Eichhornia crassipes) has garnered attention as a versatile solution for addressing water contamination challenges, particularly in the removal of organic contaminants and dyes (Cuerda-Correa et al. 2019). The versatility of this aquatic plant extends to its effectiveness in handling a range of dyes, including those that pose challenges to traditional treatment methods. In the realm of dye removal, PWH, derived from water hyacinth through a transformative process like pyrolysis, emerges as a promising adsorbent. Dyes, especially synthetic and non-biodegradable ones, pose significant challenges in wastewater treatment. Research has shown that water hyacinth adsorbents can effectively remove dyes, including cationic and anionic dyes, from wastewater. The adsorption capacity depends on factors such as pH, initial dye concentration, and contact time. Moreover, the regeneration of water hyacinth adsorbents for dye removal has been studied as a sustainable approach, promoting their reusability (Wang et al. 2021).

Water hyacinth's ability to remove dyes can be attributed to its high surface area and porous structure, allowing for the physical adsorption of dye molecules onto the plant material. Electrostatic interactions between the charged dye molecules and the surface of water hyacinth also contribute to the adsorption process (Cuerda-Correa et al. 2019). The pH of the solution can significantly affect dye removal efficiency, with different dyes showing varying adsorption behavior in response to pH changes. This pH dependency underscores the need to optimize pH conditions for specific dye removal applications.

In the realm of dye removal, water hyacinth exhibits versatility across different dye classes, showcasing effectiveness in adsorbing various types of dyes. One notable class is Azo dyes, which are characterized by the presence of the azo group (–N=N–) as their chromophore. An example of an Azo dye is Methyl Orange, and water hyacinth has demonstrated remarkable adsorption capacities for such dyes, emphasizing its potential as an eco-friendly alternative for removing synthetic dyes from industrial effluents (Mlunguza et al. 2020a, b; Shokry et al. 2020). Additionally, water hyacinth has shown effectiveness in adsorbing Anthraquinone dyes, featuring an anthraquinone nucleus as their chromophore. Alizarin, an Anthraquinone dye, is one such example. Triphenylmethane dyes, characterized by a central carbon atom bonded to three aromatic rings, include dyes like Crystal Violet, and water hyacinth has proven to be efficient in adsorbing Triphenylmethane dyes, contributing to its versatility in dye removal applications.

Furthermore, water hyacinth has demonstrated adsorption capacities for Indigo dyes, which have an indole or indoline structure. Lastly, in the category of Phthalocyanine dyes, known for their large planar, aromatic macrocyclic structure, water hyacinth has shown effectiveness in adsorbing dyes such as Copper Phthalocyanine, commonly used in blue and green pigments (Chen et al. 2019; Pan et al. 2021; Zhu et al. 2022). This diverse range of adsorption capabilities highlights water hyacinth's potential as an environmentally friendly solution for the removal of various synthetic dyes across different chemical classes from industrial effluents (Hashem et al. 2020). Moreover, the regeneration of water hyacinth adsorbents through desorption processes reinforces its status as a sustainable choice for continuous dye removal applications. In conclusion, water hyacinth presents a promising solution for the removal of dyes from wastewater, addressing the environmental challenges associated with dye pollution. Its adsorption capabilities, when optimized for specific dye types and conditions, make it a cost-effective, sustainable, and eco-friendly alternative in the field of wastewater treatment. Additionally, ongoing research into the development of efficient regeneration techniques for water hyacinth adsorbents ensures their long-term applicability in dye removal processes (Zhu et al. 2022; Ismail et al. 2019).

4 Recent Advances in Processed Water Hyacinth-Based Contaminant Removal

4.1 Insights into Future Research and Reusability of PWH Adsorbents

In recent years, the utilization of Processed Water Hyacinth (PWH) as an effective adsorbent has opened new avenues for innovative research in the field of water remediation. The unique characteristics of PWH, derived through pyrolysis, have spurred interest in exploring its untapped potential and addressing emerging challenges in contaminant removal from water sources (Hashem et al. 2020; Zubair et al. 2021; Enaime et al. 2020). As we look to the future, there is a growing emphasis on advancing research to deepen our understanding of PWH's adsorption mechanisms, optimize its performance, and expand its applicability to a broader spectrum of contaminants.

One promising aspect of future research lies in the sustainable reusability of PWH adsorbents. The capacity to regenerate and reuse PWH biochar for contaminant removal stands as a cornerstone for environmentally conscious water treatment practices. Ongoing studies are delving into innovative regeneration techniques that not only restore the adsorption capacity of PWH but also contribute to minimizing waste generation (Zubair et al. 2021; Lima et al. 2020). This focus on reusability aligns with the broader goal of developing eco-friendly and cost-effective solutions for long-term water treatment strategies. The reusability of PWH adsorbents holds significant implications for both economic and environmental sustainability. By reducing the need for frequent replacement of adsorbent materials, PWH contributes to cost savings in water treatment processes (Lima et al. 2020). Simultaneously, it aligns with the principles of the circular economy and waste reduction, promoting a holistic and eco-conscious approach to water remediation. Researchers are actively exploring diverse regeneration methods, including thermal treatment, chemical regeneration, and biological processes, to unlock the full potential of PWH adsorbents in sustainable water treatment applications.

In the pursuit of advancing water hyacinth (WH)-based contaminant removal, researchers have delved into the reusability of WH adsorbents following regeneration, offering a novel avenue for sustained water remediation (Li et al. 2021; Lima et al. 2020). This breakthrough implies that even during seasons when WH is scarce or unavailable, aqueous resources can still be effectively remediated. Furthermore, encouraging research suggests that WH can be repurposed for plant uptake, as demonstrated by the uptake of formaldehyde by plants, showcasing reusability for at least four cycles (Talaiekhozani et al. 2021; Yaashikaa et al. 2020).

However, additional studies are warranted to explore the full potential of WH adsorbents, as the range of substances WH roots can absorb, their uptake rates, and maximal concentration remain largely uncharted territory (Mlunguza et al. 2020a, b; Shokry et al. 2020). Consequently, further research, especially using hydroponic systems, is essential to comprehensively investigate the spectrum of organics that WH roots can effectively absorb. At a larger scale, the feasibility of incorporating constructed wetland systems, harnessing the capability of living WH for organics uptake, should be a focal point for exploration as a supplementary purification approach in water treatment plants. This dual focus on the reusability of PWH biochar and the exploration of living water hyacinth (WH) for organics uptake underscores the dynamic nature of research in water treatment. By combining the regenerative capabilities of PWH adsorbents with the potential of living WH systems, researchers are poised to unlock comprehensive and sustainable solutions for contaminant removal, ensuring the long-term resilience and effectiveness of water remediation practices.

4.2 Repurposing PWH as Value-Added Products

In the realm of water treatment innovation, the repurposing of Processed Water Hyacinth (PWH) extends beyond its role as a contaminant adsorbent, offering exciting prospects for the creation of value-added products. The pyrolysis process that transforms water hyacinth into biochar provides a foundation for exploring diverse applications that go beyond the traditional scope of water remediation.

One avenue for value addition involves leveraging PWH biochar as a sustainable soil amendment. The rich carbon content and porous structure of biochar enhance soil fertility and water retention, contributing to improved agricultural productivity. Studies have shown that incorporating PWH biochar into soil not only sequesters carbon but also promotes nutrient availability, ultimately enhancing crop yields (Hammam et al. 2022; Masto et al. 2013). This dual benefit aligns with the principles of circular economy and sustainable agriculture, presenting a viable solution for addressing both water pollution and soil degradation.

In the pursuit of repurposing PWH, it is crucial to consider the principles of sustainable and circular design. By transforming a water pollutant into a resource for agriculture, construction, and industry, PWH becomes a symbol of eco-conscious innovation. The development of value-added products not only addresses water pollution challenges but also contributes to the broader goals of resource efficiency and environmental sustainability. As research in this field progresses, the potential applications of PWH as a versatile, value-generating resource continue to expand, heralding a new era in the sustainable utilization of water hyacinth beyond traditional water treatment methodologies.

4.3 Expanding PWH's Removal Capabilities

Recent advancements in the application of Processed Water Hyacinth (PWH) have propelled its removal capabilities to new heights, offering promising solutions to contemporary water pollution challenges. One notable breakthrough involves the development of tailored PWH-based adsorbents with enhanced selectivity for specific contaminants. Researchers have fine-tuned the pyrolysis process to create biochar with surface modifications that optimize its affinity for target pollutants, ranging from heavy metals to complex organic compounds (Sahoo et al. 2019; Feng et al. 2021; Enaime et al. 2020). This customization of PWH biochar opens avenues for precision water treatment, where the adsorbent can be designed to selectively target and remove contaminants of concern.

Moreover, advancements in nanotechnology have led to the integration of PWH-derived nanomaterials in water treatment applications. The unique properties of nanoscale PWH materials, such as increased surface area and reactivity, enhance their adsorption capabilities (Bapat and Jaspal 2020; Nidheesh et al. 2021). This nanotechnological approach allows for the removal of contaminants at the molecular level, demonstrating the versatility of PWH in addressing even the most challenging pollutants. The integration of PWH nanomaterials into filtration and purification systems showcases the adaptability of this eco-friendly solution in modern water treatment technologies.

In addition to adsorption, recent studies have explored PWH's potential in advanced oxidation processes (AOPs) for contaminant degradation. By coupling PWH with AOPs like photocatalysis or electrochemical treatment, researchers have achieved synergistic effects that enhance the overall removal efficiency (Cuerda-Correa et al. 2019; Nidheesh et al. 2021). This integration of PWH into advanced treatment strategies represents a paradigm shift, providing a multifaceted approach to water remediation that goes beyond conventional adsorption methods.

As we witness these advancements, it becomes evident that PWH's removal capabilities are continuously evolving, driven by innovative research and interdisciplinary collaborations. The integration of nanotechnology, advanced oxidation processes, and smart technologies into PWH-based solutions marks a transformative phase in water treatment. These developments not only expand the scope of contaminants that can be effectively addressed but also pave the way for the integration of PWH into next-generation water treatment technologies that are efficient, sustainable, and tailored to meet the evolving challenges of water pollution.

5 Conclusions

In the realm of water remediation, Processed Water Hyacinth (PWH) emerges as a promising and versatile tool, showcasing a myriad of attributes that contribute to its effectiveness in removing contaminants from water sources. The evolution of water hyacinth from a notorious invasive species into a sustainable solution highlights the transformative power of innovative approaches in environmental science. The fundamental success of PWH in contaminant removal lies in its unique characteristics derived from the pyrolysis process. The resulting biochar exhibits a highly porous structure and an increased surface area, providing numerous active sites for the physical and chemical adsorption of a wide range of contaminants. The dual mechanism of physical trapping within the porous structure and chemical interactions with surface functional groups, such as hydroxyl and carboxyl, enhances the overall removal efficiency, setting PWH apart as a natural and eco-friendly adsorbent. The versatility of PWH extends beyond its adsorption capabilities, encompassing the critical aspects of regeneration and reusability. Pioneering studies explore various regeneration techniques, including thermal treatment, chemical regeneration, and biological processes, emphasizing the potential for sustained use and minimizing waste generation. The focus on reusability aligns with the principles of circular economy and eco-conscious water treatment practices, contributing to both economic and environmental sustainability. As we peer into the future, the landscape of PWH research unfolds with exciting possibilities. Insights into the adsorption mechanisms, isotherm models, and the complexity of interactions with organic pollutants lay the groundwork for continuous exploration and optimization. The reusability of PWH adsorbents opens doors to novel strategies, such as advanced regeneration processes and hybrid adsorption systems, propelling the field toward innovative and efficient water treatment solutions. Recent advances in PWH-based contaminant removal amplify its significance in addressing contemporary water pollution challenges. Tailored PWH-based adsorbents, nanomaterials, and integration into advanced oxidation processes showcase the adaptability and precision of this eco-friendly solution. The intersection of PWH with nanotechnology, artificial intelligence, and machine learning marks a transformative phase in water treatment, ensuring not only enhanced efficiency but also the seamless integration of sustainable practices. In conclusion, Processed

Water Hyacinth stands at the forefront of sustainable and effective water remediation strategies. From its inherent ability to remove a diverse array of contaminants to its pivotal role in advancing research frontiers, PWH has proven to be a beacon of innovation in environmental science. As we navigate the complexities of water pollution, the promise of PWH shines bright, offering a holistic and eco-conscious approach to safeguarding our precious water resources.

References

Abd Aziz MA, Md Isa K, Rashid RA (2017) Pneumatic jigging: influence of operating parameters on separation efficiency of solid waste materials. Waste Manag Res 35:647–655

Awe A, Opeolu B, Fatoki O, Ayanda O, Jackson V, Snyman R (2020) Preparation and characterisation of activated carbon from Vitisvinifera leaf litter and its adsorption performance for aqueous phenanthrene. Appl Biol Chem 63

Ayanda OI, Ajayi T, Asuwaju FP (2020) Eichhornia crassipes (Mart.) solms: uses, challenges, threats, and prospects. Sci World J 2020:3452172

Azhar MA, Rahman NWA, Aziz MAA, Isa KM (2021) Identification of chemical compounds from agarwood hydrosol (Aquilaria malaccensis) fruits via LC-QTOF-MS/MS analysis. IOP Conf Ser Earth Environ Sci

Aziz MAA, Isa KM, Miles NJ, Rashid RA (2019) Pneumatic jig: effect of airflow, time and pulse rates on solid particle separation. Int J Environ Sci Technol 16:11–22

Bapat S, Jaspal D (2020) Surface-modified water hyacinth (Eichhornia crassipes) over activated carbon for wastewater treatment: a comparative account. South African J Chem Suid-Afrikaanse Tydskrif Vir Chemie 73:70–80

Brown A, Adams JMM, Grasham O, Camargo-Valero M, Ross A (2020) An assessment of different integration strategies of hydrothermal carbonisation and anaerobic digestion of water hyacinth. Energies 13

Chandraseagar S, Abdulrazik AH, Abdulrahman SN, Abdaziz MA (2019) Aspen Plus simulation and optimization of industrial spent caustic wastewater treatment by wet oxidation method. IOP Conf Ser Mater Sci Eng 702:012011

Chen S, Qin C, Wang T, Chen F, Li X, Hou H, Zhou M (2019) Study on the adsorption of dyestuffs with different properties by sludge-rice husk biochar: adsorption capacity, isotherm, kinetic, thermodynamics and mechanism. J Mol Liq 285:62–74

Cuerda-Correa EM, Alexandre-Franco M, Fernández G (2019) Advanced oxidation processes for the removal of antibiotics from water. An overview. Water 12:102

Emam AA, Abo Faraha SA, Kamal FH, Gamal AM, Basseem M (2020) Modification and characterization of nano cellulose crystalline from Eichhornia crassipes using citric acid: an adsorption study. Carbohyd Polym 240:116202

Enaime G, Baçaoui A, Yaacoubi A, Lübken M (2020) Biochar for wastewater treatment—conversion technologies and applications. Appl Sci [Online] 10

Feng Q, Wang B, Chen M, Wu P, Lee X, Xing Y (2021) Invasive plants as potential sustainable feedstocks for biochar production and multiple applications: a review. Resour Conserv Recycl 164:105204

Gaurav GK, Mehmood T, Cheng L, Klemeš JJ, Shrivastava DK (2020) Water hyacinth as a biomass: a review. J Clean Prod 277:122214

Gwenzi W, Chaukura N, Noubactep C, Mukome FND (2017) Biochar-based water treatment systems as a potential low-cost and sustainable technology for clean water provision. J Environ Manag 197:732–749

Hammam AA, Mohamed ES, El-Namas AE, Abd-Elmabod SK, Badr Eldin RM (2022) Impacted application of water-hyacinth-derived biochar and organic manures on soil properties and barley growth. Sustainability 14:13096

Hashem MA, Hasan M, Momen MA, Payel S, Nur-a-Tomal MS (2020) Water hyacinth biochar for trivalent chromium adsorption from tannery wastewater. Environ Sustain Indic 5:100022

Isa KM, Kasim FH, Saad SA, Rahim MAA, Aziz MAA, Ali UFM (2017) Influence of operating parameters on biomass conversion under sub- and supercritical water conditions. Chem Eng Technol 40:537–545

Ismail NA, Abd Aziz MA, Hisyam A, Abidin MA (2021) Separation of samarium from medium rare earth mixture using multi-stage counter-current extraction. Chem Eng Commun 208:764–774

Ismail NA, Aziz MAA, Yunus MYM, Hisyam A (2019) Selection of extractant in rare earth solvent extraction system: a review. Int J Recent Technol Eng 8:728–743

Japar ANS, Aziz MAA, Razali MN (2019) Formulation of fumed silica grease from waste transformer oil as base oil. Egypt J Pet 28:91–96

Jirawattanasomkul T, Minakawa H, Likitlersuang S, Ueda T, Dai J-G, Wuttiwannasak N, Kongwang N (2021) Use of water hyacinth waste to produce fibre-reinforced polymer composites for concrete confinement: mechanical performance and environmental assessment. J Clean Prod 292:126041

Li F, He X, Srishti A, Song S, Tan HTW, Sweeney DJ, Ghosh S, Wang C-H (2021) Water hyacinth for energy and environmental applications: a review. Biores Technol 327:124809

Lima JRA, de Farias DL, Menezes THS, Oliveira RVM, Silva IAA, da Costa Cunha G, Romão LPC (2020) Potential of a magnetic hybrid material produced using water hyacinth (Eichhornia crassipes) for removal of inorganic and organic pollutants from aqueous media. J Environ Chem Eng 8:104100

Liu C, Ye J, Lin Y, Wu J, Price GW, Burton D, Wang Y (2020) Removal of Cadmium (II) using water hyacinth (Eichhornia crassipes) biochar alginate beads in aqueous solutions. Environ Pollut 264:114785

Liu L, Hu S, Shen G, Farooq U, Zhang W, Lin S, Lin K (2018) Adsorption dynamics and mechanism of aqueous sulfachloropyridazine and analogues using the root powder of recyclable long-root Eichhornia crassipes. Chemosphere 196:409–417

Madikizela LM (2021) Removal of organic pollutants in water using water hyacinth (Eichhornia crassipes). J Environ Manag 295:113153

Mamane H, Altshuler S, Sterenzon E, Vadivel VK (2020) Decolorization of dyes from textile wastewater using biochar: a review. Acta Innov 36–46

Masto RE, Kumar S, Rout TK, Sarkar P, George J, Ram LC (2013) Biochar from water hyacinth (Eichornia crassipes) and its impact on soil biological activity. CATENA 111:64–71

Milke J, Gałczyńska M, Wróbel J (2020) The importance of biological and ecological properties of phragmites Australis (Cav.) Trin. Ex Steud., in phytoremendiation of aquatic ecosystems—the review. Water [Online] 12

Mlunguza NY, Ncube S, Mahlambi PN, Chimuka L, Madikizela LM (2020a) Determination of selected antiretroviral drugs in wastewater, surface water and aquatic plants using hollow fibre liquid phase microextraction and liquid chromatography—tandem mass spectrometry. J Hazard Mater 382:121067

Mlunguza NY, Ncube S, Mahlambi PN, Chimuka L, Madikizela LM (2020b) Optimization and application of hollow fiber liquid-phase microextraction and microwave-assisted extraction for the analysis of non-steroidal anti-inflammatory drugs in aqueous and plant samples. Environ Monit Assess 192:557

Nidheesh PV, Gopinath A, Ranjith N, Praveen Akre A, Sreedharan V, Suresh Kumar M (2021) Potential role of biochar in advanced oxidation processes: a sustainable approach. Chem Eng J 405:126582

Nyamunda B, Chivhanga T, Guyo U, Chigondo F (2019) Removal of Zn (II) and Cu (II) ions from industrial wastewaters using magnetic biochar derived from water hyacinth. J Eng 2019:1–11

Pan S-Y, Dong C-D, Su J-F, Wang P-Y, Chen C-W, Chang J-S, Kim H, Huang C-P, Hung C-M (2021) The role of biochar in regulating the carbon, phosphorus, and nitrogen cycles exemplified by soil systems. Sustainability 13:5612

Razali MN, Aziz MAA, Jamin NFM, Salehan NAM (2018) Modification of bitumen using polyacrylic wig waste. In: AIP conference proceedings

Razali MN, Mohd Isa SNE, Md Salehan NA, Musa M, Abd Aziz MA, Nour AH, Yunus RM (2020) Formulation of emulsified modification bitumen from industrial wastes. Indones J Chem 20:96–104

Sahoo D, Awasthi A, Dhyani V, Biswas B, Kumar J, Reddy YS, Adarsh VP, Puthiyamadam A, Mallapureddy KK, Sukumaran RK, Ummalyma SB, Bhaskar T (2019) Value-addition of water hyacinth and para grass through pyrolysis and hydrothermal liquefaction. Carbon Resour Convers 2:233–241

Salahuddin N, Abdelwahab M, Akelah A, Elnagar M (2021) Adsorption of Congo red and crystal violet dyes onto cellulose extracted from Egyptian water hyacinth. Nat Hazards 105:1–20

Samsami S, Mohamadizaniani M, Sarrafzadeh M-H, Rene ER, Firoozbahr M (2020) Recent advances in the treatment of dye-containing wastewater from textile industries: overview and perspectives. Process Saf Environ Prot 143:138–163

Sharma R, Saini H, Paul DR, Chaudhary S, Nehra SP (2021) Removal of organic dyes from wastewater using Eichhornia crassipes: a potential phytoremediation option. Environ Sci Pollut Res 28:7116–7122

Shokry H, Elkady M, Salama E (2020) Eco-friendly magnetic activated carbon nano-hybrid for facile oil spills separation. Sci Rep 10:10265

Singh AK, Chandra R (2019) Pollutants released from the pulp paper industry: aquatic toxicity and their health hazards. Aquat Toxicol 211:202–216

Sonu K, Sogani M, Syed Z, Dongre A, Sharma G (2020) Enhanced decolorization and treatment of textile dye wastewater through adsorption on acid modified corncob derived biochar. Chem Select 5

Talaiekhozani A, Rezania S, Kim K-H, Sanaye R, Amani AM (2021) Recent advances in photocatalytic removal of organic and inorganic pollutants in air. J Clean Prod 278:123895

Tkaczyk A, Mitrowska K, Posyniak A (2020) Synthetic organic dyes as contaminants of the aquatic environment and their implications for ecosystems: a review. Sci Total Environ 717:137222

Wang Q, Jiang F, Ouyang XK, Yang LY, Wang Y (2021) Adsorption of Pb(II) from aqueous solution by Mussel shell-based adsorbent: preparation, characterization, and adsorption performance. Materials (Basel) 14

Yaashikaa PR, Kumar PS, Varjani S, Saravanan A (2020) A critical review on the biochar production techniques, characterization, stability and applications for circular bioeconomy. Biotechnol Rep 28:e00570

Zhang C, Ma X, Chen X, Tian Y, Zhou Y, Lu X, Huang T (2020) Conversion of water hyacinth to value-added fuel via hydrothermal carbonization. Energy 197:117193

Zhu C-Y, Ye Y-W, Guo X, Cheng F (2022) Design and synthesis of carbon-based nanomaterials for electrochemical energy storage. New Carbon Mater 37:59–92

Zubair M, Ihsanullah I, Abdul Aziz H, Azmier Ahmad M, Al-Harthi MA (2021) Sustainable wastewater treatment by biochar/layered double hydroxide composites: progress, challenges, and outlook. Biores Technol 319:124128

Advancements in Constructed Wetland Technology for Enhanced Domestic Pollutants Removal: A Review

Danar Arifka Rahman

Abstract Constructed wetlands (CWs) have emerged as innovative and efficient systems to address domestic wastewater pollutants. This review discusses the different types of CWs—subsurface flow, free water surface, and hybrid systems—and their efficacy in removing organic matter, nutrients (phosphorus and nitrogen), pathogens, and emerging contaminants. CWs integrate physical, chemical, and biological mechanisms to offer a holistic and sustainable approach to wastewater treatment. Its adaptability and scalability make it highly valuable, particularly in regions without centralized treatment infrastructure. Although challenges such as land use and climate sensitivity remain, the benefits of CWs, such as reduced energy requirements, biodiversity support, and cost savings, are substantial. Incoming advances in CWs design and operation promise to further improve its efficiency and expand its use, having an important role in the reduction of global water pollution, particularly domestic pollution, and environmental sustainability. This review highlights the potential of CWs technology as a promising solution for domestic wastewater treatment.

Keywords Constructed wetlands · Domestic pollution · Environmental sustainability · Wastewater

1 Introduction

Environmental pollution has been a global problem and continues to increase with the rapid development of industry and human activities. The continuous degradation of natural ecosystems due to pollutants poses significant risks to human health and biodiversity (Gunes et al. 2021). Water pollution, particularly from domestic sources,

D. A. Rahman (✉)
Department of Biology, Faculty of Science and Technology, Study Program of Environmental Engineering, Universitas Airlangga, Kampus C Mulyorejo, Surabaya 60115, Indonesia
e-mail: danar.arifka@fst.unair.ac.id

Environmental Technology and Innovation Research Group, Universitas Airlangga, Kampus C Mulyorejo, Surabaya 60115, Indonesia

A. Z. Yaser et al. (eds.), *Controlling Environmental Pollution*,
https://doi.org/10.1007/978-981-97-8931-3_16

is one of the most critical challenges faced by many countries (Falahi et al. 2022; Werkneh 2024). As urbanization and population growth, the volume of domestic wastewater increases, requiring effective treatment methods to mitigate its impact on the environment (Lamaming et al. 2022). Innovative and practical solutions are essential to addressing the growing problem, protecting water resources and sustainable environment as one of the goals of the SDGs.

Domestic pollutants primarily originate from household activities and include a wide range of contaminants such as organic matter, nutrients (nitrogen and phosphorus), pathogens, and various chemicals including emerging pollutants (pharmaceuticals, personal care products, and household cleaning agents) (de Rozari et al. 2021; Mesquita et al. 2017; Nivala et al. 2019). According to recent studies, domestic wastewater contributes significantly to the overall pollution load in aquatic environments (Feng et al. 2020; Wu et al. 2023), with approximately 80% of wastewater worldwide being released untreated into water bodies (Mara 2003). The presence of these pollutants in water bodies can lead to eutrophication (Wen et al. 2016), the spread of waterborne diseases such as diarrhea and cholera, and disruption of aquatic ecosystems (Ibrahim et al. 2021; Priya et al. 2021; Tong et al. 2022).

Several methods are employed for the treatment of domestic wastewater, including physical, chemical, and biological processes (Rahman et al. 2022a, b). Conventional treatment plants often use a combination of these methods to achieve desired effluent quality. Physical processes such as sedimentation and filtration are effective in removing suspended solids but are limited in addressing dissolved pollutants (Lohani et al. 2020). Chemical treatments, including coagulation and chlorination, can efficiently remove contaminants but may result in the production of harmful by-products. Biological treatments, such as activated sludge systems and biofilm reactors, are widely used due to their cost-effectiveness and ability to degrade organic matter. However, these systems require significant energy input and operational expertise (Crini and Lichtfouse 2019).

Constructed wetlands (CWs) have emerged as a promising alternative for the treatment of domestic wastewater. CWs are engineered systems that mimic the functions of natural wetlands by utilizing vegetation, soil, and microbial processes to remove pollutants. These systems offer a sustainable and low-cost solution for wastewater treatment, particularly in rural and peri-urban areas where access to centralized treatment facilities may be limited (Vymazal 2010). Constructed wetlands are highly effective in removing a wide range of domestic pollutants, including organic matter, nutrients, pathogens, and emerging contaminants. The integration of physical, chemical, and biological processes within CWs facilitates the comprehensive treatment of wastewater (Biswas & Chakraborty 2023; Dan et al. 2023; Rodriguez-Dominguez et al. 2022). The adaptability and scalability of CWs make them an essential component of sustainable water management strategies.

This review aims to provide a comprehensive overview of the application of CWs to enhance the removal of domestic pollutants, parameters, and pollutant removal mechanisms. Several studies have reported the application of CWs to treat greywater (Santos et al. 2024), while this work addresses the utilization of CWs to treat domestic wastewater in general, which includes both greywater and blackwater. In many developing countries, these types of wastewater are still often mixed in sewers (Sangwijit et al. 2022). The performance of CWs will also be discussed comprehensively in terms of its effectiveness in degrading pollutants frequently found in domestic wastewater. Furthermore, this article will discuss the performance evaluation of CWs-based treatment in treating domestic wastewater based on recent research as well as the challenges and limitations in the implementation and operation of CWs. By addressing these objectives, this article seeks to highlight the potential of constructed wetlands as a viable and sustainable solution for domestic wastewater treatment and to contribute to the ongoing efforts in mitigating water pollution.

2 Types of Constructed Wetlands

The classification of constructed wetlands (CWs) is technically determined by various criteria, but mainly based on water flow pattern. The well-known types of CWs often used to treat domestic wastewater are subsurface flow CWs (horizontal and vertical subsurface flow), free water surface flow CWs (FWS), and floating treatment wetlands (FTWs) (Parde et al. 2021; Sandoval et al. 2019; Vymazal 2010). The types of CWs used to treat domestic wastewater are shown in Fig. 1.

2.1 Subsurface Flow Constructed Wetland

Subsurface flow constructed wetlands (SSF-CWs) are designed to allow water to flow beneath the surface of a porous medium, typically gravel or sand, planted with wetland vegetation. SSF-CWs can be further divided into horizontal flow constructed wetland (HF-CW) and vertical flow constructed wetland (VF-CW) systems. HF-CW systems involve horizontal water movement through the medium, whereas VF-CW systems involve vertical water movement, typically facilitated by intermittent dosing (Gorgoglione and Torretta 2018).

HF-CW allows wastewater to flow horizontally through the substrate from an inlet to an outlet. This design provides a continuous saturated environment where anaerobic and facultative bacteria can thrive, facilitating the breakdown of organic pollutants. HF-CW systems are highly effective in removing organic matter, nitrogen, and pathogens (Mesquita et al. 2017). The systems are widely used for domestic and industrial wastewater treatment due to their ability to prevent surface exposure and thus reduce the risk of mosquito breeding and odor problems. However, the efficiency of HF-CW can be limited by the reduced oxygen availability, which can hinder the removal of certain pollutants like ammonia and organic matter under high loading conditions (Shukla et al. 2021).

On the other hand, VF-CW systems are designed such that wastewater is evenly distributed over the surface and flows vertically through the substrate layer to the

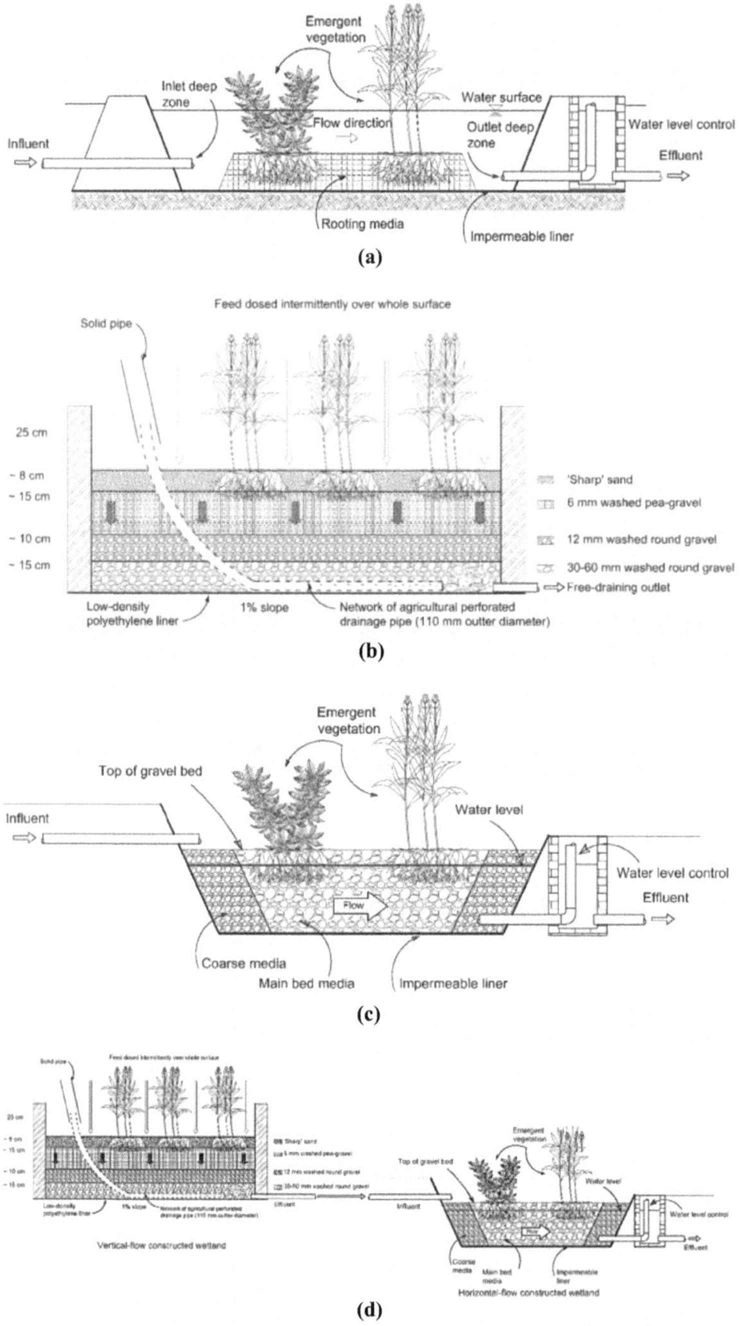

Fig. 1 Different types of CWs configuration: **a** Free water surface wetland, **b** Vertical subsurface flow wetland system, **c** Horizontal subsurface flow wetland system, and **d** Hybrid wetland system (Almuktar et al. 2018)

sewer system. The substrate usually consists of gravel or sand, which supports vegetation and provides a medium for microbial communities that degrade contaminants. VF-CW is also known for its efficient oxygen transfer, which enhances the removal of ammonia and other nitrogen compounds (David et al. 2023). Other studies have also reported high removal efficiencies for biochemical oxygen demand (BOD) and total suspended solids (TSS) due to aerobic conditions and microbial activity within the wetland (Kumar and Singh 2017; Martín et al. 2013).

2.2 Free Water Surface Constructed Wetland

Free water surface constructed wetlands (FWS-CWs) are designed to simulate natural marshes, with water flowing over the surface of the soil substrate. These systems support a diverse range of emergent vegetation and provide habitats for wildlife (Mander et al. 2014). They are particularly suitable for applications where aesthetic and ecological benefits are desired in addition to water treatment, such as in parks and nature reserves (Kumar et al. 2023). Included in free water surface wetland is floating treatment wetlands (FTWs), where plants grow on floating mats or rafts on the water surface. The roots of the plants extend into the water column, where they provide a surface for microbial biofilm development and direct uptake of nutrients (Parde et al. 2021; Shahid et al. 2018). FTWs are commonly used for the treatment of nutrient-rich waters, such as agricultural runoff and eutrophic water bodies. The main advantages of FTWs include their ability to adapt to varying water levels and their effectiveness in removing nutrients, particularly nitrogen and phosphorus, through plant uptake and microbial processes (Kumar et al. 2023).

2.3 Hybrid Constructed Wetland

Hybrid constructed wetlands combine different types of CWs to optimize treatment performance. Hybrid wetlands can leverage the advantages of each system type by integrating two or more CWs systems, such as FWS, FTWs, HF-CWs, and VF-CWs. These integrated systems are particularly effective in achieving higher pollutant removal efficiencies, including the reduction of nutrients, heavy metals, and organic contaminants. In addition, CWs can also present aerobic and anaerobic processes in a unified system (Fernandez-Fernandez et al. 2020; Vega De Lille et al. 2021; Wu et al. 2017). Hybrid CWs are often employed in settings requiring high-quality effluent, such as in sensitive ecological areas or for water reuse applications (Domscheit 2023).

3 Pollutant Removal Mechanisms

The tremendous progress in the research of artificial wetlands to treat wastewater makes the application of CWs more effective and massive. That is because of numerous advantages including cost-effectiveness, lower energy requirements, and the ability to treat a variety of pollutants (Rahman et al. 2020). This section provides a discussion of the mechanisms involved in pollutant removal in CWs, focusing on physical, chemical, and biological processes (Fig. 2).

3.1 Physical Processes

Physical processes in CWs primarily involve the removal of contaminants through mechanical means, such as the settling of suspended particles (sedimentation) and the trapping of particulate matter (filtration) in wastewater. The pollutants that are mostly removed in the physical process are TSS, organic matter, heavy metals, and nutrients, including phosphorus and nitrogen (Hassan et al. 2021; Tang et al. 2021). These processes are essential for the primary treatment of wastewater, reducing the load of solids and associated pollutants before biological and chemical treatments take over for further purification.

Sedimentation is the process by which suspended particles in the wastewater settle to the bottom of the wetland due to gravity. This process is influenced by factors such as particle size, density, and the flow rate of water through the wetland. Total suspended solids (TSS), which often include organic matter, heavy metals, and pathogens, are largely removed after the sedimentation process. In the meantime, the filtration process also removes pollutants similar to sedimentation, except that the

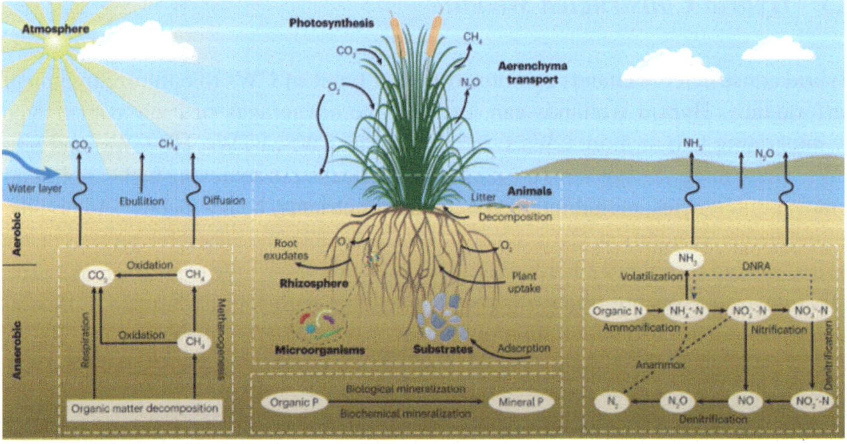

Fig. 2 Pollutant removal in constructed wetland (Wu et al. 2023)

processes are different. Filtration traps fine suspended particles and microorganisms, while chemical processes such as absorption also occur (Malyan et al. 2021; Shukla et al. 2021).

3.2 Chemical Processes

Chemical processes in CWs involve the transformation or removal of contaminants through chemical reactions. These processes include adsorption, where pollutants adhere to the surface of a substrate, in this case, it could be the media in CWs (Salama et al. 2019). In addition, there is also the process of precipitation, where soluble pollutants are converted into insoluble forms and then removed from the water. These mechanisms are integral to improving CWs treatment performance, especially for the removal of heavy metals, nutrients (nitrogen and phosphorus), and organic contaminants (Ma et al. 2017). Ge et al. (2019) reported the long-term efficiency of natural pyrite and enriched biochar in enhancing simultaneous nitrogen and phosphorus removal in CWs, highlighting the adaptability and sustainability of these systems in treating different types of wastewater.

3.3 Biological Processes

Biological processes in CWs work together with physical and chemical processes in degrading pollutants. The operation is mainly driven by microorganisms and plants that include microbial degradation, phytoremediation, and anaerobic processes. Biological processes play an important role in nutrients and organic pollutants removal (Crini and Lichtfouse 2019; Malyan et al. 2021). Microbial degradation involves bacteria, fungi, and protozoa decomposing organic pollutants into simpler compounds like carbon dioxide, water, and biomass, with bacteria in the rhizosphere of wetland plants playing a crucial role in breaking down complex organic compounds (Zhao et al. 2020). Phytoremediation, on the other hand, involves plants such as Phragmites australis and Typha latifolia absorbing contaminants through their roots, which not only removes pollutants but also supports microbial communities that further degrade these substances (Cheng et al. 2021). The roots of these plants enhance the bioavailability of pollutants to microorganisms, increasing the wetland system's overall efficiency (Zhao et al. 2019). Additionally, anaerobic digestion, facilitated by anaerobic bacteria in oxygen-free conditions, decomposes organic matter into methane and carbon dioxide, reducing the biochemical oxygen demand (BOD) in wastewater and effectively degrading pharmaceuticals and personal care products (Chandel et al. 2022; Chen et al. 2019).

Several studies have investigated that biological processes have a significant effect on pollutant removal. Lizama-Allende et al. (2021) explored the use of specific substrates and bioaugmentation with plant growth-promoting rhizobacteria can further improve pollutant removal efficiency, showcasing the potential for optimization in CWs design and operation. Furthermore, a research by Lawan et al. (2023) has highlighted the effectiveness of horizontal subsurface flow constructed wetlands in removing heavy oil and recalcitrant organic compounds from petrochemical wastewater, achieving removal efficiencies of up to 90%.

4 Constructed Wetlands Performance

The utilization of CWs has undergone significant development globally, used in both developed and developing countries. Originally designed for domestic and municipal wastewater treatment, CWs have now been proven effective in treating a diverse range of wastewater types, including industrial wastewater, mine drainage, landfill leachate, and contaminated water bodies such as rivers and lakes. Moreover, CWs have demonstrated the ability to remove heavy metals and micropollutants from wastewater, making it a versatile tool in various environmental management activities (Rodriguez-Dominguez et al. 2022; Sossalla et al. 2021).

Studies highlight the adaptability of CWs to diverse environmental conditions, including different seasons and temperature ranges. This adaptability makes CWs particularly suitable for tropical and subtropical regions, as shown in Table 1., where they can function effectively throughout the year. Reports have shown that CWs can maintain high performance in water contaminant reduction despite varying hydraulic loading levels and seasonal changes (Cui et al. 2022; Zhang et al. 2024a, b, c). In addition, recent reviews emphasize the ecological and environmental benefits of CWs. They not only provide effective wastewater treatment but also contribute to biodiversity conservation and habitat creation. The integration of CWs into urban and industrial wastewater management systems is increasingly seen as a sustainable and cost-effective approach to improve water quality and reduce water pollution (Parde et al. 2021; Wu et al. 2023). This section will discuss the performance of CWs in removing pollutants according to the reported works as shown in Table 1.

4.1 Organic Matter Removal

CWs effectively remove organic matter through various mechanisms, mainly through biological processes, which are microbial degradation and plant uptake (Aqdas and Hashmi 2023; Cheng et al. 2021). Microbial communities in the wetlands play a crucial role in degrading organic pollutants. These microorganisms, including bacteria and fungi, use the organic matter as a food source, decomposing it into simpler substances. This process is enhanced by the specific environmental conditions within the wetland, such as anaerobic and aerobic zones, which support different microbial populations and degradation pathways (Kang et al. 2023). Furthermore, various substrates, plant roots, and organic materials provide a habitat and food

Table 1 Performances of CWs for domestic pollutant removal

Site	Setup of CWs	Media	Constructed Wetland design and configuration				Initial concentration (mg/L) and removal efficiency (%)										Reference
			Plant	HRT	FR	L × W × D	BOD	COD	TSS	TDS	TP	Phosphate	TN	N-NH_4^+	Nitrate	Coliforms	
Sakon Nakhon, Thailand	FWS	Water	*H. verticillate*	28 d	–	39 cm × 41 cm	30 84	110 63	– –	– –	– –	0.23 87	– –	– –	– –	– –	Sangwijit et al. (2022)
Côte d'Ivoire, West Africa	VF-CW	Gravel + shale Gravel + laterate	*P. purpureum*	–	6.43 L/d	(0.4 × 0.4 × 0.6) m	288.6 84.7 78	481.3 83 76.9	– – –	– – –	11.8 72.4 58.4	– – –	143.6 72.2 55.5	– – –	– – –	– – –	Fatim Traoré et al. (2024)
Aligarh, India	HF-CW	Coarse gravel + fine gravel	*P. karka* *I. kashmiriana* *S. latifolia*	–	0.12 m^3/d	(1.8 × 0.6 × 0.9) m	– – – –	105.7 84.39 74.83 76.16	40.2 79.6 55.22 59.7	– – – –	16.6 70.48 60.24 62.05	– – – –	31.9 71.47 56.43 62.38	– – – –	– – – –	– – – –	Qadiri et al. (2021)
Carimate, Italy	Clarifier + VF-CW + FWS	Coarse gravel + gravel + pea gravel	*P. australis* + *S. lacustris* + *M. trifollata* + *T. minima* + *N alba* + *L. salicaria* + *Nuphar iutea* + *I. pseudacorus* + *C. riparia* + *T. latifolia* + *P. amphibia*	–	900 m^3/h	8500 m^2 + 4500 m^2	– –	294 83.2	– –	– –	4.7 64.6	– –	27.4 –	18.5 63.8	– –	– –	Masi et al. (2023)
Jarabacoa, Dominican Republic	HF-CW	Gravel	*V. zizanioides*	6.6 d	7.5 m^3/d	70 m^2	309.09 95	277.78 95	118 88	– –	– –	15.38 46	15.87 55	– –	– –	1204.82 MPN/ 100 mL 25	Pérez et al. (2024)

(continued)

Table 1 (continued)

Site	Setup of CWs	Media	Constructed Wetland design and configuration				Initial concentration (mg/L) and removal efficiency (%)										Reference
			Plant	HRT	FR	L × W × D	BOD	COD	TSS	TDS	TP	Phosphate	TN	$N\text{-}NH_4^+$	Nitrate	Coliforms	
Spain	HUSB + aerated HF-CW	Gravel	*P. australis*	2.17 d	1 m^3/d	(4 × 1.4 × 0.9) m	– –	207 97	– –	– –	– –	– –	19.5 44.7	14.1 80	– –	– –	Pascual et al. (2024)
Egypt	FWS	Water	*E. crassipes* *P. communis* *L. gibba*	2 d	11.25 m^3/d	(3 × 15 × 0.5) m	27 59.5 67.9 61.9	– – – –	35 53.7 63.1 50.7	– – – –	– – – –	2.3 26.6 24.2 21.8	– – – –	3 16.9 16.3 13.7	– – – –	– – – –	El Hawary & Shaban, (2018)
Kuwait	VF-CW	Gravel + random packing + soil	*I. cylindrical*	10.4 h	1.67 m^3/m^2.d	(60 × 60 × 80) cm	80.92 59.1	131.62 59.1	– –	– –	– –	18.93 24.7	48.87 55	- –	27.99 76.6	– –	Khajah & Ahmed, (2023)
Santa Isabel, Ecuador	VF-CW	Coarse, medium, fine gravel + fine, coarse sand	*P. australis* *C. papyrus*	1.12 d	0.6 m^3/d	(3 × 1 × 0.65) m	95.75 75.39 80.69	222.44 64.68 69.87	88.83 62.85 32.46	– – –	7.42 49.38 50	– – –	– – –	29.52 70.7 69.69	0.84 −90.29 −88.67	5.4×10^{10} MPN/100 mL 96.02 98.08	García-Ávila et al. (2019)
Jashore, Bangladesh	VF-CW	Gravel + coco fiber + medium gravel + coarse sand	*S dichotomus*	1 d	0.012 m^3/d	(30 × 30 × 40) cm	21.4 61.6	98 53.98	– –	545.67 21.22	– –	– –	– –	– –	3.2 70.25	– –	M.-A. Rahman et al. (2022a, b)

(continued)

Table 1 (continued)

Site	Setup of CWs	Media	Constructed Wetland design and configuration				Initial concentration (mg/L) and removal efficiency (%)										Reference
			Plant	HRT	FR	L × W × D	BOD	COD	TSS	TDS	TP	Phosphate	TN	N-NH_4^+	Nitrate	Coliforms	
Ghana	HF-CW	Waste clay brick Palm kernel shells	*T. geniculata*	–	80 L/d	(35 × 105 × 40) cm	158.8 70.56 39.03	267 71.89 11.92	39.2 90.13 53.79	– – –	21.67 65.18 37.37	– – –	25.6 39.96 29.63	– – –	– – –	1.33 × 10^8 CFU/100 mL 89.49 98.32	Obeng et al. (2023)
Sarnadas Rodão, Portugal	HF-CW	Gravel	*P. australis*	–	73 m^3/d	1120 m^2	– –	– –	– –	– –	– 23	– –	– 6.9	– 5.8	– –	– –	Mesquita et al. (2017)
China, Indoor (20–32 °C)	VF-CW	Gravel + gravel & straw biochar + sand	*Iris pseudacorus L*	3 d		16 cm × 40 cm	– –	50 80	– –	– –	0.75 70	– –	15 90	10 85	– –	– –	Wang et al. (2022)
Benin	FWS in facultative & maturation pond (3 stages)	Water	*Eichhornia crassipes* *Pistia stratiotes L* *Lemna minor L*	8 d	100 L/d	(1.1 × 0.9 × 0.8) m	140 93.57	228.8 92.31	700 82.86	– –	– –	8.2 79.59	96 82.71	– –	– –	– –	Akowanou et al. (2023)
Arusha, Tanzania	HF-CW	Gravel	*C. alternifolius* *C. indica* Hybrid	7 d	–	(1.5 × 0.5 × 1) m	86.67 71 70.1 72.7	156.29 79.5 77.7 77.9	– – – –	– – – –	– – – –	3.84 65.4 60.7 69.5	– – – –	– – – –	– – – –	– – – –	Karungamye et al. (2023)

(continued)

Table 1 (continued)

Site	Setup of CWs	Media	Constructed Wetland design and configuration				Initial concentration (mg/L) and removal efficiency (%)										Reference
			Plant	HRT	FR	L × W × D	BOD	COD	TSS	TDS	TP	Phosphate	TN	N-NH_4^+	Nitrate	Coliforms	
Kupang, Indonesia	VF-CW	Gravel + sand and pumice	*C. citratus*	1 d	2 L/d	10.2 cm × 55 cm	– –	499 83	1722 97.3	– –	– –	3.18 88	– –	– –	2.93 86	1813 CFU/100 mL 97	de Rozari et al. (2021)
Mytilene, Greece	VF-CW	Coarse gravel + fine gravel + vermiculite	*T. jasminoides* + *L. japonica* + *C. laevis*	–	12 L/d	(0.65 × 0.47 × 0.6) m	107 99.62	358 97.31	43 91.92	– –	5.3 45.28	– –	16.5 39.39	<2 –	– –	– –	Stefanatou et al. (2023)
Egypt	HF-CW	Gravel Shredded tire rubber chips Corrugated pieces of plastic pipes	*P. australis*	–	6.61 m^3/d	(10 × 2 × 0.65) m	232 43.45 38.69 40.4	344 43.04 38.25 49	150 48.39 45.16 60.65	– – – –	– – – –	– – – –	– – – –	– – – –	– – – –	– – – –	Zidan et al. (2015)

source for microbes, thus enhancing their activity and efficiency in biodegrading organic pollutants (Arliyani et al. 2021; Yang et al. 2021). Most studies show highly satisfactory removal of organic matter, up to 99.62% for BOD and 97.31 for COD using VF-CW system with multi-vegetation (*T. jasminoides, L. japonica, and C. laevis*) (Stefanatou et al. 2023). Meanwhile, there are some studies that showed low removal efficiency. This could be due to the release of organic compounds by the natural substrate. Another reason is due to the descent of waste residues, such as oil, to the outlet (Obeng et al. 2023).

4.2 TSS and TDS Removal

TSS removal primarily occurs through physical mechanisms such as sedimentation and filtration. Sedimentation acts to settle heavier particles by gravity, while filtration utilizes substrates and plant roots to trap particles, besides, aggregation and coagulation processes promote the formation of larger flocs, enhancing removal efficiency (Wu et al. 2015). For TDS removal, CWs utilize adsorption, where dissolved ions attach to soil surfaces and plant roots, where vegetation absorbs nutrients and other dissolved substances. Microbial activity also provides a crucial role in degrading organic dissolved solids through biodegradation and denitrification (Gorito et al. 2017; Zhang et al. 2024a, b, c). Subsurface constructed wetlands have relatively high efficiency in TSS removal, both horizontal and vertical flow, compared to free water surface wetlands (FWS). However, several studies reported that VF-CW is superior in removing TSS, even reaching >90% efficiency (de Rozari et al. 2021; Stefanatou et al. 2023). This can be seen from the design of the reactor and the flow in the system, where the vertical system is more optimal in the sedimentation and filtration process. We have discussed the effect of reactor design and flow in previous research (Rahman et al. 2022a, b). TDS removal is linearly related to TSS, but, due to its dissolved properties, the physical processes don't have a significant role in its removal, but chemical processes such as adsorption and precipitation do (Carstensen et al. 2019).

4.3 Phosphorus Removal

Phosphorus is one of the challenges in CWs systems because in large quantities it can cause eutrophication and various water quality problems in CWs systems (Shukla et al. 2021). Phosphorus removal in several studies showed adequate results in subsurface wetlands. According to El Hawary and Shaban (2018), an initial phosphorus concentration of 2.3 mg/L can be removed up to 26% with an FWS system. Khajah and Ahmed (2023), who used VF-CW also showed a similar efficiency, from an initial concentration of 18.93 mg/L, the removal was 24%. The low removal efficiency in VF-CW is due to the high flow rate given. Because at a lower flow rate, which is

1.04 m^3/m^2.d, the removal efficiency can reach 43%. Higher efficiency on SSF-CW is due to the plant's root zone penetrates deep into the substrate, thus allowing for more extensive root-surface interactions. Meanwhile, plant uptake in FWS is limited to the surface and near-surface water layers (Vymazal 2011).

4.4 Nitrogen Removal

The mechanisms of nitrogen removal in CWs are complex and multifaceted, including processes such as ammonification, nitrification, denitrification, plant uptake, and microbial assimilation (Tazkiaturrizki et al. 2018; Wu et al. 2023). Advancements in CWs design and materials, such as the incorporation of specific substrates that enhance microbial activity, have been explored to improve nitrogen removal rates. The different types of plants used also affect nitrogen removal efficiency. Qadiri et al. (2021) reported that HF-CW with *Phragmites karka* has nitrogen removal efficiency of 71.47%, followed by HF-CW setup with *Sagittaria latifolia* and *Iris kashmiriana* at 62.38% and 56.43%, respectively. However, the study conducted by García-Ávila et al. (2019) showed negative results for nitrate removal for both plants used, C. papyrus and P. australis, at −88.67% and −90.29%, respectively. This result indicates that there is a nitrification process in the system, but a lack of denitrification conditions due to well aerated VF-CW design in the study (Abdelhakeem et al. 2016; Stefanakis and Tsihrintzis 2012). Other study also reported negative result of nitrogen removal (Tsihrintzis 2017).

4.5 Pathogenic Microbe Removal

Pathogen removal in CWs occurs through various mechanisms, specifically physical (sedimentation and filtration), chemical (adsorption), and biological (microbial degradation) processes (Khalifa et al. 2020). Sedimentation and filtration are primary processes where larger pathogens settle and trap between substrates. Pathogens attach to the surface of soil particles and organic matter through adsorption, which is influenced by the physicochemical properties of the pathogen and the substrate (Cydzik-Kwiatkowska and Zielińska 2016; Cyprowski et al. 2018). Further, biofilm formation on plant roots and substrate surfaces is another crucial mechanism, which creates a complex microenvironment where pathogens are trapped and degraded (Guan et al. 2015). Several studies have shown high removal efficiency of pathogenic microbial of around 89–98% (de Rozari et al. 2021; García-Ávila et al. 2019; Obeng et al. 2023). Moreover, there are four main factors that can increase pathogen removal, which are (1) finer substrate material, (2) warmer water temperature, (3) prolonged hydraulic retention time (HRT), and (4) shallower bottom depth (Kadlec and Wallace 2008).

4.6 *Micropollutants Removal*

Micropollutants, including pharmaceuticals, personal care products, hormones, pesticides, and industrial chemicals, are often found in household wastewater. These substances are commonly used in households and then discharged into the sewer through activities such as bathing, washing, cleaning, and toilet flushing. CWs are effective in removing various micropollutants through the synergy of physical, chemical, and biological processes. Adsorption on substrates and biofilms is the main mechanism, while microbial degradation breaks down complex organic compounds into simpler and less harmful forms. The roots of wetland plants can also facilitate the uptake and metabolism of micropollutants (Chatterjee and Shah 2023). Sossalla et al. (2021) reported that CWs could degrade micropollutants, such as caffeine, ibuprofen, naproxen, benzotriazole, diclofenac, and acesulfame, by 94 –> 96%. This shows the remarkably high effectiveness of CWs in degrading micropollutants.

5 Constructed Wetlands Sustainability

The environmental sustainability of CWs is underlined by its low-energy requirements and its ability to support biodiversity. CWs is a passive system that does not require the extensive mechanical aeration and energy inputs typical of conventional wastewater treatment plants. Integration of CWs with innovative materials can further enhance its performance in pollutant removal (Zhang et al. 2024a, b, c), making it a strong option for sustainable water management. CWs are globally utilized, including in countries with high population density, including China and Indonesia. China's CWs are extensively developed across large areas with detailed national guidelines (Zhang et al. 2024a, b, c), while Indonesia's CWs are implemented in specific projects and small urban areas, especially in coastal and tourism areas (Perdana et al. 2020), but China's CWs face significant challenges for technical adaptations due to cold climate temperatures. However, although reports have shown that CWs have great performance and are even technically possible to treat both greywater and blackwater in a single constructed wetland (Sangwijit et al. 2022), in the context of domestic wastewater, the best practice often involves using separate systems (Abdel-Shafy et al. 2017; Collivignarelli et al. 2020). This approach allows for customized treatment processes that address the specific characteristics of each wastewater type more effectively, ensuring higher efficiency and better quality of the treated effluent.

From a socio-economic perspective, CWs offer a cost-effective alternative to traditional wastewater treatment methods (Sharma et al. 2024). The construction and operational costs of CWs are generally lower, making them accessible to communities with limited financial resources. CWs can also be integrated into urban landscapes, providing green spaces that enhance urban resilience and sustainability. Despite their benefits, the implementation of CWs faces challenges, mainly related to land requirements and climate adaptability. Effective CWs require a considerable amount of land, which can be a limiting factor in densely populated areas, and there is also the challenge of plant saturation over time. The method for disposing of plants in constructed wetlands is to cut the plants at the base, so that the remaining root system can continue to provide some ecological benefits until it decomposes naturally. The harvested biomass can then be used in several ways, such as composting to produce organic fertilizer, incineration to produce energy, or, if contaminated with pollutants, safe disposal in landfills designed to handle hazardous waste. In addition, the performance of CWs can be affected by climatic conditions, requiring careful design and management to ensure year-round efficiency and optimize its function.

6 Conclusion

Advancements in constructed wetlands (CWs) technology have shown significant potential in enhancing the removal of domestic pollutants from wastewater. The integration of physical, chemical, and biological processes in CWs provides a comprehensive treatment solution that is sustainable and cost-effective. Various types of CWs, including subsurface flow, free water surface, and hybrid systems, have demonstrated high efficiency in removing organic matter, nutrients, pathogens, and emerging contaminants. The adaptability and scalability of CWs make it an important component in modern water management strategies, especially in areas that lack centralized treatment facilities. Saturated plants in constructed wetlands can be disposed by cutting the plant's base and utilizing the biomass in several ways, such as composting to produce organic fertilizer, burning to produce energy, or, if contaminated with pollutants, safe disposal in landfills designed to handle hazardous waste. Despite the challenges associated with land requirements and climatic conditions, the benefits of CWs in supporting biodiversity, reducing energy consumption, and improving water quality underscore its importance as a viable solution for domestic wastewater treatment. Continued research and innovation in CWs design and operation will further improve their performance and expand their application, contributing to global efforts in reducing water pollution and promoting environmental sustainability.

References

Abdel-Shafy HI, El-Khateeb MA, Shehata M (2017) Blackwater treatment via combination of sedimentation tank and hybrid wetlands for unrestricted reuse in Egypt. Desalin Water Treat 71(April):145–151. https://doi.org/10.5004/dwt.2017.20538

Abdelhakeem SG, Aboulroos SA, Kamel MM (2016) Performance of a vertical subsurface flow constructed wetland under different operational conditions. J Adv Res 7(5):803–814. https://doi.org/10.1016/J.JARE.2015.12.002

Akowanou AVO, Deguenon HEJ, Balogoun KC, Daouda MMA, Aina MP (2023) The combined effect of three floating macrophytes in domestic wastewater treatment. Sci Afr 20:e01630. https://doi.org/10.1016/j.sciaf.2023.e01630

Almuktar SAAAN, Abed SN, Scholz M (2018) Wetlands for wastewater treatment and subsequent recycling of treated effluent: a review. Environ Sci Pollut Res 25(24):23595–23623. https://doi.org/10.1007/s11356-018-2629-3

Aqdas A, Hashmi I (2023) Role of water hyacinth (Eichhornia crassipes) in integrated constructed wetlands: a review on its phytoremediation potential. Int J Environ Sci Technol 20(2):2259–2266. https://doi.org/10.1007/s13762-022-04181-0

Arliyani I, Tangahu BV, Mangkoedihardjo S (2021) Selection of plants for constructed wetlands based on climate and area in the interest of processing pollutant parameters on leachate: a review. IOP Conf Ser Earth Environ Sci 835(1). https://doi.org/10.1088/1755-1315/835/1/012003

Biswas A, Chakraborty S (2023) Organics and coliform removal from low strength domestic wastewater using integrated constructed wetland–microbial fuel cell reactor with bioelectricity generation. J Clean Prod 408:137204. https://doi.org/10.1016/J.JCLEPRO.2023.137204

Carstensen MV, Larsen SE, Kjærgaard C, Hoffmann CC (2019) Reducing adverse side effects by seasonally lowering nitrate removal in subsurface flow constructed wetlands. J Environ Manag 240:190–197. https://doi.org/10.1016/J.JENVMAN.2019.03.081

Chandel N, Ahuja V, Gurav R, Kumar V, Tyagi VK, Pugazhendhi A, Kumar G, Kumar D, Yang YH, Bhatia SK (2022) Progress in microalgal mediated bioremediation systems for the removal of antibiotics and pharmaceuticals from wastewater. Sci Total Environ 825:153895. https://doi.org/10.1016/J.SCITOTENV.2022.153895

Chatterjee A, Shah MP (2023) Constructed wetland for metals: removal mechanisms and analytical challenges BT-recent trends in constructed wetlands for industrial wastewater treatment. In: Shah MP (ed.). Springer Nature Singapore, pp 31–41. https://doi.org/10.1007/978-981-99-2564-3_2

Chen P, Blaney L, Cagnetta G, Huang J, Wang B, Wang Y, Deng S, Yu G (2019) Degradation of ofloxacin by perylene diimide supramolecular nanofiber sunlight-driven photocatalysis. Environ Sci Technol 53(3):1564–1575. https://doi.org/10.1021/acs.est.8b05827

Cheng YX, Chen J, Wu D, Liu YS, Yang YQ, He LX, Ye P, Zhao JL, Liu SS, Yang B, Ying GG (2021) Highly enhanced biodegradation of pharmaceutical and personal care products in a novel tidal flow constructed wetland with baffle and plants. Water Res 193:116870. https://doi.org/10.1016/J.WATRES.2021.116870

Collivignarelli MC, Carnevale Miino M, Gomez FH, Torretta V, Rada EC, Sorlini S (2020) Horizontal flow constructed wetland for greywater treatment and reuse: an experimental case. Int J Environ Res Public Health 17(7). https://doi.org/10.3390/ijerph17072317

Crini G, Lichtfouse E (2019) Advantages and disadvantages of techniques used for wastewater treatment. Environ Chem Lett 17(1):145–155. https://doi.org/10.1007/s10311-018-0785-9

Cui X, Wang J, Wang X, Khan MB, Lu M, Khan KY, Song Y, He Z, Yang X, Yan B, Chen G (2022) Biochar from constructed wetland biomass waste: a review of its potential and challenges. Chemosphere 287:132259. https://doi.org/10.1016/J.CHEMOSPHERE.2021.132259

Cydzik-Kwiatkowska A, Zielińska M (2016) Bacterial communities in full-scale wastewater treatment systems. World J Microbiol Biotechnol 32(4):66. https://doi.org/10.1007/s11274-016-2012-9

Cyprowski M, Stobnicka-Kupiec A, Ławniczek-Wałczyk A, Bakal-Kijek A, Gołofit-Szymczak M, Górny RL (2018) Anaerobic bacteria in wastewater treatment plant. Int Arch Occup Environ Health 91(5):571–579. https://doi.org/10.1007/s00420-018-1307-6

Dan A, Deng Y-Y, Guo Q-M, Jiang Y, Chen C-X (2023) A three-year study on the treatment of domestic-industrial mixed wastewater using a full-scale hybrid constructed wetland. Environ Sci Pollut Res 30(11):31256–31267. https://doi.org/10.1007/s11356-022-23936-3

David G, Rana MS, Saxena S, Sharma S, Pant D, Prajapati SK (2023) A review on design, operation, and maintenance of constructed wetlands for removal of nutrients and emerging contaminants. Int J Environ Sci Technol 20(8):9249–9270. https://doi.org/10.1007/s13762-022-04442-y

de Rozari P, Krisnayanti DS, Refli YKV, Atie MRR (2021) The use of pumice amended with sand media for domestic wastewater treatment in vertical flow constructed wetlands planted with lemongrass (Cymbopogon citratus). Heliyon, 7(7):e07423. https://doi.org/10.1016/j.heliyon.2021.e07423

Domscheit E (2023) Constructed wetlands and resource protection BT-wastewater management and technologies. In: Debik E, Bahadir M, Haarstrick A (eds.). Springer Nature Switzerland, pp 3–11. https://doi.org/10.1007/978-3-031-36298-9_1

El Hawary A, Shaban M (2018) Improving drainage water quality: constructed wetlands-performance assessment using multivariate and cost analysis. Water Sci 32(2):301–317. https://doi.org/10.1016/j.wsj.2018.07.001

Falahi OAAL, Abdullah SRS, Hasan HA, Othman AR, Ewadh HM, Kurniawan SB, Imron MF (2022) Occurrence of pharmaceuticals and personal care products in domestic wastewater, available treatment technologies, and potential treatment using constructed wetland: a review. Process Saf Environ Prot 168:1067–1088. https://doi.org/10.1016/J.PSEP.2022.10.082

Fatim Traoré N, Pétémanagnan Ouattara J-M, Michaël Zahui F, Cyrille Beda AJ, Messou A (2024) Assessment of secondary metabolites in Pennisetum purpureum planted into constructed wetlands using shale and laterite as substrate for wastewater treatment. Heliyon 10(13):e33284. https://doi.org/10.1016/j.heliyon.2024.e33284

Feng L, Liu Y, Zhang J, Li C, Wu H (2020) Dynamic variation in nitrogen removal of constructed wetlands modified by biochar for treating secondary livestock effluent under varying oxygen supplying conditions. J Environ Manag 260:110152. https://doi.org/10.1016/J.JENVMAN.2020.110152

Fernandez-Fernandez MI, de la Vega PTM, Jaramillo-Morán MA, Garrido M (2020) Hybrid constructed wetland to improve organic matter and nutrient removal. Water 12(7). https://doi.org/10.3390/w12072023

García-Ávila F, Patiño-Chávez J, Zhinín-Chimbo F, Donoso-Moscoso S, Flores del Pino L, Avilés-Añazco A (2019). Performance of Phragmites Australis and Cyperus Papyrus in the treatment of municipal wastewater by vertical flow subsurface constructed wetlands. Int Soil Water Conserv Res 7(3):286–296. https://doi.org/10.1016/j.iswcr.2019.04.001

Ge Z, Wei D, Zhang J, Hu J, Liu Z, Li R (2019) Natural pyrite to enhance simultaneous long-term nitrogen and phosphorus removal in constructed wetland: three years of pilot study. Water Res 148:153–161. https://doi.org/10.1016/J.WATRES.2018.10.037

Gorgoglione A, Torretta V (2018) Sustainable management and successful application of constructed wetlands: a critical review. Sustainability 10(11). https://doi.org/10.3390/su10113910

Gorito AM, Ribeiro AR, Almeida CMR, Silva AMT (2017) A review on the application of constructed wetlands for the removal of priority substances and contaminants of emerging concern listed in recently launched EU legislation. Environ Pollut 227:428–443. https://doi.org/10.1016/J.ENVPOL.2017.04.060

Guan W, Yin M, He T, Xie S (2015) Influence of substrate type on microbial community structure in vertical-flow constructed wetland treating polluted river water. Environ Sci Pollut Res 22(20):16202–16209. https://doi.org/10.1007/s11356-015-5160-9

Gunes K, Masi F, Ayaz S, Tuncsiper B, Besiktas M (2021) Domestic wastewater and surface runoff treatment implementations by constructed wetlands for Turkey: 25 years of experience. Ecol Eng 170:106369. https://doi.org/10.1016/J.ECOLENG.2021.106369

Hassan I, Chowdhury SR, Prihartato PK, Razzak SA (2021) Wastewater treatment using constructed wetland: current trends and future potential. Processes 9(11). https://doi.org/10.3390/pr9111917

Ibrahim Y, Ouda M, Kadadou D, Banat F, Naddeo V, Alsafar H, Yousef AF, Barceló D, Hasan SW (2021) Detection and removal of waterborne enteric viruses from wastewater: a comprehensive review. J Environ Chem Eng 9(4):105613. https://doi.org/10.1016/J.JECE.2021.105613

Kadlec RH, Wallace S (2008) Treatment Wetlands (2nd Editio). CRC Press. https://doi.org/10.1201/9781420012514

Kang Y, Ma H, Jing Z, Zhu C, Li Y, Wu H, Dai P, Guo Z, Zhang J (2023) Enhanced benzofluoranthrene removal in constructed wetlands with iron- modified biochar: mediated by dissolved organic matter and microbial response. J Hazard Mater 443:130322. https://doi.org/10.1016/J.JHAZMAT.2022.130322

Karungamye P, Rugaika A, Mtei K, Machunda R (2023) The removal of ciprofloxacin from synthetic wastewater in constructed wetland. HydroResearch 6:138–146. https://doi.org/10.1016/j.hydres.2023.04.001

Khajah M, Ahmed ME (2023) Performance evaluation of a pilot wetland system for wastewater treatment. J Eng Res. https://doi.org/10.1016/j.jer.2023.11.027

Khalifa ME, El-Reash YGA, Ahmed MI, Rizk FW (2020) Effect of media variation on the removal efficiency of pollutants from domestic wastewater in constructed wetland systems. Ecol Eng 143:105668. https://doi.org/10.1016/J.ECOLENG.2019.105668

Kumar M, Singh R (2017) Performance evaluation of semi continuous vertical flow constructed wetlands (SC-VF-CWs) for municipal wastewater treatment. Biores Technol 232:321–330. https://doi.org/10.1016/J.BIORTECH.2017.02.026

Kumar S, Sangwan V, Kumar M, Deswal S (2023) A survey on constructed wetland publications in the past three decades. Environ Monit Assess 195(8):992. https://doi.org/10.1007/s10661-023-11516-y

Lamaming J, Saalah S, Rajin M, Ismail NM, Yaser AZ (2022) A review on bamboo as an adsorbent for removal of pollutants for wastewater treatment. Int J Chem Eng. https://doi.org/10.1155/2022/7218759

Lawan MS, Kumar R, Rashid J, Barakat MA (2023) Recent advancements in the treatment of petroleum refinery wastewater. Water 15(20). https://doi.org/10.3390/w15203676

Lizama-Allende K, Ayala J, Jaque I, Echeverría P (2021) The removal of arsenic and metals from highly acidic water in horizontal subsurface flow constructed wetlands with alternative supporting media. J Hazard Mater 408:124832. https://doi.org/10.1016/J.JHAZMAT.2020.124832

Lohani SP, Khanal SN, Bakke R (2020) A simple anaerobic and filtration combined system for domestic wastewater treatment. Water-Energy Nexus 3:41–45. https://doi.org/10.1016/j.wen.2020.03.004

Ma N, Wang W, Gao J, Chen J (2017) Removal of cadmium in subsurface vertical flow constructed wetlands planted with Iris sibirica in the low-temperature season. Ecol Eng 109:48–56. https://doi.org/10.1016/J.ECOLENG.2017.09.008

Malyan SK, Yadav S, Sonkar V, Goyal VC, Singh O, Singh R (2021) Mechanistic understanding of the pollutant removal and transformation processes in the constructed wetland system. Water Environ Res 93(10):1882–1909. https://doi.org/10.1002/wer.1599

Mander Ü, Dotro G, Ebie Y, Towprayoon S, Chiemchaisri C, Nogueira SF, Jamsranjav B, Kasak K, Truu J, Tournebize J, Mitsch WJ (2014) Greenhouse gas emission in constructed wetlands for wastewater treatment: a review. Ecol Eng 66:19–35. https://doi.org/10.1016/J.ECOLENG.2013.12.006

Mara D (2003) Domestic wastewater treatment in developing countries. 1st ed. Routledge. https://doi.org/10.4324/9781849771023

Martín M, Gargallo S, Hernández-Crespo C, Oliver N (2013) Phosphorus and nitrogen removal from tertiary treated urban wastewaters by a vertical flow constructed wetland. Ecol Eng 61:34–42. https://doi.org/10.1016/J.ECOLENG.2013.09.046

Masi F, Sarti C, Cincinelli A, Bresciani R, Martinuzzi N, Bernasconi M, Rizzo A (2023) Constructed wetlands for the treatment of combined sewer overflow upstream of centralized wastewater treatment plants. Ecol Eng 193:107008. https://doi.org/10.1016/j.ecoleng.2023.107008

Mesquita MC, Albuquerque A, Amaral L, Nogueira R (2017) Seasonal variation of nutrient removal in a full-scale horizontal constructed wetland. Energy Procedia 136:225–232. https://doi.org/10.1016/J.EGYPRO.2017.10.246

Nivala J, Boog J, Headley T, Aubron T, Wallace S, Brix H, Mothes S, van Afferden M, Müller RA (2019) Side-by-side comparison of 15 pilot-scale conventional and intensified subsurface flow wetlands for treatment of domestic wastewater. Sci Total Environ 658:1500–1513. https://doi.org/10.1016/J.SCITOTENV.2018.12.165

Obeng E, Essandoh HMK, Akodwaa-Boadi K, Taylor TS, Kusi I, Appiah-Effah E (2023) Recycled clay bricks and palm kernel shell as constructed wetland substrate for wastewater treatment: an engineered closed circuit circular economy approach. Environ Technol Innov 32:103324. https://doi.org/10.1016/j.eti.2023.103324

Parde D, Patwa A, Shukla A, Vijay R, Killedar DJ, Kumar R (2021) A review of constructed wetland on type, treatment and technology of wastewater. Environ Technol Innov 21:101261. https://doi.org/10.1016/J.ETI.2020.101261

Pascual A, Álvarez JA, de la Varga D, Arias CA, Van Oirschot D, Kilian R, Soto M (2024) Horizontal flow aerated constructed wetlands for municipal wastewater treatment: the influence of bed depth. Sci Total Environ 908:168257. https://doi.org/10.1016/j.scitotenv.2023.168257

Perdana MC, Hadisusanto S, Purnama ILS (2020) Implementation of a full-scale constructed wetland to treat greywater from tourism in Suluban Uluwatu Beach, Bali, Indonesia. Heliyon 6(10):e05038. https://doi.org/10.1016/J.HELIYON.2020.E05038

Pérez Y, Vargas E, García-Cortés D, Hernández W, Checo H, Jáuregui-Haza U (2024) Efficiency and effectiveness of systems for the treatment of domestic wastewater based on subsurface flow constructed wetlands in Jarabacoa, Dominican Republic. Water Sci Eng 17(2):118–128. https://doi.org/10.1016/j.wse.2023.08.004

Priya AK, Pachaiappan R, Kumar PS, Jalil AA, Vo DVN, Rajendran S (2021) The war using microbes: a sustainable approach for wastewater management. Environ Pollut 275:116598. https://doi.org/10.1016/J.ENVPOL.2021.116598

Qadiri RZZ, Gani KM, Zaid A, Aalam T, Kazmi AA, Khalil N (2021) Comparative evaluation of the macrophytes in the constructed wetlands for the treatment of combined wastewater (greywater and septic tank effluent) in a sub-tropical region. Environ Chall 5:100265. https://doi.org/10.1016/j.envc.2021.100265

Rahman DA, Priambodo EW, Caturputranto TW, Wahyudianto FE (2022) Kinetics of pollutants removal in wetlands influenced by retention time and number of plants using *Cyperus alternifolius*. J Ecol Eng 23(12):37–43. https://doi.org/10.12911/22998993/154848

Rahman M-A, Rahaman M-H, Yasmeen S, Rahman MM, Rabbi FM, Shuvo OR, Usamah (2022) Phytoremediation potential of Schumannianthus dichotomus in vertical subsurface flow constructed wetland. Environ Chall 9:100631. https://doi.org/10.1016/j.envc.2022.100631

Rahman ME, Bin Halmi MI, Bin Abd Samad MY, Uddin MK, Mahmud K, Abd Shukor MY, Sheikh Abdullah SR, Shamsuzzaman SM (2020) Design, operation and optimization of constructed wetland for removal of pollutant. Int J Environ Res Public Health 17(22). https://doi.org/10.3390/ijerph17228339

Rodriguez-Dominguez MA, Bonefeld BE, Ambye-Jensen M, Brix H, Arias CA (2022) The use of treatment wetlands plants for protein and cellulose valorization in biorefinery platform. Sci Total Environ 810:152376. https://doi.org/10.1016/j.scitotenv.2021.152376

Salama E-S, Roh H-S, Dev S, Khan MA, Abou-Shanab RAI, Chang SW, Jeon B-H (2019) Algae as a green technology for heavy metals removal from various wastewater. World J Microbiol Biotechnol 35(5):75. https://doi.org/10.1007/s11274-019-2648-3

Sandoval L, Zamora-Castro SA, Vidal-Álvarez M, Marín-Muñiz JL (2019) Role of wetland plants and use of ornamental flowering plants in constructed wetlands for wastewater treatment: a review. Appl Sci 9(4). https://doi.org/10.3390/app9040685

Sangwijit C, Phoosuwan N, Sangwijit P, Thakong W, Saikhao L (2022) Assessment of growth ability and removal efficiency of the Hydrilla (Hydrilla verticillata) in wastewater. Egypt J Aquat Res 48(3):199–204. https://doi.org/10.1016/j.ejar.2021.11.004

Santos J, Rodrigues S, Magalhães M, Rodrigues K, Pereira L, Marinho G (2024) A state-of-the-art review (2019–2023) on constructed wetlands for greywater treatment and reuse. Environ Chall 16(May). https://doi.org/10.1016/j.envc.2024.100973

Shahid MJ, Arslan M, Ali S, Siddique M, Afzal M (2018) Floating wetlands: a sustainable tool for wastewater treatment. CLEAN–Soil, Air, Water, 46(10):1800120. https://doi.org/10.1002/clen.201800120

Sharma S, Asolekar SR, Thakur VK, Asokan P (2024) Unlocking the potential: a paradigm-shifting approach for valorizing lignocellulosic waste biomass of constructed wetland enabling environmental and societal sustainability. Ind Crops Prod 207:117709. https://doi.org/10.1016/j.indcrop.2023.117709

Shukla R, Gupta D, Singh G, Mishra VK (2021) Performance of horizontal flow constructed wetland for secondary treatment of domestic wastewater in a remote tribal area of Central India. Sustainable Environment Research 31(1):13. https://doi.org/10.1186/s42834-021-00087-7

Sossalla NA, Nivala J, Reemtsma T, Schlichting R, König M, Forquet N, van Afferden M, Müller RA, Escher BI (2021) Removal of micropollutants and biological effects by conventional and intensified constructed wetlands treating municipal wastewater. Water Res 201:117349. https://doi.org/10.1016/j.watres.2021.117349

Stefanakis AI, Tsihrintzis VA (2012) Effects of loading, resting period, temperature, porous media, vegetation and aeration on performance of pilot-scale vertical flow constructed wetlands. Chem Eng J 181–182:416–430. https://doi.org/10.1016/J.CEJ.2011.11.108

Stefanatou A, Schiza S, Petousi I, Rizzo A, Masi F, Stasinakis AS, Fyllas N, Fountoulakis MS (2023) Use of climbing and ornamental plants in vertical flow constructed wetlands treating greywater. J Water Process Eng 53:103832. https://doi.org/10.1016/j.jwpe.2023.103832

Tang Z, Wood J, Smith D, Thapa A, Aryal N (2021) A review on constructed treatment wetlands for removal of pollutants in the agricultural runoff. In Sustainability 13(24). https://doi.org/10.3390/su132413578

Tazkiaturrizki T, Soewondo P, Handajani M (2018) Nitrogen removal on recycling water process of wastewater treatment plant effluent using subsurface horizontal wetland with continuous feed. IOP Conf Ser Earth Environ Sci 106(1). https://doi.org/10.1088/1755-1315/106/1/012078

Tong Y, Wang X, Elser JJ (2022) Unintended nutrient imbalance induced by wastewater effluent inputs to receiving water and its ecological consequences. Front Environ Sci Eng 16(11):149. https://doi.org/10.1007/s11783-022-1584-x

Tsihrintzis VA (2017) The use of vertical flow constructed wetlands in wastewater treatment. Water Resour Manag 31(10):3245–3270. https://doi.org/10.1007/s11269-017-1710-x

Vega De Lille MI, Hernández Cardona MA, Tzakum Xicum YA, Giácoman-Vallejos G, Quintal-Franco CA (2021) Hybrid constructed wetlands system for domestic wastewater treatment under tropical climate: Effect of recirculation strategies on nitrogen removal. Ecol Eng 166:106243. https://doi.org/10.1016/J.ECOLENG.2021.106243

Vymazal J (2010) Constructed wetlands for wastewater treatment. Water 2(3):530–549. https://doi.org/10.3390/w2030530

Vymazal J (2011) Plants used in constructed wetlands with horizontal subsurface flow: a review. Hydrobiologia 674(1):133–156. https://doi.org/10.1007/s10750-011-0738-9

Wang H, Xu J, Sheng L, Teng H (2022) Study on treatment of city tail water by constructed wetland with corn straw biochar substrate. Environ Technol Innov 28:102855. https://doi.org/10.1016/j.eti.2022.102855

Wen S, Liu H, He H, Luo L, Li X, Zeng G, Zhou Z, Lou W, Yang C (2016) Treatment of anaerobically digested swine wastewater by Rhodobacter blasticus and Rhodobacter capsulatus. Biores Technol 222:33–38. https://doi.org/10.1016/J.BIORTECH.2016.09.102

Werkneh AA (2024) Decentralized constructed wetlands for domestic wastewater treatment in developing countries: field-scale case studies, overall performance and removal mechanisms. J Water Process Eng 57(April 2023):104710. https://doi.org/10.1016/j.jwpe.2023.104710

Wu H, Wang R, Yan P, Wu S, Chen Z, Zhao Y, Cheng C, Hu Z, Zhuang L, Guo Z, Xie H, Zhang J (2023) Constructed wetlands for pollution control. Nature Reviews Earth and Environment 4(4):218–234. https://doi.org/10.1038/s43017-023-00395-z

Wu H, Zhang J, Ngo HH, Guo W, Hu Z, Liang S, Fan J, Liu H (2015) A review on the sustainability of constructed wetlands for wastewater treatment: design and operation. Biores Technol 175:594–601. https://doi.org/10.1016/J.BIORTECH.2014.10.068

Wu Y, Han R, Yang X, Zhang Y, Zhang R (2017) Long-term performance of an integrated constructed wetland for advanced treatment of mixed wastewater. Ecol Eng 99:91–98. https://doi.org/10.1016/J.ECOLENG.2016.11.032

Yang C, Fu T, Wang H, Chen R, Wang B, He T, Pi Y, Zhou J, Liang T, Chen M (2021) Removal of organic pollutants by effluent recirculation constructed wetlands system treating landfill leachate. Environ Technol Innov 24:101843. https://doi.org/10.1016/J.ETI.2021.101843

Zhang J, Jiang Y, Zhang H, Feng D, Bu H, Li L, Lu S (2024a) A critical review of characteristics of domestic wastewater and key treatment techniques in Chinese villages. Sci Total Environ 927(January):172155. https://doi.org/10.1016/j.scitotenv.2024.172155

Zhang S, Shen C, Zhang F, Wei K, Shan S, Zhao Y, Man YB, Wong MH, Zhang J (2024b) Microplastics removal mechanisms in constructed wetlands and their impacts on nutrient (nitrogen, phosphorus and carbon) removal: a critical review. Sci Total Environ 918:170654. https://doi.org/10.1016/J.SCITOTENV.2024.170654

Zhang Z, Hou Y, Li P, Wang J (2024c) Enhancement of geotechnical properties of loess using nano clay and nano-iron oxide. Environ Earth Sci 83(13):418. https://doi.org/10.1007/s12665-024-11716-w

Zhao Y, Mao W, Pang L, Li R, Li S (2020) Influence of Phragmites communis and Zizania aquatica on rhizosphere soil enzyme activity and bacterial community structure in a surface flow constructed wetland treating secondary domestic effluent in China. Environ Sci Pollut Res 27(21):26141–26152. https://doi.org/10.1007/s11356-020-08904-z

Zhao Y, Song X, Cao X, Wang Y, Zhao Z, Si Z, Yuan S (2019) Modified solid carbon sources with nitrate adsorption capability combined with nZVI improve the denitrification performance of constructed wetlands. Biores Technol 294:122189. https://doi.org/10.1016/J.BIORTECH.2019.122189

Zidan ARA, El-Gamal MM, Rashed AA, El-Hady Eid MAA (2015) Wastewater treatment in horizontal subsurface flow constructed wetlands using different media (setup stage). Water Sci 29(1):26–35. https://doi.org/10.1016/j.wsj.2015.02.003

Effect of Different Carbon Sources on Biosurfactant Production by Bacteria Isolated from Hydrocarbon-Contaminated Soil

Usman Ali Bukar, Abdullahi Hassan Kawo, Binta Muhammad Aminu, Aminu Yusuf Fardami, Husaini Hauwa, HarunaYahaya Ismail, Ruqayyah Abbas Usman, and Aminu Aliyu

Abstract Prospecting for microorganisms and substrates for the synthesis of significant amounts of biosurfactants is becoming more and more prominent due to the rising demand for these substances for use in environmental and industrial applications. This study was carried out to evaluate the effect of different carbon sources on biosurfactant production using bacteria isolated from hydrocarbon-contaminated soil in Hotoro Nasarawa Local Government, Kano State, Nigeria. The physicochemical properties of the hydrocarbon-contaminated soil as well as cultural, morphological, and biochemical characteristics of the isolates were analysed using standard procedures. Methods for screening the isolates for biosurfactant production that include blood haemolysis, drop collapse, oil displacement, and emulsification index tests were employed. The result of physicochemical characteristics of the hydrocarbon-contaminated soil samples from this study revealed that the pH of the all soil samples were weakly acidic from 6.80 to 7.06. The temperature was ranged between 38 and 41 °C. The viable heterotrophic bacterial count of all the samples ranges from 1.56 to 2.78×10^6 cfu/g. Four (4) bacterial genera were biochemically identified as *Bacillus*

U. A. Bukar (✉) · H. Ismail
Department of Microbiology, University of Maiduguri, Maiduguri, Borno State, Nigeria
e-mail: uthmaniyyer@gmail.com

H. Ismail
e-mail: hyismail@unimaid.edu.ng

A. H. Kawo · B. M. Aminu · R. A. Usman
Department of Microbiology, Bayero University Kano, Rimin Gata, Kano State, Nigeria
e-mail: ahkawo.mcb@buk.edu.ng

A. Y. Fardami
Department of Microbiology, Usmanu Danfodiyo University, Gidan Alkali, Sokoto State, Nigeria

H. Hauwa
Department of Microbiology, Sokoto State University, Dangeda, Sokoto State, Nigeria

A. Aliyu
Department of Microbiology, Federal University Gusau, Gusau, Zamfara State, Nigeria

A. Z. Yaser et al. (eds.), *Controlling Environmental Pollution*,
https://doi.org/10.1007/978-981-97-8931-3_17

subtilis, Bacillus subtilis, Bacillus lichenoformis, and *Bacillus* sp. The four (4) identified bacterial isolates were all positive for haemolysis test, drop collapse and oil displacement test. The most potent isolate during the screening was *Bacillus subtilis* which had the highest biosurfactant production ability of 0.83 ± 0.07 and 1–90 ± 0.01 mg/ml for used engine oil and diesel as carbon sources, respectively. It can be concluded that the production of biosurfactant from the diesel substrate was higher than the yield from the used engine oil substrate.

Keywords Biosurfactant · Bacteria · Carbon source · Diesel · Used engine oil

1 Introduction

Surfactants are amphipathic compounds that act primarily at fluid phase interfaces, such as the oil–water or air–water interfaces (Adamu et al. 2015). Lowering the surface and interfacial tensions of liquid at interfaces is one of their many properties as surface-active agents. In essence, surfactants, also known as "surface-active agents," are chemical substances that lessen the surface tension of a liquid, the interfacial tension between two liquids, or the tension between a liquid and a solid (Aparna et al. 2011). Secondary metabolites known as biosurfactants are produced by microbial cells that are either growing or at rest (bacteria, fungi and yeast). The substances can either be metabolized within the cell or released into the surrounding environment as a component of the cell membrane (Mulligan 2005).

Biosurfactants are surface-active substances produced by microbes that have many characteristics with chemical surfactants. They gather at interfaces between fluids of various polarities, such as water and hydrocarbons, and they lessen the interfacial surface tension (Arku et al. 2011). Glycolipids (sophorose lipid, rhamnose lipid), hydroxylated and cross-linked fatty acids, polysaccharides lipid complexes, lipoprotein-lipopeptides, and phospholipids make up the majority of them (Banat et al. 2008). According to Kaya et al. (2014), biosurfactant is a biological substance that plays significant intracellular (regulatory molecules, food stores, etc.), extracellular (cell hydrophobicity and mycelium formation), and intercellular (quorum sensing, biofilm formation, antagonism, and microbial community structure) roles in the physiology and ecology of microorganisms (Kaya et al. 2014).

The benefits of biologically synthesized surfactants include their biodegradability, low toxicity, biocompatibility, and digestibility, availability of raw materials, acceptable production economics, specificity, and effectiveness, to name a few. These attributes support their acceptance in a variety of applications (Kosaric 2001). Biosurfactant-producing microorganisms are abundant in nature and can be found in both freshwater and saltwater as well as on land and in the sea (soil, sediment, and sludge). They also survive in a variety of temperatures, pH levels, and salinities and can be found in harsh situations (like oil reservoirs, for example) (Bezza and Chirwa 2015). Additionally, they can be isolated from settings that haven't been disrupted and in which they play physiological roles that don't involve solubilizing hydrophobic contaminants, like antibacterial activity, biofilm formation or processes of motility, and surface colonization (Hamme et al. 2006).

Biosurfactant-producing microorganisms are abundant in nature and can be found in both freshwater and saltwater as well as on land and in the sea (soil, sediment, and sludge). They also survive in a variety of temperatures, pH levels, and salinities and can be found in harsh situations (like oil reservoirs, for example) (Bezza and Chirwa 2015). Additionally, they can be isolated from settings that haven't been disrupted and in which they play physiological roles that don't involve solubilizing hydrophobic contaminants, like antibacterial activity, biofilm formation or processes of motility, and surface colonization (Hamme et al. 2006).

Since olive oil is an insoluble in water carbon source, it produced the finest emulsifying rhamnolipid. Furthermore, the growth-supported rhamnolipid synthesis medium could only be made using nitrogen sources like sodium nitrate and ammonium nitrate (Hamme et al. 2006). With glucose and hexadecane as carbon sources, the yield and emulsification index, however, were constrained because of inhibition brought on by the pH drop. This disease was most likely brought on by the creation of secondary acid metabolites such as uronic acid (Abouseoud et al. 2008). The present study was, therefore, undertaken to assess the effect of used engine oil and diesel as carbon sources on the biosurfactant production by the most potent bacterial isolate isolated from hydrocarbon-contaminated soil.

2 Materials and Methods

2.1 Study Area

Kano is located in the North West of Nigeria and the Metropolitan Kano has 499 square kilometres (193 square miles) with a population of 2,828,861. The headquarters of Kano State is Kano city which is 12° 0′ 0.0000" N and 8° 31′ 0.0012" E (National Population Census 2006) (Fig. 1).

2.2 Sample Collection

Soil samples were collected from hydrocarbon-contaminated soil in Hotoro metropolis, Nasarawa Local Government, Kano State, Nigeria. The samples were collected in sterile polythene bag and were immediately transported to the Postgraduate Laboratory of the Department of Microbiology, Bayero University Kano, Kano State, Nigeria. The soil samples were collected from four (4) different locations; Three (3) of which were from hydrocarbon-contaminated soil and the one was from natural soil within Hotoro, Nasarawa LGA Kano State.

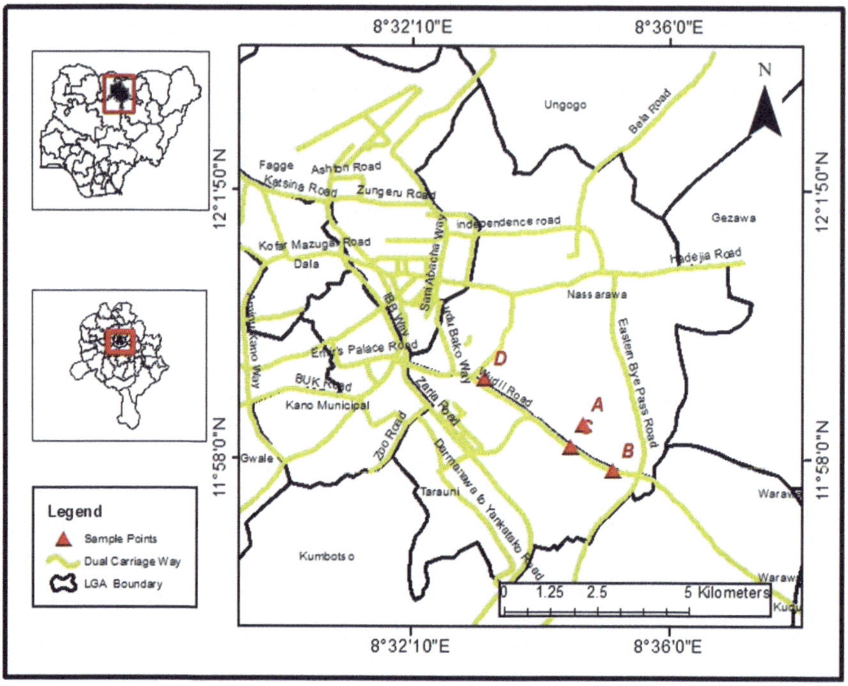

Fig. 1 Map of sampling area (A, B, C, and D)

2.3 *Physicochemical Analysis of the Soil Samples*

Determination of pH

The soil pH was determined according to the method of International Institute of Tropical Agriculture (IITA) (International Institute for Tropical Agriculture (IITA) 1979). Twenty (20) grams of air-dried soil were sieved and placed in 50 ml capacity beaker. Twenty (20) millilitres of distilled water was added and allowed to stand for 30 min, while stirring occasionally with a glass rod. The pH meter was calibrated with buffer of pH 7.0 before use. The electrode of the pH meter was inserted into the partly settled suspension and the reading on the pH meter was noted and recorded accordingly in triplicate and the mean was taken.

Determination of Temperature

Temperature was determined at the point of sample collection by dipping the bulb of mercury-in-glass thermometer into the soil and readings were recorded using the procedure of Association of Official Analytical Chemists (AOAC) (Association of Official Analytical Chemists (AOAC) 1990).

Determination of Electrical Conductivity

Soil water suspension was prepared using a soil to water ratio of 1:2.5. The contents were allowed to equilibrate at room temperature for 30 min and the electrical conductivity was recorded using EC metre (model-4070, Amber Science, USA) after standardizing with 0.01 N KCl solution at 25 °C (Page et al. 1982).

Determination of Moisture Content

An empty crucible was weighed (W_0) and 2 g of soil was added and weighed again (W_1). Samples were dried in hot air oven at 105 °C until constant weight was achieved (W_2). Both the container and the dried sample were weighed again (International Institute for Tropical Agriculture (IITA) 1979). The moisture content was calculated as follows:

$$\%Moisture = \frac{\text{W1} - \text{W2}}{\text{W1} - \text{W0}} \times 100$$

Determination of Organic Carbon Content

One gram of soil sample was weighed in duplicates and transferred to 250 ml Erlenmeyer flask. Ten millilitres of potassium dichromate solution and 20 ml of concentrated H_2SO_4 were added and the contents of each flask were shaken gently until properly mixed. One hundred millilitres of distilled water were added and allowed to stand for 30 min. This was followed by adding 4 drops of methyl red indicator and titrating against 0.5N ferrous sulphate solution until the colour changed from blue to red (Page et al. 1982). The percentage of carbon was calculated according to the formula: % Organic Carbon = Me K2Cr2O7–Me $FeSO_4$ × 100 Weight (Wet soil–dry soil).

Determination of Nitrogen Content of the Soil

Total nitrogen was determined by Macro-Kjeldahl digestion method of Juo (1979). Five grams of soil sample were weighed into a 500 ml Macro-Kjeldahl flask and 20 ml of distilled water was added. The content was swirled for five minutes and allowed to stand for 30 min. One tablet of mercury catalyst and 10 g of K_2SO_4 were added and 30 ml of conc. H_2SO_4 was added through an automated pipette. The content of the flask was heated gently in the digestion stand. After cooling, 100 ml of distilled water was added and transferred into another clean macro-Kjeldahl flask (750 ml) and the sand residue washed four times with 50 ml of distilled water. All the washings were transferred into the same flask. Fifty millilitres (50 ml) H_3BO_3 indicator solution was added into 500 ml Erlenmeyer flask, which was placed under the condenser of distillation apparatus and 150 ml of 10N NaOH was introduced. This was followed by distillation. For the condenser to remain cool (30^0C) and prevent frothing, sufficient cold water was allowed to flow through the condenser. Ammonium was determined in the distillate by treating against 0.01N standard H_2SO_4 using a 25 ml burette graduated at 0.1 ml interval. The colour changed at the end point from green to pink.

Percentage nitrogen was calculated using the formula:

$$\%Nitrogen = N \times 0.014 \times Vd \times 10 \times 100$$

$$\mathrm{A} \times \text{ Weight of sample}$$

where: N = Normality of acid, Vd = Volume of the digest, A = Aliquot of digest

Determination of Soil Particle Size

The hydrometer method for the mechanical analysis of particle size distribution presented by Gee and Bauder (1986) and adopted by Adamu et al. (2015) was used. Air dried soil was sieved using 425 μm pore size sieve. Fifty one gram (51 g) was transferred into 1 L shaken to mix. Fifty millilitres (50 ml) of 5% sodium hexametaphosphate was added followed by 100 ml of distilled water. The soil suspension was stirred thoroughly for 15 min and transferred into a cylinder containing hydrometer. Distilled water was added to the lower blue line of the cylinder. The volume changed to 1130 ml and the hydrometer was removed. The top of the cylinder was covered with hand and inverted several times until all soil was in suspension. The cylinder was placed on a flat surface and time was noted (Gee and Bauder 1986). Hydrometer was placed in the suspension and the first reading (H1) was noted immediately after 40 s. Subsequently, the temperature (T1) was recorded after the hydrometer was removed. The suspension was allowed to stand for 3 h and the second reading was taken for hydrometer (H2) and the temperature (T2), respectively.

The percentage sand, clay, and silt were calculated thus: (Gee and Bauder 1986).
Sand = 100.0-(H1 + 0.3(T1-20)-2.0)2.
Clay = H2 + 0.3(T2-20)-2.0
Silt = 100.0-(% sand + % clay).

Enumeration of Bacterial Loads in the Soil Samples

A stock solution for serial dilution was made by dispensing 1 g of soil into 100 ml of distilled water and shaken thoroughly and transferring one (1) ml into a test tube containing 9 ml of sterile distilled water and subsequently making a serial dilution of up to 10^5. Using the spread plate method technique, 0.1 ml of the suspension from dilution of 10^3, 10^4, and 10^5 were plated aseptically on a prepared Nutrient Agar (NA) and incubated at 30 °C for 24 h. The results were determined by multiplying the number of counts with the dilution used and expressed as colony forming units per gram (cfu/g) of soil and the isolates were also subcultured in bushnell haas agar (Kiran et al. 2010).

2.4 *Characterization of Bacterial Isolates*

Gram's Staining

Gram's staining was carried out as described by Harley and Prescott (2002). Smear of the bacterial isolate was made on a clean grease-free glass slide using a drop of water. The smear was allowed to air dry and then passed over a flame in order to be fixed. After fixing, the smear was covered with a primary dye (i.e., crystal violet) for one minute and washed with water. The slide was covered with Lugol's iodine and washed after one minute. Ethanol was used to rapidly decolorize the smear and washed with water. This was followed by the application of safranin and left for 30 s and later washed with water. Back of the slide was cleaned with a cotton wool and allowed to air dry. The slides were examined using oil immersion objective lens (× 100) of the microscope. Bluish or purplish colour indicates Gram positive while red or pinkish colour indicates Gram negative bacteria.

Spore Staining

A smear was made on a slide and heat fixed. Malachite green (5%) solution was applied and heated until steam rises and allowed to cool, and washed gently with cold water. The smear was counterstained with 0.5% safranin for 30 s and washed with water. The slide was blot dried and was examined under oil immersion objective lens for the presence of spores. Spores stained green while vegetative cells stained red (Ochei and Kolhatkar 2000).

Starch Hydrolysis Test

Cultures of bacteria were grown on starch agar. The colonies were flooded with iodine. Clear area observed around the isolate indicated positive result (Harley and Prescott 2002).

Voges-Proskauer Test

To the culture in above, 0.6 ml 5% α-napthol solution was added and shaken. The test tubes were sloped and examined after 15 min. A red coloration indicated a positive VP reaction (Ochei and Kolhatkar 2000).

Urease Production Test

Slants of urea medium in universal bottles were inoculated with a loopful of the isolates by streaking. These were incubated at 37 °C for 4 days and examined daily (Harley and Prescott 2002).

Screening of Isolates for Biosurfactant Production

All the pure bacterial isolates were subjected to screening for their ability to produce biosurfactants as described by Kosaric (2001).

Blood Haemolysis Test

Isolates were inoculated on blood agar plates containing 5% (v/v) sheep blood and incubated at room temperature for 24 h. Haemolytic activity was detected as the occurrence of define clear zone around a colony (Fardami et al. 2022).

Drop Collapse Test

A drop of used engine oil was placed on a grease-free slide, and one drop of the cell-free supernatant was placed at the Centre of the oil drop. The shape of drop on the oil surface was observed for flattening. The collapse of the drop was due to the reduction of interfacial tension between the liquid drops and the hydrophobic surface of the oil. Additionally, the time required by the oil drop to collapse was recorded (Fardami et al. 2022).

Oil Spreading Test

Oil spreading technique was carried out according to the method described by Maneerat and Phetrong (Maneerat and Phetrong 2007). Forty millilitres of distilled water were added to Petri dishes followed by the addition of 100 μL of diesel on the surface of the water. Then, one drop of the supernatant was put on the used engine oil and diesel surface, respectively. The diameter of the clear zone on oil surface was measured using a metre ruler. The time taken to achieve the spread was also noted.

Emulsification Capacity Test

Emulsification activity of the biosurfactants were carried out in accordance of a procedure of Fardami et al. (2022). Four ml (4 ml) of used engine oil was added to an equal amount of cell-free supernatant and Vortexed at 500 rpm for 10 min. After 24 h, the height of the stable emulsion layer was measured using a metre rule. The emulsification index (E_{24}) was calculated as the ratio of the height of the emulsion layer and the total height of the liquid. As given by the expression:

$$E24 = \frac{\text{h emulsion}}{\text{h total}} \times 100$$

where: E_{24} is emulsion index after 24 h, h emulsion is the height of the emulsion layer, h total is the total height of the liquid.

Biosurfactant Production

The carbon sources that were use are used engine oil and diesel. The media that were used for the experiment is a mineral medium then the medium was supplemented with 2% engine oil and 2% diesel, respectively, for the growth of the bacteria, as the sole source of carbon and energy. One (1) ml of the isolates was pipetted into each of the mineral media using a sterile syringe. The inoculated flasks were plugged with cotton wool and incubated at 37 °C on a shaker operating at 120 rpm for 5 days. After the incubation period, the broth of each isolate was centrifuged at 6000 rpm for 10 min, and the supernatants were filtered to obtain cell-free supernatants (Priya and Usharani 2009). The biosurfactant concentration was measured indirectly by measuring the absorbance of crude biosurfactant at wavelength of 500 nm. Absorbance was measured using spectrophotometer Rahman et al. (2002).

3 Results and Discussion

3.1 Physical and Chemical Properties of the Soil Samples

Table 1 shows the result for the physical and chemical characteristics of the soil from the sampling sites A, B, and C as well as the control site D soil (agricultural soil) that has no hydrocarbon contamination. The pH and temperature of the soil in the site A, B, and C are slightly the same. The soil sample from the sites A and D had the highest value of pH recorded 7.06 while site C had lowest value recorded 6.80. The soil sample in the site A had the highest value of temperature recorded 40°C while the soil samples of sites B and C had average temperature of 38°C and 39°C, respectively. The temperature of the soil samples in the control site D was recorded 41°C. The soil samples from site B had the highest average of the percentage of organic carbon 13.72% while sites A and C had 8.88 and 8.77%, respectively. The texture of the soil from the sampling sites A and C was Loamy Sandy (LS) while site B had sand class and site D had sandy loamy class each as determined using soil triangle (Table 1).

The lithosphere, or layer of rock that covers the globe, is sandwiched between the atmosphere and the soil. Along with being a component of the biosphere, it also

Table 1 Physicochemical characteristics of the soil samples

Parameters	Site A (Mean ± S.D.)	Site B (Mean ± S.D.)	Site C (Mean ± S.D.)	Control soil (Mean ± S.D.)
pH	7.06 ± 0.03	6.92 ± 0.07	6.80 ± 0.04	7.06 ± 0.01
Temperature (°C)	40.2 ± 1.60	38.0 ± 0.80	39.6 ± 0.40	41 ± 0.80
Moisture (%)	1.92 ± 0.05	1.87 ± 0.07	2.10 ± 0.20	2.42 ± 0.91
Organic Carbon (%)	8.88 ± 0.13	13.72 ± 0.06	8.77 ± 0.007	0.70 ± 0.13
Nitrogen (%)	0.04 ± 0.01	0.13 ± 0.01	0.04 ± 0.008	0.14 ± 0.004
Phosphorus (mg/kg)	11.18 ± 0.12	12.76 ± 0.25	11.27 ± 0.07	5.62 ± 0.05
Potassium (mg/kg)	0.17 ± 0.06	0.05 ± 0.01	0.07 ± 0.07	0.04 ± 0.01
Copper	2.64 ± 0.31	2.40 ± 0.09	2.54 ± 0.20	2.40 ± 0.09
CEC (cmol/kg)	3.47 ± 0.43	3.28 ± 0.14	3.38 ± 0.28	1.44 ± 0.38
EC (ds/m)	0.17 ± 0.06	0.19 ± 0.01	0.40 ± 0.002	0.07 ± 0.007
Sand (%)	84	92	86	64
Silt (%)	12	4	10	20
Clay (%)	4	4	4	16
Textural Class	Loamy Sandy	Sandy	Loamy Sandy	Sandy Loamy

Key: S.D. = Standard deviation

creates the boundary between the lithosphere and water bodies (hydrosphere). The soil is known to be home to a variety of microorganisms, most of which are bacteria, and they make use of the soil's various substrates. The types and qualities of soil affect the different bacterial species that exist there. Due to the fact that both physical and chemical characteristics influence microbial development and activity as well as soil production, the physicochemical research of a given area is crucial.

In this study, physical and chemical properties of the soil were evaluated based on a number of criteria, including pH, electrical conductivity, texture, moisture, and temperature. The pH of the soil affects how acidic, alkaline, or neutral the soil is, and this has an impact on the persistence of bacteria in the soil. The pH results of this study, which ranges from 6.80 to 7.06 and was fairly neutral, are comparable to those obtained by Jemimah et al. (2018), who worked on the isolation and identification of bacteria in petroleum hydrocarbon-polluted soils in the north-west of Bayelsa State, Nigeria. More so, this result was in conformism with the findings of Akeredolu and Akinnibosun's (2017), who had a pH range of 6.5–6.7 who analysed petroleum-product-contaminated soil of NNPC depot in Benin City, Nigeria.

Temperature has a significant impact on microbial activity. Microorganisms are categorized into three types based on their preferred temperature range: mesophiles, which prefer moderate temperatures, and thermophiles, which prefer extremely cold temperatures (heat-loving microbes). However, warmth does cause an increase in enzyme activity. But if the temperature rises too much, the enzyme's activity decreases and the protein itself denatures. Because each bacterial species has specific growth temperature requirements that are mostly influenced by the temperature requirements of its enzymes, temperature is a major indicator of the bacterial activity in contaminated soil. The temperature variations in this study are consistent with those reported by Saidu et al. (2020), which vary from 35 to 45 °C.

Another element that controls bacterial activity is organic carbon, which was once thought to be a crucial indicator of the soil's quality. The range of organic carbon in this study's results, 0.70 to 13.72, is in agreement with (Akeredolu and Akinnibosun 2017) findings, which focused on bacterial isolation and physicochemical analyses of petroleum-product-contaminated soil from NNPC/PPMC depot, Benin City, Nigeria, and found 2.37–8.19 organic carbon.

The most important element that plants can get from the soil is nitrogen, which is a growth inhibitor for plants (Gorde and Jadhav 2013). The percentage of nitrogen recovered in this investigation ranged between 0.04% and 0.14%. This effect may be caused by anthropogenic activity, according to research by Salami et al. (2002). Potassium is a key component in numerous biological processes and is essential for the growth of all living things. The amount of potassium found in this investigation was consistent with the findings of Osakwe (2014). Every living cell contains the most significant element, phosphorus, which is also one of the most significant micronutrients required for plant growth. The result of Phosphorus levels of this study which ranges from 5.62 to 12.76 mg/kg and this had agreed with the finding of Osakwe (2014).

Table 2 Mean heterotrophic bacterial counts from the sampling sites of hydrocarbon-contaminated soils

Sampling site	Mean bacterial colony count (cfu/g) ± S.D
A	$2.78 \times 10^6 \pm 0.13$
B	$2.24 \times 10^6 \pm 0.10$
C	$2.46 \times 10^6 \pm 0.09$
D (Control Site)	$1.56 \times 10^7 \pm 0.04$

Key: S.D. = Standard deviation

The physical and chemical variability of soil contributes to its high level of microbial diversity. Sand, clay, and silt content ratios add together to form soil texture, All of the samples in this study yielded percentage results range for sand (64–84%), silt (4–20%), and clay (4–16%) which is consistent with the findings of Jemimah et al. (2018) and Hazim and Al-Ani (2019). The bio-physical characteristics of soil are influenced by the relatively stable characteristic known as soil texture. The retention capacity of the soil is determined by the proportions of various textures, the analysis of textural class of the soil samples in this study revealed that all the samples are were loamy sandy expect site B which is Sandy and its shows high percentage sand (92%) (Table 1), which is also consistent with the findings of Jemimah et al. (2018) and Hazim and Al-Ani (2019).

The overall heterotrophic bacterial count in soil samples ranged from 1.56 to 2.78 X106cfu/g (Table 2). The results of findings by Monger et al. (2022) on the impact of semi-natural broadleaf woods and grassland on soil characteristics and flood discharge are consistent with the current findings.

3.2 *Bacteriological Characteristics of the Soil Samples*

The viable aerobic heterotrophic bacterial counts are presented in Table 2. The highest mean count of 2.78×10^6 CFU/g was observed at site A, while lowest bacterial counts recorded 1.56×10^7 CFU/g at site D. The morphological and biochemical characteristics of the isolates were presented in Table 3. A total of Four (4) bacterial isolates were identified from the soil samples. Four (4) bacterial isolates were identified, each have different frequency of occurrence. The Four (4) bacterial genera are *Bacillus subtilis, Bacillus subtilis, Bacillus lichenoformis,* and *Bacillus* sp. (Table 3).

The isolation of *Bacillus subtilis* in this investigation corresponds to that of Oyewole et al. (2020) who reported the discovery of *Bacillus subtilis* from a diesel-contaminated soil sample. Al-Dhabaan(Al-Dhabaan 2018) reported the isolation and identification of *Bacillus subtilis* and *Bacillus* sp. from oil-polluted soil in Dhahran, Saudi Arabia. Nimrat et al. (2020), reported a similar detection of *Bacillus subtilis* in hydrocarbon-contaminated soil. The presence of *Bacillus* in soil may be due to its capacity to tolerate adverse environmental conditions due to its ability to generate spores.

Table 3 Morphological and biochemical characterization of isolates obtained from soil of sampling sites

Code	Sha	GR	Spo	Cat	Oxd	MR	Ure	Ind	VP	Gl	Lac	Cit	Organism
B2	Rod	+	+	+	+	–	–	–	+	+	+	+	*Bacillus subtilis*
B5	Rod	+	+	+	–	–	–	–	+	+	–	+	*Bacillus* sp.
C2	Rod	+	+	+	+	–	–	–	+	+	+	+	*Bacillus lichineformis*
D9	Rod	+	+	+	+	–	–	–	+	+	+	+	*Bacillus subtilis*

Key: Sp-spore; GR-Gram Reaction, Cat-Catalase, Oxd-Oxidase, MR-Methyl Red Reaction, Ure-Urease, Id-Indole, Cit-Citrase, Gl-glucose, Lac-Lactose, VP-Voges-Proskauertest

Bacillus licheniformis is another Gram-positive *Bacilli* bacteria identified from the current study. Ours was the least common genus found in this investigation, and its isolation matched that of Rabani (2020) who recovered *Bacillus licheniformis* from petroleum-contaminated soil.

3.3 Screening of the Isolates for Biosurfactant Production

Screening of the isolates for the production of biosurfactant was carried out based on blood haemolysis, oil drop collapse, oil spreading, and emulsification index tests. The Four (4) identified genus of bacterial isolates were first subjected to blood haemolysis test and all the ten isolates were β-haemolysis positive and it was observed that the supernatant of all isolates was positive for drop collapse and oil spreading tests, and *Bacillus subtilis* was the most potent isolate (Table 4).

In this study, a biosurfactant screening approach was used to detect the isolates' potential to produce biosurfactants (*Bacillus subtilis, Bacillus lichinoformis,* and *Bacillus sp.*). All of the isolates tested positive in the blood haemolysis test (Table 4). Several researchers, including Ibrahim et al. (2020); Youssef et al. (Youssef et al.

Table 4 Result of screening of the bacterial isolates

Isolate	Haemolysis	E_{24} (%)	Drop collapse Result time	Oil spreading Result diameter
A	+VE	61	+VE 1.5 m	+ VE 3.3 cm
B	+VE	73	+VE 1 m	+ VE 3.0 cm
C	+VE	66	+VE 1 m	+ VE 2.5 cm
D	+VE	65	+VE 1 m	+ VE 2.4 cm

Key: The letters A, B, C, and D stand for *Bacillus* sp., *Bacillus subtilis, Bacillus lichenformis,* and *Bacillus subtilis,* respectively.

2004); and Plaza et al. (2006) have employed this technique to test bacteria for their potential to create biosurfactants.

Mulligan (Banat et al. 2014) proposed using blood agar lysis as a preliminary screening approach for biosurfactant production. The organisms produced haemolytic activity as a marker of biosurfactant production due to the presence of translucent clearing zones in the blood agar plates as haemolytic activity. Confirmatory detection of biosurfactants by the bacterial isolates was also made possible through the use of techniques such as oil spreading technique, drop collapse assay, and emulsification test. Several authors Banat et al. (2014); Youssef et al. (2004), Walter et al. (2010)) have suggested the use of this confirmatory technique for screening microorganisms with potential to produce biosurfactants. The drop collapse technique is a qualitative method commonly used to measure the surface tension of liquids. The method relies on the destabilization of liquid droplets by surfactants, described as surface-active compounds Walter et al. (2010). Thus, the presence of biosurfactants causes the liquid droplets to spread or collapse over a hydrophobic surface due to the interfacial tension between the liquid drop and the hydrophobic surface. Emulsification is a functional property of biosurfactants which refers to the dispersion of one liquid phase in another, causing the mixing of two immiscible liquids (Inès and Dhouha 2015). Measuring the emulsification activity is one of the indirect methods used for screening possible biosurfactant production by microorganisms. The results of oil drop collapse, oil displacement test, and emulsification (Table 4) in this present study were all positive in their individual response. This had agreed with the findings of Ibrahim et al. (2020), Joshi et al. (2008) and Ibrahim et al. (2013).

3.4 Effect of Different Carbon Source on Biosurfactant Production by **Bacillus Subtilis**

Table 5 and Fig. 2 shows the effect of used engine oil and diesel as carbon sources on the biosurfactants production by *Bacillus subtilis* (B2), the effect of carbon sources on the production of biosurfactants was determined at different concentrationsranginge from 1 to 5% (v/v). As the carbon sources concentration increases biosurfactant production increases until the optimum activity for used engine oil 0.83 ± 0.07 mg/ml and for diesel 1.90 ± 0.01 at 3% v/v, respectively. Biosurfactant production then decreases at 4 and 5% concentration (Table 5). Optimum production of biosurfactant by this organism was observed on the 5th day. However, the highest yield was recorded when diesel was used as carbon source with significant difference ($p \leq 0.05$) between the two carbon sources.

The production of biosurfactants by microorganisms has been a subject of increasing interest in recent years, especially due to their increasing potential application. In the present study, result showed that *Bacillus subtilis* yielded a higher amount of biosurfactants when diesel was used and compared to used engine oil as carbon

Table 5 Effect of different carbon source on biosurfactant production by *Bacillus subtilis*

Concentrations (%)	Used engine oil (Mean ± S.D.)	Diesel (Mean ± S.D.)
1	0.64 ± 0.08	1.64 ± 0.08
2	0.74 ± 0.06	1.46 ± 0.003
3	0.83 ± 0.07	1.90 ± 0.01
4	0.61 ± 0.06	1.16 ± 0.03
5	0.31 ± 0.04	0.44 ± 0.01

Key: S.D. = standard deviation

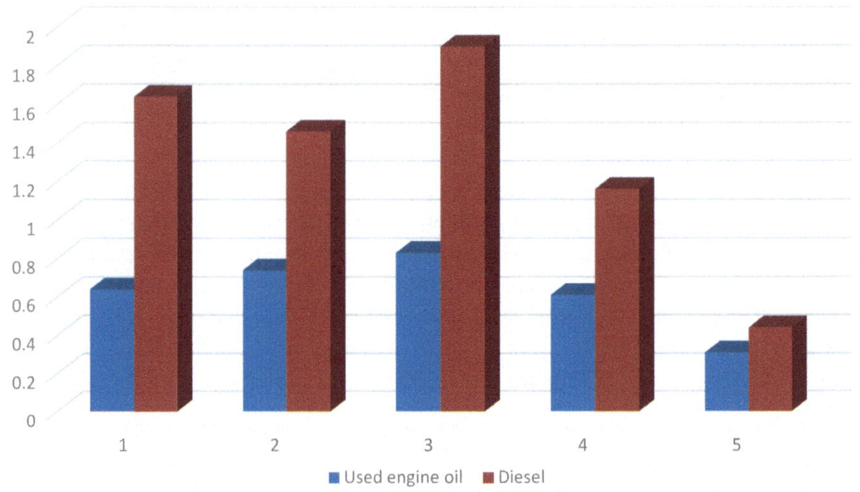

Fig. 2 Effect of used engine oil and diesel as carbon source on biosurfactant production by *Bacillus subtilis*

sources at different concentrations. However, the quality and quantity of biosurfactant production are affected and influenced by the nature of the carbon source. In this present study found that diesel substrate as a better carbon source for biosurfactant production. The result of this research is in conformity with the finding of Desai and Banat (1997) reported diesel to be a good source of carbon substrate for biosurfactant production. It also agreed with the finding of Abdulsalam et al. (2016) who worked on comparative studies on the biosurfactant production capacity of *Bacillus subtilis* and *Pseudomonas aeruginosa* using engine oil and diesel, respectively, as Substrate in Samaru Zaria, Kaduna State Nigeria.

From the result of this study, 0.83 mg/ml biosurfactant was produced by *Bacillus subtilis* using 3% engine oil as carbon source and this is in conformity with the finding of Abdulsalam et al. (2016) who recorded 0.54 mg/ml of biosurfactant produced by Bacillus subtilis using engine oil as carbon source too. This result also agreed with Pradhan et al. (2022) with regard to biosurfactant production using engine oil as carbon source. The production of biosurfactant using diesel as a carbon source was recorded as 1.90 mg/ml and this result had slightly differed with the finding of Abdulsalam et al. (2016) who recorded 0.63 mg/ml. The slight difference might be due to the fact that the production of the biosurfactant was achieved by two different bacterial species. This result also agreed with Chandran and Das (2010) with regard to biosurfactant production using diesel oil as carbon source and all findings of study for both used engine oil and diesel carbon sources is differ with the finding of Jain et al. (2013) who evaluated several carbon sources which are all soluble in water and found highest yield with starch (10.1 g/L) followed by sucrose (5.1 g/L), xylose (3.25 g/L), galactose (3.1 g/L), glucose (2.75 g/L), and fructose (2.62 g/L). The difference might be due to use of water soluble carbon sources.

4 Conclusion

The physicochemical characteristics of the hydrocarbon-contaminated soil samples from this study revealed that the pH of all soil samples was neutral ranging from 6.80 to 7.06. The average temperature was also 38 to 41 °C. Four (4) bacterial genera were biochemically identified as *Bacillus subtilis, Bacillus subtilis, Bacillus lichenoformis, Bacillus* sp.. *Bacillus subtilis* had the highest frequency of occurrence. The most potent isolate is *Bacillus subtilis* which had highest result of production ability. Based on the two carbon sources used in this research it can be understood that diesel substrate is better than used engine oil for higher yield of biosurfactant produced.

References

Abdulsalam OA, Tijjani MB, Aliyu MS, Garba I, Wada-Kura A (2016) Comparative studies on the biosurfactant production capacity of bacillus subtilis and Pseudomonas Aeruginosa using engine oil and diesel respectively as substrate. Nigerian J Basic Appl Sci 24(2):102–108

Abouseoud M, Maachi R, AmraneA, Boudergua, S. Nabi, A (2008) Evaluation of different carbon and nitrogen sources in production of biosurfactant by *Pseudomonas fluorescens*. Desalination 223:143–151

Adamu A, Ijah UJ, Riskuwa ML, Ismail HY, Ibrahim UB (2015) Isolation of biosurfactant producing bacteria from tannery effluents in Sokoto Metropolis, Nigeria. Int J Inno Sci Eng Technol 2(1):366–373

Akeredolu DO, Akinnibosun FI (2017) Isolation of bacteria and physicochemical analyses of petroleum-products contaminated soil from NNPC/PPMC depot, Benin City. Nigeria 2(4):124–127

Al-Dhabaan FA (2018) Morphological, biochemical and molecular identification of petroleum hydrocarbons biodegradation bacteria isolated from oil polluted soil in Dhahran, Saud Arabia. Saudi J Biol Sci 26:1247–1252

Aparna A, Srinikethan G, Smitha H (2011) Effect of addition of biosurfactant produced by *Pseudomonas spp.* on biodegradation of crude oil. In: 2nd International conference on environmental science and technology 2011, IPCBEE 6: V1-71–V1-75. IACSIT Press, Singapore.

Arku AY, Musa SM, Mofoke AL, Dibal JM (2011) Re-examining raw effluents from Nigerian Bottling Company Maiduguri for crop irrigation. J App Phytotechnol Environ Sanit 1(1):43–49
Association of Official Analytical Chemists (AOAC) (1990) Methods of Analysis. 12th Edition, AOAC, (1990) Washington DC. American Public Health, USA
Banat IM, Makkar RS, Cameotra SS (2008) Potential commercial applications of microbial surfactants. Appl Microbiol Biotechnol 53(5):495–508
Banat IM, Satpute SK, Cameotra SS, Patil R, Nyayanit NV (2014) Cost effective technologies and renewable substrates for biosurfactants' production. Front Microbiol 5:697
Bezza FA, Chirwa EMN (2015) Production and applications of lipopeptide biosurfactant for bioremediation and oil recovery by Bacillus subtilis CN2. Biochem Engr J 101:168–178
Chandran P, Das N (2010) Biosurfactant production and diesel oil degradation by yeast species Trichosporon asahii isolated from petroleum hydrocarbon contaminated soil. Int J Eng Sci Technol 2(12):6942–6953
Desai JD, Banat IM (1997) Microbial production of surfactants and their commercial potential. Microbiol Mol Biol Rev 61:47–64
Fardami AY, Kawo AH, Yahaya S, Riskuwa-Shehu ML, Lawal I, Ismail HY (2022) Isolation and Screening of Biosurfactant-producing Bacteria from Hydrocarbon-contaminated Soil in Kano Metropolis, Nigeria. J Biochem Microbiol Biotechnol 10(1):52 57
Gee GW, Bauder JW (1986) Particle size analysis. In: Klute A (ed) Methods of soil analysis part 1. American Society of Agronomy. Wisconsin, USA, pp 383–411
Gorde SP, Jadhav V (2013) Assessment of water and soil quality parameters: a review. Int J Eng Res Appl 3:2029–2035
Harley JP, Prescott LM (2002) Laboratory exercises in microbiology, 5th edn. McGraw-Hill Companies, New York, United State of America
Hazim RN, Al-Ani MA (2019) Effect of petroleum hydrocarbons contamination on soil microorganisms and biodegradation. Rafidain J Sci 28(1):13–22
Ibrahim ML, Ijah UJJ, Manga SB, Bilbis L (2013) Production and partial characterization of biosurfactant produced by crude oil degrading bacteria. Int Biodeterio Biodegra 81:28–34. https://doi.org/10.1016/j.ibiod.2012.11.012
Ibrahim S, Abdul Khalil K, Zahri KNM, Gomez-Fuentes,Convey C, Zulkharnain A, Sabri S, Alias SA, González-Rocha G, Ahmad SA (2020). Biosurfactant production and growth kinetics studies of the waste canola oil-degrading bacterium *Rhodococcus erythropolis* AQ5-07 from Antarctica. Molecule 25(17):3878
Inès M, Dhouha G (2015) Glycolipid biosurfactants: Potential-related biomedical and Biotechnological applications. Carbohydrate Res 416:59–69
International Institute for Tropical Agriculture (IITA) (1979) Selected methods for soil and plant analysis. Oyo road, Ibadan, Nigeria, pp 102–214
Jain RM, Mody K, Joshi N, Mishra A, Jha B (2013) Production and structural characterization of biosurfactant produced by an alkaliphilic bacterium, Klebsiella sp.: Evaluation of different carbon sources. Colloids Surf B Biointerfaces 108:199–204
Jemimah TB, Jonathan AA, Ojeniyi SO, Joseph TO (2018) Isolation and identification of bacteria in petroleum hydrocarbons polluted soils in North-West Bayelsa State, Nigeria. Covenant J Phys Life Sci 1(2):1–13
Joshi S, Bharucha C, Jha S, Yadav S, Nerurkar A, Desai AJ (2008) Biosurfactant production using molasses and whey under thermophilic conditions. Biores Technol 99:195–199
Juo ASR (1979) Selected methods for soil and plant analysis. International Institute of Tropical Agriculture manual, p 136
Kaya T, Aslım B, Kariptaş E (2014) Production of biosurfactant by *Pseudomonas* spp. isolated from industrial waste in Turkey. Turkey J Biol 38:307–317
Kiran SG, Thomas TA, Selvin J, Sabarathnam B, Lipton AP (2010) Optimization and characterization of a new lipopeptide biosurfactant produced by marine Brevibacteriumaureum MSA13 in solid state culture. Bior Technol 101:2389–2396

Kosaric N (2001) Biosurfactants and their application for soil bioremediation. Food Technol Biotechnol 39(4):295–304

Maneerat S, Phetrong K (2007) Isolation of biosurfactant-producing marine bacteria and characteristics of selected biosurfactant. Songklanakarin J Sci Technol 29(3):781–791

Monge FV, Spracklen DJ, Kirkby M, Schofield L (2022) The impact of semi-natural broadleaf woodland and pasture on soil properties and flood discharge. Hydrol Process 36(1):14453

Said RM, Muhammed GA, Peter SG (2020) Isolation and molecular identification of hydrocarbon degrading bacteria from contaminated soil in mechanic Village Dutse, Jigawa State. Int J Microbiol Biotechnol 5(1):28–33

Mulligan CN (2005) Environmental applications for biosurfactants. Env Pollu 133(2):183–198

National Population Census, National Census Provisional Results. National Population Commission, Abuja, Nigeria (2006)

Nimrat S, Lookchanb S, Boonthai T, Vuthiphandchai V (2020) Isolation, optimisation and gasoline biodegradation by lipopeptide-producing Bacillus subtilis SE1. J Sci Soc Thail 195–205

Ochei J, Kolhatkar A (2000) Medical laboratory science, theory and practice. Tata McGraw Hill Publishers, New Delhi, India

Osakwe SA (2014) Evaluation of physicochemical characteristics of soils in the flood disaster affected areas of Isoko region of delta state, Nigeria. IOSR J Appl Chem (IOSR-JAC). Ver I. 7(5):24–31

Oyewole OA, Zobeashia SLT, Oladoja OE, Musa IO, Terhemba IT (2020) Isolation of bacteria from diesel contaminated soil for diesel remediation. J Bio-Sci 28:33–41

Page AL, Miller RH, Keey DR (1982) Methods of soil analysis-chemical and microbiological properties part-2, 2nd edn. American Society of Agronomy, vol 9. Madison, Wisconsin, USA

Plaza GA, Zjawiony I, Banat IM (2006) Use of different methods for detection of Thermophilic biosurfactant producing bacteria from hydrocarbon-contaminated and bioremediated soils. J Petrol Sci Eng 50:71–77. https://doi.org/10.1016/j.petrol.2005.10.005

Pradhan K, Jujaru M, Gaur S, Kumar S, Jain A (2022) Biosurfactant production by free and immobilized cells of pseudomonas aeruginosa using waste engine oil: a comparative analysis

Priya T, Usharani G (2009) Comparative study for biosurfactant production by using *Bacillus subtilis* and *Pseudomonas aeruginosa*. Botany Res Int 2(4):284–287.

Rabani MS, Habib A, Gupta MK (2020) Polycyclic aromatic hydrocarbons: toxic effects and their bioremediation strategies. Biorem Biotechnol l(4):65–105. (Springer Cham)

Rahman KSM, Rahman TJ, McClean S, Marchant R, Banat IM (2002) Rhamnolipid biosurfactant production by strains of Pseudomonas aeruginosa using low-cost raw materials. Biotechnol Progr 18(6):1277–1281

Salami AT, Farounbi AI, Muoghalu JI (2002) Effects of cement production on vegetation in a part of Southwestern Nigeria. Tanz J Sci 28(2):70–82

Van Hamme JD, Singh A, Ward OP (2006) Physiological aspects: part 1 in a series of papers devoted to surfactants in microbiology and biotechnology. Biotechnol Adv 24(6):604–620

Walter V, Syldatk C, Hausmann R (2010) Screening concepts for the isolation of biosurfactant producing microorganisms. Biosurfactants 1–13

Youssef NH, Duncan KE, Nagle DP, Savage KN, Knapp RM (2004) Comparison of methods in detection of biosurfactant production by diverse microorganisms. J Microbiol Meth 56:339–347. https://doi.org/10.1016/j.mimet.2003.11.00

Utilization of Organic Solid Waste for Hydrocarbon Contaminated Soil Remediation Through Co-composting: A Review

Rizkiy Amaliyah Barakwan

Abstract Hydrocarbons are one of the most contaminants that are difficult to degrade and toxic. Soil pollution by hydrocarbons can be caused by crude oil contamination from petroleum production activities. Therefore, hydrocarbon contaminated soil requires removal technology that is safe for the environment (biological methods). Co-composting is a biological method that can be applied to remediate hydrocarbon contaminated soil. The principle of this method is the addition of nutrients to increase and stimulate the activity of indigenous microorganisms that can decompose hydrocarbons. Co-composting is carried out by mixing polluted soil with organic waste. Co-composting is carried out with a certain composition to obtain the best amount of nutrients for microorganisms (C/N ratio). The types of organic waste added also influence this process. Co-composting has been proven to be able to remove hydrocarbons from polluted soil based on several previous studies.

Keywords Co-composting · Hydrocarbon · Organic waste · Soil remediation · Toxic chemicals

1 Introduction

Soil pollution by hydrocarbon compounds can be caused by various activities, the most common industrial activities of which are petroleum and coal mining (Tran et al. 2021; Sayara and Sanchez 2020). Exploration and exploitation of crude oil which includes drilling, refining, and transportation result in soil pollution by total petroleum hydrocarbons (TPHs) compounds (Parnian et al. 2022). Meanwhile, coal mining causes soil pollution by poly aromatic hydrocarbons (PAHs) compounds (Sayara and Sanchez 2020). Soil pollution by various hydrocarbon compounds occurs due to numerous spills from these various activities continuously. Soil pollution by

R. A. Barakwan (✉)
Environmental Engineering Study Program, Department of Biology, Faculty of Science and Technology, Universitas Airlangga, Surabaya, Indonesia
e-mail: rizkiy.amaliyah@fst.unair.ac.id

A. Z. Yaser et al. (eds.), *Controlling Environmental Pollution*,
https://doi.org/10.1007/978-981-97-8931-3_18

TPHs and PAHs often occurs in developing countries from traditional mining which is still carried out in a simple way, traditional and without good security (Sari et al. 2019).

Hydrocarbons are one of the most common contaminants that generally require remediation because they are carcinogenic organic pollutants closely related to human health and indications of water pollution (Abdelhafeez et al. 2022). Hydrocarbon contamination in soils is toxic, difficult to degrade, and will disrupt the growth of plants and other organisms that grow in it. Crude oil has a hydrocarbon component which is a persistent organic pollutants (POPs) (Yap and Hadibarata 2022). The crude oil contaminated soil that contains hydrocarbon components is a big problem because it is difficult to be degraded by microorganisms (Yap and Hadibarata 2022). This is due to its hydrophobic substances that are difficult to dissolve in water (Abdelhafeez et al. 2022). Based on its content, Minister of Environment Decree No. 128 Years 2003 stated that the crude oil contaminated soil was included in the hazardous and toxic waste (B3) category. This causes crude oil contaminated soil is one of the environmental problems that until now always arise serious attention.

Currently, the most popular technology for remediation of polluted soil containing various toxic contaminant compounds is biological technology or bioremediation techniques. Bioremediation is widely applied because it is inexpensive and environmental friendly, if it compared to physical and chemical methods (Ramadan et al. 2018). One of the bioremediation techniques for removing pollutants from contaminated soil is the co-composting method. Co-composting is carried out by mixing contaminated soil with biodegradable organic materials to add nutrients and encourage the population and activity of microorganisms that are able to remove toxic pollutants, for example, toxic organic compounds that are difficult to decompose, hydrocarbons, etc. (Parnian et al. 2022). The advantages of the co-composting method are low capital and operational costs, simple operation and design, and high efficiency in processing. The co-composting method also has the advantage of improving soil properties and reusing biodegradable organic waste so that it can reduce the volume of waste generation (Lim and Jusoh 2024a; Trihadiningrum et al. 2018).

Various studies have been carried out to prove that co-composting of hydrocarbon contaminated soil using organic materials is able to remove hydrocarbon levels. The results of co-composting of PAHs contaminated soil with organic waste showed PAHs removal of 80% (Lim and Jusoh 2024b). The reduction of hydrocarbons in contaminated soil reached 98.8% after a co-composting process with the mixture of contaminated soil and organic waste (a local sugarcane sugar factory, mixed urea, sugar, and compost) during 3 months of operation (Parnian et al. 2022). Sari et al. (2022) stated that the best composition of hydrocarbon removal in contaminated soil from co-composting process was polluted soil with organic waste (yard waste and rumen waste) (1:1, w/w) with intermittent addition of organic waste. TPH degradation was enhanced until 77.1% using co-composting of the contaminated soil amended with compost (200 kg/soil) and biochar (50 kg/soil) during 60 days of co-composting (Deebika et al. 2021). Ye et al. (2019) stated that co-composting of contaminated wetland soil and a mixture of organic materials (chopped straw, wheat bran, rotten vegetables) with a composition of 11:2:3:8 and biochar addition was able to reduce PAHs pollutants up to 87.5%. Therefore, this review aims to explain that co-composting process can be an alternative method for remediating hydrocarbons contaminated soil.

2 Hydrocarbon Compounds

Hydrocarbons are organic compounds chemically composed of hydrogen and carbon. Based on the chemical structure and number of carbon chains, hydrocarbons can be classified into several types, namely aliphatic, cycloalkane, aromatic, and poly aromatic compounds (Baldan et al. 2015). Therefore, these hydrocarbons have different physical and chemical properties. These types of hydrocarbons can be separated based on differences in volatility through fractional distillation, where the higher the number of C atoms, the lower the volatility (Varjani 2017). Meanwhile, the biodegradability of hydrocarbons depends on the amount of carbon, structure, aromaticity, and the ability to bind to the soil (Ossai et al. 2022a).

2.1 Types and Characteristics of Hydrocarbon Compounds

The types of hydrocarbon compounds in crude oil can be classified as aliphatic hydrocarbons, cycloalkane, aromatic hydrocarbons, and poly aromatic hydrocarbons (Pandolfo et al. 2023; Adedeji et al. 2022).

- Aliphatic Compounds

The rings of atoms carbon of aliphatic hydrocarbons are arranged in a linear way, branching, or circular closed (alicyclic). Aliphatic is also divided become a number of groups, namely:

a. Alkanes (paraffins) which have saturated C atomic bonds.

Alkanes are bonded saturated aliphatic hydrocarbons single and stable to chemical reactions with empirical formula of C_nH_{2n+2}. Alkanes are petroleum hydrocarbons which very easily degraded. However, alkanes in the range C5 to C10 are inhibitor in hydrocarbon degradation process. In high concentrations, this compound is toxic, because it can tear lipid membranes in microorganism cells.

b. Alkenes (olefins) are unsaturated aliphatic hydrocarbons which have at least one double bond with empirical formula of C_nH_{2n}.
c. Alkynes are unsaturated aliphatic hydrocarbons that have at least one triple bond with empirical formula of C_nH_{2n-2}.

- Cycloalkane Compounds

Cycloalkanes are alicyclic (cyclic ring) hydrocarbons single and multiple with the empirical formula of C_nH_{2n}. This compound is very stable but more reactive than alkane. As with alkane compounds, the more the number of C atoms, the higher the Sg (specific gravity), and the boiling point. Degradation of cycloalkanes usually oxidizes at the terminal methyl group and becomes alcohol.

- Aromatic Compounds

Aromatic hydrocarbons are formed from one benzene molecule where 6 atoms are arranged like a ring with single and double bonds (Pandolfo et al. 2023a). Aromatic hydrocarbons have high volatility and low solubility. Aromatics hydrocarbon content also determines the level of toxicity of the crude oil. According to Pandolfo et al. (2023a), ring benzene hydrocarbons are contained in many aromatic hydrocarbons. Examples are: benzene, toluene, ethylbenzene, and xylene which are often referred to as BTEX compounds. Aromatic hydrocarbon compounds are difficult to degrade and can produce intermediate compounds that are not desired. Basic methods of microbial attack on component aromatic ringed single form dihydrodiol compounds. Dihydrodiol is oxidized to form alkyl catechol which is an intermediate compound. The result of the oxidation of ring breakdown is the formation of aldehydes and acids that are ready for use by microorganisms for cell and energy synthesis.

- Poly Aromatic Compounds

Poly aromatic compounds are also called polycyclic aromatic hydrocarbons (PAHs) which consist of several fused aromatic compounds, for example, naphthalene, acenaphthene, and fluorene. PAHs are carcinogenic, the more the number of aromatic molecules combined, the more difficult for PAH compounds to decompose.

The ease level of hydrocarbons degradation can be seen in Fig. 1.

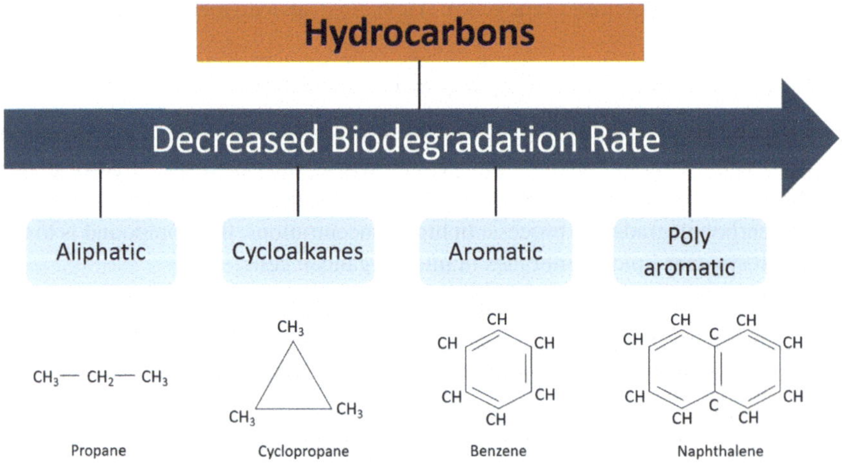

Fig. 1 Sequence of hydrocarbons degradation levels

2.2 Potential of Hydrocarbon Pollution in Soil

Soil pollution is a situation where natural/artificial chemical materials enter and change the natural soil structure (Abdelhafeez et al. 2022). Based on Government Regulation no. 150/2000, it is stated that the destruction of soil for biomass production is a change in the basic properties of the soil that exceeds the criteria standard of soil damage. Exploration and exploitation activities of petroleum which include drilling, production, refining, and transportation have the potential to cause soil pollution. It is caused by oil spills from these various activities. Petroleum contains hydrocarbon components which are nonbiodegradable and toxic organic compounds (Abdelhafeez et al. 2022; Tran et al. 2021). Hydrocarbon contamination of soil which is difficult to degrade and is toxic will be interfered the growth of plants and other organisms that grow inside the soil.

Hydrocarbons in the contaminated soil are usually in the non-aqueous phase liquid (NAPL). NAPL will undergo multiphase transport (diffusion and dispersion), mass transfer (volatilization and sorption), and biodegradation processes. The distribution of hydrocarbons in soil media depends on the properties of the hydrocarbons compound (e.g., solubility, hydrophobicity, and volatility), soil characteristics (e.g., air content, soil texture, pH, and porosity), and environmental conditions (e.g., soil air levels, weather conditions, and biological activity) (Balseiro-Romero et al. 2018; Ossai et al. 2022b).

Van Gestel et al. (2003) stated that soil pollution by hydrocarbons can be caused by diesel spills during the production and transportation processes. Moretto et al. (2005) stated that PAH contaminated soil occurred in the industrial area of Porto Marghera, Venice, Italy. Wang et al. (2011) stated that there was soil pollution in the Delta Yellow River, China because this area is a petroleum production area. For many years, there has been hydrocarbons contamination in this area due to production and transportation of petroleum, oil spills, and pipeline leaks in the soil. Baldan et al. (2015) stated that hydrocarbons contaminated soil resulting from continuous diesel fuel spills over a long period of time, namely 20–25 years happen in Northern Eastern Italy.

Soil pollution by hydrocarbons also occurs in a number of areas in Indonesia. Soil pollution by hydrocarbons occurs around PT CPI Minas oil mine, Riau due to crude oil spills during the drilling, production, and transportation processes (Karwati 2009). Ali (2012) mentions cases of soil pollution due to hydrocarbons, for example, at PT. UNILEVER Jakarta covering an area of 2.2 ha, at PT. CALTEX covering an area of 8 ha, crude oil pipe leak at PT. CONNOCO PHILLIPS 300 m long, and more other cases of pollution. Juliani and Rahman (2011) stated that soil pollution by oil sludge was a result of petroleum mining activities by PT Pertamina Cilacap. Sari et al. (2018) reported that the soil at Wonocolo public oil fields in Indonesia has been polluted by hydrocarbons in the concentration of 52,328.14–107,189.63 μg/g.

3 Co-composting Process

Co-composting is a decomposition process carried out by microorganisms on biodegradable organic waste. Initially, composting technology as a method for contaminated soil remediation did not receive much attention. Although, this technology is applicable in removing contaminated soil using organic compounds for several years (Yap and Hadibarata 2022). Research on remediation of contaminated soil using the composting method is mostly carried out on soil with small concentrations of contaminants. However, based on research by Sari et al. (2019), composting has the potential to remediate large concentrations of contaminated soil. Co-composting is a method of composting organic waste with contaminated soil (Trihadiningrum et al. 2018). Co-composting can also be interpreted as the use of other types of waste together with the waste usually used in the composting process to improve the process while reducing the disposal of other waste (Sayara and Sanchez 2020). The addition of organic matter to contaminated soil with the toxic compounds is useful as a source of co-substrates, nutrients, and microorganisms, improving soil structure and the soil's ability to retain water (Yap and Hadibarata 2022). Co-composting as a bioremediation method has been considered as an appropriate method for bioremediation of contaminated soil to reduce contaminants and improve soil structure and content. The choice of the co-composting method also makes a sustainable contribution to the reuse of biodegradable organic waste which is rich in nutrients and microorganisms (Sayara and Sanchez 2020).

4 Types of Co-composting Methods

According to Tchobanoglous et al. (1993), there are three methods of co-composting windrow system, aerated static pile system, and in vessel system. The windrow systems is commonly used in the application of crude oil contaminated soil remediation using the co-composting method (Yap and Hadibarata 2022). It was caused windrow system is the simplest and cheapest method of making compost. The raw material is organic material stacked lengthwise with a height of 0.6–1 m and a width of 2–5 m with the length of 40–50 m. This method utilizes natural air circulation, so it is influenced by dimensions of the compost raw material pile, humidity, and space pore. The disadvantage of this method is that it requires a large area. According to Yap and Hadibarata (2022) and Tchobanoglous et al. (1993), co-composting can be divided based on the process, namely the aerobic co-composting and anaerobic co-composting.

a. Aerobic Co-composting Method

Aerobic co-composting is a composting method using oxygen gas in the biological process for convert organic waste into stable materials like humus. A number of application on aerobic composting technique is garden waste, household waste, and co-composting with sewage sludge (Tchobanoglous et al. 1993). The reaction that occur in aerobic composting according to Tchobanoglous et al. (1993), namely:

$$Organic\,matter + O_2 + nutrients + microorganisms \rightarrow Compost + new\,cells + CO_2 + H_2O + NO_3 + SO_4^{2-} + heat$$

b. Anaerobic Co-composting Method

Anaerobic co-composting has more complex processes than aerobics composting. Anaerobic co-composting produces more many benefits than aerobic co-composting because of the renewable energy production in the form of methane gas formation (Tchobanoglous et al. 1993). Reactions that occur in anaerobic co-composting according to Tchobanoglous et al. (1993), namely:

$$Organic\,matter + H_2O + nutrients \rightarrow Compost + cells\,new + CO_2 + CH_4 + NH_3 + H_2S + heat$$

5 Hydrocarbon Decomposition Mechanism

The decomposition of hydrocarbons by microorganisms in the co-composting process is included in the bioenergetic or catabolism process. Bioenergetics or catabolism is part of the process metabolism that can produce chemical energy. This process occurs when organic molecules are broken down into simpler molecules (Parnian et al. 2022).

In the process of decomposing hydrocarbons, first of all, hydrocarbons are oxidized by aerobic bacteria. Terminal alkyl group oxidized to form primary alcohol. The general reaction is as follows:

$$Substrate + H\,(Alkanes,\,primary\,alcohol) + O_2 + AH_2 \rightarrow Substrate - OH + H_2O + A\,(hydroxylase)$$

The primary alcohol formed will be further oxidized to aldehyde. This compound is then oxidized to become fatty acids by dehydrogenase which requires NAD. Next, the fatty acids run into a β-oxidation process to then enter the Krebs cycle (Trihadiningrum 2012). Krebs Cycle produces CO_2, CoA, $NADH_2$, $FADH_2$, and ATP. Next on the electron transport system, H atoms coming out of acids organics in the Krebs Cycle are removed by the dehydrogenase enzyme through nucleotide compounds (NAD and NADP), flavoproteins, and cytochromes containing iron (Fe). The final product of hydrocarbon aerobic decomposition is water and ATP (Trihadiningrum 2012).

Ramavandi et al. (2018) stated that aliphatic hydrocarbon degradation mostly occurs aerobically. The process begins with the entering of oxygen into the hydrocarbon assisted by the enzyme oxygenase. The oxygenase enzyme attacks terminal methyl group to convert it to alcohol. Alcohol is oxidized to aldehyde and then becomes carboxylate acid. Carboxylic acids are further degraded through beta oxidase and then enter the Krebs cycle.

6 Hydrocarbon Removal in Contaminated Soil Using Co-composting Processes

6.1 Decomposition Mechanism of Hydrocarbons

In the recovery of crude oil contaminated soil, there are two technique approaches that can be used are bioaugmentation and biostimulation. Bioaugmentation is adding decomposing microorganisms to complement existing microbial populations. Biostimulation is adding nutrients to stimulate the growth of indigenous hydrocarbon decomposers microorganisms. Biostimulation method with additional nutrients in the remediating process of crude oil contaminated soil is assumed to be able to stimulate the growth of soil microbes. Within a certain time, recovery with this technique is possible reducing crude oil concentration. Natural microbial growth in polluted soil will degrade hydrocarbons in soil contaminated with crude oil (Handrianto et al. 2012; Lim and Jusoh 2024a).

Nutrients are elements or compounds that can used to form new cell and as energy source for cell activities (Trihadiningrum 2012). Microorganisms need nutrients for resources energy and the needs of biosynthesis. The energy that microorganisms are needed to carry out all cells activities originates from sunlight or oxidation of chemical compounds. C atoms are needed by all species of organisms because they form very essential cellular components. All organic compounds required by organisms contain C atoms, neither necessary for metabolism (carbohydrate, proteins, fat) nor which necessary to form cell components. All organisms need nitrogen to form acids aminos, nucleotides, and vitamins. Certain bacteria types use proteins or polypeptides as a source of N. Oxygen is required to form most organic molecules, such as nucleotides, amino acids, carbohydrates, etc. The need for this element can be obtained from organic nutrients or free oxygen (Trihadiningrum 2012).

Organic waste which is added to hydrocarbon contaminated soil in the co-composting process plays a role as a material to stimulate indigenous bacterial activity in the crude oil contaminated soil and provide additional levels of nutrients in the soil. Addition of nutrients especially the nutrient levels of N, P, K in crude oil contaminated soil will increase the concentration of nutrient levels in the soil, so that nutrient levels in the soil are sufficient (Sari et al. 2018). Increasing the concentration of soil nutrient levels can stimulate the growth and proliferation of microbes, one of them is hydrocarbonoclastic bacteria (Trihadiningrum et al. 2018).

In the bacterial metabolism, N nutrient is used as a building for protein, nucleic acids, and coenzymes. P nutrient is used as a constituent of nucleic acids, phospholipids, and coenzymes. Hara K is used as a cofactor for several enzymes (Trihadiningrum 2012). C nutrient is used by bacteria as a constituent of cell macromolecules, for example, proteins, carbohydrates, nucleic acids, and lipids. These carbon-containing molecules are involved in the process of metabolism. If the nutrition for this bacteria is enough for proliferation, it will increase the number of bacteria. The increase of the bacteria will optimize the crude oil hydrocarbon degradation process, so the reduction of hydrocarbons concentration is more optimal. The addition of organic waste also has a function to improve soil structure and tissue (loose and retain water), so that it can support plant growth and other organisms life in the soil, and increase the pores in the soil thereby increasing aeration (Sayara and Sanchez 2020).

Biodegradation of hydrocarbon pollutants in contaminated soil can work in the presence of surfactants. This is because surfactants are composed of hydrophilic and hydrophobic groups that can reduce surface tension and separate hydrocarbons from the matrix and soil particles. Surfactants can be formed naturally from biological processes during the co-composting process of organic waste, such as lignocellulose from plants, which is referred to as compost humic acid-like (cHALs). cHALs are produced during the formation of humus. In composting biodegradable organic waste, cHALs are generally formed in the 20-40th days of co-composting. cHALs consist of plant polymers, hydrophilic fractions, and ionic compounds (such as carboxylic acids, aldehydes, alcohols, etc.). Based on the nature of its compounds, cHALs have characteristics like a surfactant. Therefore, cHALs are able to increase the solubility of non-polar organic pollutants (Sari et al. 2018; Trihadiningrum et al. 2018; Sari et al. 2022). Sari et al. (2018) reported that a composition of 50% crude oil contaminated soil and 50% of a mixture biodegradable waste (yard waste and rumen residue) is recommended for co-composting crude oil contaminated soil that produced cHALs in the range of 7.35–38.01%.

Hydrocarbonoclastic microorganisms are the types of microorganisms that can degrade hydrocarbon compounds in the co-composting process. Microorganisms that can oxidize hydrocarbon compounds, including *Pseudomonas fluorescens, P. oleovorans, Acinobacter calcoaceticus, Candida lipolytica, Arthrobacter paraffineus, Corybacterium glutamicum, Bacillus sp., and B. cereus* (Sari et al. 2018, 2022; Arce et al. 2022; Pandolfo et al 2023b).

6.2 Co-composting Reactor

Trihadiningrum et al. (2018) used a laboratory scale experiment for aerobic in-vessel co-composting of hydrocarbons contaminated soil with a mixture of yard waste and rumen residue. Co-composting was conducted using glass reactors. The glass reactors have dimensions of 14.9 × 26.9 cm with a volume of 3.5 L which can be seen in Fig. 2. The reactor will be filled with a mixture of hydrocarbons contaminated soil samples and biodegradable organic waste a total of 1 kg with a thickness of 23 cm.

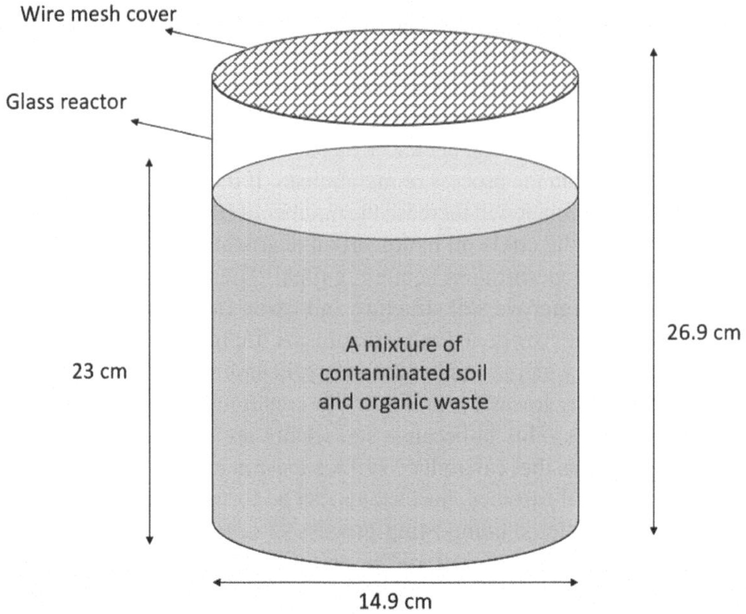

Fig. 2 Laboratory scale of co-composting reactor

Wire mesh cover was used as a cover to protect the top of the glass reactor which is open to foreign objects or animals that can break down the compost and helps in periodic manual aeration.

7 Factors Affecting Co-composting Process

Factors influencing the composting process can be divided into two types, namely based on the composting composition mixture and based on the composting process (Bernal et al. 2009). Bernal et al. (2009) says that the factors that influence the composting process based on a mixture of composting compositions that is:

- Nutrient Balance (C/N ratio)

Nutrient balance in the co-composting process can be defined as the C/N ratio (organic carbon/organic nitrogen ratio). Microorganisms require an energy source (degradable organic C) and N for their growth and activities. The sufficient and appropriate C/N ratio in the co-composting process is in the range of 25-50 because microorganisms require 30°C per unit N. The high C/N ratio can cause co-composting process to be very slow due to the excess of degradable substrate for microorganisms. Small C/N ratio indicates an excessive amount of N per degradable C and excessive amounts of inorganic N that can be lost due to the evaporation of ammonia or release from co-composting mass. A low C/N ratio can improve by adding the bulking agents to provide degradable organic C (Yap and Hadibarata 2022).

- pH Value

The pH range of 6.7–9 is effective for microbial activity during the co-composting process (Bernal et al. 2009). According to Yap and Hadibarata (2022), the optimum pH of the co-composting process is in the range of 5–7. The value of pH is not a main factor in co-composting if the material compost has a pH value in the range of 6.7–9. The pH value is very important in controlling the N degradation due to the evaporation of ammonia. Decreasing in N occurs when the pH values >7.5. Addition of sulfur compounds (S) is carried out to avoid high pH values during co-composting (Mari et al. 2005).

- Particle Size

Particle size and distribution are very important in the surface area balance for the growth of microorganisms and maintain sufficient porosity for aeration. The larger the particle size the smaller the surface areas for growth of microorganisms. Compost with large particle sizes did not decompose optimally due to the inside part of the particulate matter was difficult for microorganisms to reach (Bernal et al. 2009). Neither is the small particle size is not effective because it results in small porosity and meeting. The optimum particle size in the co-composting is 25–75 mmm (Tchobanoglous et al. 1993).

- Porosity

The percentage of porosity in small-scale co-composting is in the range of 35–50%. Higher porosity than 50% can result in a low temperature in the compost pile because energy can remove heat generation (Bernal et al. 2009).

- Humidity

The optimum humidity in co-composting is in the range of 50-60% (Gajalakshmi and Abbasi, 2008). If compost humidity exceeds 60%, then the movement of O_2 is disturbed and causes the process to change anaerobic.

Bernal et al. (2009) said that the factors that influence composting based on process composting are:

- O_2 Requirement

Aeration is one of the important factors in co-composting process. Optimum aeration can control temperature, reduce humidity and the amount of CO_2 excess, and provide O_2 for biological processes (Bernal et al. 2009). The optimum O_2 concentration is in the range of 15–20% (Miller, 1992). Controlling aeration must be able to maintain the compost temperature below 60–65 °C which ensures sufficient O_2 availability.

- Temperature

Changes in temperature values indicate a microbial activity and types of co-composting process. The optimum temperature in the co-composting process is

40–65 °C (Bernal et al. 2009). Temperatures above 55 °C are required for removing pathogenic microorganisms. If the temperature exceeds the thermophilic decomposer range, it can result in damage to the co-composting process. On temperatures above 63 °C, microorganism activity decreases quickly because the optimum thermophilic temperature has been established exceeded and microbial activity will continue to decline up to a temperature of 72 °C. In the temperature range of >50 °C, it is very good for the decomposition process (Yap and Hadibarata 2022). High reduction of the temperature in co-composting can be done in several ways, namely by controlling size and form of the waste and also cooling increase by stirring the compost (Yap and Hadibarata 2022).

- Water Content

The optimum water content in the aerobic co-composting process is in the range of 50-60%. Water content can be controlledby mixing components or adding water. If the compost water content is below 40%, it can cause the composting process to be slow (Tchobanoglous et al. 1993).

8 Case Studies

The following case studies or previous researches have been carried out to prove that co-composting of contaminated soil using organic materials can reduce pollutants contained in Table 1.

9 Conclusions

The co-composting process has been proven to be an alternative in the remediation of hydrocarbon contaminated soil. The main factor that influences this process is the amount of nutrients for the microorganisms in the compost reactor. The nutrients are influenced by the composition (the amount of organic waste and contaminated soil) and the types of organic waste which determines the efficiency of this method.

Table 1 Previous research on co-composting of contaminated soil

No	Previous research	References
1	The optimum TPHs removal (65%) was obtained in the co-composting of sludge: compost (1:0.5) during 63 days of co-composting	Ramavandi et al. (2018)
2	Co-composting of contaminated soil from ship demolition sites in Tanjung Jati, Bangkalan Regency, Madura Island with kitchens and slaughterhouses waste can remove hydrocarbon pollutants until 67.03% for 60 days of co-composting	Syawlia and Titah (2018)
3	In-vessel composting of oil contaminated soil amended with mixed yard waste and rumen waste of 50:50 w/w has removed hydrocarbon of 45.26% within 80 days of co-composting process	Trihadiningrum et al. (2018)
4	Cr(VI)-contaminated soil can be co-composted with the mixture of biomass solid waste (cow dung). The concentration of Cr (VI) reduced from 75 to 10 mg/L during 50 days of co-composting	Chen et al. (2018)
5	Degradation of PAHs in contaminated wetland soil reached 87.5% using co-composting of organic materials (chopped straw, wheat bran, rotten vegetables, and contaminated soil at a weight ratio of 11:2:3:8) and biochar addition	Ye et al. (2019)
6	The best composition that can improve the quality of mine soil (Increasing C/N ratio) is 25% soil: 75% compost of cow dung during 30 days of co-composting	Natanael et al. (2020)
7	Removal of PAHs in contaminated soil reached 72–92% using co-composting of organic waste with the composition of 1:1 (soil: organic waste)	Sayara and Sanchez (2020)
8	Biodegradation using R. communis plant and co-composting (contaminated soil: compost = 3:1) have reduced TPH in the soil by 97.7% during 30 days of co-composting	Madubun et al. (2020)
9	The co-composting process of 50% sludge from refinery of palm oil WWTP and 50% organic matter (yard waste, sawdust, and small wood chips) was the best for reducing Pb up to 60.51% for 28 days of co-composting	Rahmawati et al. (2020)
10	Aerobic co-composting remediation of TPH contaminated soil and organic waste was applied to treat TPH contamination (as high as 380,000 mg kg − 1) until 70–99% of removal efficiency during 30–150 days of co-composting	Tran et al. (2021)
11	The contaminated soil amended with compost (200 kg/soil) and biochar (50 kg/soil) enhanced the degradation of TPH by 77.1% during 60 days of co-composting	Deebika et al. (2021)
12	The removal of hydrocarbon from crude oil contaminated soil using co-composting of poultry manure and wood chips was 77–99% for 53 days of co-composting	Yap and Hadibarata (2022)
13	Co-composting process reduced TPH concentration in petroleum contaminated soil using organic waste (a local sugarcane sugar factory, mixed urea, sugar, and compost) during 3 months of operation (9.89 to 0.109 g/kg)	Parnian et al. (2022)

(continued)

Table 1 (continued)

No	Previous research	References
14	The best composition for co-composting to assist the degradation of hydrocarbon is polluted soil with organic waste (yard waste and rumen waste) (1:1, w/w) with intermittent addition of organic waste every 40 days	Sari et al. (2022)
15	PAHs were removed until 80% of using co-composting of PAHs contaminated soil with organic waste	Lim and Jusoh (2024b)

References

Abdelhafeez IA, El-Tohamy SA, Abdul-Malik MA, Abdel-Raheem SAA, El-Dars FMS (2022) A review on green remediation techniques for hydrocarbons and heavy metals contaminated soil. Curr Chem Lett 11:43–62

Adedeji JA, Tetteh EK, Amankwa MO, Sackey DA, Frimpong SO, Armah EK, Rathilal S, Mohammadi AH, Chetty M (2022) Microbial bioremediation and biodegradation of petroleum products—a mini review. Appl Sci 12:12212

Ali M (2012) Overview of the bioremediation process through oil contaminated soil testing. UPN Press, Surabaya

Arce DCR, Salas MM, Garcia DR (2022) Hydrocarbon-degrading bacteria in Colombia: systematic review. Biodegradation 33:99–116

Baldan E, Basaglia M, Fontana F, Shapleigh JP, Casella S (2015) Development, assesment, and evaluation of a biopile for hydrocarbons soil remediation. Int Biodeterior Biodegrad 98:66–72

Balseiro-Romero M, Monterroso C, Casares JJ (2018) Environmental fate of petroleum hydrocarbons in soil: review of multiphase transport, mass transfer, and natural at

Bernal MP, Alburquerque JA, Moral R (2009) Composting of animal manures and chemical criteria for compost maturity assessment a review. Biores Technol 100:5444–5453

Chen H, Dou J, Xu H (2018) Remediation of Cr(VI)-contaminated soil with co-composting of three different biomass solid wastes. J Soils Sediments 18:897–905. https://doi.org/10.1007/s11368-017-1811-4

Deebika P, Sheela AM, Ilamathi R (2021) Biochar and compost-based phytoremediation of crude oil contaminated soil. Indian J Sci Technol 14(3):220–228

Gajalakshmi S, Abbasi SA (2008) Solidwaste management by composting: state of the art. Environ Sci Technol 38:311–400

Handrianto P, Rahayu SY, Yuliani (2012) Bioremediation technology in overcoming hydrocarbon polluted soil. Prosiding Seminar Nasional Kimia Unesa 2012, Surabaya: FMIPA Universitas Negeri Surabaya, 25 February 2012, hal. 22–30. (inated with petroleum hydrocarbon: a review. Environ Technol Innov 17:100526).

Juliani A, Rahman F (2011) Bioremediation of oil sludge with the addition of compost as a bulking agent and additional nutrient source. Jurnal Sains Dan Teknologi Lingkungan 3(1):1–18

Karwati (2009) Hydrocarbon degradation in soil contaminated with petroleum with isolates A10 and D8. Bogor: Department of Chemistry, Faculty of Mathematics and Science

Lim AHJ, Jusoh MNH (2024a) Remediation of contaminated soil by polycyclic aromatic hydrocarbons using composting. Sustain Environ Insight 1(1):22–31. https://doi.org/10.53623/sein.v1i1.389

Lim AHJ, Jusoh MNH (2024b) Remediation of contaminated soil by polycyclic aromatic hydrocarbons using composting. Sustain Environ Insight 1(1):22–31

Madubun FH, Amin R, Rahyuni D (2020) Phytoremediation of hydrocarbon-contaminated soils using Jarak Kepyar (Ricinus Communis L) and compost. BioLink: Jurnal Biologi Lingkungan, Industri Dan Kesehatan 7(1):38–52. https://doi.org/10.31289/biolink.v7i1.3328

Mari I, Ehaliotis C, Kotsou M, Chatzipavlidis I, Georgakakis D (2005) Use of sulfur to control pH in composts derived from olive processing by products. Compost Sci Uti. 13:281–287

Miller FC (1992) Composting as a process based on the control of ecologically selective factors. Metting FB Jr (ed) Soil microbial ecology, applications in agricultural and environmental management. Marcel Dekker, Inc,, New York, Hal 515–544

Moretto LM, Silvestri S, Ugo P, Zorzi G, Abbondanzi F, Baiocchi C, Lacondini A (2005) Polycyclic aromatic hydrocarbons degradation by composting in a soot-contaminated alkaline soil. J Hazard Mater 126:141–148

Natanael D, Firmansyah M, Mizwar A (2020) Physical and chemical soil properties improvement on post-diamond mining soil in Cempaka using composting method based on cow dung. JTAM Teknik Lingkungan Universitas Lambung Mangkurat 3(1):25–36

Ossai IC, Ahmed A, Hassan A, Hamid FS (2020a) Remediation of soil and water contaminated with petroleum hydrocarbon: a review. Environ Technol Innov 17:100526

Ossai IC, Ahmed A, Hassan A, Hamid FS (2020b) Remediation of soil and water contam

Pandolfo E, Caracciolo AB, Rolando L (2023a) Recent advances in bacterial degradation of hydrocarbons. Water 15:375

Pondolfo E, Caracciolo AB, Rolando L (2023b) Recent advances in bacterial degradation of hydrocarbons. Water 15:375

Parnian A, Anosheh HP, Furze JN, Prasad MNV, Race M, Hulisz P, Ferraro A (2022) Full-scale bioremediation of petroleum-contaminated soils via integration of co-composting. J Soils Sediments 22:2209–2218. https://doi.org/10.1007/s11368-022-03229-5

Rahmawati A, Mizwar A, Prihatini NS (2020) Leading metal (Pb) and zinc (Zn) in sludge wastewater treatment plant and boiler ash industrial refinery and biodiesel oil palm oil with composting. JTAM Teknik Lingkungan Universitas Lambung Mangkurat 3(1):15–23

Ramadan BS, Sari GL, Rosmalina RT, Effendi AJ, Hadrah. (2018) An overview of electrokinetic soil flushing and its effect on bioremediation of hydrocarbon contaminated soil. J Environ Manag 218:309–321

Ramavandi B, Ghafarizadeh F, Alavi N, Babaei AA, Ahmadi M (2018) Biotreatment of total petroleum hydrocarbons from an oily sludge using co-composting approach. Soil Sediment Contam 27(6):524–537. https://doi.org/10.1080/15320383.2018.1489371

Sari GL, Trihadiningrum Y, Ni'matuzahroh (2018) Petroleum hydrocarbon pollution in soil and surface water by public oil fields in Wonocolo Sub-district, Indonesia. J Ecol Eng 19(2):184-193

Sari GL, Trihadiningrum Y, Ni'matuzzahroh (2019) Bioremediation of petroleum hydrocarbons in crude oil contaminated soil from Wonocolo public oilfields using aerobic composting with yard waste and rumen residue amendments. J Sustain Dev Energy, Water Environ Syst 7(3):482-492

Sari GL, Trihadiningrum Y, Ni'matuzahroh, Putra SA, Kasasiah A, Alim MS (2022). Biosurfactant produced by indigenous bacteria during composting process of crude oil polluted soil: properties and role. J Ecol Eng 23(4):297–314. https://doi.org/10.12911/22998993/146693

Sayara T, Sanchez A (2020) Bioremediation of PAH-contaminated soils: process enhancement through composting/compost. Appl Sci 10(3684):1–20. https://doi.org/10.3390/app10113684

Syawlia RM, Titah HS (2018) Removal of hydrocarbons from contaminated soils using bioremediation by aerobic co-composting methods at ship dismantle locations. J Ecol Eng 22(6):181–190. https://doi.org/10.12911/22998993/137441

Tchobanoglous G, Theisen H, Vigil SA (1993) Integrated solid waste management. Mcgraw-Hill, New York. (Tenuation Processes. Pedosphere 28:833–847)

Tran HT, Lin C, Bui XT, Ngo HH, Cheruiyot NK, Hoang HG, Vu CT (2021) Aerobic composting remediation of petroleum hydrocarbon-contaminated soil. Current and future perspectives. Sci Total Environ 753:142250. https://doi.org/10.1016/j.scitotenv.2020.142250

Trihadiningrum Y (2012) Environmental microbiology. ITSPress, Surabaya

Trihadiningrum Y, Barakwan RA, Sari GL, Pandebesie ES, Warmadewanthi IDAA (2018) Hydrocarbon removal in oil-contaminated soil using in-vessel composting with yard waste and rumen waste. J Hazard Toxic Radioact Waste 22(4):04018020. https://doi.org/10.1061/(ASCE)HZ.2153-5515.0000412

Van Gestel K, Mergaert J, Swings J, Coosemans J, Ryckeboer J (2003) Bioremediation of diesel oil-contaminated soil by composting with biowaste. Environ Pollut 125:361–368

Varjani SJ (2017) Microbial degradation of petroleum hydrocarbons. Bioresour Technol 223:277–286

Wang Z, Xu Y, Zhao J, Li F, Gao D, Xing B (2011) Remediation of petroleum contaminated soil through composting and rizosphere degradation. J Hazard Mater 190:677–685

Yap CN, Hadibarata T (2022) Remediation of contaminated soil with crude oil by composting. J Civil Eng Sci Technol 13(1):49–58. https://doi.org/10.33736/jcest.4511.2022

Ye S, Zeng G, Wu H, Liang J, Zhang C, Dai J, Xiong W, Song B, Wu S, Yu J (2019) The effects of activated biochar addition on remediation efficiency of co- composting with contaminated wetland soil. Resour Conserv Recycl 140:278–285. https://doi.org/10.1016/j.resconrec.2018.10.004

Mesophilic Composting of Vegetable Waste Using a Passive Aerated Static Bioreactor: Simultaneous Production of Compost and Leachate

Mariani Rajin, Sariah Saalah, Elora Shyakilla Atinin, Shafina Nazera Mohd Shafizan, Mohd Azmier Ahmad, and Abu Zahrim Yaser

Abstract The performance of vegetable waste composting was assessed using passive aerated static bioreactors at both laboratory and pilot scales. The new formulation tested at the lab scale showed improved compost quality compared to our previous study (Murshid et al., in Int J Recycl Org Waste Agric 13(3):1–12, 2024), prompting a move to pilot-scale experiments. Pilot-scale trials were conducted with Reactor A (featuring four perforated pipes) and Reactor B (equipped with a single perforated pipe). Reactor A exhibited marginally greater efficiency than Reactor B, attributable to enhanced aeration from the additional pipes, resulting in reduced final compost moisture content and increased phosphorus and potassium levels. Each reactor at the pilot scale produced approximately 200 L of leachate, which, based on its quality, has potential for use as a growth medium for plants.

Keywords Kundasang · Circular economy · In-vessel composting · Biodegradation · Chicken manure

1 Introduction

The Kundasang Fresh Market in Sabah is a significant source of organic waste in the Ranau district, generating a total of 13,147.50 kg of organic waste over two weeks, with a generation rate of 6.35 kg per premise per day (Yaacob et al. 2019). Vegetable

M. Rajin · S. Saalah · E. S. Atinin · S. N. M. Shafizan · A. Z. Yaser (✉)
Faculty of Engineering, Universiti Malaysia Sabah, Jalan UMS, 88400 Kota Kinabalu, Sabah, Malaysia
e-mail: zahrim@ums.edu.my

M. A. Ahmad
School of Chemical Engineering, Universiti Sains Malaysia, Engineering Campus, 14300 Nibong Tebal, Penang, Malaysia

A. Z. Yaser et al. (eds.), *Controlling Environmental Pollution*,
https://doi.org/10.1007/978-981-97-8931-3_19

waste, rich in water, nutrients, and microorganisms (Zhang et al. 2023), presents environmental risks if not managed properly. This waste is disposed of through landfilling, which has detrimental ecological impacts. The substantial volume of vegetable waste produced at Kundasang highlights the potential of composting as an effective waste management solution (Sun et al. 2024; Torrijos et al. 2021; Zhang et al. 2023). Composting can transform vegetable waste into valuable nutrients that enhance soil quality and promote plant growth. Additionally, it helps address pollution issues associated with vegetable waste accumulation. Furthermore, the leachate produced during the composting process, containing organic compounds, nitrogen, and phosphorus, can improve crop yield and soil fertility, making it a valuable resource for plant growth and soil enhancement (Ahamad Sanadi et al. 2021; Medvidović et al. 2023; Romero et al. 2013).

Mesophilic composting, which occurs at temperatures between 20 and 45 °C, supports rapid bacterial and fungal growth, generating heat and achieving higher decomposition rates compared to thermophilic composting (Adler and Sikora 2005; Alkarimiah and Suja 2019; Tang et al. 2007). Nitrogen losses are expected to be lower in mesophilic systems because volatile nitrogen loss increases with temperature and airflow, which are higher in thermophilic systems (Hong et al. 1998; Kuroda et al. 1996). Moreover, mesophilic conditions preserve more nutrients compared to thermophilic composting (50–65 °C) (Adler & Sikora, 2005; Paradelo et al. 2013). According to Semitela et al. (2019), mesophilic composting is more energy-efficient and can effectively utilize passive aeration. Passive aerated static pile composting involves placing organic waste in piles with perforated pipes that allow air to flow passively through the pile, thus providing oxygen for aerobic decomposition (Brodie et al. 2000; Jacob et al. 2022; Vu et al. 2015). This method has several advantages, including reduced labor costs compared to turned windrow systems and the capacity to handle large volumes of material with less energy requirement (Brodie et al. 2000; Jacob et al. 2022). In our previous study, Murshid et al. (2024) developed an initial compost mixture composed of vegetable waste, rice husk, and chicken manure in a ratio of 2:1:0.03. This mixture was subjected to passive aeration and turned every 5 days. The results indicated nutrient levels in the compost with N at 0.6%, P at 0.04%, and K at 0.2%. However, the manual turning required during the composting process was labor-intensive. Therefore, this study aims to explore a new compost mixture formulation utilizing a static passive aerated process. The formulation was first optimized using a lab-scale bioreactor. Following the identification of the optimal mixture, it was tested in pilot-scale reactors. These reactors were equipped with different configurations of perforated pipes to assess how various aeration methods affect temperature, moisture content, decomposition rate, and compost quality. The objective is to determine the most effective aeration configuration for efficient composting on a pilot scale.

2 Materials and Methods

2.1 Composting Materials

Both experimental scales maintained an 8:1:1 ratio of vegetable waste, rice husk, and chicken manure for the mixture of raw materials. At the lab scale, three reactors were prepared. Each reactor was loaded with 1.55 kg of vegetable waste, 0.19 kg of rice husk, and 0.19 kg of chicken manure. For the pilot-scale study, two reactors, Reactor A and Reactor B, were each filled with 600 kg of vegetable waste, 75 kg of rice husk, and 75 kg of chicken manure. The composting duration for the lab-scale experiment was 40 days, whereas it was extended to 60 days for the pilot scale.

The vegetable waste for the lab scale was taken from Penampang, Sabah. Rice husks and chicken manure were taken from a previous study (Murshid et al. 2024). For the pilot scale, the vegetable waste was gathered from vendors at the Kundasang Vegetable Fresh Market. Rice husks were acquired from Kg Matan while the chicken manure was procured from the closest poultry farm to Kundasang town (Murshid et al. 2024). To maximize the surface area of the vegetable waste, it was shredded into small pieces, approximately 5 cm in size, before being mixed with rice husk and chicken manure.

2.2 Composting Process and Sampling

The lab-scale composting process was conducted in a Polyethylene Terephthalate (PET) reactor with a capacity of 5 L and a thickness of 0.5 cm. The effective volume of the reactor was 4.0 ± 0.2 L, as illustrated in Figs. 1 and 2. Figure 1 depicts the reactor equipped with containers beneath it to prevent leachate from re-entering. Leachate was gathered through 9 small holes (5 mm in diameter) positioned at the bottom of the reactor. The study was conducted in triplicate.

The pilot-scale composting process was carried out at the Kundasang Composting Site, Sabah, Malaysia, utilizing two brick reactors with a thickness of 12 cm: Reactor A and Reactor B. Each reactor had dimensions of 154 cm in length, 162 cm in height, and 154 cm in width. Four perforated pipes were installed in the center of Reactor A as depicted in Fig. 4. Conversely, Reactor B, illustrated in Fig. 5, was equipped with a single perforated pipe. These installations were designed to facilitate natural ventilation and ensure adequate oxygen supply. Compost samples were collected from six different points within each reactor. The samples from each reactor were then combined and homogenized. At the bottom of the reactor, a nine-hole system facilitates the drainage of leachate, with a catchment container provided for this purpose. Further explanation can be obtained from Murshid et al. (2024) (Figs. 3, 4, 5 and 6).

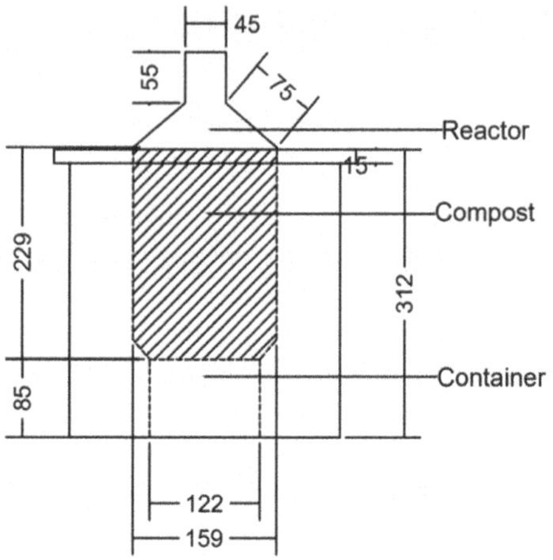

Fig. 1 Side view of the lab-scale reactor (measurement in mm)

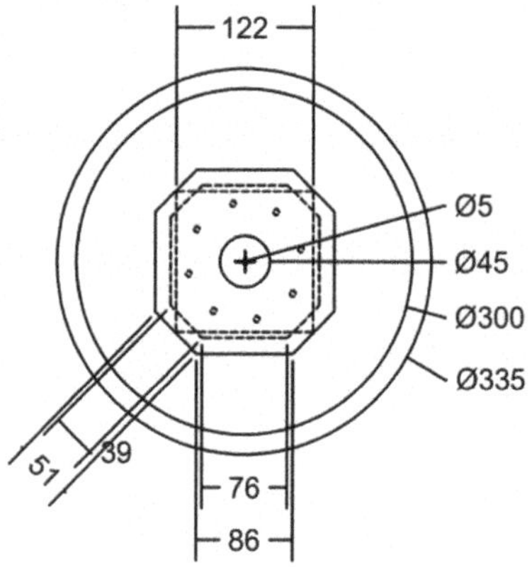

Fig. 2 Plan view of the lab-scale reactor (measurement in mm)

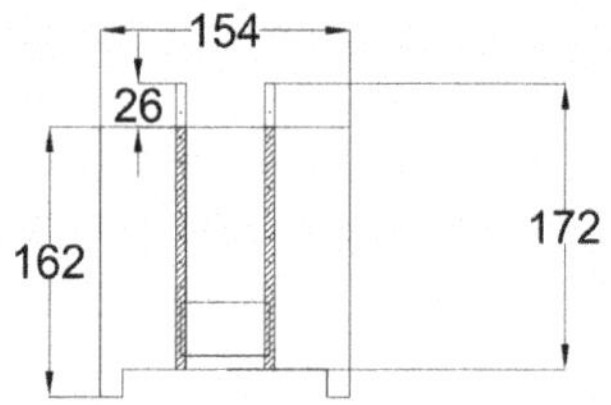

Fig. 3 Front View of Reactor A (measurement in cm)

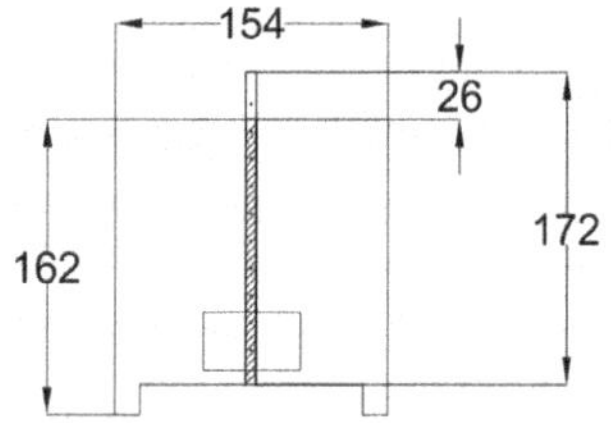

Fig. 4 Front View of Reactor B (measurement in cm)

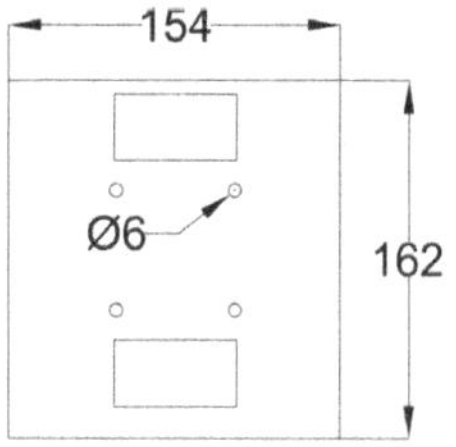

Fig. 5 Top View of Reactor A (measurement in cm)

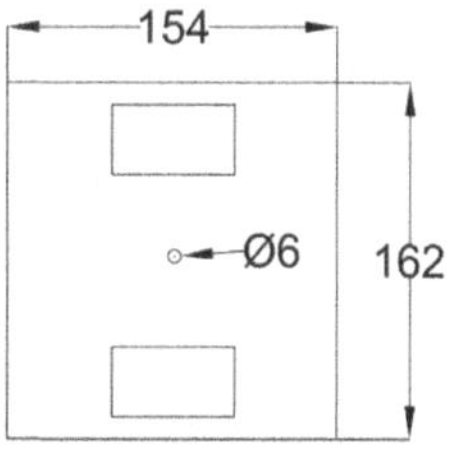

Fig. 6 Top View of Reactor B (measurement in cm)

2.3 Chemical Analysis

The pH, electrical conductivity (EC), total organic carbon (TOC), and total Kjeldahl nitrogen (TN), as well as the C: N ratio, were measured for samples collected from both the raw materials and the initial compost. Total phosphorus (TP), total potassium (TK), and total Kjeldahl nitrogen (TN) were measured for samples from the final composts. Additionally, the compost leachate was analyzed for Chemical Oxygen Demand (COD), Nitrate, Nitrite, Ammonium ion, Phosphate, and Potassium. The pH and EC were determined for a slurry consisting of 10:100 (W/V) compost: distilled water using an Eutech PC 450 pH/EC meter. TN was analyzed by using in house method based on MS 417: Part 3: 1994 while the TP and TK were analyzed by using ICP-OES. The TOC value was determined by heating the oven-dried compost sample to 550 °C for 4 h in a furnace (Faithful Instrument). Equation 1 was used to calculate the VS, while Eq. 2 was used to calculate the TOC.

$$Volatile\,solid(\%) = 100 - Ash(\%) \tag{1}$$

$$\text{TOC } (\%) = \frac{\text{Volatile solid } (\%)}{1.8} \tag{2}$$

The C to N ratio was determined by dividing the TOC value by the TN value, as TOC/TN. COD and Potassium were analyzed by using APHA 5220 D, 2012, and APHA 3120 B, respectively. Ammonium ion, Nitrate, Nitrite, and Phosphate were assessed by using HACH 8308, 8039,8507, and 8048, respectively.

2.4 Physical Analysis

The compost mass for both scales was measured at the beginning and end of the composting process. For the laboratory scale, leachate mass was recorded daily from day 1 to day 15, while for the pilot scale, leachate production was measured on day 60. Wet density calculations were performed at both the initial and final stages of the composting process. Dry density calculations were conducted after determining the moisture content of the compost. Moisture content (MC) was assessed by drying the samples in an incubator at 105 °C for 24 h. Ambient and compost temperatures were recorded daily at the laboratory scale, with the probe placed in the center of the compost pile.

2.5 Gas Emissions

The gas emissions were calculated by using the initial composition of the compost. The higher initial moisture content of the compost in this study may limit oxygen penetration, reducing aeration in the pile. As a result, the composting process might favor anaerobic conditions. Thus, the CH_4 emissions were calculated by using Eq. 4 (Metcalf and Eddy 2004):

$$C_vH_wO_xN_yS_z + \left(v - \frac{w}{4} + \frac{x}{2} + \frac{3y}{4} + \frac{z}{2}\right) \times H_2O \rightarrow \left(\frac{v}{2} + \frac{w}{8} + \frac{x}{4} + \frac{3y}{8} + \frac{z}{4}\right)CH_4 + \left(\frac{v}{2} - \frac{w}{8} + \frac{x}{4} + \frac{3y}{8} + \frac{z}{4}\right)CO_2 + yNH_3 + zH_2S \quad (4)$$

The composition of the compost was analyzed by CHNS Element Analyzer. From the composition of the compost, the CH_4 gas emissions can be calculated. According to Bachman (2019), the oxygen percentage can be calculated if an ultimate analysis is carried out on an oven-dried sample. The equation is as follows (Bachman 2019):

$$O\% = 100\% - C\% - H\% - N\% - S\% - ash\% \quad (5)$$

From Eq. 5 oxygen percentage of vegetable waste and compost was calculated. The data was tabulated in Table 1. The data in Table 1 is used to calculate gas emissions of both the laboratory scale and the pilot scale.

Table 1 Composition percentage of vegetable waste and initial compost for laboratory scale

Sample	C%	H%	N%	S%	O%	Ash%	
Initial compost	23.0	2.1	1.9	0.2	42.1[a]	30.4[b]	This study
Vegetable waste	39.4	7.0	3.2	0.6	43.3[a]	6.6[b]	This study
Vegetable waste	48.0	6.4	2.6	0.4	37.6	NA	Akinshilo (2019)
Vegetable waste	38.4	6.7	1.3	0.3	53.0	0.39	Sui et al. (2021)
Cabbage and rice husk	43.0	5.5	0.7	NA	49.0	NA	Chang et al. (2006)
Mixed vegetables and fruits	35.4	4.7	2.7	0.3	19.3	37.6	Rijo et al. (2023)

[a] Calculated using Eq. 5, [b] Ash at temperature 550 ^{0}C, NA = Not Available

2.6 Statistical Analysis

Statistical analysis and data processing, which involved calculating mean and standard deviation values, were conducted using Microsoft Excel software, 2024. Two-tailed independent t-test was used to compare the sets of data.

3 Results and Discussion

3.1 Lab-Scale Composting Performance

Temperature is a crucial factor in tracking composting efficiency, as it influences biological reaction rates, microbial population dynamics, and the physicochemical properties of compost (Antizar-Ladislao et al. 2005). In this study, the temperature data obtained from the lab-scale experiments are compared with the average temperature recorded in our previous study (Murshid et al. 2024).

Based on Fig. 7, the temperature in the lab scale was lower than the pilot scale (Murshid et al. 2024), aligning with previous research (Hu et al. 2009). According to Hu et al. (2009), the higher temperature in the pilot scale is due to the larger volume of the composting mass, which retains heat more effectively as well as better insulation that helps maintain higher internal temperatures.

Table 2 compares the lab-scale compost quality with findings from other studies. It is noteworthy that the 29% reduction in moisture content achieved in this study is similar to that reported by Bhatia et al. (2012) for composting of vegetable waste, sawdust, and cattle manure. The pH of the compost rises from 5.5 to 7.5, which is consistent with the findings of Bhatia et al. (2019). This pH increase is attributed to ammonization and gas emission during biodegradation (Wang et al. 2024). The EC value in this study is 1.2 mS/cm, which is safe for soil amendment as it is below the threshold of 4 mS/cm (Ajaweed et al. 2022). The nitrogen content dropped from 1.5 to 1.0%, which might be due to excess ammonia volatilization Chang et al. (2006). The ideal C/N ratio for composting is 25–30 (Choi 1999, as cited by Azim

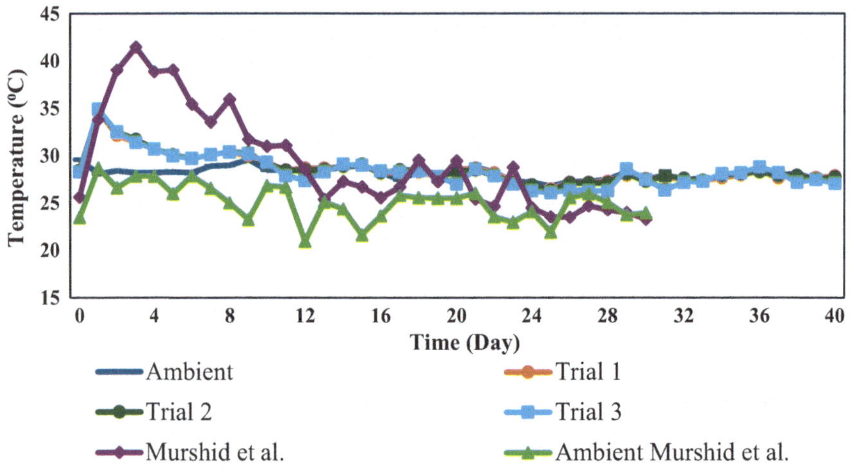

Fig. 7 Temperature profile during lab-scale composting

et al. 2018). In this study, the C/N ratio increased from 25.6 to 33.7 which might be due to the low degradation of carbon during the composting process as well as high ammonia volatilization. However, OM loss was 34.62%, which falls within the acceptable range of 30–70% as specified by TMECC (2002).

At the lab scale, the compost experienced a 50% reduction in mass as shown in Fig. 8. In contrast, Reactor A and B at the pilot scale achieved more substantial reductions of 76% and 81%, respectively. These results are consistent with the findings of Nenciu et al. (2022), that composted 2,235 kg of fruits and vegetables, which reported a cumulative reduction rate of mass exceeding 88%. The lower reduction of mass observed in the lab scale might be attributed to a lower biodegradation rate,

Table 2 Lab-scale compost quality

Parameters	This study		Other studies		References
	Initial	Final	Initial	Final	
MC, %	79	50	70	40	Bhatia et al. (2012)
pH	5.5	7.6	6.5	7.8	Bhatia et al. (2012)
EC, mS/cm	1.6	1.2	5.1	3.3	Bhatia et al. (2012)
TOC, %	38.7	33.1	34.8	28.1	Bhatia et al. (2012)
N, %	1.5	1.0	2.0	0.6	Chang et al. (2006)
P, %	NA	0.3	NA	0.4	Tratsch et al. (2019)
K, %	NA	0.3	NA	3	Tratsch et al. (2019)
C/N	25.6	33.7	26.5	14.3	Bhatia et al. (2012)
OM loss (%)	34.6		39.6		Bhatia et al. (2012)

NA = Not Available

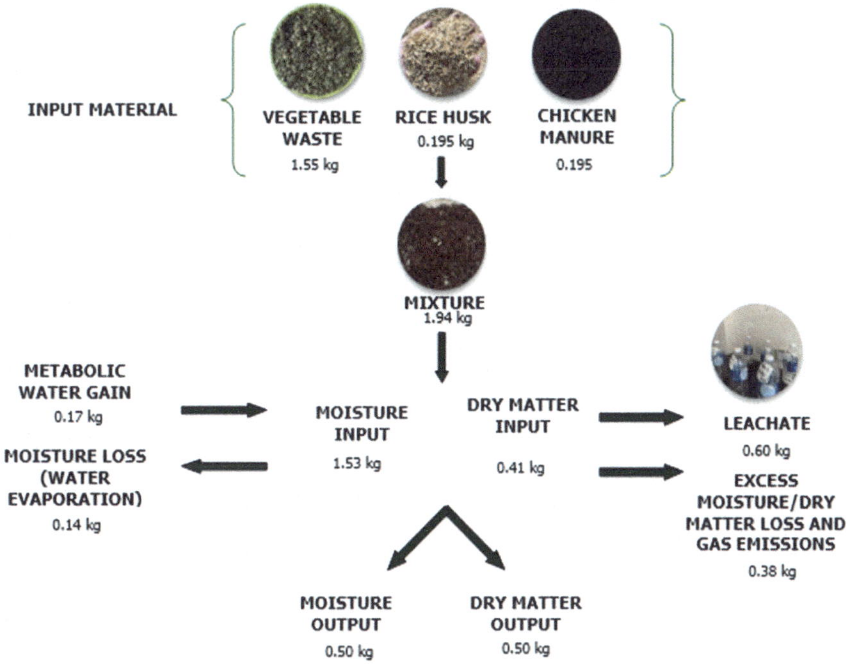

Fig. 8 Total mass balance of vegetable waste passive aerated composting system

likely caused by lower compost mixture temperature. In a lab-scale experiment, the emissions of methane (CH_4) and carbon dioxide (CO_2) were estimated using Eq. 4, resulting in weights of 0.06 kg for CH_4 and 0.04 kg for CO_2, respectively.

3.2 Pilot-Scale Composting Performance

The final compost quality in this pilot-scale study was compared with data obtained from a previous study of Murshid et al. (2024). The previous study utilized the same feedstock with a different weight ratio of 2:1:0.03, resulting in a total wet basis weight of 769 kg.

Based on Table 3, both the pilot scale and Murshid et al. (2024) achieved the ideal moisture content for finished compost, ranging from 40 to 50% (TMECC, 2002). The final pH and EC in this study are slightly higher than Murshid et al. (2024) but still within the acceptable compost range: pH 5.0–8.5 and EC < 4 mS/cm. In this study, the N, P, and K contents were as follows: Reactor A had 1.5% N, 1.3% P, and 1.9% K, while Reactor B had 2.2% N, 0.2% P, and 0.8% K. The nutrient content observed in this study is higher compared to those reported in Murshid et al. (2024). One reason for the lower nitrogen in Murshid et al. (2024) might be due to the

Table 3 Pilot-scale compost quality

Parameter	Reactor A	Reactor B	Murshid et al. (2024)	Accepted range
MC, %	40.2	48.8	62.0	40–50 (TMECC 2002)
pH	8.5	8.3	7.6	5.0–8.5 (TMECC 2002)
EC, mS/cm	3.8	4.0	0.2	<4 (Ajaweed et al. 2022)
TOC, %	37.5	33.9	45.0	NA
N, %	1.5	2.2	0.6	0.5–2.5 (TMECC 2002)
P, %	1.3	0.2	0.04	NA
K, %	1.9	0.8	0.2	NA
C/N	20.6	21.6	77.0	<20 (TMECC 2002)
OM loss (%)	40	46.3	12.0	30–70 (TMECC 2002)

NA = Not Available

lower proportion of chicken manure (1%) used in the initial compost formulation. Furthermore, the frequent turning of the compost every 5 days, as practiced by Murshid et al. (2024), may lead to ammonium loss (Ma et al. 2022). The average C/N ratio for both reactors decreased from 25 to 21, indicating minimal impact from the perforated pipe. This study observed a better C/N ratio and higher organic matter loss compared to Murshid et al. (2024). Based on the results, the compost quality obtained in this study demonstrates superior parameters compared to those of our previous study (Murshid et al. 2024).

3.3 Leachate Quality of Lab and Pilot Scale

The leachate produced from the lab scale was collected and measured every five days. The average leachate production from all trials is 0.3 L/kg. It is expected that leachate production in the pilot scale will be higher than the lab scale, i.e. more than 0.3 L/kg. The leachate quality for both scales was evaluated by using several key parameters such as pH, electrical conductivity, chemical oxygen demand, and nutrient content, as tabulated in Table 4.

The levels of electrical conductivity (EC), chemical oxygen demand (COD), and ammonium ions in the pilot scale are significantly higher compared to the lab-scale findings due to a higher biodegradation rate. The reading for both scales exhibits similar values with the reported value in Sall et al. (2019) except for pH, nitrate, and Phosphorus (P), with P being slightly higher. This elevated P value might be attributed to young leachate characteristics (Ángel López-Ramírez et al. 2019). Elevated ammonium ion concentration in pilot-scale leachate might be due to a higher biodegradation rate which leads to ammonification (Sall et al. 2019).

Table 4 Leachate quality of lab- and pilot-scale composting

Parameters	Lab scale (This study)	Pilot scale (This study)	Sall et al. (2019)
pH	7.7	7.3	5.4
EC, mS/cm	16.4	28.6	13.6
COD, mg/ L	3667.0	9724.0	NA
Ammonium ion, mg/L	127.0	1393.5	491.0
Nitrate, mg/L	105.9	114.9	3.5
Nitrite, mg/L	0.9	0.6	NA
P, mg/L	34.0	41.7	183.0
K, mg/L	3980.0	4460.0	3381.0

NA = Not Available

4 Conclusion

The vegetable waste composting operation was successfully demonstrated in a lab- and pilot-scale setting. It is interesting that the new compost formulation tested at the lab scale outperformed previous study mixtures. On the other hand, Reactor A, with its enhanced aeration due to additional perforated pipes, showed better performance than Reactor B, evidenced by reduced moisture content and increased nutrient levels. The leachate produced holds promise as a valuable growth medium due to its rich nutrient content. Overall, this study demonstrates the effectiveness of the new compost formulation and techniques, leading to improved biodegradation and nutrient profiles. For the Kundasang community, these methods offer a sustainable solution for waste management and agriculture by converting vegetable waste into high-quality compost and nutrient-rich leachate. To ensure long-term success, the community should consider diversifying funding sources, implementing sustainable practices, and exploring alternative revenue streams, such as selling compost and liquid fertilizer. These steps will enhance project sustainability, support agricultural growth, and foster financial independence.

Acknowledgements The authors would like to thank Universiti Malaysia Sabah for funding this work (DLV 2302).

References

Adler PR, Sikora LJ (2005) Mesophilic composting of arctic char manure. Compost Sci Util 13(1). https://doi.org/10.1080/1065657X.2005.10702215

Akinshilo A (2019) Energy potential from municipal solid waste (MSW) for a developing metropolis. J Therm Eng 5(6):196–204. https://doi.org/10.18186/thermal.654322

Ahamad Sanadi NFB, Ibrahim N, Ong PY, Klemeš JJ, Li C, Lee CT (2021) Dilution rate of compost leachate from different biowaste for the fertigation of vegetables. J Environ Manag 295. https://doi.org/10.1016/j.jenvman.2021.113010

Ajaweed AN, Hassan FM, Hyder NH (2022) Evaluation of physio-chemical characteristics of bio fertilizer produced from organic solid waste using composting bins. Sustainability 14(8):4738. https://doi.org/10.3390/su14084738

Alkarimiah R, Suja F (2019) Effects of technical factors towards achieving the thermophilic temperature stage in composting process and the benefits of closed rector system compared to conventional method–a mini review. Appl Ecol Environ Res 17(4). https://doi.org/10.15666/aeer/1704_99799996

Ángel López-Ramírez M, Pirene Castellanos-Onorio O, Alberto Susunaga-Miranda M, Lango-Reynoso F, del Refugio Castañeda-Chávez M, Montoya-Mendoza J (2019) Treatment of leachates of a controlled landfill in Veracruz by using the Fenton method. Nat Environ Pollut Technol Int Q Sci J 18:1–8. www.neptjournal.com

Antizar-Ladislao B, Lopez-Real J, Beck AJ (2005) Laboratory studies of the remediation of polycyclic aromatic hydrocarbon contaminated soil by in-vessel composting. Waste Manag 25(3). https://doi.org/10.1016/j.wasman.2005.01.009

Azim K, Soudi B, Boukhari S, Perissol C, Roussos S, Thami Alami I (2018) Composting parameters and compost quality: a literature review. Organ Agric 8(2). https://doi.org/10.1007/s13165-017-0180-z

Bachmann RT (2019) Re: How to calculate the oxygen content (% by weight) of biomass having high ash content from the ultimate analysis results?. https://www.researchgate.net/post/How-to-calculate-the-oxygen-content-by-weight-of-biomass-having-high-ash-content-from-the-ultimate-analysis-results

Bhatia A, Ali M, Sahoo J, Madan S, Pathania R, Ahmed N, Kazmi AA (2012) Microbial diversity during rotary drum and windrow pile composting. J Basic Microbiol 52(1):5–15. https://doi.org/10.1002/jobm.201100077

Brodie HL, Carr LE, Condon P (2000) A comparison of static pile and turned windrow methods for poultry litter compost production. Compost Sci Util 8(3). https://doi.org/10.1080/1065657X.2000.10701990

Chang JI, Tsai JJ, Wu KH (2006) Composting of vegetable waste. Waste Manag Res J Sustain Circ Econ 24(4):354–362. https://doi.org/10.1177/0734242X06065727

Hong JH, Keener HM, Elwell DL (1998) Preliminary study of the effect of continuous and intermittent aeration on composting hog manure amended with sawdust. Compost Sci Util 6(3):74–88. https://doi.org/10.1080/1065657x.1998.10701933

Hu Z, Lane R, Wen Z (2009) Composting clam processing wastes in a laboratory- and pilot-scale in-vessel system. Waste Manag 29(1):180–185. https://doi.org/10.1016/j.wasman.2008.02.016

Jacob G, Dienerowitz F, Jaschke N (2022) Spatial mathematical modeling of static compost piles with heat recovery. Detritus 20. https://doi.org/10.31025/2611-4135/2022.15213

Kuroda K, Osada T, Yonaga M, Kanematu A, Nitta T, Mouri S, Kojima T (1996) Emissions of malodorous compounds and greenhouse gases from composting swine feces. Bioresour Technol 56(2–3):265–271. https://doi.org/10.1016/0960-8524(96)00047-8

Ma Q, Li Y, Xue J, Cheng D, Li Z (2022) Effects of turning frequency on ammonia emission during the composting of chicken manure and soybean straw. Molecules 27(2):472. https://doi.org/10.3390/molecules27020472

Metcalf and Eddy (2004) Wastewater engineering: treatment and Reuse. 4th Edition, McGraw-Hill, New York

Medvidović NV, Vrsalović L, Svilović S, Bilušić A, Jozić D (2023) Electrocoagulation treatment of compost leachate using aluminium alloy, carbon steel and zinc anode. Appl Surf Sci Advan 15:100404. https://doi.org/10.1016/j.apsadv.2023.100404

Murshid N, Yaser AZ, Lamaming J, Rajin M, Saalah S, Taliban M (2024) Feasibility study of pilot scale vegetable waste composting project for Kundasang community's waste management program. Int J Recycl Org Waste Agric 13(3):1–12. https://doi.org/10.57647/j.ijrowa.2024.1303.27

Nenciu F, Stanciulescu I, Vlad H, Gabur A, Turcu OL, Apostol T, Vladut VN, Cocarta DM, Stan C (2022) Decentralized processing performance of fruit and vegetable waste discarded from retail, using an automated thermophilic composting technology. Sustainability 14(5):2835. https://doi.org/10.3390/su14052835

Paradelo R, Moldes AB, Barral MT (2012) Evolution of organic matter during the mesophilic composting of lignocellulosic winery wastes. J Environ Manage 116:18–26. https://doi.org/10.1016/j.jenvman.2012.12.001

Rijo B, Dias APS, Saksiwi ND, Pereira M, Zăvoianu R, Pavel OD, Ferreira O, Santos RGD (2023) Biofuels from pyrolysis of third-generation biomass from household and garden waste composting bin: kinetics analysis. React 4(2):295–310. https://doi.org/10.3390/reactions4020018

Romero C, Ramos P, Costa C, Márquez MC (2013) Raw and digested municipal waste compost leachate as potential fertilizer: comparison with a commercial fertilizer. J Clean Prod 59:73–78. https://doi.org/10.1016/j.jclepro.2013.06.044

Sall PM, Antoun H, Chalifour F-P, Beauchamp CJ (2019) Potential use of leachate from composted fruit and vegetable waste as fertilizer for corn. Cogent Food Agric 5(1):1580180. https://doi.org/10.1080/23311932.2019.1580180

Sun S, Guo C, Wang J, Ren L, Qu J, Guan Q, Dou N, Zhang J, Chen Q, Wang Q, Wang J, Li J, Gao Z, Zhou B (2024) Effect of initial moisture content, resulting from different ratios of vegetable waste to maize straw, on compost was mediated by composting temperatures and microbial communities at low temperatures. Chemosphere 357:141808. https://doi.org/10.1016/j.chemosphere.2024.141808

Semitela S, Pirra A, Braga FG (2019) Impact of mesophilic co-composting conditions on the quality of substrates produced from winery waste activated sludge and grape stalks: lab-scale and pilot-scale studies. Bioresour Technol 289. https://doi.org/10.1016/j.biortech.2019.121622

Sui W, Li S, Zhou X, Dou Z, Liu R, Wu T, Jia H, Wang G (2021) Potential hydrothermal-humification of vegetable wastes by steam explosion and structural characteristics of humified fractions. Molecules 26(13):3841. https://doi.org/10.3390/molecules26133841

Tang JC, Shibata A, Zhou Q, Katayama A (2007) Effect of temperature on reaction rate and microbial community in composting of cattle manure with rice straw. J Biosci Bioeng 104(4). https://doi.org/10.1263/jbb.104.321

TMECC (2002) Test methods for the examination of composting and composts. Composting Council Research and Education Foundation. Holbrook, New York. CDROM Only

Torrijos V, Calvo Dopico D, Soto M (2021) Integration of food waste composting and vegetable gardens in a university campus. J Clean Prod 315:128175. https://doi.org/10.1016/j.jclepro.2021.128175

Tratsch MVM, Ceretta CA, da Silva LS, Ferreira PAA, Brunetto G (2019) Composition and mineralization of organic compost derived from composting of fruit and vegetable waste. Revista Ceres 66(4):307–315. https://doi.org/10.1590/0034-737x201966040009

Vu QD, De Neergaard A, Tran TD, Hoang HTT, Vu VTK, Jensen LS (2015) Greenhouse gas emissions from passive composting of manure and digestate with crop residues and biochar on small-scale livestock farms in Vietnam. Environ Technol (United Kingdom) 36(23). https://doi.org/10.1080/09593330.2014.960475

Yaacob NNF, Abd Manaf L, Ash'aari ZH (2019) Quantifying the organic waste generated from the fresh market in Kundasang town, Sabah. Plan Malays J 17(10). https://doi.org/10.21837/pmjournal.v17.i10.633

Zhang T, Li H, Yan T, Shaheen SM, Niu Y, Xie S, Zhang Y, Abdelrahman H, Ali EF, Bolan NS, Rinklebe J (2023) Organic matter stabilization and phosphorus activation during vegetable waste composting: multivariate and multiscale investigation. Sci Total Environ 891:164608. https://doi.org/10.1016/j.scitotenv.2023.164608

Utilization of Spent Mushroom Compost (SMC) from *Volvariella Volvacea* Cultivation as a Plant Substrate for *Capsicum Annuum*

Afiq Aizan Hamsun and Noor Azrimi Umor

Abstract Spent mushroom compost (SMC) largely accumulates after the end of mushroom harvesting and must be properly managed other than by incineration or landfill. This study sought to determine the potential use of SMC from the cultivation of *Volvariella volvacea* as a plant substrate due to its high nutrient content. *V. volvacea* is cultivated for 35–40 days using empty fruit bunches (EFB) fiber as substrate. The SMC is treated with hot water before being used for plant cultivation. *Capsicum annum* is also cultivated using two other substrates for comparison, black soil and coco peat for 75 days without fertilizer addition. The growth of the plant, such as height, number of leaves, and pH of the substrates, is measured. The changes in chemical bonding of SMC, black soil, and coco peat were analyzed by Fourier transform infrared spectroscopy (FTIR). From the results, black soil has the highest average plant height and plant weight, 12.6 cm and 4.0 g, respectively. In addition, SMC has the highest leaf number with an average of 7.2, while coco peat has the highest root length with an average of 12.22 cm. The composition of substrates, such as primary amine, alkane, and monosubstituted alkene, decreased significantly after the cultivation process, which was analyzed by FTIR analysis. Therefore, the use of SMC as a growing medium for plant cultures is reasonable because it provides sufficient nutrients for *Capsicum annuum.*

Keywords Spent mushroom compost · *Volvariella volvacea* · *Capsicum annuum*

1 Introduction

Mushrooms are a nutritious vegetable that is high in protein, minerals, and vitamins, low in calories, and high in unsaturated fats (Wiafe-Kwagyan and Odamtten 2018). The consumption of mushrooms is increasing over the years due to the increasing

A. A. Hamsun · N. A. Umor (✉)
School of Biology, Faculty of Applied Science, University Technology Mara (UiTM), Negeri Sembilan Branch, Kuala Pilah Campus (Beting), Pekan Parit Tinggi, 72000 Kuala Pilah, Negeri Sembilan, Malaysia
e-mail: noorazrimi@uitm.edu.my

A. Z. Yaser et al. (eds.), *Controlling Environmental Pollution*,
https://doi.org/10.1007/978-981-97-8931-3_20

demand associated with the dietary style. However, the production of mushrooms also generated waste. For every kilogram of mushrooms harvested, approximately 5 kg of SMC is produced (Williams et al. 2001). The annual production of SMC is estimated to be about 170–204 billion kilograms in 2013 (Zisopoulos et al. 2016).

SMC is characterized as contains a high concentration of neutral detergent fiber (NDF), acid detergent fiber (ADF), hemicellulose, cellulose, lignin, carbohydrate, crude protein (CP), ether extract (EE), ash, dry matter (DM), calcium (Ca), and phosphorus (P) (Mohd Hanafi et al. 2018). Usually, most mushroom growers simply burned or dumped the SMC into the ground, or they simply scattered it on the ground (Umor et al. 2021). These SMC can be used as fertilizer, soil conditioner, for horticulture, and in the renewable energy sector, including direct use as fuel pellets or use as feedstock for biofuel production (Umor et al. 2021). The SMC contains sufficient nutrients and generally nontoxic ingredients for plant growth, such as calcium, protein, ash, and nitrogen, so that it can be used as biofertilizer for growing other crops (Sendi et al. 2013). In addition, SMC has the potential to improve soil structure and increase dry matter production due to its high organic matter content and availability of important plant nutrients (Jordan et al. 2008). The high use of synthetic pesticides and mineral fertilizers in agricultural and horticultural crops contibute to high cost thus increased the need for alternative and cheaper organic fertilizers and compost (Jasińska 2018). SMC provide an option as a cheap and abundance material for plant substrate (Gounani et al. 2011). However a pre-treatment process may applied prior use (Priadi and Saskiawan 2018).

In this study, SMC of empty fruit bunch fiber from *Volvariella volvacea* cultivation was tested as a plant substrate and compared with commercial organic substrate for growth of *Capsicum annum*. *Volvariella volvacea* is known as hot type mushroom that grown in hot climate and largely cultivated in China, India, and south east Asia (Singh et al. 2018).The use of oil palm empty fruit bunch (EFB) fiber for cultivation of this mushroom has been reported in Indonesia, Malaysia, and Thailand (Umor et al. 2020; Triyono et al. 2019; Nur Sakinah et al. 2019). It is expected SMC of EFB based poses various important plant nutrients, high cation exchange capacity, slow mineralization, and high organic matter content to support the growth of vegetables.

2 Materials and Methods

2.1 SMC Collection

After the final harvest, the spent mushroom compost was reused as the main substrate for the cultivation of the *Capsicum annuum* (Fig. 1). First, the SMC was sun-dried for five days to lower its moisture content and eliminate unwanted bio-deterioration agents including maggots, insects, and others (Wiafe-Kwagyan and Odamtten 2018) Then, the SMC was pasteurized by submerging the substrates in boiling water for one to two hours to reduce molds, bacteria, and contamination.

Fig. 1 SMC after the pre-treatment process

2.2 Capsicum Annuum *Cultivation*

Sweet pepper (*Capsicum annuum*) will be used as a test plant because it is one of the most preferred for the short-term cultivation of garden plants. The cultivation of *Capsicum annuum* was conducted for 60 days with three different types of plant substrates. This experiment consisted of the following treatments: black soil (A), coco peat (B), and SMC (C), which include 5 plants per treatment. The seeds were obtained from the fresh bell pepper and were submerged in water for one night before they were ready to be planted. The black soil and coco peat substrates were purchased from the local store in Kuala Pilah town. After that, the seeds of pepper were initially sown in a seed tray filled with seedbed, watered using a spray, and left for 4–5 days before the plants can be transferred into polyethylene bag. The weight of substrates filled in the polyethylene bag was standardized to 800 g for each substrate. The experiment was done with total of 15 units of plant. Layout of the experiment was presented in Fig. 2.

The plants were maintained under natural light conditions and watered two times a day. In this study, the plant height, number of leaves, root length, plant weight,

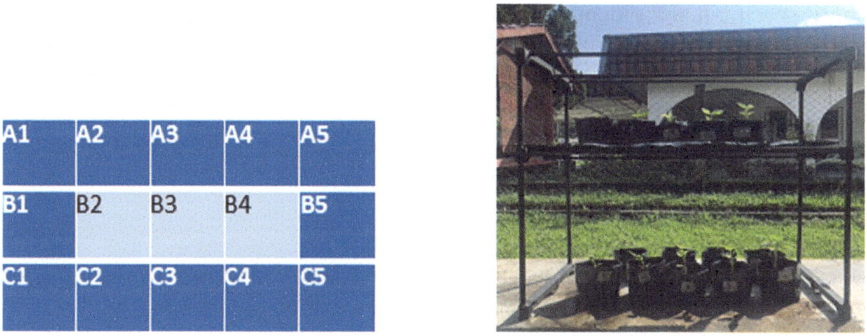

Fig. 2 Layout of the experiment unit

and the pH of the soil were recorded at specific intervals throughout the cultivation process.

2.2.1 Plant Growth Characterization

The plant height was estimated by using a meter ruler placed firmly from the surface of the substrate to the highest leaf tip of the plant in centimeters (Wiafe-Kwagyan and Odamtten 2018). Initial plant height was the same for all samples approximately after transfer from the seedling bed to the substrate. The height of the plant was measured every week. The number of leaves was counted manually per week during the 8 weeks of the growth period to determine the average growth changes between every different substrate. The substrate will be removed completely from the plant and the length of the root will be measured using a ruler.

2.3 PH Value

The pH value of each treatment was measured using a pH meter. 5 g of media was mixed with 25 ml of distilled water, agitated for 30 min, and left standing for 24 h before the pH value can be measured (Jahedur Rahman et al. 2020).

2.4 FTIR Analysis

FTIR analysis of the samples was done using Spectrum 100 Spectrometer Perkin Elmer, to determine changes in the component. FTIR spectra of each sample were recorded in the range of 4000–600 cm $^{-1}$ at 4 cm^{-1} resolution and averaged over 32 scans. Before analysis, the samples were ground into powder and kept at a low temperature of 22∘C. This analysis uses a simple and fast qualitative technique using standard IR spectra. It helps identify the functional group(s) of the samples' components, which can provide insightful observation of compositional changes or degradation (Bradley 2015).

2.5 Statistical Analysis

Data for these experiments were analyzed using IBM SPSS Statistic 26. Means comparison with Tukey's significant test at a level of $\alpha = 0.05$ was performed to determine the significant difference between the treatments.

3 Results and Discussion

Plant Height

Table 1 shows the effects of different substrates on plant height of *Capsicum annuum*. As an indicator of the plant growth, plant height, number of leaves, and root length were monitored. Sweet pepper's vegetative development was greatly influenced by different mixtures of growing substrates, cultivars, and their interactions (Jahedur Rahman et al. 2020).

Figure 3 shows the plant height curves of the three substrates. Based on the 8 weeks average, black soil treatment showed the highest average value of plant height and at the end of the experiment compared to the other two treatments. Highest plant height was obtained from black soil and coco peat with average of 12.6 and 10.66 cm, respectively, and lowest plant height was related to SMC with an average of 10.8 cm (Table 1). Since ($P > 0.05$), there is no significant difference of plant height among the treatments. These results revealed that SMC is comparable to plant substrates and can support plant growth very similar to other commercial substrates. The superiority of black soil could be attributed to appropriate aeration, water retention capacity, low bulk density, and biostability as compared to other treatments (Jahedur Rahman et al. 2020).

Number of Leaves

Table 2 shows the number of leaves of *Capsicum annuum* grown in different substrate samples. Leaf development follows a basic, flexible program and is adjusted according to the species, developmental stage, and environmental circumstances. The data collected on the investigated *Capsicum annuum* plants during the experiment showed that the highest average number of leaves was obtained from the SMC treatment by average of 7.2 leaves. Through the final week of the study, coco peat and black soil showed the similar average number of leaves count both by an average of 7.0.

The number of leaves of *Capsicum* plants grown in three different substrates were not significantly different ($P > 0.05$). This shows that SMC is high in organic matter and a good source of macro and micro nutrients for leaf development comparable to what the other substrates can provide (Wiafe-Kwagyan and Odamtten 2018). Based on previous researches, limited resources and water content in the substrate influenced the formation and development of leaf (Bar and Ori 2014).

Root Length and Plant Weight

Table 3 shows the effects of different substrates on root length and plant weight. Since ($P > 0.05$), there is no significant difference of root length among the treatments. The temperature, moisture capacity, and aeration of growing media can affect root growth, water, and mineral uptake (Gungor and Yildirim 2013). Coco peat sample showed the highest average root length while the lowest average root length was obtained from SMC treatment. It is possibly due to low water holding capacity of coco peat that causes poor air–water relationship, thus affecting oxygen diffusion to roots and its development (Jahedur Rahman et al. 2020). While the SMC has a

Table 1 Effects of different substrates on plant height of *Capsicum annuum*

Sample	Mean/SD Plant Height (cm)							
	Week 1	Week 2	Week 3	Week 4	Week 5	Week 6	Week 7	Week 8
Black soil	3.28 ± 0.54	4.12 ± 0.77	4.98 ± 1.47	6.58 ± 1.86	7.76 ± 1.90	8.60 ± 1.66	9.90 ± 2.11	12.62 ± 3.48
Coco peat	3.44 ± 0.17	3.78 ± 0.33	4.3 ± 0.61	4.96 ± 1.01	5.98 ± 0.85	7.0 ± 1.22	8.66 ± 1.23	10.66 ± 1.30
SMC	2.34 ± 0.33	3.3 ± 0.43	4.16 ± 0.61	4.58 ± 0.62	5.86 ± 1.04	7.0 ± 1.13	8.82 ± 2.20	10.8 ± 2.99

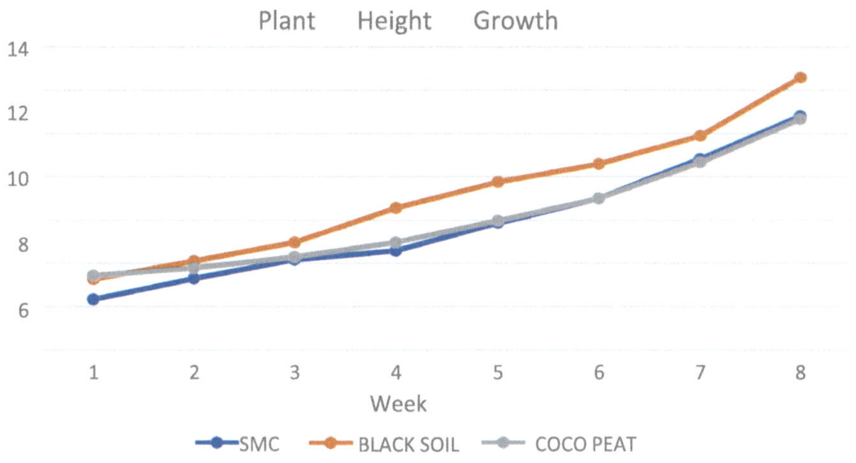

Fig. 3 Plant height growth curves for black soil, coco peat, and SMC

Table 2 Effects of different substrates on number of leaves of *Capsicum annuum*

Sample			Mean/SD leaves count					
	Week 1	Week 2	Week 3	Week 4	Week 5	Week 6	Week 7	Week 8
Black soil	2.20 ± 0.45	2.20 ± 0.45	3.40 ± 0.89	3.80 ± 0.45	4.40 ± 0.55	5.00 ± 0.71	5.80 ± 0.84	7.00 ±1.22
Coco peat	2.00 ± 0.00	2.40 ± 0.89	3.80 ± 0.45	4.00 ± 0.00	4.40 ± 0.55	4.80 ± 0.45	6.00 ± 0.71	7.00 ± 1.00
SMC	2.40 ± 0.55	2.40 ± 0.55	4.00 ± 0.00	4.00 ± 0.00	5.00 ±0.71	5.60 ± 1.14	6.40 ± 0.55	7.20 ± 0.84

higher water holding capacity making the shorter growth of root developed. Figure 4 shows the comparison of the root length of 5 plants from each substrate.

Based on Table 4, the highest plant weight was obtained from black soil treatment with an average of 4.0 g while the coco peat and SMC show more or less the same average of plant weight of 3.8 g and 3.75 g. Since ($P > 0.05$), there is no significant difference of plant weight among all treatments. Higher plant weight of the plants could be attributed to high water availability and proper aeration of water in the plants (Jahedur Rahman et al. 2020).

Table 4 shows the pH of sample before and after the cultivation of *Capsicum annuum*. Most of the plant species and cultivars require a different pH range for optimal development, however, the optimum pH of soilless substrate for adequate availability of key plant nutrients is about 6.0. The initial pH is a significant parameter of every developing substrate composition since it strongly influences the availability and indicates the intrinsic nutritional status of the substrates (16]. The results revealed that the initial pH of SMC and black soil were relatively higher than the optimum pH for the plant growth while coco peat had a lower initial pH of 5.5. SMC also

Table 3 Effects of different substrates on root length and plant weight of *Capsicum annuum*

Variable	Black soil						Coco peat						SMC					
	1	2	3	4	5	**Mean**	1	2	3	4	5	**Mean**	1	2	3	4	5	**Mean**
Root length (cm)	12.2	17.0	5.2	14.1	7.5	**11.2**	12.0	11.8	12.3	11.0	14.0	**12.22**	4.0	6.5	4.2	6.5	8.9	**5.3**
Plant weight (g)	5	4	4	3	4	**4**	3	5	4	3	4	**3.8**	3	4	3	5	5	**3.75**

Fig. 4 Root length of Capsicum plants from different substrates

Table 4 pH difference between raw substrate and after the cultivation of *Capsicum annuum.*

Substrates	Before		After
Black soil	7.2		7.6
Coco peat	5.5		5.4
SMC	7.5		8.3

possesses calcium carbonate, which acts as a short-term buffer for acidic waters and raises soil pH. Aluminium, zinc, and copper ions might accumulate to dangerous levels at very low pH values, although chemical binding can take place at pH values over 7.5 (Jahedur Rahman et al. 2020). All of these occurrences might result in nutritional deficiency in the plants.

FTIR Analysis

Fig. 5 FTIR analysis for raw samples of black soil (A1), coco peat (B2), SMC (C1)

Based on Figs. 5 and 6, the compositions of contents in all treatments were reduced significantly after the cultivation process. For coco peat samples, around peak 3416, the graph shows that the content of primary amine was highly reduced after the cultivation (Bradley 2015). The other contents found in the samples include alkane, monosubstituted alkene, carbonyl, and hydroxyl group. O–H stretching was noticed around peak 3855 in SMC after-used sample, indicating the presence of carbonyl group, and C-H stretching was also found in a raw sample of black soil which might belong to alkene (Bradley 2015).

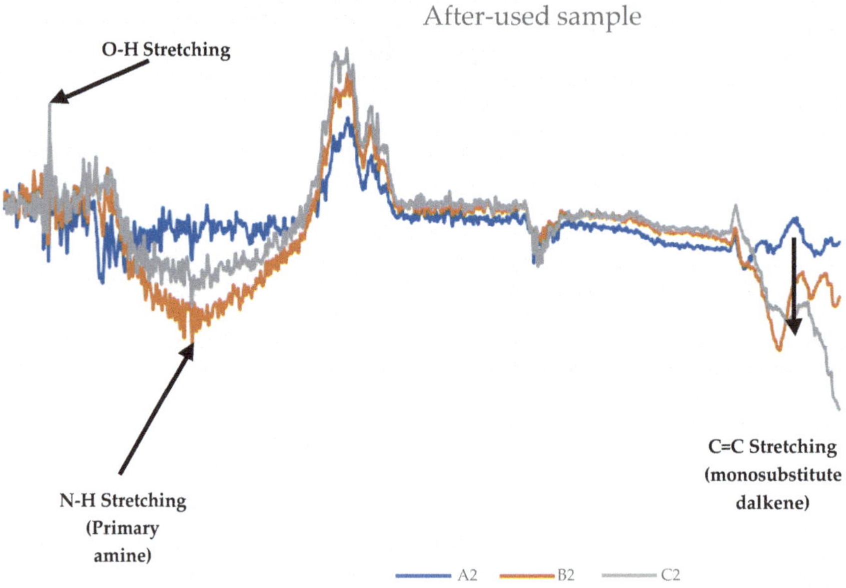

Fig. 6 FTIR analysis for after-used samples of black soil (A2), coco peat (B2), and SMC (C2)

4 Conclusion

In conclusion, the SMC is a good and suitable substrate for supporting the growth of *Capsicum annuum* plants. Based on the findings obtained from the study, black soil growing media showed the better results in the vegetative growth of the *Capsicum* plants than coco peat and SMC. Black soil showed the highest plant height and weight while coco peat showed the highest root length of 12.22 cm. Besides, SMC has the highest leaves count average of 7.2 leaves. Overall composition content of each substrate was significantly reduced which is analyzed by FTIR analysis. For further recommendation, the experiment should be performed over a longer period of time to determine and observe the other growth of the plant such as yield, moisture content, and leaf chlorophyll content. Therefore, it is strongly advised to use and employ SMC in the field experiment of the cultivation of *Capsicum annuum* and any other vegetable plants as preliminary results suggest that it can perform almost similar to the conventional substrate such as black soil and coco peat.

Acknowledgements The authors gratefully acknowledge the host institution Universiti Teknologi MARA Cawangan Negeri Sembilan, Kuala Pilah Campus for research facilities.

References

Bar M, Ori N (2014) Leaf development and morphogenesis. Co. Biol 141:4219–4230

Bradley M (2015) FTIR basic organic functional group reference chart. In: Advancing materials; thermo fisher scientific. https://www.thermofisher.com/blog/materials/a-gift-for-you-an-ftir-basic-organic-functional-group-reference-chart/

Gounani Z, Riahi H, Sharifi K (2011) Impact of using leached spent mushroom compost as a partial growing media for horticultural plants. J Plant Nutr 34:337–344. https://doi.org/10.1080/01904167.2011.536876

Gungor F, Yildirim E (2013) Effect of different growing media on quality, growth and yield of pepper (*Capsicum annuum L.*) under greenhouse_conditions. Pakistan J Botany 45(5):1605–1608

Jahedur Rahman M, Shahjahan Ali M, Islam F, Zannat Zakia M (2020) Physicochemical properties of different organic substrate mixtures and their effects on growth and yield of sweet pepper (*Capsicum annuum L.*) cultivars. http://www.scirea.org/journal/Biology

Jasińska A (2018) Spent mushroom compost (SMC)–Retrieved added value product closing loop in agricultural production. Acta Agraria Debreceniensis 185–202. https://doi.org/10.34101/actaagrar/150/1715

Jordan SN, Mullen GJ, Murphy MC (2008) Composition variability of spent mushroom compost in Ireland. Biores Technol 99(2):411–418. https://doi.org/10.1016/j.biortech.2006.12.012

Mohd Hanafi FH, Rezania S, Mat Taib S, Md Din MF, Yamauchi M, Sakamoto M, Hara H, Park J, Ebrahimi SS (2018) Environmentally sustainable applications of agro-based spent mushroom substrate (SMS): an overview. J Mater Cycles Waste

Nur Sakinah MJ, Misran A, Mahmud TMM, Abdullah S (2019) A review: production and postharvest management of Volvariella volvacea. Int Food Res J 26(2):367–376

Priadi D, Saskiawan I (2018) The utilization of spent oyster mushroom substrates into compost and its effect on the growth of Indian mustard (Brassica juncea (L.) Czern.) in the screenhouse. Int J Agric Technol 14(3):351–362

Sendi H, Mohamed MTM, Anwar MP, Saud HM (2013) Spent mushroom waste as a media replacement for Peat Moss In Kai-Lan *(Brassica oleracea* var *Alboglabra*) production. Sci World J. https://doi.org/10.1155/2013/258562

Singh K, Dey I, Kumar S (2018) Paddy straw mushroom (Volvariella Spp.): a natural scavengers who help in malnutrition and environment protection. Int J Microbiol Res 10(5):1183. https://doi.org/10.9735/0975-5276.10.5.1183-1185

Triyono S, Haryanto A, Telaumbanua M, Dermiyati, Lumbanraja J, To F (2019) Cultivation of straw mushroom (*Volvariella volvacea*) on oil palm empty fruit bunch growth medium. Int J Recycl Organic Waste Agric 8(4):381–392. https://doi.org/10.1007/s40093-019-0259-5

Umor NA, Abdullah S, Mohamad A, Ismail S, Ismail SI, Misran A (2020) Challenges and current state-of-art of the *Volvariella volvacea* cultivation using agriculture waste: a brief review. In: Advances in waste processing technology. Springer Singapore, pp. 145–156. https://doi.org/10.1007/978-981-15-4821-5_9

Umor NA, Ismail S, Abdullah S, Huzaifah MHR, Huzir NM, Mahmood NN, Zahrim AY (2021) Zero waste management of spent mushroom compost. J Mater Cycles Waste Manag 23(5):1726–1736. Springer Japan. https://doi.org/10.1007/s10163-021-01250-3

Wiafe-Kwagyan M, Odamtten GT (2018) Use of *Pleurotuseous* strain P-31 spent mushroom compost (SMC) as soil conditioner on the growth and yield performance of *Capsicum annuum L.* and *Solanum lycopersicon L.* Seedlings under greenhouse conditions in Ghana. Trop Life Sci Res 29(1):173–194. https://doi.org/10.21315/tlsr2018.29.1.12

Williams BC, McMullan JT, McCahey S (2001) An initial assessment of spent mushroom compost as a potential energy feedstock. Bioresour Technol 79(3):227–230

Zisopoulos FK, Becerra Ramírez HA, Van Der Goot AJ, Boom RM (2016) A resource efficiency assessment of the industrial mushroom production chain: the influence of data variability. J Clean Prod 126:394–408. https://doi.org/10.1016/j.jclepro.2016.03.066

Limonite as a Feed Additive Reduces Methane Emissions During Ruminal In Vitro Fermentation

Mohd Huzairi Mohd Zainudin and Toshinari Maeda

Abstract Livestock farming is a significant contributor to anthropogenic greenhouse gas emissions. A major source of these emissions is the methane produced through enteric fermentation, which occurs during the digestive process of ruminant animals. Feed additives have been identified as an effective way to reduce enteric methane emissions. However, not all feed additives are practical or widely available. Therefore, alternative additives need to be identified for local or regional use. This study aimed to evaluate the effect of limonite and tannin on enteric methane generation during in vitro rumen fermentation. The experiment involved adding different concentrations (2 mg, 10 mg and 20 mg) of limonite and tannin to rice straw (200 mg) in rumen liquor. A control treatment without additives was also included. The results showed that the treatment with 2 mg, 10 mg and 20 mg of limonite significantly reduced methane production by 80%, 91%, and 94%, respectively, compared to the tannin treatment, which only reduced methane by 59%, 78% and 87%. This suggests that limonite could be a more effective additive for reducing methane emissions. Limonite has been used in livestock production to improve animal growth performance, primarily due to its high iron and mineral content such as calcium and manganese. This study suggests that limonite could also reduce enteric methane emissions in addition to enhancing animal productivity.

Keywords Enteric fermentation · Ruminant · Limonite · Methane reduction

M. H. M. Zainudin (✉)
Laboratory of Sustainable Animal Production and Biodiversity, Institute of Tropical Agriculture and Food Security, Universiti Putra Malaysia, 43400 Serdang, Selangor, Malaysia
e-mail: mohdhuzairi@upm.edu.my

T. Maeda
Department of Biological Function and Engineering, Graduate School of Life Science and System Engineering, Kyushu Institute of Technology, 2-4 Hibikino-cho, Wakatmastu-ku, Kitakyushu-shi, Fukuoka 808-0196, Japan

A. Z. Yaser et al. (eds.), *Controlling Environmental Pollution*,
https://doi.org/10.1007/978-981-97-8931-3_21

1 Introduction

Methane is a powerful greenhouse gas that contributes to global warming. Its global potential is 21 times greater than carbon dioxide, and it accounts for 16% of total global greenhouse gas emissions. In the EU-15, agricultural emissions of methane have been estimated at 10.2 million tons per year, making it the largest source. Two-thirds of these emissions come from enteric fermentation, and one-third come from livestock manure. The rumen fermentation of feeds is the main source of methane from enteric fermentation. The production of methane in the rumen is affected by the molar percentage of different volatile fatty acids produced during fermentation (Moss et al. 2003).

The rumen is the largest compartment of the complex stomach, occupying the left half of the abdominal cavity. It is filled with digested and partially digested feed material, which is swallowed and regurgitated multiple times. Mechanical mastication and enzymatic action break down hard fibrous feeds into smaller pieces. Ruminant animals utilize anabolic nutrients instead of producing methane, making it important to develop a strategy for mitigating enteric methane emissions that benefit both the animals and the environment (Agarwal et al. 2015). Numerous in vitro and in vivo studies have demonstrated that supplementing animal feed with secondary compounds such as condensed tannins (CT) and iron can effectively lower methanogenesis. This is because tannins are toxic to certain types of rumen fiber-degrading bacteria and methanogenic archaea, thereby reducing the production of methane in the rumen. Limonite is a type of ferric hydroxide that can absorb various gases and harmful substances. It has been proven to be effective in multiple fields. Limonite found in Mt. Aso, Japan, is safe for animals to consume due to its unique mineral composition. It contains a combination of minerals such as iron, calcium, and magnesium. These minerals have been used in livestock farming, pet dietary supplements, and water purification (JAPAN LIMONITE Co. 1960). Tannins are secondary plant metabolites that have anti-microbial effects. They can form complexes in the rumen, affecting hydrogen production and utilization, consequently reducing methane formation in the rumen (Harun 2015).

As far as we are concerned, no previous studies have investigated the use of limonite as a feed additive for methane gas production during in vitro rumen fermentation. Therefore, the proposed experiment aims to determine the effects of limonite on the production of methane and other greenhouse gases during in vitro rumen fermentation. Additionally, the study aims to compare the effectiveness of limonite and tannins in reducing ruminal gas production.

2 Ruminant Nutrition in the Tropics, Feed Additives and Digestive System

The amount of feed required for maintenance varies based on factors such as animal species, age, growth rate, and climate. Irrespective of the purpose of feeding an animal, a part of the feed is utilized to support the animal's vital body processes necessary for survival. This proportion includes the amount of feed necessary to maintain all of the animal's essential tissues when it is not growing, working, or producing any product (Andbook 2009). In the tropics, animal production is influenced by the length and reliability of the wet season, which determines the nature of the animal production enterprise.

The primary feed for ruminant production in Malaysia comprises native grasses and shrubs. However, these tropical forages have lower digestibility values and become fibrous earlier compared to their temperate counterparts. The nitrogen content of these grasses is marginally above 1% (Bakrie et al. 1996). As a result, the quality of the feed is poor, and during the dry season, the supply becomes insufficient for the animals. In addition to these issues, local cattle and buffalo have low intake capacities and a nutrient imbalance. All these factors suggest that native forages can only support low levels of ruminant production in Malaysia.

According to experimental results, local Kedah-Kelantan cattle, and their crossbreeds, grazing on native grasses under farmers' conditions, gained only 170–200 g per day (Liang and Samiyah 1988). In the tropics, ruminant nutritional requirements are different compared to animals in other climates. Temperature, precipitation, and seasonal changes significantly affect the energy maintenance required by the ruminants to survive. Ruminants in tropical countries can adapt and survive in high environmental temperatures and high precipitation of rain in the tropics. Therefore, providing the animals with free access to water throughout the year helps them regulate their body temperature (Harun 2015).

2.1 Rice Straw

Rice straw is a part of the rice plant (Oryza sativa L.) that remains after the grains are harvested. It can be used for various purposes such as burning and leaving on the field before the next plowing, plowing down as a soil improver, or using it as feed for livestock. However, rice straw is not a nutritious feed and must compete with other cereal grain straws like wheat and barley. Despite being a marginal feed, in 1997, most of the rice straw harvested was used for animal feed (Kadam et al. 2000). Rice straw is a significant source of feed for ruminants in many tropical countries, especially during the dry seasons, as it is abundantly available.

Rice straw is an agricultural by-product that is composed of cellulose (37.4%), hemicellulose (44.9%), lignin (4.9%), and silica ash (13.1%) (El-Chaghaby et al.

2018). While it contains a high concentration of fiber (such as cellulose, hemicellulose, and lignin), it has low crude protein content and digestibility from a nutritive value point of view. Previous studies have shown that feeding rice straw as a sole diet to ruminants is not enough to supply nutrients required for production, but rather it only provides nutrients for a maintenance level of production. Therefore, alternative ways to improve its utilization have been sought, such as ammoniation with urea or supplementing it with protein-rich feeds like leaves from tree legumes or shrubs (Kusmartono 2007).

It is expected that the amount of straw will increase along with the progress in the rice production system. In 2008, Asia produced about 620 million tons of rice straw, and now the global production of rice straw is approximately 731 million tons per year. Since 1970, two basic approaches have been used to process the collection of field crop residues. These are post-harvest collection, where residues are collected after the crop is harvested, and simultaneous collection of crop and residue. After harvest, residues can be collected using either the small or medium baling system or the large bales system. However, collecting and managing rice straws is still a major challenge, both for environmental and economic reasons. Rice straw should be collected and used to produce energy or be managed under different options, which becomes unfeasible over long distances due to the low density of rice straw (Sandro 2016).

2.2 *Limonite*

Limonite, also known as the "Lantern mine", is formed when water containing a lot of iron, such as swamp water or shallow ocean water, comes into contact with the air and sedimentation occurs. The resulting precipitate is known as limonite. Limonite was formed in the mountains of Aso, Japan, after three large eruptions occurred 300 thousand years ago, 150 thousand years ago, and 90 thousand years ago. These eruptions caused the formation of a large crater lake between Mt. Aso and Japan. The outer ring of the crater can be used for farming and animal husbandry as a mineral supplement. In 1974, a large amount of red iron oxide, known as "Iron Tan", was discovered in the stone chambers and sarcophagi in the Yayoi ancient burial zone around Aso Town, Japan. This iron oxide was produced by limonite acting on the limestone present in the area. When limonite is heated, the iron component turns reddish-brown and becomes unstable. Limonite, or ferric hydroxide, can be obtained by digesting freshly precipitated ferrous oxide for three years. It is highly effective in absorbing various gases and harmful substances and can be used in a variety of applications, such as water purification, livestock feed additives, and pet dietary supplements (JAPAN LIMONITE Co. 1960).

Limonite contains iron, calcium, and magnesium in a balanced manner and has been found to have many beneficial properties. Its absorption power is demonstrated in all fields, making it a valuable resource for a wide range of applications. Limonite ores are a mixed type of minerals. As a result, nickel is mainly hosted in goethite

(FeNi)O(OH) by substituting for iron or absorption, while quartz is the major non-metallic mineral. The limonitic layer, which includes limonite and part of the saprolite zone, has a higher amount of iron and a relatively lower amount of nickel and magnesia.

2.3 Tannin

Tannins are phenolic compounds found in plants, with molecular weights ranging from 500 to 3000 Da. They occur widely in a variety of plants like monocots, dicots, and ferns, and are commonly consumed by ruminants (Chowdhury et al. 2004). More than 8000 different tannins have been detected in plant cells' free and bound forms. Generally, tannins are commonly found in plant species from tropical or arid/semi-arid areas. Terrestrial plant tannins can be divided into four major groups, which are gallotannins, ellagitannins, proanthocyanidins (condensed tannins), and complex tannins.

Tannins have various applications in medicine, food, and beverage. The effectiveness of tannins depends on their concentration and whether they act as a complexing or precipitating agent (Paridah et al. 2016). Tannins also possess anti-microbial properties and can form complexes with proteins in the rumen, which can affect hydrogen production and utilization, and subsequently, methane production. The effect of tannins varies depending on the type and content present in the animal's diet. While tannins may have beneficial effects, such as reducing methanogenesis, they can also have undesirable effects, such as reducing voluntary dry matter intake and nutrient digestibility, particularly of fiber (Lima et al. 2019). The presence of tannins negatively affects feed intake, digestibility, and production. However, the extent of the effect of tannins depends on the type of tannins present in the animal's diet and its tolerance toward the plant's secondary compounds (Harun 2015).

2.4 Rumen

The rumen is a specialized chamber in the gastrointestinal tract of animals. It plays host to a complex microbial ecosystem that is responsible for extracting nutrients from a variety of agricultural by-products, which are otherwise considered poor-quality lignocellulose. Rumen is the largest compartment of the stomach and occupies the left half of the abdominal cavity. It is a spacious compartment that contains both digested and partially digested feed material. The feed is swallowed and then regurgitated multiple times, and a combination of mechanical mastication and enzymatic action on the hard fiber-containing feeds results in breakdown into smaller pieces. The rumen is divided into different sacs by various pillars, which are folds of the ruminal wall thickened with additional muscle fibers.

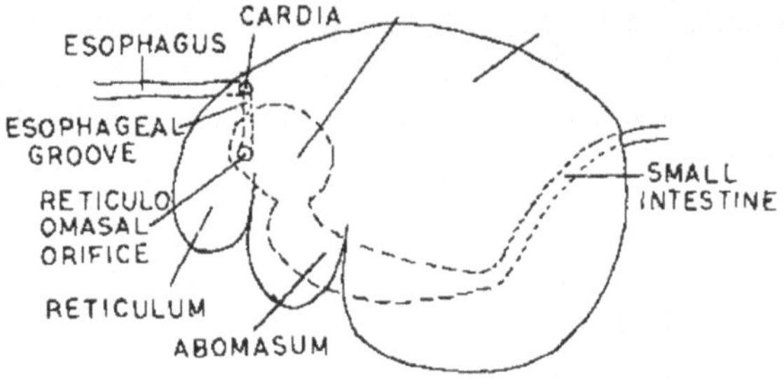

Fig. 1 Diagram of ruminant stomach

The rumen in ruminants is separated from the reticulum by the rumino-reticular fold (Agarwal et al. 2015). This is where ingested food is retained and undergoes extensive microbial digestion and fermentation. During feeding, mastication comminutes the food into particles that vary in volume from one to 103 mm^3 and as long as 10 cm. Young, lush, green feeds are comminuted more than older and drier vegetation. Ruminants further reduce the particle size by rumination, especially with dry feeds. The digesta in the reticulum is regurgitated and rechewed before returning to the rumen. Only particles with a volume of about 5 mm^3 or less can pass from the reticulum into the next compartment, which is the omasum (Hungate 2003). The rumen evolved from being a storage space for the ingested plant feed to become an almost ideal fermentation chamber (Michaiowski 2005). The extended surface area of the rumen facilitates the absorption of fermentation products, particularly volatile fatty acids (VFA) (Zhou et al. 2015) (Fig. 1).

2.5 Rumen Microbial Ecosystem

A healthy rumen is essential for optimum rumen ecology, which includes maintaining a proper balance of microbial population, pH, rumen substrates, and the end products of fermentation activity (Harun 2015). The rumen, which is a four-chambered stomach found in grazing animals, is home to a complex network of microorganisms ranging from archaea to protozoa (Choudhury et al. 2015). These microorganisms work together in a synergistic relationship to extract energy from food and produce lignocellulolytic enzymes, which aid in digestion (Krause et al. 2003). When the rumen ecosystem is disturbed or the microorganisms are absent, it can have a cascading effect on animal health and productivity.

Ruminants rely on the large quantities of plant material present in their rumen, along with the dense microbial population (over 10^{10} cells/g) colonizing the area, to meet up to 80% of their energy requirements from the end products of carbohydrate fermentation (Michaiowski 2005). The rumen ecosystem is classified into three domains of microbes—Bacteria (bacteria), Archaea (methanogens), and Eucarya (protozoa and fungi)—according to Woese's classification. The rumen constantly receives and passes material and has a 10–13% dry matter content. The temperature is typically regulated between 38 and 41 °C, and the pH range is between 5.5 and 6.9, though it is one of the most variable factors in the rumen environment. Complex feedstuffs are broken down into simple sugars, transported, and fermented by other members of the microbial population, which, in turn, produce branched-chain fatty acids, vitamins, or other cofactors for other bacteria responsible for breaking down feedstuffs (Puniya et al. 2015).

2.6 *Rumen Fermentation*

Herbivores consume plant structural carbohydrates such as lignin, cellulose, and hemicellulose. However, they are unable to utilize these compounds themselves because they lack the respective hydrolytic enzymes to break them down. Therefore, symbiotic microorganisms are established in their digestive tracts to help break down these compounds into energy that can be used by both the microorganisms and the host animal. The natural rumen microflora in thc herbivore's digestive system consists of strict anaerobes since anaerobic conditions are dominant in the rumen (Puniya et al. 2015). The rumen ecosystem's microbial ecology is incredibly complex. The system's ability to effectively convert complex carbohydrates to fermentable sugars is partly due to the efficient disposal of H_2 by reducing CO_2 to methane by methanogens. Although methane production can be inhibited for short periods, the system can easily revert to its initial levels of methane production through various adaptive mechanisms (McAllister and Newbold 2008).

2.7 *Methane Gas*

Methane gas is a potent greenhouse gas that has grown 2.5-fold over three centuries, primarily due to agricultural expansion. Ruminant livestock farming is a leading contributor to this growth, as it generates and emits methane during digestion, also known as enteric fermentation (Lassey 2007). Livestock production accounts for a significant 25% of the anthropogenic methane released from enteric fermentation, particularly from ruminant animals (Lesschen et al. 2011; Mustapha et al. 2017). Methane is produced by methanogenic archaea, a specialized group of microbes found in anaerobic environments, including the rumen. These methanogens use primarily H_2 and CO_2 as substrates to produce methane, which fills an important functional niche in the ecosystem.

During fermentation, ruminant animals utilize carbon sources for their nutrition and milk production (Beauchemin et al. 2008; Mustapha et al. 2017). However, this process also produces methane gas, contributing to approximately 17% of total global anthropogenic CH_4 production, making it a significant environmental issue. Several nutritional approaches have been proposed to mitigate the production of ruminant methane (CH_4) emissions, such as supplementing feed with essential oils, nisin, and condensed tannins. However, these methods are often temporary as the rumen microbiota can adapt to changes in the ecosystem created by the mitigating agents. Additionally, some of these agents can also inhibit the growth of important rumen microorganisms, such as peptidase-producing and cellulolytic bacteria, which can lead to lower fiber digestion in the rumen. Therefore, it is important to continue exploring new and effective methods to reduce ruminant methane emissions while ensuring the animals' health and well-being (Candyrine et al. 2018).

3 Materials and Methods

3.1 Animal, Sample, and Rumen Liquor Sampling

This study conducted experiments using different combinations of rice straw as a substrate mixed with varying concentrations of limonite obtained from JAPAN LIMONITE Co., LTD and tannin. The experiment included a control group with unmixed rice straws. All treatments were mixed with 40:60 concentrate and rice straw, and different concentrations of limonite and tannin were added. Samples from each treatment were dried in an oven at 60 °C, ground, and passed through a 1 mm sieve. The sample was weighed to approximately 200 mg before the start of the experiment. The rumen liquor was obtained from a healthy fistulated cow provided by Ladang 15, Universiti Putra Malaysia. The liquor was strained through 4 layers of muslin cloth and kept in an anaerobic environment in a flask.

3.2 Buffer and Mineral Solutions Preparations

The following is the method used to prepare a buffer solution for in vitro studies as per Menke and Steingass (1988). Firstly, 39.0 g of sodium hydrogen carbonate ($NaHCO_3$) was weighed and dissolved in 600 ml of distilled water. The solution was then made up to 1 L. Secondly, a micro-mineral solution was prepared by weighing and dissolving 13.2 g calcium chloride ($CaCl_2{\cdot}2H_2O$), 10.0 g manganese chloride ($MnCl_2{\cdot}4H_2O$), 1.0 g cobalt chloride ($CoCl_2{\cdot}6H_2O$), and 8.0 g iron chloride ($FeCl_3{\cdot}6H_2O$) in 50 ml of distilled water. The solution was then made up to 100 ml with distilled water. Thirdly, a macro-mineral solution was prepared by weighing and dissolving 5.7 g disodium hydrogen phosphate (Na_2HPO_4), 6.2 g potassium dihydrogen phosphate (KH_2PO_4), and 0.6 g magnesium sulfate ($MgSO_4{\cdot}7H_2O$) in 500 ml of distilled water. The solution was then made up to 1 L with distilled water. Fourthly, an anaerobic indicator solution was prepared by weighing 100 mg of resazurin and dissolving it in 100 ml of distilled water. Finally, the reducing solution was prepared by measuring 4 ml of sodium hydroxide (1M NaOH) weighing 625 mg of sodium sulfide ($Na_2S{\cdot}9H_2O$) and adding them to 95 ml of distilled water.

3.3 Preparation of Media

400 ml of distilled water, 0.1 ml of micro-mineral solution, 200 ml of buffer solution, 200 ml of macro-solution, and 1 ml of resazurin solution were mixed in a flask. This mixture was then continuously flushed with carbon dioxide gas and 40 ml of reducing solution was added. The flushing process continued until the mixture changed from bluish to reddish and finally to colorless. After that, rumen fluid was added to the buffer medium mixture in a 1:2 ratio. For in vitro studies approximately 30 ml of buffer media mixture was in each serum bottle.

3.4 Incubation Procedure

For each sample, 12 serum bottles were incubated for 10 days. Twelve treatments, including the standards, were tested in triplicate. For each treatment, 30 ml of rumen liquor was measured and supplemented with a mixture of rice straw and either 1, 5, or 10% limonite or tannin. The serum bottles for each treatment were then tightly sealed with butyl rubber stoppers, crimped, and sparged with carbon dioxide gas for 2 min to create an anaerobic condition. Anaerobic digestion was carried out to understand the impact of limonite and tannin on methane production. The serum bottles were incubated at 37 °C and shaken at 120 rpm for 10 days. The gas produced was collected over water using Dalton's Law every alternate day for a total of 10 days. After 10 days of incubation, the media inside the serum bottles was collected for pH analysis. The samples were stored at −20 °C for further analysis.

3.5 Gases Produced Analysis

The methane, hydrogen, and carbon dioxide volume was analyzed during the incubation using a gas chromatograph. The gas chromatography was set to a current of 100 mA and a temperature of 50 °C. A gas sample of 50 μl was injected into the gas chromatograph, and nitrogen was used as the carrier gas. The identified peak was compared to the standard methane, hydrogen, and carbon dioxide prepared. Each gas sample was injected at least three times as technical replicates. The amount of methane, hydrogen, and carbon dioxide produced was calculated based on a specific equation:

$$y = mx$$

$$y = \frac{m \times x \times 36 \times 10^{-3}}{50 \times 10^{-6}}$$

where,

y = nmole
m = Slope value
x = Area
36 ml = 36 ml serum bottle headspace
50 μl = Volume of gas injected.

4 Results and Discussion

4.1 Ruminal Gases Productions

Table 1 demonstrates the effect of various concentrations of limonite and tannin on ruminal gas production during in vitro rumen fermentation. The results show that the mean values for all gas production in both treatments are significantly different ($P <$ 0.0001), with limonite producing a slightly lower mean value than tannin. The results confirm that ruminal gas production (methane, hydrogen, carbon dioxide) decreases as higher concentrations are used in both treatments of limonite and tannin. This indicates that limonite produces a lower amount of ruminal gas during in vitro rumen fermentation. Overall, the study strongly suggests that adding limonite and tannin significantly reduces the production of methane, hydrogen, and carbon dioxide. The results also indicate that the addition of limonite has a far greater effect on reducing hydrogen and carbon dioxide production, and most importantly, inhibiting methane production compared to tannin. Tannin, on the other hand, has a lower effect on reducing ruminal gas production.

Methane is produced as a result of the fermentation of food within the rumen of animals, particularly during the process of methanogenesis. This methane production plays a crucial role in the disposal of metabolic hydrogen generated during microbial metabolism. Studies have shown that plants and plant extracts have a significant impact on methane production both in vivo and in vitro (Saminathan et al. 2015). Additionally, iron is an effective inhibitor of methanogens and methane production (Van Bodegom et al. 2004). Recent research has shown that limonite and tannin have a significant impact on the production of ruminal gases, as they both have similar mechanisms for inhibiting the production of methane.

It has been found that limonite can reduce acetate and hydrogen levels in the rumen below what methanogens can metabolize, according to a study by Lan and Yang (2019). However, no research has been done on the use of limonite with rumen and the production of ruminal gases during in vitro rumen fermentation. Conversely, tannin inhibits methanogen growth by reducing fiber digestion and causing hydrogen production in the rumen to decrease, according to a study by Tavendale et al. (2005). While the effect of tannin on rumen methanogens cannot be excluded, tannin also inhibits protozoal activity and some fiber degraders due to its anti-microbial activity. Additionally, tannin has the ability to interfere with the availability and efficiency of

Table 1 The effect of different concentrations of limonite and tannin during in vitro rumen fermentation

Gas	Day	Gas production (μmol)								SEM	*P*-value
		Limonite				Tannin					
		Control	2 mg	10 mg	20 mg	Control	2 mg	10 mg	20 mg		
CH_4	2	175.22 ± 12.79^{f}	151.03 ± 1.37^{hij}	148.04 ± 2.17^{ijk}	132.50 ± 1.35^{lmn}	187.55 ± 2.96^{e}	180.19 ± 6.30^{ef}	160.59 ± 16.94^{gh}	129.70 ± 9.18^{mn}	3.78	<0.0001
	4	189.09 ± 15.59^{e}	141.82 ± 2.19^{jkl}	132.28 ± 0.65^{lmn}	111.87 ± 6.85^{o}	207.42 ± 4.91^{d}	164.25 ± 2.87^{g}	137.51 ± 1.30^{klm}	106.66 ± 11.32^{opq}		<0.0001
	6	207.83 ± 0.99^{d}	135.27 ± 0.53^{lm}	101.01 ± 2.25^{pq}	79.97 ± 1.15^{r}	220.69 ± 5.26^{c}	156.82 ± 1.37^{ghi}	122.73 ± 1.04^{n}	96.83 ± 4.49^{q}		<0.0001
	8	230.55 ± 0.62^{abc}	109.89 ± 6.66^{op}	61.88 ± 3.91^{st}	49.84 ± 5.96^{u}	225.58 ± 1.13^{bc}	133.54 ± 1.37^{lm}	85.11 ± 4.63^{r}	65.00 ± 4.24^{s}		<0.0001
	10	237.07 ± 0.44^{a}	46.42 ± 0.62^{u}	21.23 ± 8.70^{vw}	13.61 ± 0.12^{vw}	236.17 ± 3.04^{ab}	97.07 ± 4.26^{q}	52.43 ± 4.85^{u}	31.09 ± 0.22^{v}		<0.0001
H_2	2	117.55 ± 3.02^{f}	104.14 ± 0.26^{hijk}	103.34 ± 1.35^{ijk}	99.55 ± 1.67^{kl}	118.46 ± 1.11^{f}	111.01 ± 1.55^{g}	109.27 ± 0.08^{gh}	107.28 ± 0.31^{ghi}	1.87	<0.0001
	4	125.45 ± 0.37^{e}	100.88 ± 0.36^{jk}	95.32 ± 6.85^{lm}	88.06 ± 1.27^{nop}	124.60 ± 0.56^{e}	107.01 ± 3.05^{ghi}	104.90 ± 0.75^{hij}	101.55 ± 2.06^{jk}		<0.0001
	6	$137.25 \pm 2.02^{\mathbf{d}}$	90.40 ± 0.70^{mno}	87.88 ± 0.09^{nop}	85.63 ± 0.23^{op}	138.27 ± 0.45^{d}	100.86 ± 2.04^{jk}	92.16 ± 3.38^{mn}	89.04 ± 0.31^{no}		<0.0001
	8	144.48 ± 1.61^{c}	64.48 ± 3.39^{rs}	61.69 ± 0.55^{t}	58.86 ± 0.85^{t}	155.65 ± 1.75^{b}	87.516 ± 10.05^{nop}	83.29 ± 1.24^{p}	77.41 ± 1.86^{q}		<0.0001
	10	158.70 ± 0.73^{b}	46.30 ± 1.46^{u}	33.82 ± 4.06^{v}	28.94 ± 0.43^{v}	173.39 ± 1.81^{a}	77.36 ± 9.20^{q}	67.51 ± 1.03^{r}	60.08 ± 0.86^{ts}		<0.0001
CO_2	2	719.76 ± 17.95^{de}	666.93 ± 1.54^{fg}	652.02 ± 1.72^{gh}	601.94 ± 23.35^{lmn}	727.43 ± 25.76^{d}	704.67 ± 18.62^{e}	672.42 ± 6.63^{f}	608.13 ± 22.09^{lm}	5.97	<0.0001
	4	759.64 ± 8.26^{c}	646.39 ± 1.48^{hi}	608.29 ± 3.17^{lm}	592.55 ± 2.30^{mnop}	764.42 ± 16.29^{c}	651.39 ± 1.44^{gh}	641.24 ± 9.19^{hij}	607.21 ± 1.31^{lm}		<0.0001
	6	785.13 ± 6.48^{b}	614.37 ± 13.00^{kl}	613.12 ± 3.44^{kl}	582.66 ± 5.81^{opqr}	800.25 ± 0.92^{ab}	632.25 ± 8.12^{ij}	627.70 ± 1.23^{kj}	597.67 ± 0.51^{lmno}		<0.0001
	8	803.49 ± 2.83^{a}	589.27 ± 0.52^{nopq}	579.79 ± 6.14^{pqrs}	568.18 ± 8.80^{rst}	807.77 ± 0.88^{a}	593.84 ± 3.05^{mnop}	583.51 ± 5.90^{opqr}	573.11 ± 4.71^{qrst}		<0.0001
	10	809.11 ± 2.38^{a}	574.78 ± 1.05^{qrs}	559.84 ± 1.05^{t}	526.05 ± 4.61^{u}	816.32 ± 0.61^{a}	589.85 ± 0.66^{nopq}	580.38 ± 4.74^{pqrs}	565.58 ± 0.62^{st}		<0.0001
Total gas produced	2	1012.54 ± 8.19^{i}	922.10 ± 0.10^{jk}	903.41 ± 5.24^{kl}	833.99 ± 26.37^{mno}	1033.44 ± 23.91^{h}	995.88 ± 10.77 i	942.27 ± 23.49^{j}	845.11 ± 31.58^{m}	7.30	<0.0001
	4	1074.18 ± 6.96^{g}	889.09 ± 4.02^{l}	835.89 ± 9.37^{mn}	792.48 ± 5.82^{q}	1096.44 ± 20.64^{f}	922.64 ± 7.36^{jk}	883.65 ± 7.14^{l}	815.42 ± 14.69^{nop}		<0.0001
	6	1130.21 ± 7.50 e	840.04 ± 11.77^{m}	802.01 ± 5.78^{pq}	748.26 ± 7.20^{s}	1159.21 ± 3.90^{d}	889.94 ± 11.53^{l}	842.60 ± 3.57^{m}	783.54 ± 4.29^{qr}		<0.0001
	8	1178.56 ± 1.84^{cd}	763.83 ± 9.53^{rs}	703.36 ± 10.60^{t}	676.89 ± 13.91^{u}	1189.00 ± 0.25^{bc}	814.90 ± 8.37^{op}	751.91 ± 2.51^{s}	715.52 ± 10.81^{t}		<0.0001
	10	1204.88 ± 2.09^{b}	667.51 ± 1.88^{u}	614.89 ± 3.59^{v}	568.61 ± 5.16^{w}	1225.88 ± 4.25^{a}	764.27 ± 5.61^{rs}	700.323 ± 0.93^{t}	656.75 ± 0.46^{u}		<0.0001

CH_4: Methane, H_2: Hydrogen, CO_2: Carbon dioxide, SEM: Standard error mean. All means are significantly different ($P < 0.0001$)

protein utilization in the digestive tract by forming complexes, according to a study by Hess et al. (2006).

The in vitro rumen fermentation involves the production of hydrogen by limonite and tannin. However, this hydrogen production leads to the inhibition of methanogenesis in the rumen. The hydrogen produced during ruminal fermentation is the primary substrate utilized by methanogens, and methanogenesis represents the most significant hydrogen sink in the rumen. Therefore, reducing hydrogen production in rumen fermentation can disrupt the methane production pathway.

A recent experiment conducted by Martinez-Fernandez et al. (2016) showed that when hydrogen production was reduced and methanogenesis was inhibited, there was a shift in rumen fermentation to the production of propionate and branched-chain fatty acids. This shift occurred because the ruminal microorganisms changed their fermentation patterns in response to decreasing their dependence on pathways that become thermodynamically less favorable. Instead, they shifted to thermodynamically more favorable pathways, as explained by Russell and Wallace (1997).

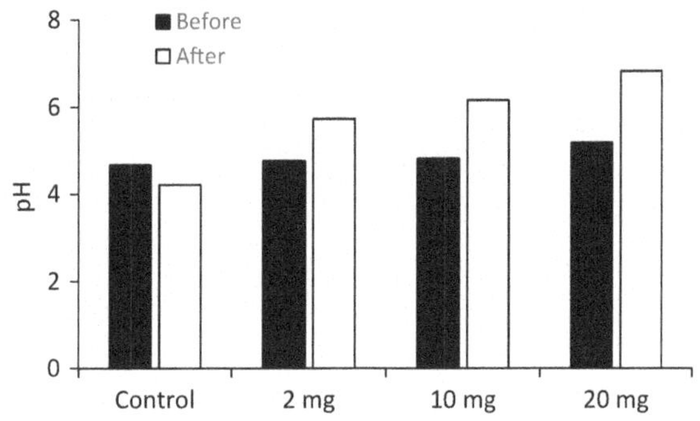

Fig. 2 pH level in treatment with limonite

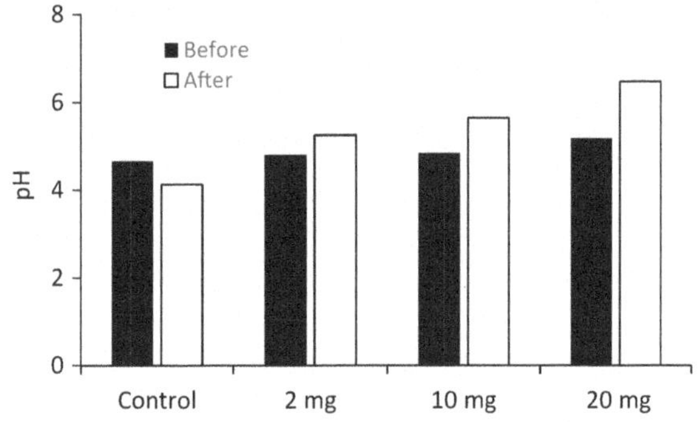

Fig. 3 pH level in treatment with tannin

4.2 pH of Rumen in Vitro Fermentation

The study examined the in vitro rumen pH levels and the impact of limonite and tannin on these levels. The results presented in Figs. 2 and 3 showed that the in vitro rumen pH levels for the control group slightly decreased by the end of the experiment. In contrast, the groups receiving 2, 10 and 20 mg of limonite and tannin displayed an increase in the in vitro rumen pH levels. This study recorded the highest pH value in the group receiving 20 mg of limonite, while the lowest value was observed in the control group which did not receive any limonite or tannin. Despite the differences in pH levels among the groups, the results indicated that the differences were only slight. This suggests that the addition of limonite and tannin had a minimal impact on in vitro rumen pH levels. Overall, the study provides valuable insights into the impact of limonite and tannin on rumen pH levels, which can have important implications for ruminant animal nutrition.

Saminathan et al. (2015) revealed that the rumen, which is a complex digestive system in cows, works best when the pH level is between 6.89 and 7.02. This pH range is vital for maintaining normal activities of the rumen such as cellulolysis and proteolysis, which break down complex food components into smaller and simpler ones. When the pH level becomes too low, it negatively affects the ion transport through the membrane, inhibits enzymes, and causes the membrane proteins to denature. This, in turn, prevents microbial growth. Methanogens, microorganisms that produce methane, can survive in pH levels below 6.0 by entering metabolic stasis, a state of reduced metabolic activity. However, if the pH level drops below 6.0, methane production in cow rumen fluid stops. Once the pH level is restored to above 6.0, methane production resumes, as observed by Kessel and Russell (1996). Overall, this study showed that the treatment with 2 mg, 10 mg, and 20 mg of limonite significantly reduced methane production by 80%, 91%, and 94%, respectively, compared to the tannin treatment, which only reduced methane by 59%, 78%, and 87% (Table 2). This suggests that limonite could be a more effective additive for reducing methane emissions. Limonite is a naturally occurring mineral rich in iron and minerals such as calcium and manganese. This mineral has been extensively used in livestock production to improve animal growth performance and overall health. Recent studies have shown that apart from its nutritional benefits, limonite could also be an effective way to reduce enteric methane emissions in livestock. Enteric methane emissions are a significant contributor to greenhouse gas emissions; reducing them could positively impact the environment. Therefore, the use of limonite in livestock production not only improves animal productivity but also has the potential to mitigate the negative impact of livestock farming on the environment.

Table 2 Summary of enteric methane reduction during in vitro rumen fermentation using limonite and tannin as an additive

	Methane, CH_4 (ml/0.2 mg DM)	% Reduction with respect to control
Control	5.3	–
Limonite		
2 mg	1.04	80
10 mg	0.48	91
20 mg	0.31	94
Tannin		
2 mg	2.17	59
10 mg	1.17	78
20 mg	0.69	87

5 Conclusion

The current study found that including limonite and tannin affects gas production during in vitro rumen fermentation, with higher concentrations of limonite leading to greater reductions in ruminal gases. The ruminal gases produced were also significantly impacted by the concentration of limonite and tannin. The study concludes that the effects of limonite and tannin on ruminal gas production and rumen fermentation parameters may differ depending on their concentration. Methane production was reduced by the inclusion of limonite and tannin due to the utilization of hydrogen, which inhibits methane production. Further research is needed to determine the characteristics of limonite and its effect on rumen methanogen, protozoal, and bacterial populations.

References

Agarwal N, Kamra DN, Chaudhary LC (2015) Rumen microbial ecosystem of domesticated ruminants. In: Rumen microbiology: from evolution to revolution. https://doi.org/10.1007/978-81-322-2401-3_2

Andbook H (2009) An animal and nutrition handbook second revision. https://umkcarnivores3.files.wordpress.com/2012/02/animal-nutrition2.pdf

Bakrie E, Liang E, Tareque AMM (1996) Ruminant nutrition and production in the tropics and subtropics. Australian Centre for International Agricultural Research, pp 45–87

Beauchemin KA, Kreuzer M, O'Mara F, McAllister TA (2008) Nutritional management for enteric methane abatement: a review. Aust J Exp Agric 48:21–27. https://doi.org/10.1071/EA07199

Candyrine SCL, Mahadzir MF, Garba S, Jahromi MF, Ebrahimi M, Goh YM et al (2018) Effects of naturally produced lovastatin on feed digestibility, rumen fermentation, microbiota, and methane emissions in goats over a 12-week treatment period. PLoS ONE 13(7):e0199840. https://doi.org/10.1371/journal.pone.0199840

Choudhury PK, Salem AZM, Jena R, Kumar S, Singh R, Puniya AK (2015) Rumen microbiology: an overview. In: Puniya AK, Singh R, Kamra DN (eds) Rumen microbiology: from evolution to revolution, pp 3–16. https://doi.org/10.1007/978-81-322-2401-3_1

Chowdhury SP, Khanna S, Verma SC, Tripathi AK (2004) Molecular diversity of tannic acid degrading bacteria isolated from tannery soil. J Appl Microbiol 97(6):1210–1219. https://doi.org/10.1111/j.1365-2672.2004.02426.x

El-Chaghaby GA, Ramis ES, Ahmad AF (2018) Rice straw and rice straw ash for the removal of brilliant green dye from wastewater. Asian J Appl Chem Res 1–9. https://doi.org/10.9734/ajacr/2018/v1i29633

Harun NLAB (2015) Effects of substituting rice straw with different levels of Leucaena leucocephala (Lam.) De Wit and Manihot esculenta Crantz leaves on rumen fermentation characteristics and microbial populations in goats (Universiti Putra Malaysia). https://doi.org/10.1145/3132847.3132886

Hess HD, Tiemann TT, Noto F, Carulla JE, Kreuzer M (2006) Strategic use of tannins as means to limit methane emission from ruminant livestock. Int Congr Ser 1293:164–167. https://doi.org/10.1016/j.ics.2006.01.010

Hungate RE (2003) The rumen microbial ecosystem. Annu Rev Ecol Syst 6(1):39–66. https://doi.org/10.1146/annurev.es.06.110175.000351

JAPAN LIMONITE Co. L (1960) History of limonite. www.limonic.co.jp. Website. http://www.limonic.co.jp/gaiyo/kaisyagaiyou01.html

Kadam KL, Forrest LH, Jacobson WA (2000) Rice straw as a lignocellulosic resource: collection, processing, transportation, and environmental aspects. Biomass Bioenerg 18(5):369–389. https://doi.org/10.1016/S0961-9534(00)00005-2

Krause DO, Denman SE, Mackie RI, Morrison M, Rae AL, Attwood GT, McSweeney CS (2003) Opportunities to improve fiber degradation in the rumen: microbiology, ecology, and genomics. FEMS Microbiol Rev 27(5):663–693. https://doi.org/10.1016/S0168-6445(03)00072-X

Kusmartono (2007) Effects of supplementing Jackfruit (Artocarpus heterophyllus L) wastes with urea or Gliricidia/cassava leaves on growth, rumen digestion and feed degradability of sheep fed on rice straw basal diet. Retrieved from livestock research for rural development website. http://www.lrrd.org/lrrd19/2/kusm19021.htm

Lan W, Yang C (2019) Ruminal methane production: associated microorganisms and the potential of applying hydrogen-utilizing bacteria for mitigation. Sci Total Environ 654:1270–1283. https://doi.org/10.1016/j.scitotenv.2018.11.180

Lassey KR (2007) Livestock methane emission: from the individual grazing animal through national inventories to the global methane cycle. Agric For Meteorol 142(2):120–132. https://doi.org/10.1016/j.agrformet.2006.03.028

Lesschen JP, van den Berg M, Westhoek HJ, et al (2011) Greenhouse gas emission profiles of European livestock sectors. Anim Feed Sci Technol 166–167:16–28. https://doi.org/10.1016/j.anifeedsci.2011.04.058

Liang JB, Samiyah MN (1988) Comparative intake, digestibility, and utilization of guinea grass by buffaloes and cattle. MARDI Res J 16(1):43–47

Lima PR, Apdini T, Freire AS, Santana AS, Moura LML, Nascimento JCS, Rodrigues RTS, Dijkstra J, Neto AFG, Queiroz MAÁ, Menezes DR (2019) Dietary supplementation with tannin and soybean oil on intake, digestibility, feeding behavior, ruminal protozoa and methane emission in sheep. Anim Feed Sci Technol 249(January):10–17. https://doi.org/10.1016/j.anifeedsci.2019.01.017

Martinez-Fernandez G, Denman SE, Yang C, Cheung J, Mitsumori M, McSweeney CS (2016) Methane inhibition alters the microbial community, hydrogen flow, and fermentation response in the rumen of cattle. Front Microbiol 7:1122. https://doi.org/10.3389/fmicb.2016.01122

Menke HH, Steingass H (1988) Estimation of the energetic feed value obtained from chemical analysis and in vitro gas production using rumen fluid. Anim Res Develop 28:7–55

McAllister TA, Newbold CJ (2008) Redirecting rumen fermentation to reduce methanogenesis. Aust J Exp Agric 48(2):7–13. https://doi.org/10.1071/EA07218

Michaiowski T (2005) Rumen protozoa in the growing domestic ruminant. Biol Grow Anim 2(C):54–74. https://doi.org/10.1016/S1877-1823(09)70036-0

Moss AR, Jouany J-P, Newbold J (2003) Methane production by ruminants: its contribution to global warming. Ann Zootech. https://doi.org/10.1051/animres:2000119

Mustapha NA, Sharuddin SS, Zainudin MHM, Ramli N, Shirai Y, Maeda T (2017) Inhibition of methane production by the palm oil industrial waste phospholine gum in a mimic enteric fermentation. J Clean Prod 165:621–629. https://doi.org/10.1016/j.jclepro.2017.07.129

Paridah M, Moradbak A, Mohamed A, Owolabi F, Taiwo A, Asniza M, Abdul Khalid SH (2016) We are IntechOpen, the world's leading publisher of open access books built by scientists, for scientists TOP 1%. Intech, i(tourism), 13. https://doi.org/10.5772/57353

Puniya AK, Singh R, Kamra DN (eds) (2015) Rumen microbiology: from evolution to revolution. https://doi.org/10.1007/978-81-322-2401-3

Russell JB, Wallace RJ (1997) Energy-yielding and energy-consuming reactions. In: Hobson PN, Stewart CS (eds) The rumen microbial ecosystem, pp 246–282. https://doi.org/10.1007/978-94-009-1453-7_6

Saminathan M, Sieo CC, Abdullah N, Wong CM, Ho YW (2015) Effects of condensed tannin fractions of different molecular weights from a Leucaena leucocephala hybrid on in vitro methane production and rumen fermentation. J Food Sci Technol 95(13):2742–2749. https://doi.org/10.1002/jsfa.7016

Sandro JM (2016) Report: rice straw collection report: rice straw collection. https://doi.org/10.13140/RG.2.1.4205.1447

Tavendale MH, Meagher LP, Pacheco D, Walker N, Attwood GT, Sivakumaran S (2005) Methane production from in vitro rumen incubations with Lotus pedunculatus and Medicago sativa, and effects of extractable condensed tannin fractions on methanogenesis. 124:403–419. https://doi.org/10.1016/j.anifeedsci.2005.04.037

Van Bodegom PM, Scholten JCM, Stams AJM (2004) Direct inhibition of methanogenesis by ferric iron. FEMS Microbiol Ecol 49(2):261–268. https://doi.org/10.1016/j.femsec.2004.03.017

Van Kessel JAS, Russell JB (1996) The effect of pH on ruminal methanogenesis. FEMS Microbiol Ecol 20(4):205–210. https://doi.org/10.1016/0168-6496(96)00030-X

Zhou M, Chen Y, Guan LL (2015) Rumen bacteria BT-rumen microbiology: from evolution to revolution. In: Puniya AK, Singh R, Kamra DN (eds). https://doi.org/10.1007/978-81-322-2401-3_6

Conclusions and Future Perspectives

Mohd Armi Abu Samah, Fazilah Ariffin, A. K. Haghi, and Abu Zahrim Yaser

Abstract The controlling of environmental pollution is a complex task that involves understanding the various forms of pollution and implementing effective strategies to mitigate their impact. Pollution occurs in many forms, such as air, water, soil, and noise. Basically, the stems from numerous sources, including industrial activities, vehicular emissions, agricultural practices, and household waste. Understanding these basics is crucial for developing comprehensive solutions to address and manage pollution effectively. Environmental pollution can refer to introducing harmful substances or energy into the environment, leading to adverse effects on ecosystems and human health. Air pollution, primarily caused by emissions from factories, vehicles, and burning fossil fuels, leads to respiratory problems and contributes to climate change. Water pollution, resulting from industrial discharge, agricultural runoff, and improper waste disposal, contaminates water bodies, affecting marine life and posing health risks to humans. Soil pollution, often due to excessive use of pesticides and industrial waste, degrades soil quality, impacting agriculture and food security (Ibrahim et al., Environmental Advances 12, 2023). Noise pollution, primarily from industrial activities and urbanization, affects human health by causing stress, hearing loss, and sleep disturbances.

M. A. A. Samah (✉)
Kulliyyah of Science, International Islamic University Malaysia, 25200 Kuantan Pahang, Malaysia
e-mail: marmi@iium.edu.my

F. Ariffin
Faculty of Science and Marine Environment, Universiti Malaysia Terengganu, 21030 Kuala Nerus, Terengganu, Malaysia

A. K. Haghi
Department of Chemistry, Institute of Molecular Sciences, University of Coimbra, Coimbra, Portugal

A. Z. Yaser
Faculty of Engineering, Universiti Malaysia Sabah, Jalan UMS, 88400 Kota Kinabalu, Sabah, Malaysia

A. Z. Yaser et al. (eds.), *Controlling Environmental Pollution*,
https://doi.org/10.1007/978-981-97-8931-3_22

The controlling of environmental pollution is a complex task that involves understanding the various forms of pollution and implementing effective strategies to mitigate their impact. Pollution occurs in many forms, such as air, water, soil, and noise. Basically, the stems from numerous sources, including industrial activities, vehicular emissions, agricultural practices, and household waste. Understanding these basics is crucial for developing comprehensive solutions to address and manage pollution effectively. Environmental pollution can refer to introducing harmful substances or energy into the environment, leading to adverse effects on ecosystems and human health. Air pollution, primarily caused by emissions from factories, vehicles, and burning fossil fuels, leads to respiratory problems and contributes to climate change. Water pollution, resulting from industrial discharge, agricultural runoff, and improper waste disposal, contaminates water bodies, affecting marine life and posing health risks to humans. Soil pollution, often due to excessive use of pesticides and industrial waste, degrades soil quality, impacting agriculture and food security (Ibrahim et al. 2023). Noise pollution, primarily from industrial activities and urbanization, affects human health by causing stress, hearing loss, and sleep disturbances.

The effectiveness of pollution control measures depends on several factors, including the technology employed, the enforcement of regulations, and the collaboration between government, industries, and the community. Technological advancements, such as cleaner production methods, pollution control devices, and green technologies, play a crucial role in reducing emissions and waste. For instance, the adoption of catalytic converters in vehicles has significantly reduced air pollution. Similarly, advanced wastewater treatment plants help in purifying industrial discharge before it reaches water bodies (Zahrim et al. 2007, 2018). However, the technology alone is not sufficient. Strong regulatory frameworks and their strict enforcement are essential to ensure compliance. Governments must implement and enforce environmental laws and standards to hold polluters accountable. Moreover, continuous evaluation and updating of policies are necessary to address emerging challenges and incorporate new scientific findings.

Public awareness and education are fundamental in managing and controlling environmental pollution. Educating the public about the causes and consequences of pollution and the importance of environmental conservation can lead to behavioral changes that reduce pollution. Awareness campaigns, educational programs, and community engagement initiatives can empower individuals and communities to take proactive steps in pollution prevention. As such, many individuals have continually inhaled particulate matter, which originates from smog in the atmosphere (Rosman et al. 2019). Schools, universities, and community organizations can play a vital role in disseminating information and fostering an environmentally conscious mindset. Simple actions, such as reducing plastic use, conserving water, and segregating waste for recycling, can collectively make a significant impact. Encouraging responsible consumer behavior, such as choosing eco-friendly products and supporting sustainable businesses, also contributes to reducing pollution.

Implementing practical solutions at individual, community, and industrial levels is essential to control environmental pollution effectively. Here are some key strategies such as reducing plastic use, plastic pollution is a major environmental issue, and promoting the use of reusable bags, bottles, and containers can significantly reduce plastic waste. Governments and businesses can also implement policies to reduce single-use plastics and encourage the use of biodegradable alternatives.

Effective Waste Management: Proper waste management practices, such as recycling, using food waste as compost (Razali et al. 2024), and safe disposal of hazardous waste, are crucial. Establishing efficient waste collection and processing systems can prevent waste from contaminating the environment. Community-led initiatives, such as clean-up drives and recycling programs, can enhance waste management efforts. Knowledge of solid waste generation and composition is necessary for accurate decision-making in terms of engineering design, planning, and forecasting for the management strategy of urban waste (Samah et al. 2019). Using Renewable Energy: Transitioning to renewable energy sources, such as solar, wind, and hydroelectric power, can reduce dependence on fossil fuels and lower greenhouse gas emissions. Governments can incentivize the adoption of renewable energy through subsidies, tax breaks, and supportive policies. On an individual level, installing solar panels and using energy-efficient appliances can contribute to energy conservation. The Eco-Friendly Transportation: Promoting the use of electric vehicles, public transportation, cycling, and walking can reduce air pollution from vehicular emissions. Developing infrastructure for non-motorized transport and implementing policies to limit vehicle use in urban areas can encourage the shift to greener transportation options. Sustainable Agriculture: Adopting sustainable farming practices, such as organic farming, crop rotation, and integrated pest management, can reduce soil and water pollution from chemical fertilizers and pesticides. Supporting local and organic food production can also reduce the environmental footprint of agriculture.

Tree Planting and Urban Greening: Increasing green spaces and planting trees can help absorb carbon dioxide, improve air quality, and enhance urban environments. Community tree-planting initiatives and government-led afforestation programs can contribute to carbon sequestration and biodiversity conservation. In conclusion, controlling environmental pollution requires a multifaceted approach that combines technological innovation, regulatory enforcement, public awareness, and practical solutions. Understanding the basics of pollution and its sources is the first step in developing effective strategies. Ensuring the effectiveness of these strategies involves continuous evaluation and collaboration among all stakeholders. Public awareness and education are crucial in fostering an environmentally conscious society that actively participates in pollution prevention. Practical solutions, such as reducing plastic use, effective waste management, using renewable energy, promoting eco-friendly transportation, practicing sustainable agriculture, and increasing urban greening, can collectively reduce pollution and ensure a sustainable environment for future generations. Continuous efforts and collective action are key to achieving these goals and maintaining a healthy planet.

Artificial Intelligence (AI) and machine learning (ML) are revolutionizing pollution abatement by providing sophisticated tools for monitoring, predicting, and mitigating environmental pollutants. These technologies enable real-time data analysis and pattern recognition, offering insights that traditional methods cannot match. For instance, AI-powered sensors and drones can monitor air and water quality with high precision, detecting pollutants at their sources and enabling rapid response to environmental hazards. Machine learning algorithms analyze vast amounts of environmental data to predict pollution trends and identify the most effective abatement strategies. This predictive capability allows for proactive measures, reducing the overall environmental impact. Moreover, AI-driven optimization models can enhance the efficiency of industrial processes, minimizing waste and emissions. By integrating AI and ML into pollution abatement efforts, we can develop smarter, more responsive environmental management systems that contribute to a cleaner and healthier planet.

An integrated system for pollution eradication encompasses a comprehensive approach that combines advanced technologies, regulatory frameworks, and community engagement to tackle pollution effectively. Such a system leverages IoT (Internet of Things) devices, AI, and big data analytics to create a cohesive network for monitoring and managing pollution. IoT sensors collect real-time environmental data, which is then processed by AI algorithms to detect pollution sources and predict future contamination events. This information is crucial for policymakers to develop targeted regulations and for industries to implement cleaner technologies. Furthermore, an integrated system promotes collaboration among government agencies, private sector stakeholders, and the public, ensuring that pollution control measures are universally adopted and enforced. Public awareness campaigns and community participation play vital roles in this ecosystem, fostering a culture of sustainability and environmental stewardship. By integrating technological innovations with coordinated efforts across various sectors, we can establish a robust and effective framework for eradicating pollution and safeguarding our environment for future generations.

References

Ibrahim NE, Sevakumaran V, Ariffin F (2023) Preliminary study on glyphosate-degrading bacteria isolated from agricultural soil. Environ Adv 12:100368

Razali RM, Murshidi NZ, Ghazali MSM, Ariffin F (2024) Effect of effective microorganisms in fermentation of rice husk and anchovy head using lab-scale treatment. Pertanika J Trop Agric Sci 47(1)

Rosman PS, Samah MAA, Yunus K, Hussain MRM (2019) Particulate matter (PM 2.5) at construction site: a review. Int J Recent Technol Eng (IJRTE) 8(1C2):225–259. ISSN: 2277-3878

Samah MAA, Victor D, Dawda B, Manaf LA, Ismail SNS, Hamid KBA, Yunus K, Kamarudin MKA (2019) Composition and generation of municipal solid waste (MSW). Balakong City Case Study, Malaysia. Herald NAMSCA, pp 152–158

Zahrim AY, Gilbert ML, Janaun J (2007) Treatment of pulp and paper mill effluent using photo-fenton's process. J Appl Sci 7(15):2164–2167

Zahrim AY, Azreen I, Jie SS, Yoiying C, Felijia J, Hasmilah H, Gloriana C, Khairunis I (2018) Nanoparticles enhanced coagulation of biologically digested leachate. In: Nanotechnology in water and wastewater treatment: theory and applications, pp 205–241

The manufacturer's authorised representative in the EU is Springer Nature Customer Service Centre GmbH, Europaplatz 3, 69115 Heidelberg, Germany. If you have any concerns regarding our products, please contact ProductSafety@springernature.com

Printed and bound by CPI Group (UK) Ltd, Croydon, CR0 4YY

01/10/2026

02225930-0013